# Matrix Analysis of Frame Structures

Advancing computer technology has created new opportunities for sophisticated assessment and analysis of structural performance, especially using matrix and finite element methods. This textbook employs these methods using sophisticated computational techniques through simple step-by-step processes.

It covers the fundamentals required in any approach to structural analysis, strong form, equilibrium, and compatibility and includes an introduction to virtual work principles to express equilibrium and compatibility conditions of a frame structure, making use of Tonti diagrams. It shows how to construct a master stiffness matrix using an approach based on a system without rigid body modes. It then sets out in more detail the matrix approach to structural analysis, including the construction of the master stiffness matrix.

This textbook is essential for senior undergraduates and graduate students and is also useful for consulting engineers.

# Matrix Analysis of Frame Structures

## A Modern Approach

Suchart Limkatanyu

**CRC Press**
Taylor & Francis Group
Boca Raton  London  New York

CRC Press is an imprint of the
Taylor & Francis Group, an **informa** business

Designed cover image: Canva

MATLAB® and Simulink® are trademarks of The MathWorks, Inc. and are used with permission. The MathWorks does not warrant the accuracy of the text or exercises in this book. This book's use or discussion of MATLAB® or Simulink® software or related products does not constitute endorsement or sponsorship by The MathWorks of a particular pedagogical approach or particular use of the MATLAB® and Simulink® software.

First edition published 2026
by CRC Press
4 Park Square, Milton Park, Abingdon, Oxon, OX14 4RN

and by CRC Press
2385 NW Executive Center Drive, Suite 320, Boca Raton FL 33431

CRC Press is an imprint of Informa UK Limited

*British Library Cataloguing-in-Publication Data*
A catalogue record for this book is available from the British Library

ISBN: 9781032977928 (hbk)
ISBN: 9781032976662 (pbk)
ISBN: 9781003595458 (ebk)

DOI: 10.1201/9781003595458

Typeset in Times
by Deanta Global Publishing Services, Chennai, India

# Contents

# About the Author

**Suchart Limkatanyu** is Full Professor in the Department of Civil Engineering at the Prince of Songkla University in Thailand. He received his B.Eng (with distinction) from the Prince of Songkla University in 1996 and his M.S. and Ph.D. from the University of Colorado, Boulder, in 2002. Prof. Limkatanyu's research covers a variety of topics in structural engineering and applied mechanics with emphasis on computational mechanics of solids and structures. His research interests include seismic analysis, design, and retrofitting of bridges; nonlinear analysis of reinforced concrete, prestressed concrete, and steel structures under static and dynamic loading; soil–pile interactions; and earthquake engineering.

Prof. Limkatanyu is considered one of the pioneers in formulating a consistent force-based frame model for nonlinear analyses of frame structures, and his publication on this model has been highly cited in the literature. Stemming from reinforced concrete frame models with bond interfaces, Prof. Limkatanyu developed a novel finite frame element to investigate the failure behavior of reinforced concrete beams strengthened with steel and fiber-reinforced polymer (FRP) plates, and several researchers have subsequently adopted this model as a reference. In addition to structural-engineering problems, Prof. Limkatanyu has conducted research work on geotechnical-engineering problems. Numerous soil-shallow foundation and soil–pile foundation models were proposed by him.

Recently, Prof. Limkatanyu has expanded his research interests to the field of nanotechnology and smart materials and has proposed several structural models (e.g. bar, beam, plate) with the inclusion of small-scale and surface effects inherent to nano-sized structures. These structural models have found a broad spectrum of modern applications in engineering and science such as bionanoactuators, nanosensors, nanowires, nanoprobes, and graphene sheets. Prof. Limkatanyu has over 20 years of experience of teaching Matrix Structural Analysis to graduate students. Based on his above-mentioned research work and teaching experience, he unified these two ingredients and put them together into his unique textbook on *Matrix Analysis of Frame Structures*. Furthermore, Prof. Limkatanyu also authored a mathematical book titled *Introduction to the Calculus of Variations*.

# Forward

I am pleased to write this Introduction to a brand new textbook by Dr. Suchart Limkatanyu, a former doctoral student at the University of Colorado who is presently Professor of Civil Engineering at the Prince of Songkla University in Thailand. It is entitled *Matrix Analysis of Frame Structures: A Modern Approach*. As the title suggests, the book coverage is intended to be fairly comprehensive in range. This range makes it also suitable for supporting various courses in Civil and Mechanical Engineering.

Following an overview of topics in Chapter 1, Chapter 2 presents the strong formulation of Mechanics of Material elements, including bars, and thin and thick beams. One novel feature is the use of Tonti diagrams to graphically convey the set of governing equations. In the writer's experience, such a representation helps students to understand both the strong forms and subsequent approximate (weak form) models such as those of the Finite Element Method (FEM).

Chapter 3 proceeds to the matrix formulation of structural analysis. This is a subject that historically preceded FEM by two decades but eventually merged with it in the 1950s. The focus is on equilibrium matrices at local and global levels. Chapter 4 continues with the compatibility kinematic analysis of Mechanics of Material elements. Kinematic constraints are treated with the Master–Slave method.

Chapter 5 introduces energy concepts through the concept of virtual displacements and the Principle of Virtual Work (PVW). The exposition is hierarchical in nature. Starting from the concept of particle and rigid body, it proceeds to elastic elements through the constitutive equations and emphasizes the dual nature of equilibrium and compatibility relations.

Chapter 6 discusses the construction of the master stiffness matrix from conceptual arguments, an approach based on a system without rigid body modes. Chapters 7 and 8 introduce the classical Direct Stiffness Method (DSM). By contrast, Chapter 9 goes over the flexibility approach, also called the Force Method in the literature. While the latter was traditionally covered in early books on Matrix Structural Analysis as the "dual" of the stiffness approach, it has been rarely addressed in textbooks since 1965, and thus represents a distinguishing feature of the present book.

How does the book's context fit the present status of computational structural mechanics? To answer that question, a brief historical summary is pertinent.

Contemporary structural mechanics has gone through several distinguishing periods. The fundamental theory that supports structural engineering evolved rapidly following the Industrial Revolution. It was a synergistic confluence of the older Newtonian mechanics combined with mathematical developments such as the Theory of Elasticity, and of practical engineering design and application methods embodied in Mechanics of Materials (an area known as Strength of Materials before 1970). The classical theory was essentially completed by the early twentieth century. Engineers still play by those rules, which are implemented in design codes, for typical structures.

Computational mechanics is a more recent newcomer. It started modestly with the formulation of discrete structural configurations in matrix form during the early 1930s at the National Physical Laboratory (NPL) in Teddington, UK. The initial motivation was the prediction of flutter in newer, higher speed, monoplane aircraft configurations. Matrix algebra was found to be well suited for the new electronic desk calculators then coming into use. The essential steps, grouped under the name Matrix Structural Analysis (MSA), began entering courses in Aeronautical Engineering during the late 1940s, as well as Civil Engineering in the late 1950s. It was to become one of the foundation pillars of the Finite Element Method (FEM).

FEM also had unassuming beginnings in the US aerospace industry. While several variants emerged, only one thrived. The survivor, called the Direct Stiffness Method (DSM), was developed at the Boeing Company in Seattle, Washington, in the early 1950s. It was motivated by the appearance of newer aircraft wing configurations. FEM is a synergistic combination of three ingredients. A classical one: energy methods in mechanics, which dates back to the eighteenth century, provided the underlying theory. A more recent tool: Matrix Structural Analysis provided modeling, formulation, implementation, and solution tools. And a newcomer: the commercial digital computer, which appeared in 1951, made calculations feasible.

Both MSA and FEM emerged as a result of technical challenges in evolving aircraft technology. Drivers were aero-elastic stability at NPL and delta wing configurations at Boeing. Truly, necessity is the mother of invention. The initial focus of both was dynamics and vibration; the idea of using FEM in statics came later as the method expanded to Civil Engineering. Note also that without computers, FEM would be just a curiosity, and only large aerospace companies were able to purchase mainframes in the 1950s.

The expansion of FEM in the 1960s was explosive. The method entered academia primarily through Civil and Aerospace Engineering departments, triggering a publication surge. Early developments emphasized the physical interpretation: a complex structure is divided into simpler pieces (elements), and this idealized structure is solved exactly; this is the basic idea behind backward error analysis. But the mathematical interpretation advanced rapidly since the mid-1960s and facilitated extension to nonstructural field problems such as fluid dynamics and thermomechanics.

The post-1970 years were a consolidation period in which DSM took over as the dominant version, commercial codes proliferated, and new discretization techniques such as hybrid elements and meshless methods emerged. The computer implementations tried to keep pace with advances in computer hardware and software (such as interactivity, graphics, and personal computers), as well as advances in numerical algorithms.

Since about the 1990s, three "hot" computational mechanics areas have emerged. The three "multis": multiscale, multiphysics, and multiprocessing. Multiphysics addresses the interaction of different physical behaviors, as in structures and fluids, at similar physical scales. Models that span a range of physical scales (e.g., molecular through crystal) are the framework of multiscale simulations; this area has been driven by rapid developments in advanced material design (e.g. bio- and meta-materials) as well as micro- and nano-mechanic devices. Multiprocessing refers to computational methods that use problem decomposition to achieve concurrency; a feature that is becoming feasible at the PC network level.

Mapping back the book material into the foregoing history, it is obvious that it spans a range covering structural mechanics fundamentals (Chapters 1–5) and entrenched developments in computational mechanics (Chapters 6–9). In summary, a nice balance is struck between established and evolving knowledge. Teachers, students, and researchers should encounter a wealth of educational and research material. My best wishes for success.

**Carlos Felippa**
*Professor of Aerospace Engineering Sciences*
*University of Colorado at Boulder*

# Preface

During the last 50 years, the level of structural analysis needed to assess and analyze structural performance has become increasingly sophisticated, especially with drastic advances in computer technology. The driving forces which necessitate higher requirement on structural analysis stem partly from heavy property damages and loss of human lives associated with extreme events (earthquakes, terrorist attacks, etc.). This forces structural engineers to seek to understand more advanced structural analysis in order to achieve these goals. Therefore, all analysis methods have eventually moved toward the computer-based approaches. Chief among them are the matrix and finite element methods.

This textbook has been carefully prepared with a view to meet the needs of students, practicing engineers, and researchers who wish to enhance their specialization in the matrix analysis of frame structures. In this textbook, we set out to give a coherent treatise of the underlying assumptions and concepts concerning "matrix analysis of frame structures".

Part I: Fundamental Ingredients and Basic Treatments on the Virtual Work Principles. This part consists of Chapters 1 through 5. It covers fundamental ingredients required in any approach of structural analysis and an introduction to the virtual work principles. Following an introduction presented in Chapter 1, the differential (strong) form of several frame members is presented in Chapter 2. The equilibrium and compatibility conditions of a frame member and a frame structure are discussed in Chapters 3 and 4, respectively. This part is wrapped up by Chapter 5, discussing the virtual displacement and virtual force principles as an alternative way of expressing the equilibrium and compatibility conditions, respectively.

Part II: The Stiffness Method and the Flexibility Method. This part consists of Chapters 6 through 9. It discusses the matrix approach of structural analysis. The stiffness method is presented in its rudimentary form in Chapter 6 and subsequently in its modern form in Chapter 7. The general framework of a frame analysis program based on the stiffness approach is discussed in Chapter 8. This part is wrapped up by Chapter 9, which discusses the flexibility method.

It is worth remarking that this book can be used to develop a course on Matrix Analysis of Frame Structures (e.g. 220-501: Matrix Structural Analysis offered at the Department of Civil and Environmental Engineering, Prince of Songkla University).

Finally, comments, suggestions, and questions are most welcome from the readers. They may be sent to Suchart Limkatanyu, Department of Civil Engineering, Prince of Songkla University, Songkhla, Thailand, 90110 or to suchart.l@psu.ac.th.

**Hat Yai, Songkhla,**
Thailand Suchart Limkatanyu

# Acknowledgment

This textbook would not have been finished without the knowledge and support from the following teachers who have been involved in the development of my specialization in structural engineering and structural mechanics, both at Prince of Songkla University and the University of Colorado, Boulder: Dr. Boon Chantaksinopas, Dr. Fukit Nilrat, Prof. Enrico Spacone, Prof. Carlos Felippa, Prof. Kaspar Willam, Prof. Benson Shing, Prof. Victor Saouma, Prof. Kurt Gerstle, Prof. Dan Frangopol, and Prof. Ross Corotis.

I would also like to extend my special thanks to Senior Lecturer Mr. Wiwat Sutiwipakorn and Mrs. Sammi Nagaratnam for reviewing and correcting the English of this textbook.

Finally, it is impossible to express enough thanks to my wife Prof. Dr. Benjamas Cheirsilp, my daughter Panalee Limkatanyu, my son Paesol Limkatanyu, and to my mother Mrs. Kanjanee Limkatanyu. They have endured so much to make this book possible, and their unconditional love has not gone forgotten.

# 1 Introduction

## 1.1 MATRIX STRUCTURAL ANALYSIS: ITS ESSENCE AND ROLES

Statics, Dynamics, Mechanics of Materials, and Basic Structural Analysis form a series of core structural-engineering courses in most civil engineering curricula. The last course enables students to determine reactions, deflections, and internal forces of both statically determinate and indeterminate structures under static loadings. Besides the vector approach, the scalar (energy) approach has also been introduced. Mostly, problems are confined to two-dimensional ones. Three-dimensional problems are hardly encountered in a course since emphasis is on hand calculations. However, students who wish to pursue specializations in structural engineering and structural mechanics do have to undertake more advanced structural analysis courses, such as Matrix Structural Analysis. Figure 1.1 shows a general sequence of structural engineering courses. It can be seen that Matrix Structural Analysis is only at the third tier of the long series of structural-engineering courses. It, however, provides a crucial link between undergraduate and graduate structural-engineering courses. Consequently, it is no surprise why it appears again and again in Comprehensive, Preliminary, and Qualifying Examinations in Structural Engineering and Structural Mechanics Program (e.g. at the University of Colorado at Boulder, USA).

Matrix Structural Analysis or Advanced Structural Analysis automatically forces students to get involved with computer implementation. Therefore, in this textbook, it is deemed necessary to provide equal emphasis on both the classical tools and the modern tools in structural analysis and to develop a comprehensive understanding of computer-aided analyses of structures.

In current structural engineering practice, computer softwares perform most, if not all, structural analyses. As structural engineers, one should not just use them but rather ought to try to understand what goes on inside those "black boxes" by developing enough confidence to be capable of scrutinizing and modifying them to perform certain specific tasks, and most crucially, to realize their limitations (Saouma 1997).

Matrix structural analysis also serves as an introductory course to finite element analysis and other more advanced courses. However, understanding the connection and the role of these two analyzing methods is deemed essential. The finite element method involves the analysis of two- or three-dimensional continua. Fundamental unknowns are the nodal displacements, and the internal forces are usually stresses. Even though some civil engineering structures (e.g. dams, shells) need a two- or three-dimensional continuum to model, most civil engineering structures may be considered to be composed of "stick" one-dimensional elements such as beams, girders, columns, etc. For these elements, their displacements and internal forces are somewhat more complicated/sophisticated than those involved in continuum finite elements.

## 1.2 OBJECTIVES OF STRUCTURAL ANALYSIS

Generally, the response of a structural model to a set of loadings and imposed displacements is completely defined by the nodal displacements of the model. This means that the internal displacements, deformations, and internal forces can be expressed in terms of the nodal displacements. For linearly elastic structures, relationships between these quantities are unique.

DOI: 10.1201/9781003595458-1

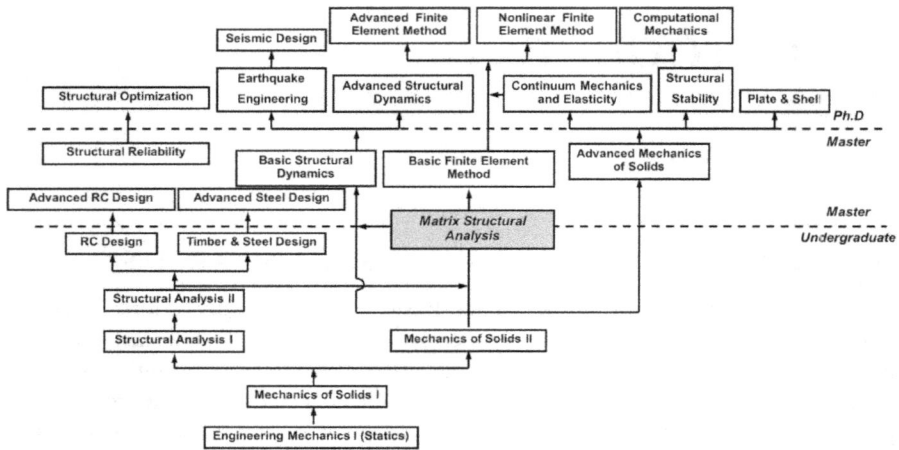

**FIGURE 1.1**   A General Sequence of Structural-Engineering Courses.

Consequently, the problem can be solved either by computing the nodal displacements first (the displacement method) or by starting off with the internal forces (the force method).

Generally, objectives of structural analysis can be summarized thus:

**Given:** A structural model with prescribed geometry and material properties under the actions of loads and predefined displacements.

**Find:** The resulting nodal displacements, member forces, internal forces, member deformations, and support reactions.

It is essential to distinguish between structural analysis and stress analysis. The main tasks of structural analysis are to determine reactions, deflections, and internal forces while that of the stress analysis are to determine the stresses at any point in the structure.

## 1.3   FUNDAMENTAL INGREDIENTS OF STRUCTURAL ANALYSIS

Equilibrium, compatibility, and stiffness (or flexibility) are the three fundamental ingredients required in every method of structural analysis.

### 1.3.1   EQUILIBRIUM

The first necessary condition required in any structural analysis approach is equilibrium. This means that the system of forces applying on the structure as a whole, or on any part, or on an infinitesimal segment of the structure must be in balance. It is often incorrectly understood that equilibrium conditions must hold only for the structure under static loadings. This is not true. In the case of a structure under dynamic loadings, equilibrium conditions must also hold, but they keep changing with respect to time. At any given instant of time, the resisting forces (elastic or inelastic) must be in equilibrium with the external forces acting on the structure. These external forces include dynamic loads as well as damping and inertia forces. Consequently, from the viewpoint of d'Alembert's principle, static equilibrium is just a special case of dynamic equilibrium but without damping and inertia forces.

A statically determinate structure is one in which equilibrium conditions are sufficient to calculate its internal forces under the given set of external loadings. This structural type belongs to an important class of structures that are widely erected in practice. Examples of statically determinate structures are: simply supported bridge girder (Figure 1.2*a*) and inverted Parker truss bridge (Figure 1.2*b*). The most noticeable advantage of statically determinate structures is their insensitivity to thermal changes or support settlements. In other words, their internal forces are not induced under such circumstances. Furthermore, statically determinate structures also play a crucial role as primary structures in the process of analyzing statically indeterminate structures by the force method.

A statically indeterminate structure is one in which equilibrium conditions are not sufficient to calculate its internal forces under the given set of external loadings. Consequently, determination of its internal forces for the given loads is not as straightforward as that in the case of a statically determinate structure. This is due to the fact that, in general, more than one internal force distribution is feasible due to its ability to redistribute the internal forces. This redistribution capability provides additional structural safety against collapses due to extreme or abnormal loadings. One other distinct feature of a statically indeterminate structure is that its internal forces could develop without any apparent loadings when it is subjected to a temperature change, creep, lack of fit of members, or support settlement. Most real structures are statically indeterminate; examples are long-span roof truss (Figure 1.3*a*) and multi-story frame building (Figure 1.3*b*).

### 1.3.2 COMPATIBILITY

Any structure will deform once its internal stresses (resultant stresses) are developed due to mechanically or non-mechanically applied loadings. As a result of internal stresses, each point of the structure displaces to its neighboring position. Generally, movements of these points differ from point to point and vary continuously within the structure. Consequently, there occur relative displacements and angular changes. Collectively, these relative displacements and angular changes are known as "deformations". It is important to note that if there are no relative displacements or relative angular changes from point to point, it will imply that the structure as a whole moves as a rigid body. Common deformations of structural members are the extension/contraction of a bar in tension/compression, the rotation of a cross-section relative to its neighboring cross-section of a beam in flexure, and the deflection of a cross-section relative to its neighboring cross-section of a beam in shear. "Strain" is generally defined as the rate of change of deformation. For a bar under tensile force, sectional axial strain is defined as the extension per unit length of the fiber; for a beam under bending moment, sectional curvature is defined as the angular deformation per unit length of the beam; and for a beam under shear force, sectional shear strain is defined as the relative deflection per unit length of the beam. These sectional strains are often known as resultant strains, and their corresponding forces are known as resultant stresses.

(*a*) Simply Supported Bridge Girder          (*b*) Inverted Parker Truss Bridge

**FIGURE 1.2**   Examples of Statically Determinate Structures.

(a) Long-Span Roof Truss                    (b) Multi-Story Frame Building

FIGURE 1.3    Examples of Statically Indeterminate Structures.

Geometrical relations between deformation and displacement could be determined purely by compatibility or kinematical consideration. Generally, compatibility or kinematical conditions are nonlinear. However, like equilibrium conditions, they could be linearized based on an assumption that both the deformation and the displacement are small. Linearization implies that the effects of structural displacement and deformation on structural equilibrium are ignored.

Members of a structure could not deform independently since they are joined together at the nodes to form the structural framework. As a whole, the overall structure displaces correspondingly to its member connectivity and constraints imposed upon by its supports. For example, nodal displacements of a truss structure can be computed only when axial deformations of all its members and the constraints imposed upon by member connectivity are considered simultaneously. This has led to the famous Williot-Mohr displacement diagram.

Another important application of kinematical consideration is on the collapse of a structure. For a frame, incipient collapse occurs when sufficient hinges have formed to cause the structure to behave as a mechanism. The yield line theory is invented based on the kinematic consideration of collapsed slabs.

### 1.3.3  STIFFNESS OR FLEXIBILITY

Relations between member forces and corresponding member deformations can be defined in terms of either member stiffness or member flexibility. These two properties are dual; stiffness indicates how stiff the member can resist the forces while flexibility indicates how compliant the member can resist deformations. Stiffness or flexibility relations can be obtained based on known material properties and proper hypotheses on stress and strain distributions in a member. For linear structural analysis, stiffnesses are constant and depend only on the geometric properties of the member, such as section depth. They are independent of the deformation level.

## 1.4  METHODS OF STRUCTURAL ANALYSIS

Generally, structural analysis methods can be classified into two main categories: displacement and force methods.

### 1.4.1  THE DISPLACEMENT METHOD

In the displacement method, the nodal displacements are treated as basic unknowns. The size of the problem depends on the degree of kinematical indeterminacy. For each nodal displacement, there is a relevant independent equilibrium equation. Accordingly, there are as many independent equilibrium equations as unknown nodal displacements. Therefore, the numbers of independent equilibrium equations are sufficient to determine the unknown nodal displacements. As the equilibrium equations are employed for the determination of unknown nodal displacements, the name "equilibrium method" is also used. Since the stiffness relations are used to express the nodal forces in terms of the nodal displacements, it is also known as the "stiffness method".

### 1.4.2 The Force Method

Just like two sides of a coin, the coexistence of the displacement method and the force method is another example of dualism. In the force method, a set of reactive and/or internal forces known as redundant forces are treated as basic unknowns. The number of selected redundant forces is equal to the degree of statical indeterminacy of the structure, and the size of the problem depends on that degree. For each redundant force, there is a relevant condition of compatibility. Consequently, there are as many compatibility equations as the unknown redundant forces. Therefore, the numbers of compatibility equations are sufficient to determine the unknown redundant forces. As the compatibility equations are employed for the determination of the unknown redundant forces, the name "compatibility method" is also used. Since the flexibility relations are used to express the nodal deformations in terms of nodal forces, it is also known as the "flexibility method".

In both methods, a system of simultaneous equations is needed to be solved. The number of simultaneous equations is dictated by the degree of kinematical indeterminacy for the displacement method and by the degree of statical indeterminacy for the force method. In structural engineering applications, the degree of kinematical and statical indeterminacies of real structures is generally large. Therefore, the task of solving a large system of simultaneous equations might appear to be impossible unless it is assisted by a computer. This is the essential aspect of this textbook. By organizing relevant unknowns in vectors, matrix algebra method can be used for the solution process. This eventually leads to the "matrix displacement" and the "matrix force" methods. The so-called "Tonti Diagram" (Figure 1.4) is used to schematically visualize the roles of the aforementioned fundamental ingredients employed in the displacement and force methods.

In practice, almost all computerized structural analyses are carried out based on the matrix displacement method. This is due to its ease of programming and implementation on computers. Therefore, this textbook will mainly focus on the displacement method. However, the matrix force method is recently experiencing a rebirth as its superiority has been discovered in the field of nonlinear frame analysis (Sae-Long et al. 2021), thanks to the pioneering works by Mahasuverachai and Powell (1982) and Zeris and Mahin (1988). The reader who has a keen interest in the force-based frame model is recommended to visit doctoral dissertations by students of Prof. F.C. Filippou (University of California, Berkeley, USA) and of Prof. E. Spacone (presently at University "G. D'Annunzio" of Chieti, Pescara, Italy, and formerly at University of Colorado, Boulder, USA).

## 1.5 COORDINATE SYSTEMS

Usually, there are two basic Cartesian coordinate systems: a local or member right-handed, orthogonal reference system $x,y,z$ and a global or structure right-handed, orthogonal reference system X,Y,Z, as shown in Figure 1.5 for two-dimensional frame. These two coordinate systems are essential in communication between a structure and its constituent members as shown in subsequent chapters.

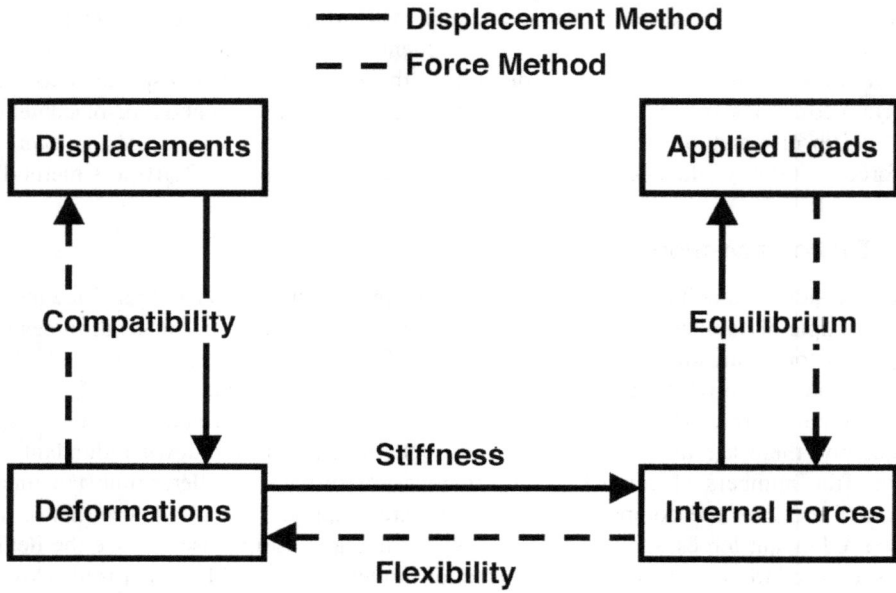

**FIGURE 1.4**   Tonti Diagram for Displacement and Force Methods.

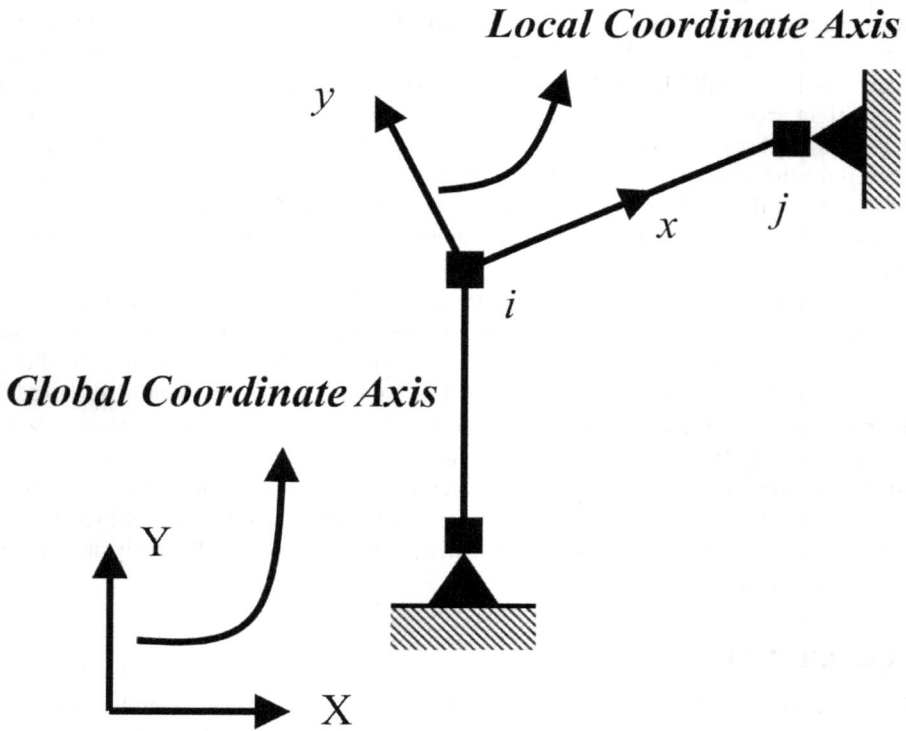

**FIGURE 1.5**   Local and Global Coordinate Axes (Plane Structure).

Even though the choice of the origin of the global coordinate system is arbitrary, the origin of the local coordinate system usually coincides with an end node of the member. The end node coinciding with the origin of the local coordinate system is named node $i$, and the node at the other end of the member is named node $j$. For a planar structure, this gives rise to a unique local $y$-axis in the plane if the local $z$-axis is co-directional with the global Z-axis, that is, pointing normal to the plane toward the reader. Nodal forces and nodal displacements are regarded as positive when directed in the positive coordinate axis. Moments and rotations in the plane are considered positive when acting counter clockwise.

## 1.6 DEGREES OF FREEDOM

Generally, the degree of freedom (DOF) of a structure is equal to the minimum number of independent components of generalized displacements of structural nodes required in positioning a displaced configuration of the structure when subjected to loads. The term "independent" implies that all DOFs must not be related to one another. The following two examples are employed to gain insight into independent degrees of freedom.

A roller support (Node $A$) on an inclined plane in Figure 1.6$a$ has three displacements: rotation $\theta$, translation $u$ along the X-axis, and translation $v$ along the Y-axis. The two translations, however, are not independent since they are kinematically constrained such that the roller support must move along the inclined plane. Consequently, there are only two independent displacements ($\theta$, and $u$ or $v$). It is also noted that the displacements have been referred to as *generalized* displacements since this term includes both rotation and translation.

A frame structure shown in Figure 1.6$b$ has three displacement degrees of freedom: rotation $\theta$, translation $u$ along the X-axis, and translation $v$ along the Y-axis when all frame members are extensible. If axial deformations of all frame members are not allowed, displacements $u$ and $v$ are constrained to be zero due to the inextensible constraint. Thus, this frame possesses only one degree of freedom, that is $\theta$. This constrained condition will be explored in a later chapter.

Depending on the structural type, there may be one or several displacements at each node. In most cases, the numbers of DOF in local and global coordinates are the same. One obvious exception is the 2-node truss member. In the local coordinates, the axial deformation is described by one translation at each node while in the global coordinates, the axial deformation is described by two and three translations for two- and three-dimensional problems, respectively, at each node.

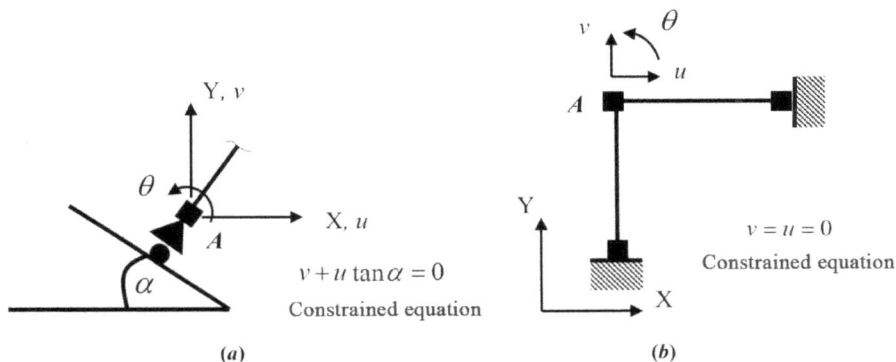

**FIGURE 1.6** Independent Displacements. (a) Inclined-Support Constraint, (b) Inextensible Constraint.

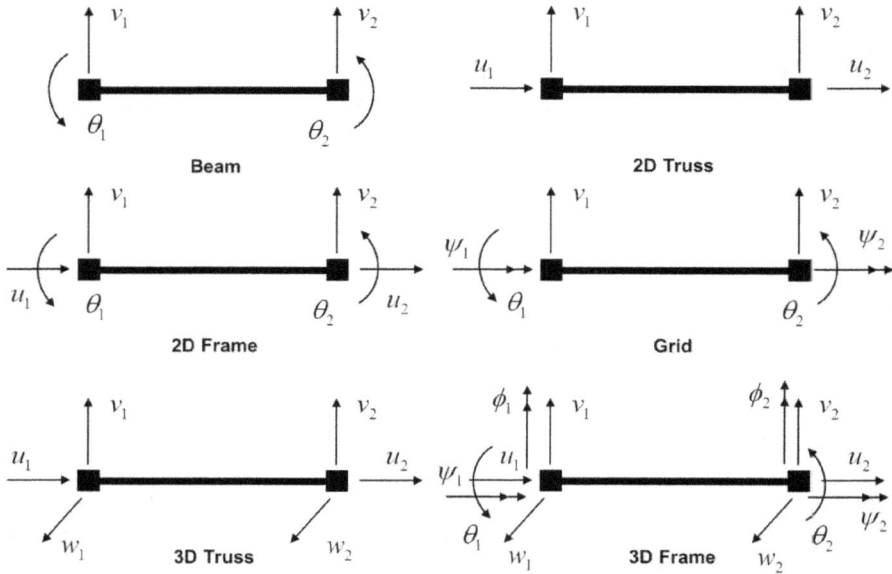

**FIGURE 1.7**   Types of Frame (Skeletal) Members.

**TABLE 1.1**
**Degrees of Freedom for Different Types of Skeletal Members**

| Type | Node 1 | Node 2 | $K_{local}$ | $K_{global}$ |
|------|--------|--------|-------------|--------------|
| 2D Beam | $F_{y1}, M_{z1}$ | $F_{y1}, M_{z1}$ | 4×4 | 4×4 |
|  | $v_1, \theta_1$ | $v_2, \theta_3$ |  |  |
| 2D Truss | $F_{x1}$ | $F_{x2}$ | 2×2 | 4×4 |
|  | $u_1$ | $u_2$ |  |  |
| 2D Frame | $F_{x1}, F_{y1}, M_{z1}$ | $F_{x2}, F_{y2}, M_{z2}$ | 6×6 | 6×6 |
|  | $u_1, v_1, \theta_1$ | $u_2, v_2, \theta_2$ |  |  |
| Grid | $T_{x1}, F_{y1}, M_{z1}$ | $T_{x2}, F_{y2}, M_{z2}$ | 6×6 | 6×6 |
|  | $\Psi_1, v_1, \theta_1$ | $\Psi_2, v_2, \theta_2$ |  |  |
| 3D Truss | $F_{x1}, F_{y1}, F_{z1}$ | $F_{x2}, F_{y2}, F_{z2}$ | 2×2 | 6×6 |
|  | $u_1, v_1, w_1$ | $u_2, v_2, w_2$ |  |  |
| 3D Frame | $F_{x1}, F_{y1}, F_{z1}, T_{x1}, M_{y1}, M_{z1}$ | $F_{x2}, F_{y2}, F_{z2}, T_{x2}, M_{y2}, M_{z2}$ | 12×12 | 12×12 |
|  | $u_1, v_1, w_1, \Psi_1, \phi_1, \theta_1$ | $u_2, v_2, w_2, \Psi_2, \phi_2, \theta_2$ |  |  |

Consequently, it is important to understand the degrees of freedom associated with the various types of *skeletal* structures as shown in Figure 1.7. Nodal displacement DOFs and the corresponding nodal force DOFs for various types of members are summarized in Table 1.1.

## 1.7   MEMBER AND STRUCTURE

Whether the matrix displacement or matrix force method is employed to analyze the structure, the analysis processes are performed at two levels; namely, member level and structure level. In this textbook, the main interests will be on the frame (skeletal) structures shown in Figure 1.8.

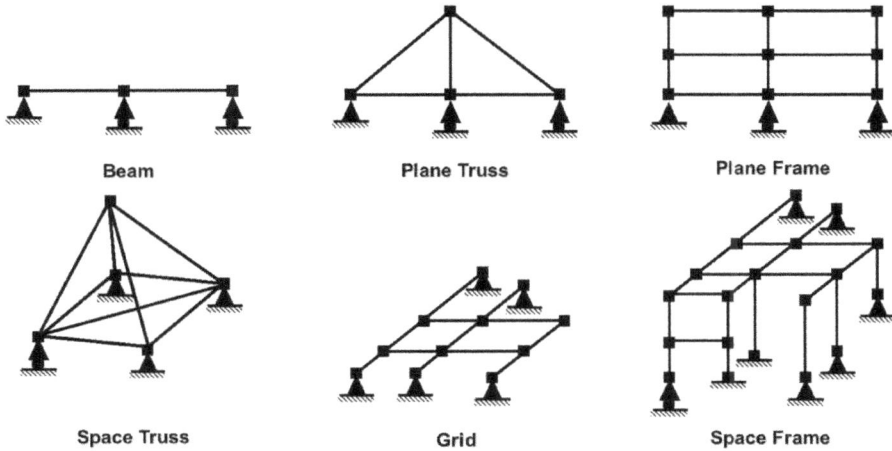

**FIGURE 1.8** Types of Frame (Skeletal) Structures.

At member level, each member (Figure 1.7) is considered independently as a free body with rigid body mode in the matrix displacement method (direct stiffness approach) and without rigid body mode in the matrix force method (flexibility approach). Member properties (e.g. stiffness or flexibility matrix, member load vector, etc.) are determined. At this level, certain quantities of member displacements or forces are treated as unknowns, and member equations are expressed in terms of these unknowns.

At structural level, structural equations gathered from the member equations are expressed in terms of the unknowns of all connected members. However, these unknowns are coupled even though they are treated independently at the member level. Consequently, there is a need for a linkage between the member and structural quantities. This could be accomplished through the member mapping vector ($LM$). This vector of a given member links the structural degree of freedom to each one of the member degree of freedom. It is best illustrated how to construct this vector through the following examples.

However, before proceeding further, it is noted that the reader should have a sufficient background on how to number members, nodes, and degree of freedom of a given structure. If this is not the case, it is advisable to consult his/her textbook on basic structural analysis.

## EXAMPLE 1.1

For the given members, nodes, and DOFs numbering systems for the truss shown in Figure 1.9, construct the mapping vectors for all members.

**Solution:** Based on the given members, nodes, and DOFs numbering systems, the mapping vector $LM$ of member 2 is

$$LM_2 = \begin{Bmatrix} 1 \\ 2 \\ 3 \\ 4 \end{Bmatrix}_{member} \begin{matrix} \Rightarrow \\ \Rightarrow \\ \Rightarrow \\ \Rightarrow \end{matrix} \begin{Bmatrix} 5 \\ 6 \\ 7 \\ 8 \end{Bmatrix}_2$$

This basically indicates that member DOF 1 links to structural DOF 5, member DOF 2 links to structural DOF 6, member DOF 3 links to structural DOF 7, and member DOF 4 links to

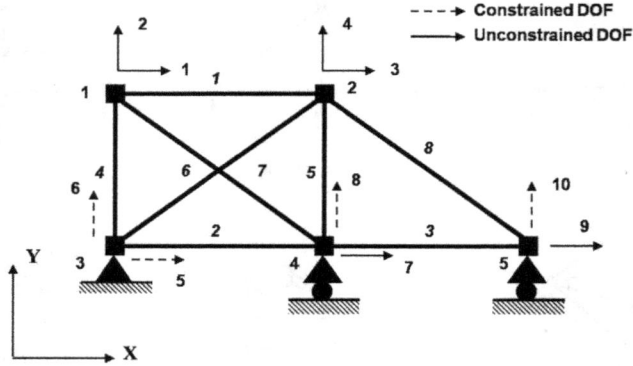

**FIGURE 1.9**   Example 1.1.

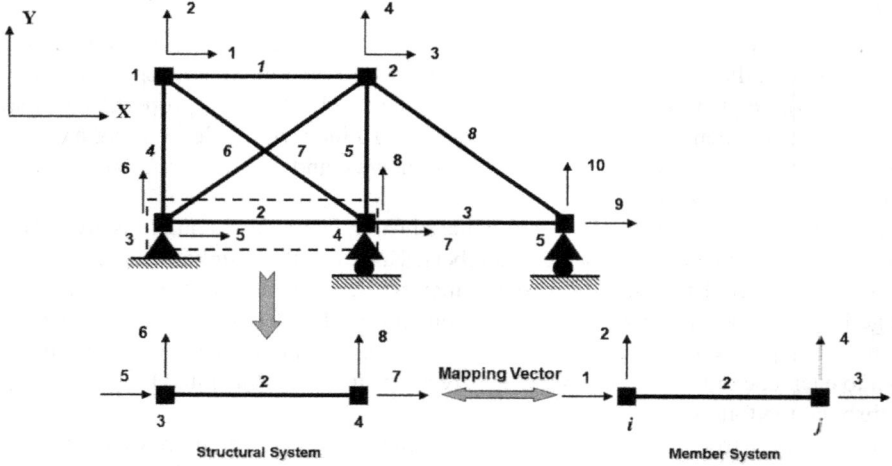

**FIGURE 1.10**   Example 1.1 (Continued).

structural DOF 8. The process of constructing the mapping vector $LM_2$ is schematically presented in Figure 1.10.

Following the same approach, the mapping vectors of other members are

$$LM_1 = \begin{Bmatrix} 1 \\ 2 \\ 3 \\ 4 \end{Bmatrix}_1 \; ; \; LM_3 = \begin{Bmatrix} 7 \\ 8 \\ 9 \\ 10 \end{Bmatrix}_3 \; ; \; LM_4 = \begin{Bmatrix} 5 \\ 6 \\ 1 \\ 2 \end{Bmatrix}_4 \; ; \; LM_5 = \begin{Bmatrix} 7 \\ 8 \\ 3 \\ 4 \end{Bmatrix}_5 \; ; \; LM_6 = \begin{Bmatrix} 5 \\ 6 \\ 3 \\ 4 \end{Bmatrix}_6 \; ; \; LM_7 = \begin{Bmatrix} 1 \\ 2 \\ 7 \\ 8 \end{Bmatrix}_7 \; ; \; LM_8 = \begin{Bmatrix} 3 \\ 4 \\ 9 \\ 10 \end{Bmatrix}_8 \; ;$$

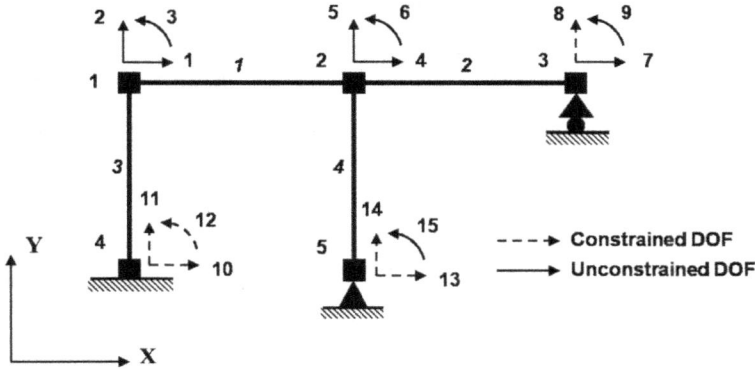

**FIGURE 1.11**   Example 1.2.

**EXAMPLE 1.2**

For the given members, nodes, and DOFs numbering systems for the frame shown in Figure 1.11, construct the mapping vectors for all members.

**Solution:** Based on the given members, nodes, and DOFs numbering systems, the mapping vector $LM$ of member 1 is

$$LM_1 = \begin{Bmatrix} 1 \\ 2 \\ 3 \\ 4 \\ 5 \\ 6 \end{Bmatrix}_{member} \begin{matrix} \Rightarrow \\ \Rightarrow \\ \Rightarrow \\ \Rightarrow \\ \Rightarrow \\ \Rightarrow \end{matrix} \begin{Bmatrix} 1 \\ 2 \\ 3 \\ 4 \\ 5 \\ 6 \end{Bmatrix}_1$$

This basically indicates that member DOF 1 links to structural DOF 1, member DOF 2 links to structural DOF 2, member DOF 3 links to structural DOF 3, member DOF 4 links to structural DOF 4, member DOF 5 links to structural DOF 5, and member DOF 6 links to structural DOF 6. The process of constructing the mapping vector $LM_1$ is schematically presented in Figure 1.12.

Following this approach, the mapping vectors of other members are

$$LM_2 = \begin{Bmatrix} 4 \\ 5 \\ 6 \\ 7 \\ 8 \\ 9 \end{Bmatrix}_2 \; ; \; LM_3 = \begin{Bmatrix} 10 \\ 11 \\ 12 \\ 1 \\ 2 \\ 3 \end{Bmatrix}_3 \; ; \; LM_4 = \begin{Bmatrix} 13 \\ 14 \\ 15 \\ 4 \\ 5 \\ 6 \end{Bmatrix}_4$$

## 1.8   MATRIX VS. FINITE ELEMENT METHODS

Even though the main focus in this textbook is on matrix methods of structural analysis, it is worth introducing the general idea of the finite element method. To a certain extent, the reader

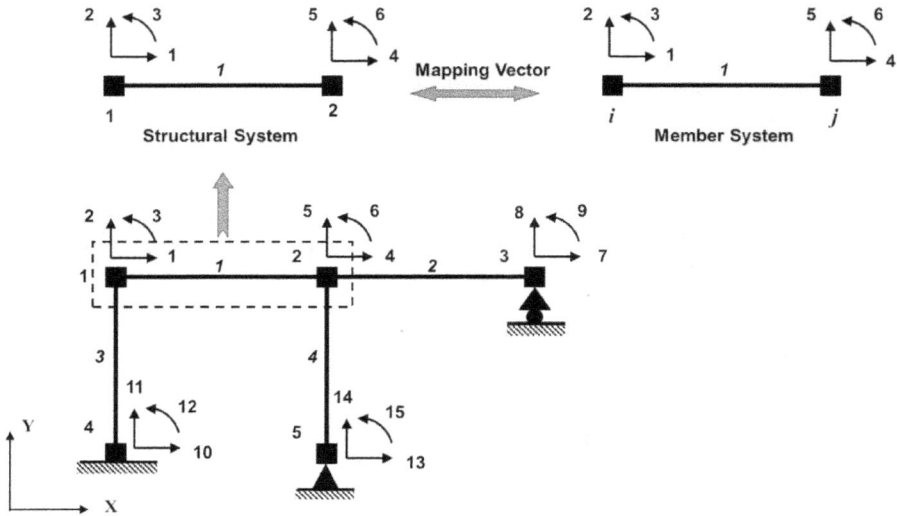

**FIGURE 1.12** Example 1.2 (Continued).

might have heard the term "finite element method" due to its popularity among academic and industrial communities. Matrix methods that will mainly be pursued in this textbook can be used to analyze only structures consisting of beams, frames, and trusses. In general, only a single frame member (shown in Figure 1.7) is sufficient to represent each structural component since its member force-displacement relations can be derived in an "exact" manner with certain underlying assumptions (e.g. prismatic member, homogeneous material, etc.). However, this single-frame member representation becomes invalid for certain types of members, for example: tapered beams (see Figure 1.13a) or beams on elastic foundations. In these circumstances, one needs to discretize the beam member into a finite number of simple beam elements as shown in Figure 1.13b. The heart of the finite element method stems from the realization that finding a "*good*" solution over the entire domain (e.g. a tapered beam) may be quite awkward and hardly possible. Consequently, the whole domain is discretized into several sub-domains (elements), and an approximate solution is constructed only inside each element domain based on energy principles and interpolation concepts. This is feasible since solution fields are separately assumed for each element domain and integral over the entire solution domain, usually occurring in the expression for energy or work, can be regarded as the sum of the integrals over the various element domains. At the limit, as the number of elements approaches infinity, the approximate solution should eventually converge to the true solution. For continuous structures (e.g. dams, flat slabs) which cannot be modeled by frame elements, two- or three-dimensional elements such as

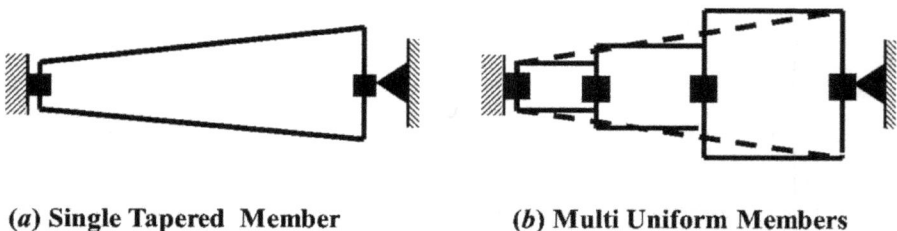

(*a*) **Single Tapered Member**          (*b*) **Multi Uniform Members**

**FIGURE 1.13** Structural Idealization of the Tapered Beam.

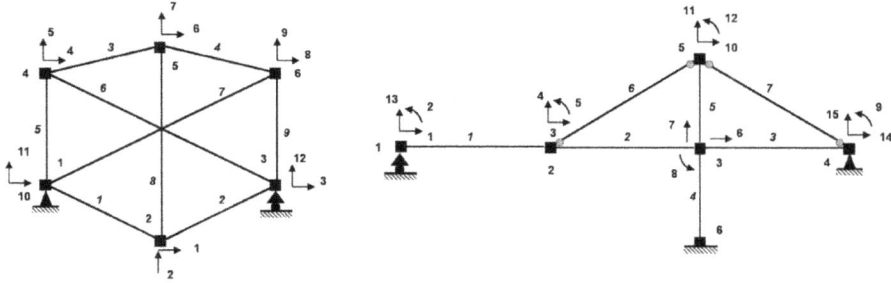

**FIGURE 1.14** Problem 1.1.

triangular or tetrahedral elements are required to model these structures, and this subject will be pursued in a course on the Finite Element Method (see Figure 1.1).

## 1.9 EXERCISE

**Problem 1.1:** For the given members, nodes, and DOFs numbering systems for the structures shown in Figure 1.14, construct the mapping vectors for all members.

## REFERENCES

Kanchi, M.B. 1993. *Matrix methods of structural analysis* (2nd Edition). Wiley Eastern Limited.

Mahasuverachai, M. and G.H. Powell. 1982. *Inelastic analysis of piping and tubular structures* (EERC Report 82/27). Earthquake Engineering Research Center, University of California.

Millais, M. 2017. *Building structures: Understanding the basics* (3rd Edition). Routledge.

Norris, C.H., J.B. Wilbur, and S. Utku. 1984. *Elementary structural analysis* (3rd Edition). McGraw-Hill Inc.

Sae-Long, W., S. Limkatanyu, C. Hansapinyo, W. Prachasaree, J. Rungamornrat, and M. Kwon. 2021. Nonlinear flexibility-based beam element on Winkler-Pasternak foundation. *Geomechanics and Engineering* 24(4): 371–388.

Saouma, V.E. 1997. *CVEN 5525 course notes: Matrix structural analysis.* Department of Civil, Environmental, and Architectural Engineering, University of Colorado.

Tonti, E. 1976. The reason for analogies between physical theories. *Applied Mathematical Modeling* 1(1): 37–50.

Wang, C.K. 1983. *Intermediate structural analysis.* McGraw-Hill Inc.

Zeris, C.A. and S.A. Mahin. 1988. Analysis of reinforced concrete beam-columns under uniaxial excitation. *ASCE Journal of Structural Engineering* 114(4): 804–820.

# 2 Frame Theories

## 2.1 INTRODUCTION

A frame-wall building shown in Figure 2.1a is composed of beams, columns, masonry infills, and structural wall. Various types of frame (skeletal) members presented in Chapter 1 can be combined to model this frame-wall building as shown in Figure 2.1b. Beams can be modeled by Euler-Bernoulli beam member, columns can be modeled by Euler-Bernoulli frame member, masonry infills can be modeled by bar member, and structural wall can be modeled by Timoshenko frame member. The main objective of this chapter is to derive the governing differential equations of these skeletal members. It is worth mentioning that along each frame member, there are regions where conventional beam theories (Euler-Bernoulli or Timoshenko beam theory) are applicable, usually away from beam-column connection joints and loading points, and regions where conventional beam theories are not applicable (typically beam-column connection joints, corbels, etc.). The former regions are named *B-Regions* where *B* stands for beam, and the latter regions are named *D-Regions* where *D* stands for discontinuity, disturbance, or detail and are necessarily discretized by continuum finite element models. More comprehensive details on *B* and *D* regions are presented in a notable paper by Schlaich et al. (1987).

This chapter focuses only on the *B*-regions and derives the governing differential equations of the classical Euler-Bernoulli and Timoshenko beam theories.

To save the best for last, a bar member instead of beams is first discussed since the governing differential equations of this member type is simpler than those of the beams. It is noted that all governing differential equations presented in this chapter are derived based on the assumption that deformations and displacements are small; hence equilibrium equations are enforced with respect to the undeformed configuration. From now on, the terms "bar" and "beam" are used to replace the term "frame" in subsequent sections for the sake of relevance.

## 2.2 BAR MEMBER

The free body diagram of an infinitesimal portion of a frame member under axial load is shown in Figure 2.2. The axial equilibrium condition in the bar leads to the following equation:

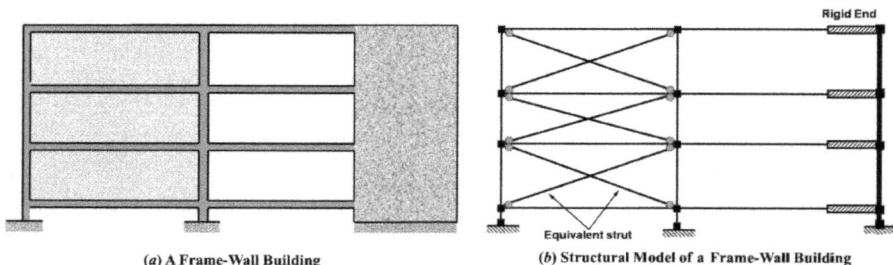

(a) A Frame-Wall Building

(b) Structural Model of a Frame-Wall Building

**FIGURE 2.1** A Frame-Wall and Its Structural Model.

DOI: 10.1201/9781003595458-2

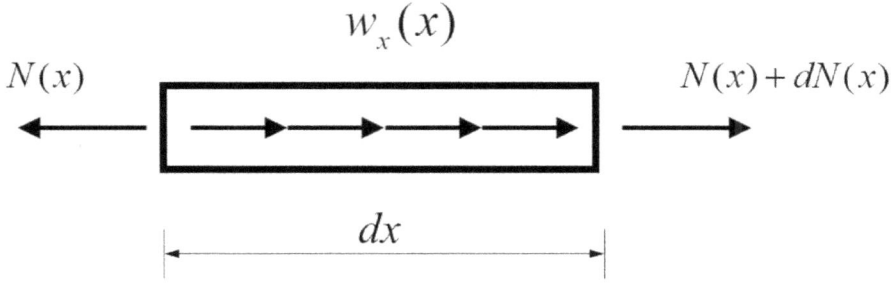

**FIGURE 2.2**   Infinitesimal Segment of a Bar.

*Equilibrium:*

$$-N(x) + N(x) + dN(x) + w_x(x)dx = 0 \quad \Rightarrow \quad \frac{dN(x)}{dx} = -w_x(x) \qquad (2.1)$$

where $N(x)$ is the axial force in the bar and $w_x(x)$ is the axially distributed load. It is noted that the axial force $N(x)$ can be regarded as the resultant stress since it is defined as an integral of the axial stress $\sigma_{xx}(x)$ over the bar sectional area $A$:

$$N(x) = \int_A \sigma_{xx}(x)dA \qquad (2.2)$$

Based on the small displacement assumption, the axial strain $\varepsilon_0(x)$ is related to the axial displacement $u_0(x)$ through the following compatibility equation:

*Compatibility:*

$$\varepsilon_0(x) = \frac{du_0(x)}{dx} \qquad (2.3)$$

The strain measure defined in Eq. (2.3) is commonly known as "uniaxial engineering strain".

Relation between the axial force and the axial strain is defined through the constitutive law:

*Constitutive law:*

$$N(x) = EA(x)\varepsilon_0(x) \qquad (2.4)$$

where $EA(x)$ is the axial rigidity of the bar.

Substituting Eqs. (2.3) and (2.4) into (2.1) leads to the governing differential equation:

*Governing differential equation:*

$$\frac{d}{dx}\left[EA(x)\frac{du_0(x)}{dx}\right] = -w_x(x) \qquad (2.5)$$

In the simple case of constant axial rigidity $EA$, the governing differential Eq. (2.5) can be simplified as:

$$EA(x) = EA \quad \Rightarrow \quad EA\frac{d^2u_0(x)}{dx^2} = -w_x(x) \qquad (2.6)$$

The general form of the solution to the governing differential Eq. (2.5) is:

$$u_0(x) = u_0^h(x) + u_0^p(x) \tag{2.7}$$

where $u_0^h(x)$ is the homogeneous solution which represents the axial displacement in the absence of member load, while $u_0^p(x)$ is the particular solution which represents the axial displacement due to the presence of member load. It is assumed that the reader should have a sufficient background in engineering mathematics. If this is not the case, it is advisable to consult his/her textbook on engineering mathematics (e.g. Kreyszig 2011).

The governing differential equation of the bar is of second order in the unknown displacement field $u_0(x)$. In addition to the governing differential equation, boundary conditions are necessarily added in order to determine the solution to the problem. Boundary conditions can be classified into two categories: *Essential* and *Natural*. The former is also known as the Dirichlet boundary type, while later is also known as the Neumann boundary type. Both boundary conditions are physically interpreted, respectively, as constraints on displacements (geometric) and forces at the member ends. Figure 2.3 provides the general representation of solution domain $\Omega$, essential boundary $\Gamma_u$, and natural boundary $\Gamma_t$ for one- and two-dimensional bodies.

Mathematically speaking, one is seeking for the function $u_0(x)$, which possesses the following characteristics:

1. The axial force field $N(x)$ computed from the axial displacement field $u_0(x)$ through compatibility (2.3) and constitutive (2.4) relations must satisfy the differential equilibrium Eq. (2.1) in the pointwise sense throughout the solution domain $\Omega\,(0 < x < L)$.
2. It must satisfy the specified boundary conditions at the member ends (at $x=0$ and $x=L$).

It is noted that, for a given degree of freedom, only one of the two boundary conditions can be prescribed. In the case of the bar, one has either

$$u_0(0) = u_0 \text{ or } N(0) = N_0 \tag{2.8}$$

and either

$$u_0(L) = u_L \text{ or } N(L) = N_L \tag{2.9}$$

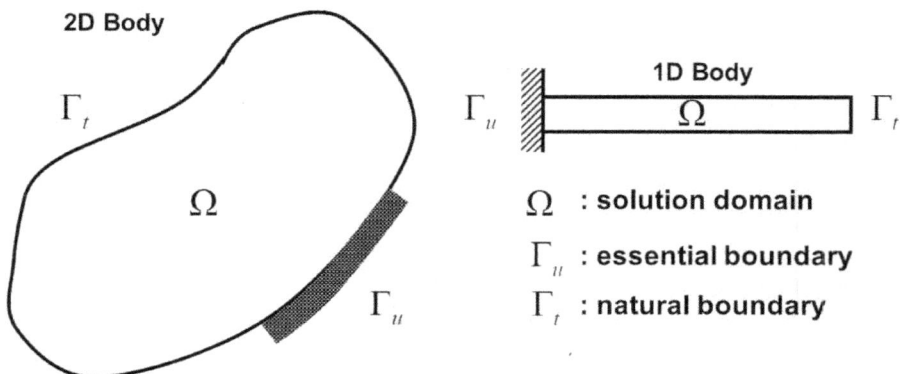

**2D Body**

$\Gamma_t$

$\Omega$

$\Gamma_u$

**1D Body**

$\Gamma_u$    $\Omega$    $\Gamma_t$

$\Omega$ : solution domain

$\Gamma_u$ : essential boundary

$\Gamma_t$ : natural boundary

**FIGURE 2.3** Illustrations of Main Solution Domain $\Omega$, Essential Boundary Conditions $\Gamma_u$, and Natural Boundary Conditions $\Gamma_t$ ($\Gamma_u \cup \Gamma_t = \Gamma$).

In other words, either force or displacement can be prescribed at the same boundary. Figure 2.4 shows all four possible combinations of boundary constraints. However, only the first three (Figure 2.4a, b, and c) are kinematically admissible, while the last one in Figure 2.4d is kinematically inadmissible since it does not inhibit the bar from translating as a rigid body.

It is worth mentioning that an elegant way to represent together the governing equations ((2.1), (2.3), and (2.4)) of the bar problem and its boundary conditions (essential and natural) is via the so-called "Tonti's Diagram" shown in Figure 2.5. This diagram can also be used to present the

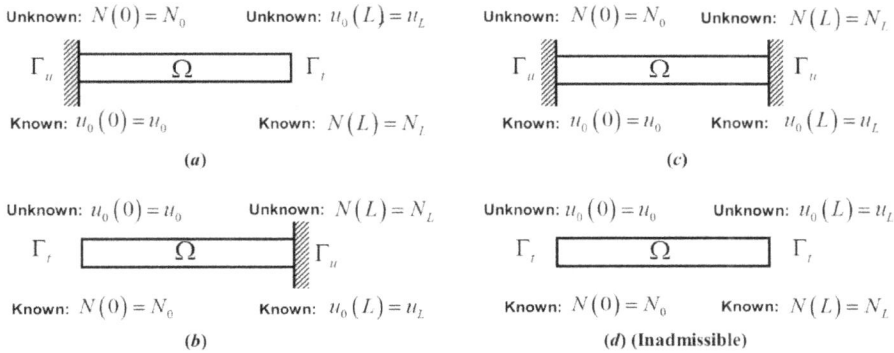

FIGURE 2.4 Four Possible Combinations of Boundary Constraints: Bar Problem.

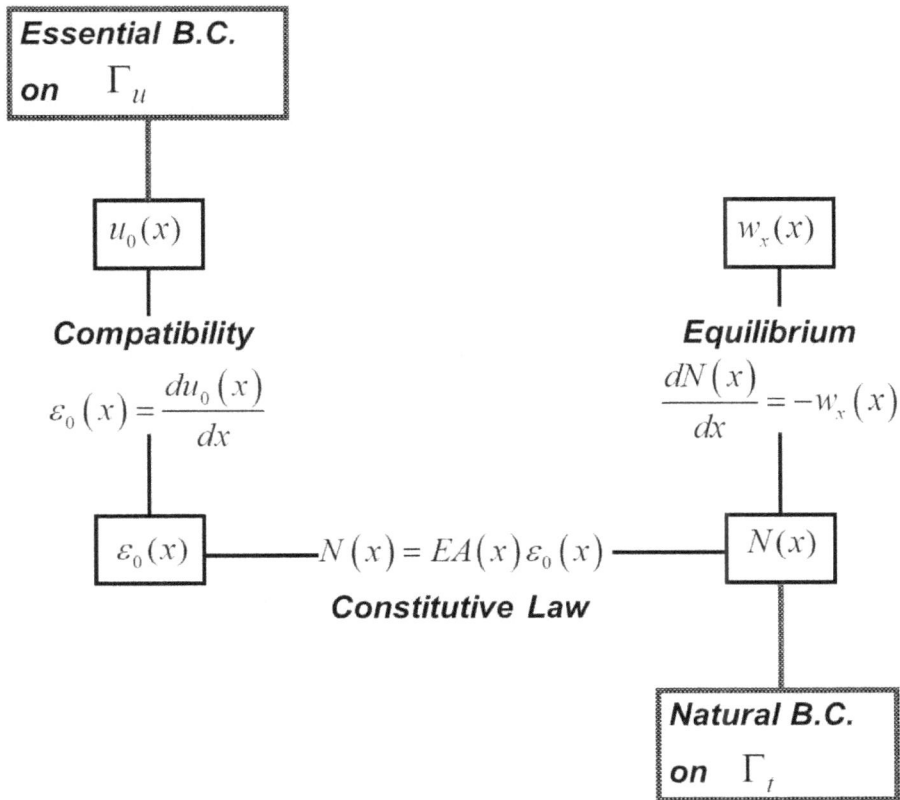

FIGURE 2.5 Tonti's Diagram for Bar Problem + Boundary Conditions.

governing differential equations of other field problems (e.g. Electromagnetism, Fluid Dynamics, Quantum Mechanics).

## EXAMPLE 2.1

Starting from the problem differential equation, find the axial displacement and axial force distributions along the constant cross-section column under its own weight and point load at the free end shown in Figure 2.6.

**Solution:** In the case of constant axial rigidity $EA$, the governing differential equation is:

$$EA\frac{d^2 u_0(x)}{dx^2} = w_x(x) = A\rho \quad \Rightarrow \quad \frac{d^2 u_0(x)}{dx^2} = \frac{\rho}{E}$$

The solution is in the form of

$$u_0(x) = u_0^h(x) + u_0^p(x)$$

where $u_0^h(x)$ is the homogeneous solution and can be defined as:

$$u_0^h(x) = C_1 x + C_2$$

$u_0^p(x)$ is the particular solution and can be defined as:

$$u_0^p(x) = \frac{\rho}{2E} x^2$$

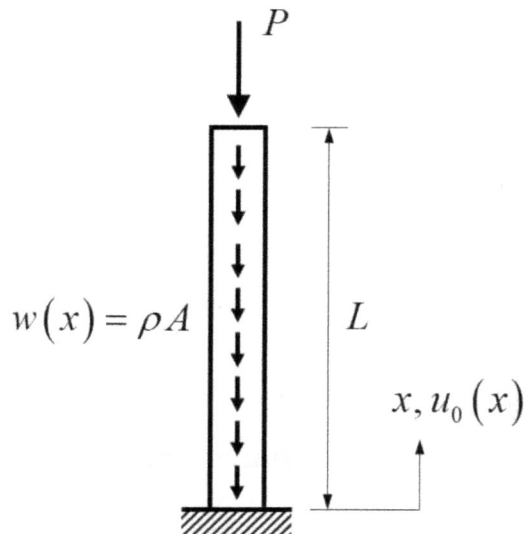

**FIGURE 2.6**  Example 2.1.

Total solution $u_0(x)$ is:

$$u_0(x) = u_0^h(x) + u_0^p(x) = C_1 x + C_2 + \frac{\rho}{2E} x^2$$

Constants $C_1$ and $C_2$ are determined from the boundary conditions.
Boundary Conditions:
Essential Boundary Condition on $\Gamma_u$:

$$u_0(0) = 0 \quad \Rightarrow \quad C_2 = 0$$

Natural Boundary Condition on $\Gamma_t$:

$$N(L) = EA \frac{du_0(x)}{dx}\bigg|_{x=L} = -P \quad \Rightarrow \quad C_1 = -\frac{\rho L}{E} - \frac{P}{EA}$$

It is observed that boundary conditions only affect the homogeneous solution.
The axial displacement field is:

$$u_0(x) = -\frac{\rho}{E}\left(Lx - \frac{x^2}{2}\right) - \frac{Px}{AE} \quad \Rightarrow \qquad \textit{Quadratic}$$

The axial strain field is:

$$\varepsilon_0(x) = \frac{du_0(x)}{dx} = -\frac{\rho}{E}(L-x) - \frac{P}{EA} \quad \Rightarrow \qquad \textit{Linear}$$

The axial force field is:

$$N(x) = EA\varepsilon_0(x) = -\rho A(L-x) - P \quad \Rightarrow \qquad \textit{Linear}$$

It is left to the reader to verify both global and local equilibriums.

## 2.3   BEAM MEMBER (EULER-BERNOULLI AND TIMOSHENKO BEAM THEORIES)

Two widely used beam theories are discussed here: the Euler-Bernoulli and the Timoshenko beams. The Euler-Bernoulli beam considers only flexural deformation, while the Timoshenko beam deals with both flexural and shear deformations. Before pursuing further the beam theories, certain assumptions on beam theories should be pointed out:

1. *Plane Symmetry*: The longitudinal axis is straight, and the cross-section of the beam has a longitudinal plane of symmetry. The resultant of the transverse loads acting on each section lies on that plane. The support conditions are also symmetric about this plane. This results in symmetrical bending.

2. *Cross-Section*: The beam cross-section is either constant or varies gradually (tapered beam).
3. *Geometrical Linearity*: Transverse displacements, rotations, and deformations are small so that the assumptions of infinitesimal deformations are valid.
4. *Material Linearity*: The material is assumed to be elastic and isotropic. Heterogeneous beams fabricated with several isotropic materials, such as reinforced concrete, are excluded.

### 2.3.1 EULER-BERNOULLI BEAM THEORY

The fundamental hypothesis of the Euler-Bernoulli beam theory is that "Plane sections remain plane and normal to the longitudinal axis". This assumption is associated with the name of Kirchhoff for plate and with Love for shell. The kinematics description and deformation definition of cross-section points for the Euler-Bernoulli beam are presented in Figure 2.7 and Figure 2.8, respectively. The cross-section ab is normal to the longitudinal axis of the undeformed beam. In the deformed configuration, the deformed cross-section a′b′ remains plane and is normal to the longitudinal axis. This simply implies that the axial displacement $u(x,y)$ and vertical displacement $v(x,y)$ fields at any point with a distance $y$ from the longitudinal axis can be written as:

$$u(x,y) = u_0(x) - y\frac{dv_0(x)}{dx} \text{ and } v(x,y) = v_0(x) \tag{2.10}$$

The corresponding deformations are:
   The axial strain (engineering strain):

$$\varepsilon(x,y) = \frac{du(x,y)}{dx} = \frac{du_0(x)}{dx} - y\frac{d^2v_0(x)}{dx^2} = \varepsilon_0(x) - y\kappa_0(x) \tag{2.11}$$

The shear strain:

$$\gamma_0(x) = \frac{dv(x,y)}{dx} + \frac{du(x,y)}{dy} = \frac{dv_0(x)}{dx} - \frac{dv_0(x)}{dx} = 0 \tag{2.12}$$

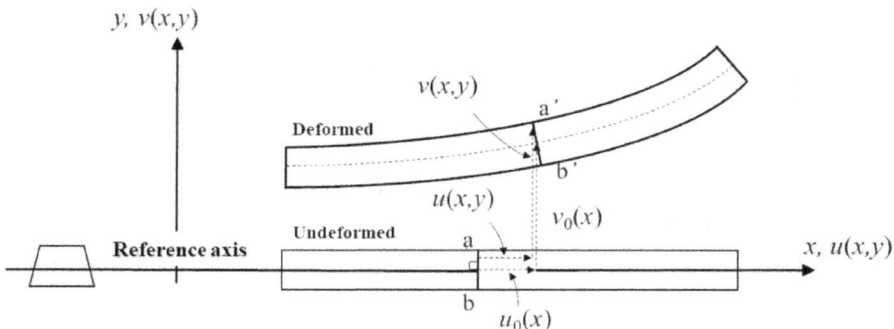

**FIGURE 2.7**   Kinematics Description of the Cross-Section Points for the Euler-Bernoulli Beam.

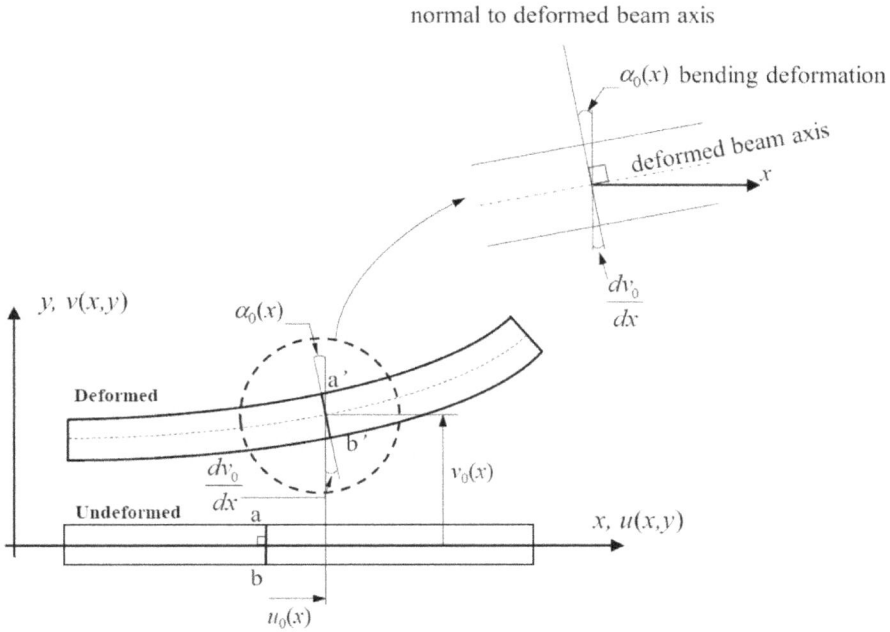

**FIGURE 2.8**   Sectional Deformation of the Euler-Bernoulli Beam.

where $\varepsilon_0(x) = du_0(x)/dx$ is the strain at the reference axis and $\kappa_0(x) = d^2v_0(x)/dx^2$ is the section curvature.

It is noted that Eq. (2.12) simply implies that based on the kinematical assumption shown in Figure 2.7, the sectional shear strain vanishes.

*Equilibrium:*

Free body diagram of an infinitesimal portion of a beam member under transverse load is shown in Figure 2.9. Vertical equilibrium condition of the beam leads to the following relation:

$$\frac{dV(x)}{dx} + w_y(x) = 0 \tag{2.13}$$

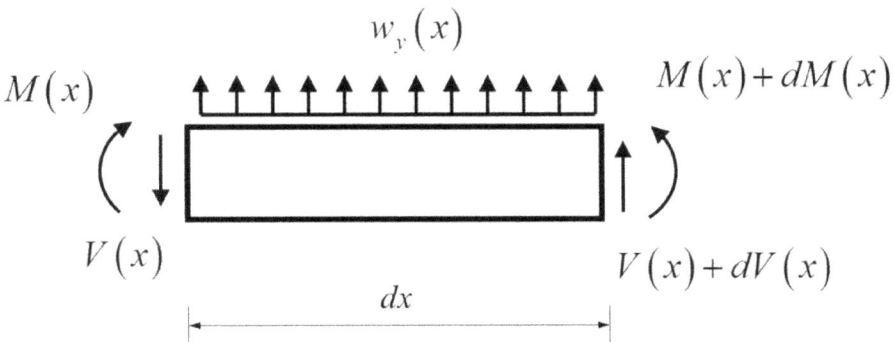

**FIGURE 2.9**   Infinitesimal Segment of a Beam.

where $V(x)$ is the shear force acting on the beam section, and $w_y(x)$ is the vertical distributed load. Moment equilibrium yields the following relation:

$$\frac{dM(x)}{dx} + V(x) = 0 \tag{2.14}$$

where $M(x)$ is the bending moment acting on the beam section.

It is noted that the bending moment $M(x)$ can be regarded as the resultant stress since it is defined as an integral of the axial stress $\sigma_{xx}(x)$ over the beam sectional area $A$:

$$M(x) = -\int_A y\sigma_{xx}(x)dA \tag{2.15}$$

Based on the Euler-Bernoulli beam hypothesis, the effects of shear force $V(x)$ on beam behavior are not considered. This is a reasonable approximation for members with medium to large span-to-depth ratios. As a result, the shear force $V(x)$ should be removed from Eq. (2.14). This can be accomplished by differentiating Eq. (2.14) with respect to $x$ and combining it with Eq. (2.13) to obtain:

$$\frac{d^2M(x)}{dx^2} - w_y(x) = 0 \tag{2.16}$$

Based on the small displacement assumption, the sectional curvature $\kappa_0(x)$ is related to the transverse displacement $v_0(x)$ through the following compatibility equation:

*Compatibility:*

$$\kappa_0(x) = \frac{d^2v_0(x)}{dx^2} \tag{2.17}$$

*Constitutive law:*

The relation between the sectional moment and sectional curvature is defined through the constitutive law:

$$M(x) = IE(x)\kappa_0(x) \tag{2.18}$$

where $IE(x)$ is the flexural rigidity of the beam.

Substituting Eqs. (2.17) and (2.18) into (2.16) leads to the governing differential equation:

*Governing differential equation:*

$$\frac{d^2}{dx^2}\left[IE(x)\frac{d^2v_0(x)}{dx^2}\right] = w_y(x) \tag{2.19}$$

In the simple case of constant flexural rigidity $IE$, the governing differential Eq. (2.19) can be simplified as:

$$IE(x) = IE \quad \Rightarrow \quad IE\frac{d^4v_0(x)}{dx^4} = w_y(x) \tag{2.20}$$

The governing differential equation of the Euler-Bernoulli beam is of fourth order in the unknown field $v_0(x)$. In addition to the above governing differential equation, boundary conditions are necessarily added in order to solve the problem. The essential and natural boundary conditions are physically interpreted as constraints on displacements (transverse displacement $v_0(x)$ and rotation $v_0'(x)$) and forces (shear $V(x)$ and moment $M(x)$) at the member ends, respectively.

Mathematically speaking, one is seeking for the function $v_0(x)$, which possesses the following attributes:

1. The bending moment field $M(x)$ computed from the transverse displacement field $v_0(x)$ through compatibility (2.17) and constitutive (2.18) relations must satisfy the differential equilibrium Eq. (2.16) in the point-wise sense throughout the solution domain $\Omega \left(0 < x < L\right)$.
2. It must satisfy the specified boundary conditions at the member ends (at $x=0$ and $x=L$).

Unlike the bar problem, in the beam problem, both essential and natural conditions can be prescribed at the same boundary as shown, for example in Figure 2.10.

At the left end ($x=0$), one can specify:

$$v_0\left(0\right)=0 \quad \Rightarrow \quad Essential \, and \, M\left(0\right)=M_0 \quad \Rightarrow \quad Natural \tag{2.21}$$

Similarly, at the right end ($x=L$), one can prescribe:

$$\theta\left(L\right)=v_0'\left(L\right)=0 \quad \Rightarrow \quad Essential \, and \, V\left(L\right)=V_0 \quad \Rightarrow \quad Natural \tag{2.22}$$

However, for a given degree of freedom, only one of the two boundary conditions can be prescribed.

Figure 2.11 shows all kinematically admissible boundary constraints of a beam. These boundary constraints are admissible in the sense that they prevent the beam from translating and rotating as a rigid body. It is also observed that at least two essential boundary conditions must be prescribed. Furthermore, the beam becomes statically indeterminate if more than two essential boundary constraints are present.

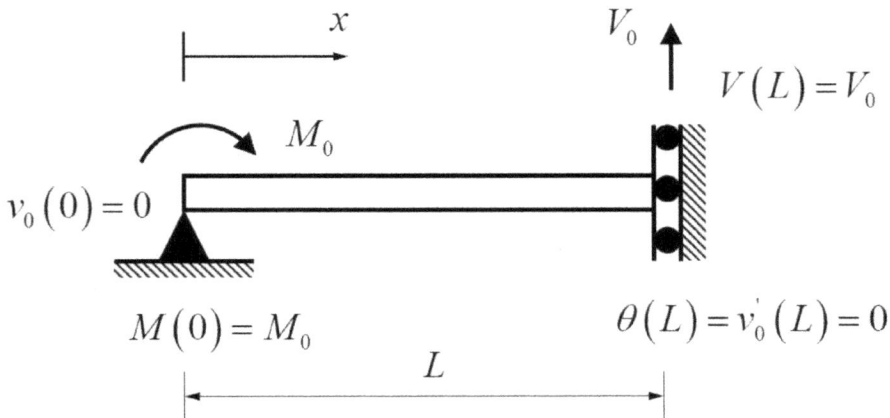

**FIGURE 2.10** An Example of Mixed Boundary Conditions for the Euler-Bernoulli Beam.

Known: $\theta(0), V(0)$          Known: $v_0(L), M(L)$

Known: $v_0(0), M(0)$          Known: $v_0(L), M(L)$

Ω                    Ω

Unknown: $V(0), \theta(0)$     Unknown: $V(L), \theta(L)$     Unknown: $M(0), v_0(0)$     Unknown: $V(L), \theta(L)$

(*a*) simply-supported beam                          (*b*) shear release-roller beam

Known: $v_0(0), \theta(0)$          Known: $v_0(L), M(L)$     Known: $v_0(0), \theta(0)$          Known: $v_0(L), \theta(L)$

Ω                    Ω

Unknown: $V(0), M(0)$     Unknown: $V(L), \theta(L)$     Unknown: $V(0), M(0)$     Unknown: $V(L), M(L)$

(*c*) propped cantilever beam                          (*d*) fixed-fixed end beam

Known: $v_0(0), \theta(0)$          Known: $V(L), M(L)$

Ω

Unknown: $V(0), M(0)$     Unknown: $v_0(L), \theta(L)$

(*e*) cantilever beam

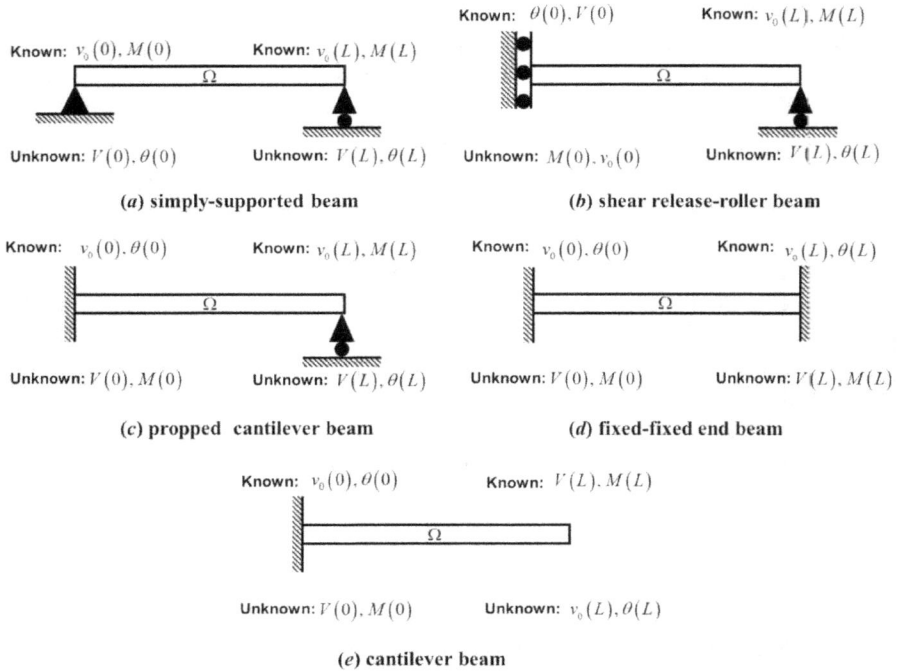

**FIGURE 2.11**  Kinematically Admissible Constraints of the Euler-Bernoulli Beam.

The governing equations ((2.16), (2.17), and (2.18)) together with boundary conditions (essential and natural) form the boundary-value problem for establishing the vertical displacement field of a beam and are schematically represented through Tonti's Diagram of Figure 2.12.

## EXAMPLE 2.2

Starting from problem differential equation, find the transverse displacement, rotation, bending moment, and shear force distributions along a prismatic Euler-Bernoulli propped cantilever beam subjected to a uniformly distributed load, as shown in Figure 2.13.

**Solution:** In the case of constant flexural rigidity *IE*, the governing differential equation is:

$$IE\frac{d^4 v_0(x)}{dx^4} = w_y(x) = -w_{y0}$$

Boundary conditions:

(1) $v_0(0)=0$: Essential

(2) $\theta(0) = \dfrac{dv_0}{dx}\bigg|_{x=0} = 0$ : Essential

(3) $M(L) = 0 \quad \Rightarrow \quad IE\dfrac{d^2 v_0}{dx^2}\bigg|_{x=L} = 0 \quad \Rightarrow \quad \dfrac{d^2 v_0}{dx^2}\bigg|_{x=L} = 0$ : Natural

(4) $v(L) = 0$ : Essential

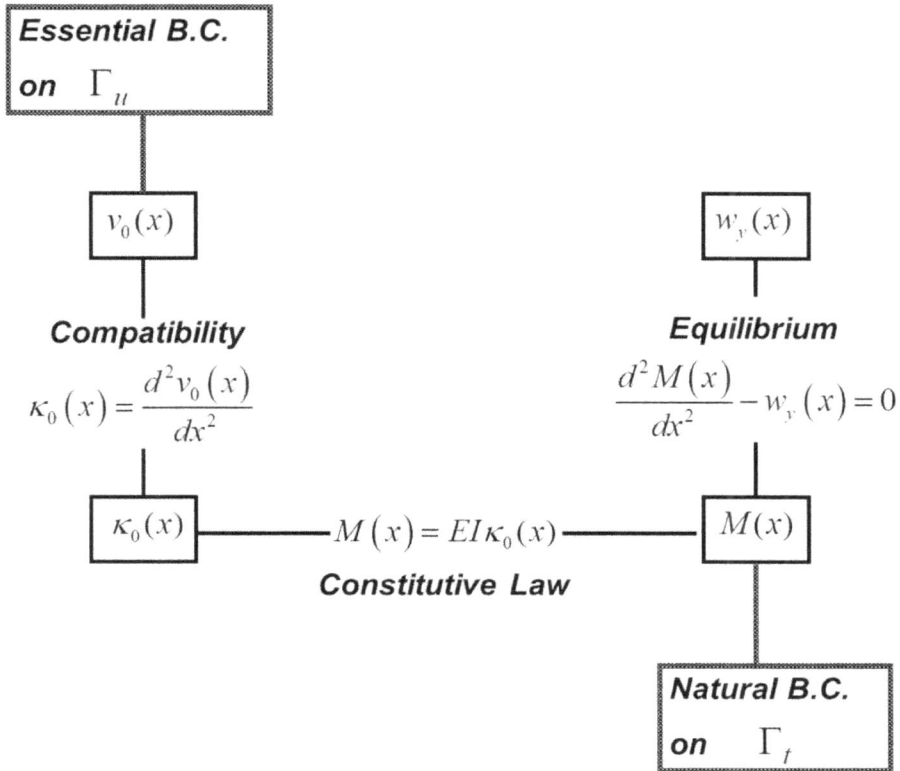

**FIGURE 2.12** Tonti's Diagram for the Euler-Bernoulli Beam Problem + Boundary Conditions.

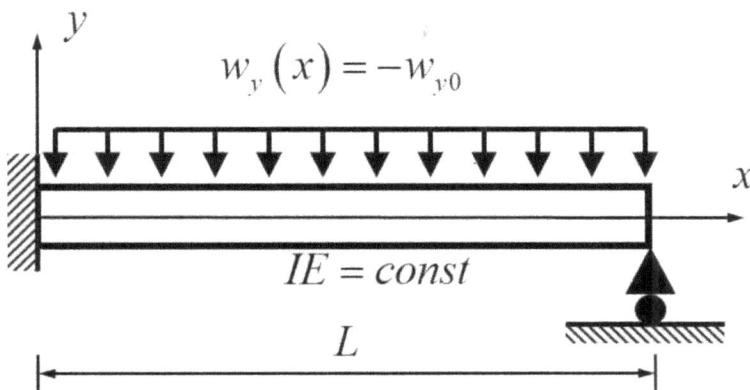

**FIGURE 2.13** Example 2.2.

The solution is in the form of

$$v_0(x) = v_0^h(x) + v_0^p(x)$$

Homogenous solution $v_0^h(x)$:

$$IE \frac{d^4 v_0(x)}{dx^4} = 0 \quad \Rightarrow \quad v_0^h(x) = C_1 + C_2 x + C_3 x^2 + C_4 x^3$$

Particular solution $v_0^p(x)$:

$$IE \frac{d^4 v_0(x)}{dx^4} = w_y(x) = -w_{y0} \quad \Rightarrow \quad v_0^p(x) = -\frac{w_{y0}}{24IE} x^4$$

The total solution is:

$$v_0(x) = C_1 + C_2 x + C_3 x^2 + C_4 x^3 - \frac{w_{y0}}{24IE} x^4$$

Constants $C_1$, $C_2$, $C_3$, and $C_4$ can be found from the four boundary conditions:

(1)    $v_0(0) = 0$:   $C_1 = 0$

(2)    $\theta(0) = \dfrac{dv_0}{dx}\bigg|_{x=0} = 0$:   $C_2 = 0$

(3)    $\dfrac{d^2 v_0}{dx^2}\bigg|_{x=L} = 0$:   $2C_3 + 6C_4 L - \dfrac{w_{y0}}{2IE} L^2 = 0$

(4)    $v_0(L) = 0$:   $C_3 L^2 + C_4 L^3 - \dfrac{w_{y0} L^4}{24IE} = 0$

From (3) and (4), $C_3 = -\dfrac{w_{y0} L^2}{16IE}$ and $C_4 = \dfrac{5 w_{y0} L}{48IE}$

The complete solution for this problem is

$$v_0(x) = -\frac{w_{y0} L^2}{16IE} x^2 + \frac{5 w_{y0} L}{48IE} x^3 - \frac{w_{y0}}{24IE} x^4$$

The bending moment along the beam is

$$M(x) = IE \frac{d^2 v_0(x)}{dx^2} = w_{y0}\left( -\frac{L^2}{8} + \frac{5}{8} Lx - \frac{x^2}{2} \right)$$

The shear force along the beam is

$$V(x) = -IE \frac{d^3 v_0(x)}{dx^3} = w_{y0}\left( \frac{5L}{8} - x \right)$$

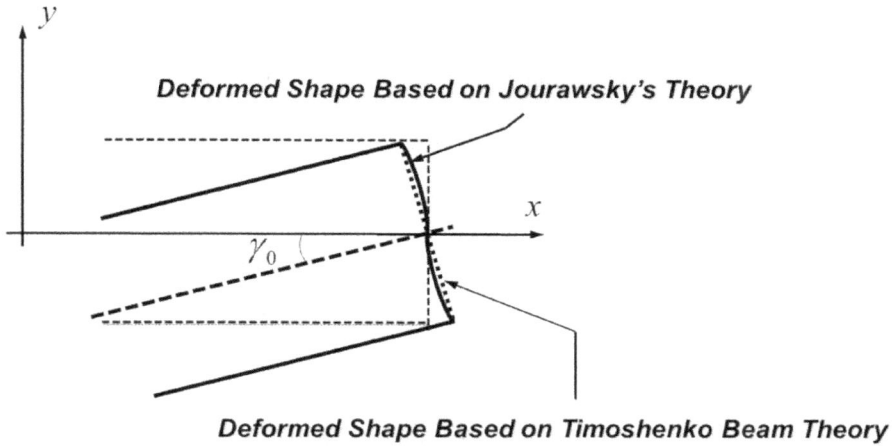

**FIGURE 2.14** Cross-Section Deformations: Jourawsky's Theory vs. Timoshenko Beam Theory.

It is left to the reader to verify both global and local equilibriums.

## 2.3.2 TIMOSHENKO BEAM THEORY

### 2.3.2.1 Theoretical Background

The Timoshenko beam theory describes a simplification of more accurate beam theory accounting for the effects of shear deformations. The fundamental hypothesis of the Timoshenko beam theory is that "***Plane sections remain plane but lose normality to the longitudinal beam axis due to shear deformation***". This assumption is associated with the name of Mindlin and Reissner (among others) for plate and with Naghdi for shell. The implication of such a hypothesis is shown in Figure 2.14 and results in a constant shear deformation (stress) over the cross-section (Spacone 2003). This is in opposition to what happens following Jourawsky's theory. The comparison between the associated constant shear stress distribution and that computed based on Jourawsky's theory is shown in Figure 2.15 (Spacone 2003).

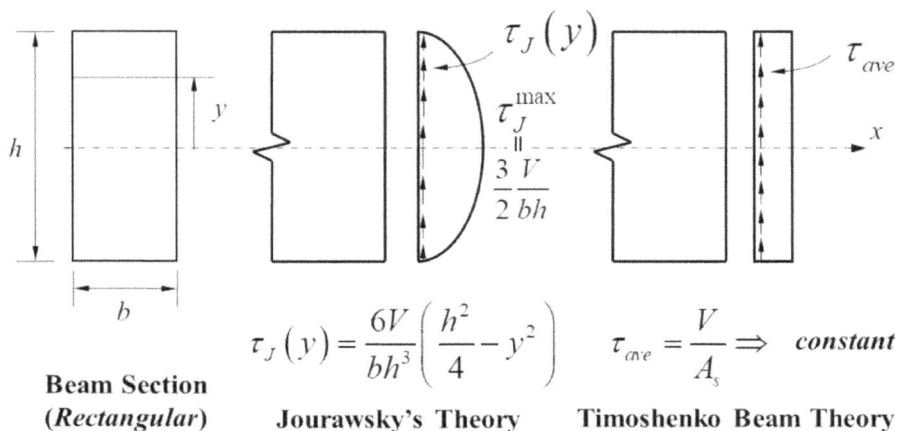

$$\tau_J(y) = \frac{6V}{bh^3}\left(\frac{h^2}{4} - y^2\right) \qquad \tau_{ave} = \frac{V}{A_s} \implies constant$$

**Beam Section**     **Jourawsky's Theory**     **Timoshenko Beam Theory**
***(Rectangular)***

**FIGURE 2.15** Shear Stress Distributions in Rectangular Cross-Section: Jourawsky's Theory vs. Timoshenko Beam Theory.

The heart of Timoshenko's argument is as follows: The shear force $V(x)$ is computed from the integral of the shear stresses over the sectional area $A$.

In the linear elastic case with constant shear modulus $G$,

$$V(x) = \int_A \tau_{xy}(x) dA = \int_A G(x) \gamma_{xy}(x) dA \Rightarrow If\ G(x) = G \Rightarrow V(x) = G \int_A \gamma_{xy}(x) dA \quad (2.23)$$

If $\gamma_0$ is the shear strain at the centroidal axis, the following relation is obtained:

$$V \le GA\gamma_0 \quad (2.24)$$

If the equality of resulting shear forces is enforced, the shear correction factor $k$ can be computed:

$$V = \int_A G\gamma_{xy} dA = (GA\gamma_0)k; \quad where\ k \le 1 \quad (2.25)$$

The shear correction factor $k$ is typically called the Timoshenko shear coefficient which depends on the cross-section shape and is defined for each type of cross-section. If one calls $A_s = kA$ the shear area, then

$$V = GA_s\gamma_0 \quad (2.26)$$

Values of $k$ for a couple of simple cross-sections are:

$$k = 5/6\ for\ Rectangular\ Section\ and\ k = 9/10\ for\ Circular\ Section \quad (2.27)$$

More thorough discussion and derivation of the Timoshenko shear coefficient can be found in Timoshenko and Gere (1972).

The total vertical displacement of the beam reference axis can be decomposed into flexural deflection and shear deflection, as shown in Figure 2.16.

$$v_0(x) = v_f(x) + v_s(x) \quad (2.28)$$

where $v_f(x)$ is the deflection due to flexure only, while $v_s(x)$ is the deflection due to shear only.

Based on Eq. (2.28) and Figure 2.16, one has:

$$\theta(x) = \frac{dv_0(x)}{dx} = \frac{dv_f(x)}{dx} + \frac{dv_s(x)}{dx} \quad (2.29)$$

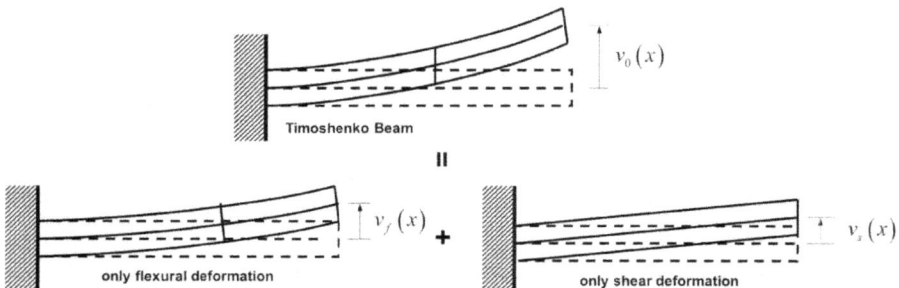

**FIGURE 2.16** Displacement Decomposition of Timoshenko Beam.

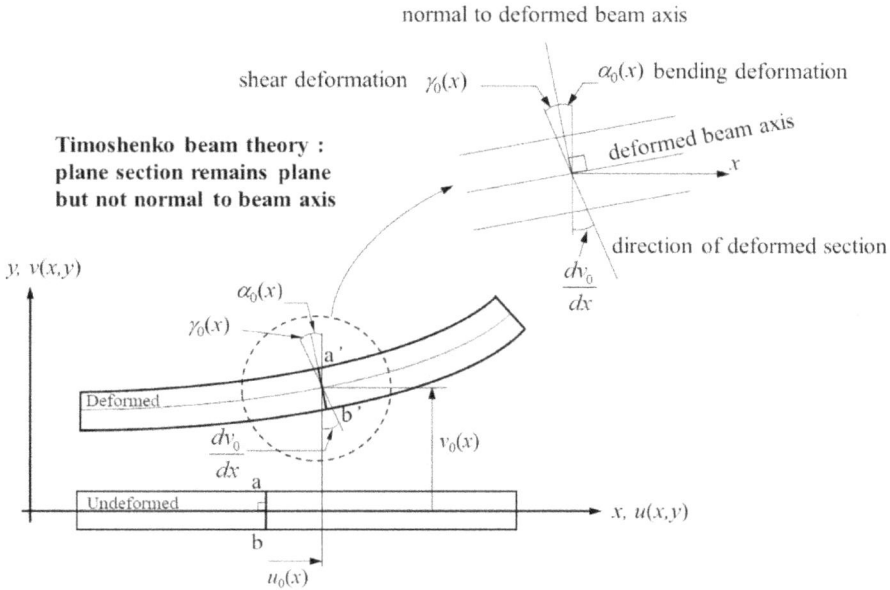

**FIGURE 2.17**   Kinematics Description of Cross-section Points for the Timoshenko Beam.

$$\theta(x) = \alpha_0(x) + \gamma_0(x) \tag{2.30}$$

Equation (2.30) infers that there are two contributions to the total section rotation $\theta(x)$; one is from the flexural rotation $\alpha_0(x) = dv_f(x)/dx$ and the other from the shear rotation $\gamma_0(x) = dv_s(x)/dx$.

From the kinematics description shown in Figure 2.17, one can observe that the deformed cross-section a′b′ remains plane but rotates with a total angle of $dv_0(x)/dx$. The difference between $dv_0(x)/dx$ and $\alpha_0(x)$ (flexural rotation) defines the section shear deformation $\gamma_0(x)$.

Based on Figure 2.17, the axial displacement $u(x,y)$ and vertical displacement $v(x,y)$ fields can be written as:

$$u(x,y) = u_0(x) - y\alpha_0(x) \text{ and } v(x,y) = v_0(x) \tag{2.31}$$

The corresponding deformations are:

$$\varepsilon(x,y) = \frac{du(x,y)}{dx} = \frac{du_0(x)}{dx} - y\frac{d\alpha_0(x)}{dx}$$
$$\gamma_0(x) = \frac{du(x,y)}{dy} + \frac{dv(x,y)}{dx} = -\alpha_0(x) + \frac{dv_0(x)}{dx} \tag{2.32}$$

The generalized section deformations are:

*Sectional axial strain at reference axis:*

$$\varepsilon_0(x) = \frac{du_0(x)}{dx} \tag{2.33}$$

*Sectional curvature:*

$$\kappa_0(x) = \frac{d\alpha_0(x)}{dx} \tag{2.34}$$

*Sectional shear strain:*

$$\gamma_0(x) = \frac{dv_0(x)}{dx} - \alpha_0(x) \tag{2.35}$$

From the constitutive relations for normal and shear stresses:

$$\sigma_{xx}(x) = E(x)\varepsilon_{xx}(x)$$
$$\tau_{xy}(x) = G(x)\gamma_{xy}(x) \tag{2.36}$$

The generalized section forces are:

*Sectional axial force at reference axis:*

$$N(x) = \int_A \sigma_{xx}(x)dA = EA(x)\frac{du_0(x)}{dx} = EA(x)\varepsilon_0(x) \tag{2.37}$$

*Sectional bending moment:*

$$M(x) = -\int_A \sigma_{xx}(x)ydA = IE(x)\frac{d\alpha_0(x)}{dx} = IE(x)\kappa_0(x) \tag{2.38}$$

*Sectional shear force:*

$$V(x) = \int_A \tau_{xy}(x)dA = GA_s(x)\gamma_0(x) \tag{2.39}$$

It is essential to point out that the generalized section deformations ((2.33), (2.34), and (2.35)) are the conjugate pairs of the generalized section forces ((2.37), (2.38), and (2.39)) based on the virtual work principle.

### 2.3.2.2 Governing Differential Equations

*Equilibrium:*

The free body diagram of an infinitesimal portion of a frame member under transverse load is shown in Figure 2.9. Vertical and moment equilibriums lead to the following equations:

$$\frac{dV(x)}{dx} + w_y(x) = 0 \text{ and } \frac{dM(x)}{dx} + V(x) = 0 \tag{2.40}$$

Based on the kinematical assumption of the Timoshenko beam theory, the sectional deformations ($\kappa_0(x)$ and $\gamma_0(x)$) are related to the transverse displacement $v_0(x)$ and flexural section-rotation $\alpha_0(x)$ through the following compatibility equations:

*Compatibility:*

$$\kappa_0(x) = \frac{d\alpha_0(x)}{dx}$$

$$\gamma_0(x) = \frac{dv_0(x)}{dx} - \alpha_0(x)$$

(2.41)

*Constitutive laws:*

The sectional constitutive laws for flexure and shear are:

$$M(x) = IE(x)\kappa_0(x) \text{ and } V(x) = GA_s(x)\gamma_0(x)$$

(2.42)

Substituting Eqs. (2.41) and (2.42) into (2.40) leads to the governing differential equation as:

*Governing differential equations:*

$$\frac{d}{dx}\left[GA_s(x)\left[\frac{dv_0(x)}{dx} - \alpha_0(x)\right]\right] + w_y(x) = 0$$

$$GA_s(x)\left[\frac{dv_0(x)}{dx} - \alpha_0(x)\right] + \frac{d}{dx}\left[IE(x)\frac{d\alpha_0(x)}{dx}\right] = 0$$

(2.43)

There are two coupled differential equations, which govern the Timoshenko beam. Both of them are of second order for the unknown fields $v_0(x)$ and $\alpha_0(x)$.

In the simple case of constant flexural rigidity $IE$ and shear rigidity $GA_s$, the governing differential Eq. (2.43) can be simplified as:

$$GA_s\left[\frac{d^2v_0(x)}{dx^2} - \frac{d\alpha_0(x)}{dx}\right] + w_y(x) = 0$$

$$GA_s\left[\frac{dv_0(x)}{dx} - \alpha_0(x)\right] + IE\frac{d^2\alpha_0(x)}{dx^2} = 0$$

(2.44)

Similar to the Euler-Bernoulli beam, in addition to the above governing differential equations, boundary conditions are necessarily supplied in order to solve the problem. The essential and natural boundary conditions are physically interpreted as constraints on displacements (transverse displacement and flexural rotation) and forces (shear and moment) at the member ends, respectively.

Mathematically speaking, one is seeking for the functions $v_0(x)$ and $\alpha_0(x)$, which possess the following attributes:

1. The sectional force fields ($M(x)$ and $V(x)$) computed from the displacement fields ($v_0(x)$ and $\alpha_0(x)$) through compatibility (2.41) and constitutive (2.42) relations must satisfy the differential equilibrium Eq. (2.40) in the point-wise sense throughout the solution domain $\Omega$ $(0 < x < L)$.
2. It must satisfy the specified boundary conditions at the member ends (at $x = 0$ and $x = L$).

Like in the Euler-Bernoulli beam problem, both essential and natural conditions can be prescribed at the same boundary, as shown in Figure 2.18 for example.

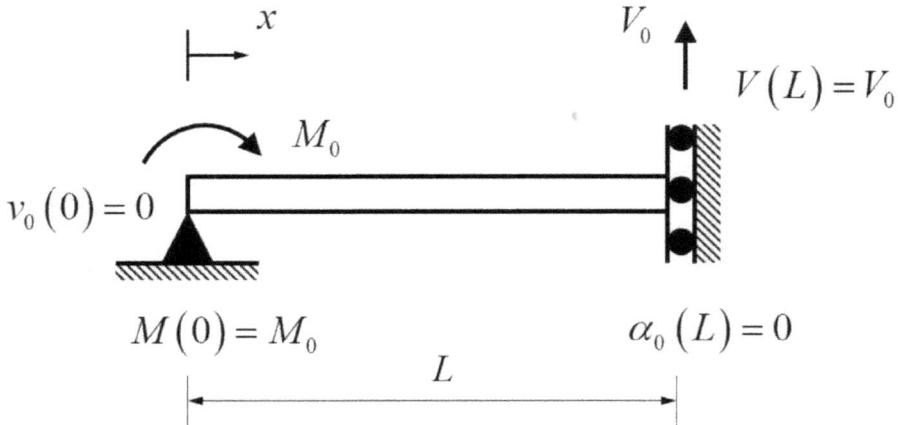

**FIGURE 2.18**   An Example of Mixed Boundary Conditions for the Timoshenko Beam.

At the left end ($x = 0$), one can specify:

$$v_0(0) = 0 \quad \Rightarrow \quad \textit{Essential} \text{ and } M(0) = M_0 \quad \Rightarrow \quad \textit{Natural} \qquad (2.45)$$

Similarly, at the right end ($x = L$), one can prescribe:

$$\alpha_0(L) = 0 \quad \Rightarrow \quad \textit{Essential} \text{ and } V(L) = V_0 \quad \Rightarrow \quad \textit{Natural} \qquad (2.46)$$

However, for a given degree of freedom, only one of the two boundary conditions can be prescribed.

Figure 2.19 shows all the kinematically admissible boundary constraints of the beam. These boundary constraints are admissible in the sense that they prevent the beam from translating and rotating as a rigid body. There are four boundary conditions (essential and natural) necessarily required to obtain the complete solution for a given problem. It is also important to note that at least two of these must be essential; otherwise, one must have a mechanism.

The governing equations ((2.40), (2.41), and (2.42)) together with the boundary conditions (essential and natural) form the boundary-value problem for establishing the vertical displacement and rotational fields of a beam and are schematically shown through Tonti's diagram of Figure 2.20.

## EXAMPLE 2.3

Starting from problem differential equation, find the displacement and force distributions along the cantilever Timoshenko beam under tip load shown in Figure 2.21.

**Solution:** In the case of constant flexural $IE$ and shear $GA_s$ rigidities, the governing differential equations are:

$$GA_s \left[ \frac{d^2 v_0(x)}{dx^2} - \frac{d\alpha_0(x)}{dx} \right] = 0 \qquad (a)$$

$$GA_s \left[ \frac{dv_0(x)}{dx} - \alpha_0(x) \right] + IE \frac{d^2 \alpha_0(x)}{dx^2} = 0 \qquad (b)$$

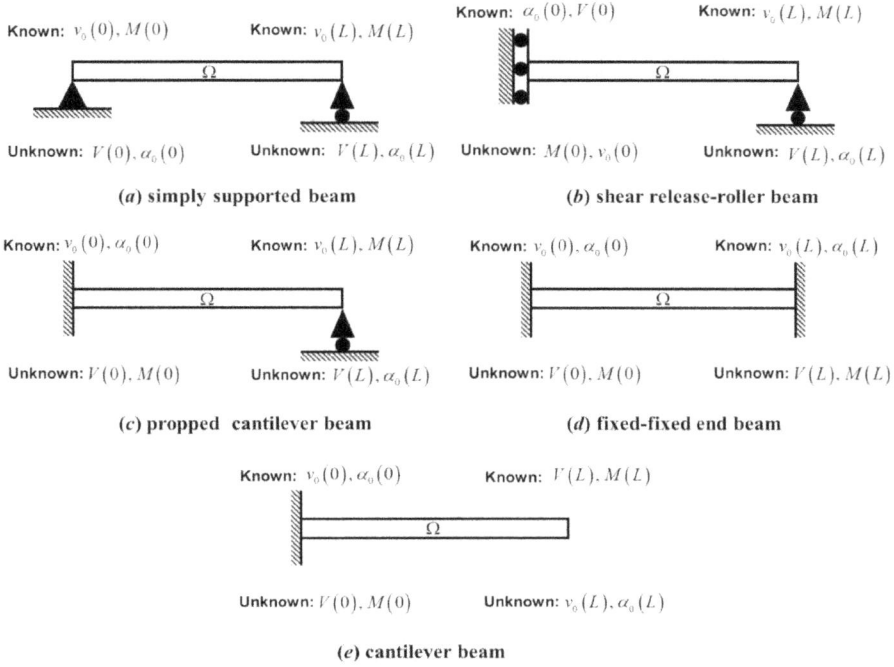

**Known:** $v_0(0), M(0)$      **Known:** $v_0(L), M(L)$      **Known:** $\alpha_0(0), V(0)$      **Known:** $v_0(L), M(L)$

**Unknown:** $V(0), \alpha_0(0)$      **Unknown:** $V(L), \alpha_0(L)$      **Unknown:** $M(0), v_0(0)$      **Unknown:** $V(L), \alpha_0(L)$

(*a*) simply supported beam          (*b*) shear release-roller beam

**Known:** $v_0(0), \alpha_0(0)$      **Known:** $v_0(L), M(L)$      **Known:** $v_0(0), \alpha_0(0)$      **Known:** $v_0(L), \alpha_0(L)$

**Unknown:** $V(0), M(0)$      **Unknown:** $V(L), \alpha_0(L)$      **Unknown:** $V(0), M(0)$      **Unknown:** $V(L), M(L)$

(*c*) propped  cantilever beam          (*d*) fixed-fixed end beam

**Known:** $v_0(0), \alpha_0(0)$      **Known:** $V(L), M(L)$

**Unknown:** $V(0), M(0)$      **Unknown:** $v_0(L), \alpha_0(L)$

(*e*) cantilever beam

**FIGURE 2.19**  Kinematically Admissible Constraints of the Timoshenko Beam.

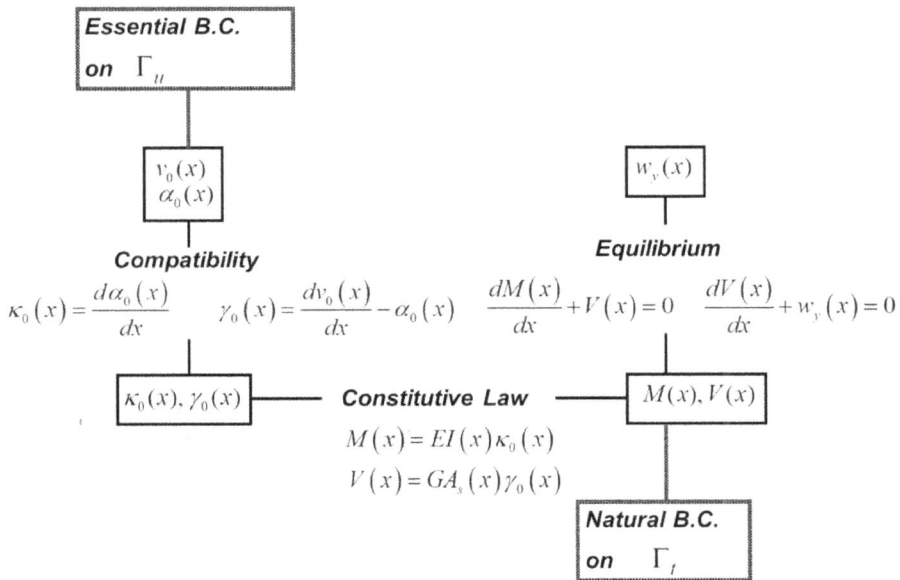

**Essential B.C.**

**on** $\Gamma_u$

$v_0(x)$
$\alpha_0(x)$

$w_y(x)$

**Compatibility**

$$\kappa_0(x) = \frac{d\alpha_0(x)}{dx} \qquad \gamma_0(x) = \frac{dv_0(x)}{dx} - \alpha_0(x)$$

**Equilibrium**

$$\frac{dM(x)}{dx} + V(x) = 0 \qquad \frac{dV(x)}{dx} + w_y(x) = 0$$

$\kappa_0(x), \gamma_0(x)$      **Constitutive Law**      $M(x), V(x)$

$$M(x) = EI(x)\kappa_0(x)$$
$$V(x) = GA_s(x)\gamma_0(x)$$

**Natural B.C.**

**on** $\Gamma_t$

**FIGURE 2.20**  Tonti's Diagram for the Timoshenko Beam Problem + Boundary Conditions.

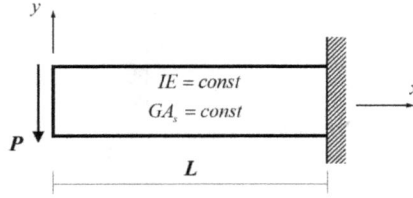

**FIGURE 2.21**   Example 2.3.

Boundary Conditions:

(1) $M(0) = 0$ : Natural

$$IE \frac{d\alpha_0}{dx}\bigg|_{x=0} = 0 \quad \Rightarrow \quad \frac{d\alpha_0}{dx}\bigg|_{x=0} = 0$$

(2) $V(0) = P$ : Natural

$$GA_s \gamma_0(0) = GA_s \left[\frac{dv_0}{dx}\bigg|_{x=0} - \alpha_0(0)\right] = P$$

(3) $\alpha_0(L) = 0$ : Essential

(4) $v_0(L) = 0$ : Essential

From the second governing equation (b),

$$\frac{dv_0(x)}{dx} = \alpha_0(x) - \frac{IE}{GA_s} \frac{d^2\alpha_0(x)}{dx^2} \tag{c}$$

Substitute (c) into the first governing equation (a), the following relation is obtained:

$$\frac{d^3\alpha_0(x)}{dx^3} = 0 \tag{d}$$

Integrating (d) leads to:

$$\alpha_0(x) = C_1 + C_2 x + C_3 x^2 \tag{e}$$

Substituting (e) into (c) results in:

$$\frac{dv_0(x)}{dx} = -\frac{2IE}{GA_s} C_3 + C_1 + C_2 x + C_3 x^2 \tag{f}$$

Integrating $(f)$ with respect to $x$:

$$v_0(x) = C_4 + \left( C_1 - \frac{2IE}{GA_s} C_3 \right) x + \frac{C_2 x^2}{2} + \frac{C_3 x^3}{3} \tag{g}$$

Imposing the boundary conditions, one can determine the integration constants:

(1) $\left. \dfrac{d\alpha_0}{dx} \right|_{x=0} = 0 \Rightarrow C_2 = 0$

(2) $GA_s \left[ \left. \dfrac{dv_0}{dx} \right|_{x=0} - \alpha_0(0) \right] = P \Rightarrow C_3 = -\dfrac{P}{2IE}$

(3) $\alpha_0(L) = 0 \Rightarrow C_1 = \dfrac{PL^2}{2IE}$

(4) $v_0(L) = 0 \Rightarrow C_4 = -\dfrac{PL^3}{3IE} - \dfrac{PL}{GA_s}$

The vertical displacement field is:

$$v_0(x) = \frac{P}{GA_s}(x - L) - \frac{PL^3}{3IE} + \frac{PL^2}{2IE} x - \frac{P}{6IE} x^3 \tag{h}$$

The flexural rotation field is:

$$\alpha_0(x) = \frac{PL^2}{2IE} - \frac{P}{2IE} x^2 \tag{i}$$

In Eq. $(h)$, the contribution of shear deformation is

$$v_s(x) = \frac{P}{GA_s}(x - L) \tag{j}$$

and the contribution of flexural deformations is

$$v_f(x) = -\frac{PL^3}{3IE} + \frac{PL^2}{2IE} x - \frac{P}{6IE} x^3 \tag{k}$$

The total tip deflection is

$$v_0(0) = -\frac{PL}{GA_s} - \frac{PL^3}{3IE} = \underbrace{\frac{PL}{GA_s}}_{Shear} + \underbrace{\frac{PL^3}{3IE}}_{Flexure} \quad \downarrow$$

The flexural rotations at the beam ends are:

$$\alpha_0(0) = \frac{PL^2}{2IE} \text{ and } \alpha_0(L) = 0$$

The total rotation is:

$$\theta(x) = \frac{dv_0(x)}{dx} = \frac{P}{GA_s} + \frac{PL^2}{2IE} - \frac{Px^2}{2IE} \qquad (l)$$

The total rotations at beam ends are:

$$\theta(0) = \underbrace{\frac{P}{GA_s}}_{Shear} + \underbrace{\frac{PL^2}{2IE}}_{Flexure} \quad \text{and} \quad \theta(L) = \underbrace{\frac{P}{GA_s}}_{Shear}$$

It is interesting to observe that at the fixed end, only the flexural rotation $\alpha_0$ is zero. This is clearly shown in Figure 2.16.

The shear strain field is:

$$\gamma_0(x) = \frac{dv_0(x)}{dx} - \alpha_0(x) = \frac{P}{GA_s} \qquad (m)$$

The bending moment along the beam axis is:

$$M(x) = IE\frac{d\alpha_0(x)}{dx} = -Px \qquad (n)$$

## EXAMPLE 2.4

Derive the stiffness matrix $\mathbf{K}_{T\text{-}Beam}$ of a beam member from the governing differential equation for a simply supported prismatic Timoshenko beam as shown in Figure 2.22.

**Solution:** For the case of a Timoshenko beam member with constant flexural rigidity $IE$ and constant shear rigidity $GA_s$, the beam kinematics (displacement system) is completely defined by the end (nodal) displacements ($\alpha_1$ and $\alpha_2$) and the beam kinetics (force system) is entirely described by the end (nodal) forces ($M_1$ and $M_2$). This simply means that for known nodal displacements and nodal forces, the entire beam behavior is completely defined. It is essential to recall from basic courses on structural analysis that the Timoshenko beam member stiffness matrix $\mathbf{K}_{T\text{-}Beam}$ relating the nodal forces and nodal displacements is:

$$\mathbf{P} = \mathbf{K}_{T-Beam}\mathbf{U}$$

$$\mathbf{P} = \lfloor M_1 \quad M_2 \rfloor^T \text{ and } \mathbf{U} = \lfloor \alpha_1 \quad \alpha_2 \rfloor^T$$

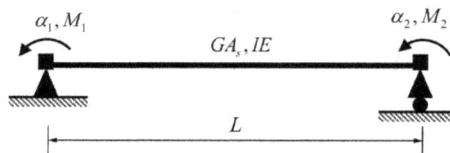

**FIGURE 2.22** Example 2.4.

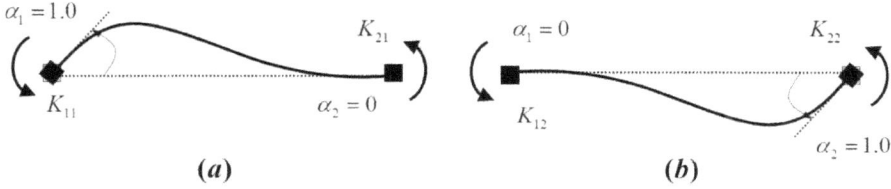

**FIGURE 2.23**   Example 2.4 (Continued).

$$\mathbf{K}_{T\text{-}Beam} = \begin{bmatrix} K_{11} & K_{12} \\ K_{21} & K_{22} \end{bmatrix}$$

This beam member stiffness matrix can also be obtained from the governing differential equation.

It is noted that the stiffness influence coefficient $K_{ij}$ corresponds to the reaction at degree of freedom $i$ due to a unit displacement at degree of freedom $j$.

Column 1 of $\mathbf{K}_{T\text{-}Beam}$ is obtained by imposing the following essential boundary conditions as shown in Figure 2.23$a$:

$$v_0(0) = 0, \ \alpha_0(0) = 1.0, \ v_0(L) = 0, \text{ and } \alpha_0(L) = 0$$

It is recalled that:

$$GA_s \left[ \frac{d^2 v_0(x)}{dx^2} - \frac{d\alpha_0(x)}{dx} \right] = 0 \qquad (a)$$

$$GA_s \left[ \frac{dv_0(x)}{dx} - \alpha_0(x) \right] + IE \frac{d^2\alpha_0(x)}{dx^2} = 0 \qquad (b)$$

From the second governing equation ($b$),

$$\frac{dv_0(x)}{dx} = \alpha_0(x) - \frac{IE}{GA_s} \frac{d^2\alpha_0(x)}{dx^2} \qquad (c)$$

Substituting ($c$) into the first governing equation ($a$), the following relation is obtained:

$$\frac{d^3\alpha_0(x)}{dx^3} = 0 \qquad (d)$$

Integrating ($d$) leads to:

$$\alpha_0(x) = C_1 + C_2 x + C_3 x^2 \qquad (e)$$

Substituting ($e$) into ($c$) results in:

$$\frac{dv_0(x)}{dx} = -\frac{2IE}{GA_s} C_3 + C_1 + C_2 x + C_3 x^2 \qquad (f)$$

Integrating $(f)$ with respect to $x$:

$$v_0(x) = C_4 + \left(C_1 - \frac{2IE}{GA_s}C_3\right)x + \frac{C_2 x^2}{2} + \frac{C_3 x^3}{3} \tag{g}$$

Boundary Conditions:

(1) $v_0(0) = 0 \quad \Rightarrow \quad C_4 = 0$ : Essential

(2) $\alpha_0(0) = 1.0 \quad \Rightarrow \quad C_1 = 1.0$ : Essential

(3) $v_0(L) = 0 \quad \Rightarrow \quad \left(-\frac{2IE}{GA_s}C_3 + 1\right) + \frac{C_2 L^2}{2} + \frac{C_3 L^3}{3} = 0$ : Essential

(4) $\alpha_0(L) = 0 \quad \Rightarrow \quad 1 + C_2 L + C_3 L^2 = 0$ : Essential

$$\Rightarrow \quad C_3 = \frac{1}{2\left(\dfrac{L^2}{6} + \dfrac{2IE}{GA_s}\right)} \quad \text{and} \quad C_2 = -\frac{1}{L} - \frac{L}{2\left(\dfrac{L^2}{6} + \dfrac{2IE}{GA_s}\right)}$$

$$\alpha_0(x) = 1 - \left(\frac{1}{L} + \frac{L}{2\left(\dfrac{L^2}{6} + \dfrac{2IE}{GA_s}\right)}\right)x + \frac{1}{2\left(\dfrac{L^2}{6} + \dfrac{2IE}{GA_s}\right)}x^2$$

It follows that:

$$M(x) = IE\frac{d\alpha_0(x)}{dx} = IE\left[-\frac{1}{L} - \frac{L}{2\left(\dfrac{L^2}{6} + \dfrac{2IE}{GA_s}\right)} + \frac{1}{\dfrac{L^2}{6} + \dfrac{2IE}{GA_s}}x\right]$$

Thus:

$$K_{11} = M_1 = -M(0) = IE\left[\frac{1}{L} + \frac{L}{2\left(\dfrac{L^2}{6} + \dfrac{2IE}{GA_s}\right)}\right] \quad \text{and}$$

$$K_{21} = M_2 = M(L) = IE\left[-\frac{1}{L} + \frac{L}{2\left(\dfrac{L^2}{6} + \dfrac{2IE}{GA_s}\right)}\right]$$

$\Phi = 12IE/GA_s L^2$ is defined as a dimensionless measure of the flexure-to-shear rigidity relation. Therefore, the above expression can be simplified to

$$K_{11} = \frac{(4+\Phi)IE}{L(1+\Phi)} \quad \text{and} \quad K_{21} = \frac{(2-\Phi)IE}{L(1+\Phi)}$$

Column 2 of $\mathbf{K}_{T\text{-}Beam}$ is obtained by imposing the following essential boundary conditions as shown in Figure 2.23*b*:

$$v_0(0) = 0, \; \alpha_0(0) = 0, \; v_0(L) = 0, \text{ and } \alpha_0(L) = 1.0$$

Following the same procedure results in:

$$K_{12} = \frac{(2-\Phi)IE}{L(1+\Phi)} \text{ and } K_{22} = \frac{(4+\Phi)IE}{L(1+\Phi)}$$

Consequently, the stiffness matrix is:

$$\mathbf{K}_{T\text{-}Beam} = \frac{IE}{L(1+\Phi)}\begin{bmatrix} (4+\Phi) & (2-\Phi) \\ (2-\Phi) & (4+\Phi) \end{bmatrix}$$

If the beam is slender, the shear deformation will be very small compared to the flexural deformation, then $\Phi \to 0$ ($GA_s$ is large compared to $IE$) and the stiffness matrix of the Timoshenko beam eventually degenerates into that of the Euler-Bernoulli beam (Spacone 2003).

$$\Phi = 0 \quad \Rightarrow \quad \mathbf{K}_{EB\text{-}Beam} = \mathbf{K}_{T\text{-}Beam} = \frac{IE}{L}\begin{bmatrix} 4 & 2 \\ 2 & 4 \end{bmatrix}$$

## 2.4 EXERCISES

**Problem 2.1:** Starting from the problem differential equation, find the reactions for the prismatic bar shown in Figure 2.24*a*.

**Problem 2.2:** Derive the expression for the transverse displacement, rotation, shear force, and bending moment fields of a prismatic Euler-Bernoulli beam as shown in Figure 2.24*b*.

**Problem 2.3:** Starting from the problem differential equation, find the transverse displacement, rotation, bending moment, and shear force distributions along a prismatic Timoshenko propped cantilever beam subjected to a uniformly distributed load as shown in Figure 2.24*c*.

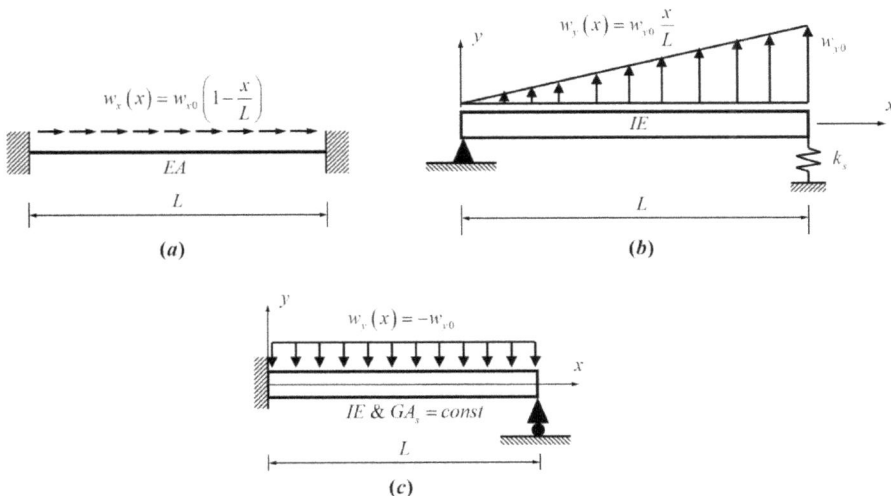

**FIGURE 2.24** Problems 2.1–2.3.

**TABLE 2.1**

**Summary of Euler-Bernoulli and Timoshenko Beam Theories**

| | **EULER-BERNOULLI BEAM THEORY** | **TIMOSHENKO BEAM THEORY** |
|---|---|---|

**Equilibrium**

$$\frac{dV(x)}{dx} = -w_y(x)$$

$$\frac{dM(x)}{dx} = -V(x)$$

$$\frac{d^2M}{dx^2} = w_y(x)$$

**Constitutive Law**

$$M = EI(x)\kappa_0$$

$$M = EI(x)\kappa_0$$

$$V = GA_s(x)\gamma_0$$

**Compatibility**

$$\kappa_0 = \frac{d^2v_0}{dx^2}$$

$$\kappa_0 = \frac{d\alpha_0}{dx}$$

$$\gamma_0 = \frac{dv_0}{dx} - \alpha_0$$

**Problem Differential Equation**

$$\frac{d^2}{dx^2}\left(EI(x)\frac{d^2v_0}{dx^2}\right) = w_y(x)$$

$$\frac{d}{dx}\left(GA_s(x)\left(\frac{dv_0}{dx} - \alpha_0\right)\right) = -w_y(x)$$

$$GA_s(x)\left(\frac{dv_0}{dx} - \alpha_0\right) = -\frac{d}{dx}\left(IE(x)\frac{d\alpha_0}{dx}\right)$$

$$EI = const \quad \Rightarrow \quad EI\frac{d^4v_0}{dx^4} = w_y(x)$$

$$EI = const \qquad GA_s = const$$

$$GA_s\left(\frac{d^2v_0}{dx^2} - \frac{d\alpha_0}{dx}\right) = -w_y(x)$$

$$GA_s\left(\frac{dv_0}{dx} - \alpha_0\right) = -EI\frac{d^2\alpha_0}{dx^2}$$

**Solution**

$$v_0(x)$$

$$v_0(x), \alpha_0(x)$$

## REFERENCES

Armenakas, A.E. 1991. *Modern structural analysis: The matrix method approach.* McGraw-Hill Inc.

Kanchi, M.B. 1993. *Matrix methods of structural analysis* (2nd Edition). Wiley Eastern Limited.

Kreyszig, E. 2011. *Advanced engineering mathematics* (10th Edition). John Wiley & Sons Inc.

Reddy, J.N. 2019. *An introduction to the finite element method* (4th Edition). McGraw-Hill Inc.

Schlaich, J., K. Schafer, and M. Jennewein. 1987. Toward a consistent design of structural concrete. *PCI Journal* 32(3): 74–150.

Spacone, E. 2003. *CVEN 5525 course notes: Matrix structural analysis.* Department of Civil, Environmental, and Architectural Engineering, University of Colorado.

Timoshenko, S.P. and J.M. Gere. 1972. *Mechanics of materials.* Van Nostrand Reinhold.

Tonti, E. 1976. The reason for analogies between physical theories. *Applied Mathematical Modeling* 1(1): 37–50.

# 3 Equilibrium Equations

## 3.1 INTRODUCTION

As mentioned in Chapter 1, equilibrium is one of the three fundamental ingredients required in any structural analysis approach. In this chapter, the main focus is on the different views of equilibrium (statical) considerations faced in structural analysis. There are various forms in describing equilibrium conditions. They can be expressed in forms of stresses or forces (resultant stresses). The first form is often encountered in the course on Continuum Mechanics, while the second one, discussed in Chapter 2, is found in basic courses on Mechanics of Materials and Structural Analysis. Equilibrium relations must hold for all levels of free bodies including a differential segment, a member, a part of the structure, and a whole structure.

In this textbook, only equilibrium considerations of frame structures are of interest. Equilibrium considerations of other structural types (e.g. plate or shell) are to be pursued in more advanced courses, such as the Finite Element Method.

## 3.2 EQUILIBRIUM IN A LOCAL REFERENCE SYSTEM: A FRAME MEMBER

A two-dimensional frame structure under general loadings is shown in Figure 3.1. The Cartesian reference system X-Y-Z is used to describe its geometry and global quantities. The Z-axis is normal to the structure plane and points toward the reader. The girder in the third story is taken as a generic frame member. Imaginary cutting planes are introduced to isolate this girder. As only plane structures are of interest in this textbook, each imaginary cut induces three pairs of forces, namely a pair of forces parallel to the member axis, a pair of forces perpendicular to the member axis, and a pair of moments. The isolated girder with its member loads and induced end forces is shown in Figure 3.2. In this section, the equilibrium relations for this generic frame member with respect to its local reference system are established. The Cartesian reference system $x$-$y$-$z$ is used as local reference axes for each member. Conventionally, node $i$ is chosen as a starting one while node $j$ is selected as an ending one. The positive local $x$-axis points toward node $j$ from node $i$. For plane structures, the preference is to keep the local $z$-axis coinciding with the global Z-axis pointing toward the reader. Consequently, the local $y$-axis is uniquely defined. It is assumed that the forces acting on the member side of each cut point are in the positive direction of the local

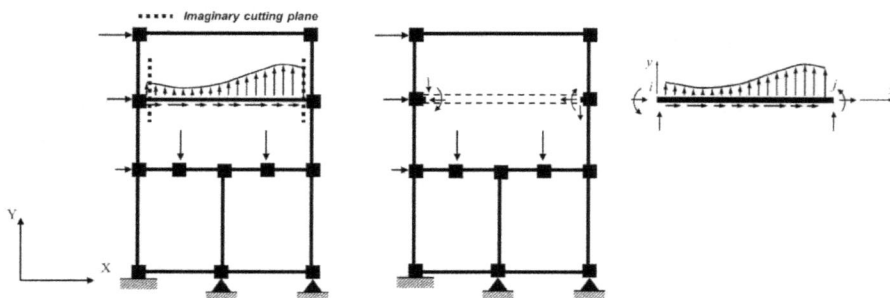

**FIGURE 3.1** A Two-Dimensional Frame Structure under General Loadings.

DOI: 10.1201/9781003595458-3

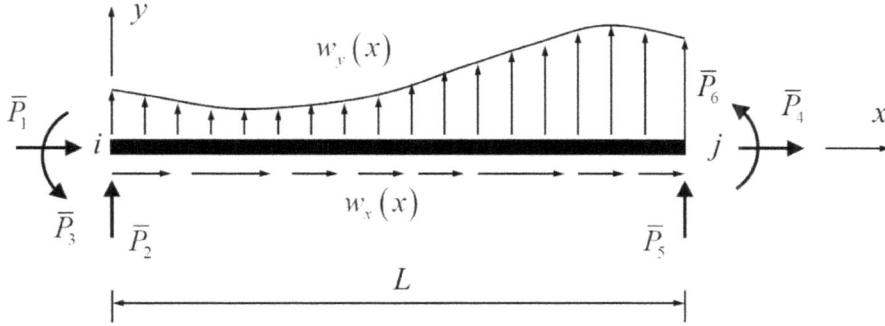

**FIGURE 3.2**  A Two-Dimensional Frame Member.

axes. According to Newton's third law of motion, the direction of the forces acting on the node side of the cut can be concluded.

The member end forces in the local reference system (Figure 3.2) can be grouped in a vector $\bar{\mathbf{P}}$ as:

$$\bar{\mathbf{P}} = \lfloor \bar{\mathbf{P}}_1 \quad \bar{\mathbf{P}}_2 \rfloor^T = \lfloor \bar{P}_1 \quad \bar{P}_2 \quad \bar{P}_3 \quad \bar{P}_4 \quad \bar{P}_5 \quad \bar{P}_6 \rfloor^T \tag{3.1}$$

The components of the end force vector $\bar{\mathbf{P}}$ are arranged in the following fashion: beginning from node $i$, one assign first the degree of freedom (DOF) in the local $x$-axis (first component), then the DOF in the local $y$-axis (second component), then the DOF about the local $z$-axis (third component), and then proceed in this fashion to node $j$ of the member. These six components form the complete set of end forces for a member free body in the local reference system. However, only three out of these six forces are independent. Consequently, one can select only three out of six as independent forces known as "basic member forces". The other three end forces are dependent and can be expressed in terms of the basic member forces. Interrelationships between the independent and dependent end forces can be determined from the equilibrium conditions of the member as a free body. Herein, the so-called "Simply Supported Beam System" shown in Figure 3.3 is employed to establish interrelationships between the independent and dependent end forces.

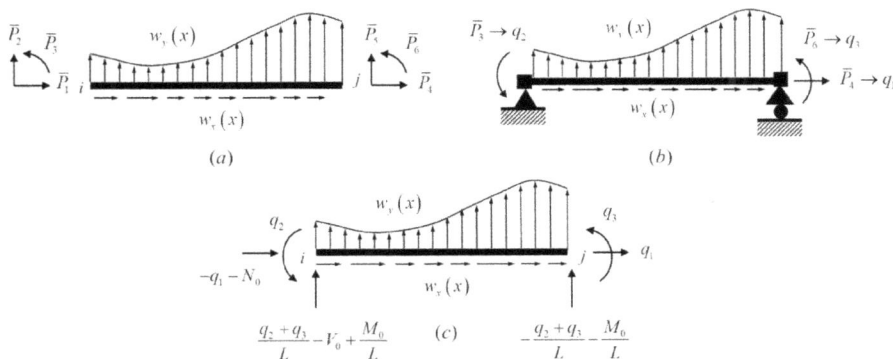

**FIGURE 3.3**  Simply Supported Beam Model: (*a*) Complete System, (*b*) Basic System, (*c*) Basic System with Reactive Forces.

Considering three equilibrium conditions of a frame member shown in Figure 3.3a, one has the following:

Summation of forces in the local x-axis:

$$\sum F_x = 0; \bar{P}_1 + \bar{P}_4 + \int_0^L w_x(x)dx = 0 \quad \Rightarrow \quad \bar{P}_1 = -\bar{P}_4 - N_0 \tag{3.2}$$

Summation of moments about node $j$ of a frame member:

$$\sum_j M_z = 0; \bar{P}_3 - \bar{P}_2 L + \bar{P}_6 - \int_0^L (L-x)w_y(x)dx = 0 \quad \Rightarrow \quad \bar{P}_2 = \frac{\bar{P}_3}{L} + \frac{\bar{P}_6}{L} - V_0 + \frac{M_0}{L} \tag{3.3}$$

Summation of moments about node $i$ of a frame member:

$$\sum_i M_z = 0; \bar{P}_3 + \bar{P}_5 L + \bar{P}_6 + \int_0^L w_y(x)xdx = 0 \quad \Rightarrow \quad \bar{P}_5 = -\frac{\bar{P}_3}{L} - \frac{\bar{P}_6}{L} - \frac{M_0}{L} \tag{3.4}$$

where

$$N_0 = \int_0^L w_x(x)dx; V_0 = \int_0^L w_y(x)dx; \text{ and } M_0 = \int_0^L w_y(x)xdx \tag{3.5}$$

Based on the three equilibrium equations ((3.2), (3.3), and (3.4)), $\bar{P}_3$, $\bar{P}_4$, and $\bar{P}_6$ are selected as the independent end forces. Therefore, $\bar{P}_1$, $\bar{P}_2$, and $\bar{P}_5$ are the dependent end forces and can be expressed in terms of the independent end forces via equilibrium consideration. It is convenient to rename the independent end forces as:

$$\bar{P}_4 \to q_1; \quad \bar{P}_3 \to q_2; \text{ and } \bar{P}_6 \to q_3 \tag{3.6}$$

Consequently, the basic member force vector **q** is defined as:

$$\mathbf{q} = \lfloor q_1 \quad q_2 \quad q_3 \rfloor^T = \lfloor \bar{P}_4 \quad \bar{P}_3 \quad \bar{P}_6 \rfloor^T \tag{3.7}$$

This basic system can be viewed as a simply supported beam shown in Figure 3.3b. This virtual supporting system merely helps the reader to select the choice of equilibrium conditions and is employed by Limkatanyu et al. (2014) to describe the basic system for the geometrically nonlinear frame model based on unification of Euler-Bernoulli-von Karman beam theory and corotational concept. However, this visualizing concept could lead to misconceptions. From Eqs. (3.2), (3.3), and (3.4), the matrix form of equilibrium equations for the simply supported beam system can be expressed as:

$$\left\{ \begin{array}{c} \bar{P}_1 \\ \bar{P}_2 \\ \bar{P}_5 \end{array} \right\} = \left[ \begin{array}{ccc} -1 & 0 & 0 \\ 0 & \dfrac{1}{L} & \dfrac{1}{L} \\ 0 & -\dfrac{1}{L} & -\dfrac{1}{L} \end{array} \right] \left\{ \begin{array}{c} q_1 \\ q_2 \\ q_3 \end{array} \right\} + \left\{ \begin{array}{c} -N_0 \\ -V_0 + \dfrac{M_0}{L} \\ -\dfrac{M_0}{L} \end{array} \right\} \tag{3.8}$$

$$\begin{Bmatrix} \bar{P}_4 \\ \bar{P}_3 \\ \bar{P}_6 \end{Bmatrix} = \begin{bmatrix} 1 & 0 & 0 \\ 0 & 1 & 0 \\ 0 & 0 & 1 \end{bmatrix} \begin{Bmatrix} q_1 \\ q_2 \\ q_3 \end{Bmatrix} + \begin{Bmatrix} 0 \\ 0 \\ 0 \end{Bmatrix} \tag{3.9}$$

and for the complete system (Figure 3.3*a* and *c*) as:

$$\begin{Bmatrix} \bar{P}_1 \\ \bar{P}_2 \\ \bar{P}_3 \\ \bar{P}_4 \\ \bar{P}_5 \\ \bar{P}_6 \end{Bmatrix} = \begin{bmatrix} -1 & 0 & 0 \\ 0 & \dfrac{1}{L} & \dfrac{1}{L} \\ 0 & 1 & 0 \\ 1 & 0 & 0 \\ 0 & -\dfrac{1}{L} & -\dfrac{1}{L} \\ 0 & 0 & 1 \end{bmatrix} \begin{Bmatrix} q_1 \\ q_2 \\ q_3 \end{Bmatrix} + \begin{Bmatrix} -N_0 \\ -V_0 + \dfrac{M_0}{L} \\ 0 \\ 0 \\ -\dfrac{M_0}{L} \\ 0 \end{Bmatrix} \tag{3.10}$$

or in a concise format as:

$$\begin{Bmatrix} \bar{\mathbf{P}}_1 \\ \bar{\mathbf{P}}_2 \end{Bmatrix} = \begin{bmatrix} \bar{\mathbf{b}}_{1B} \\ \bar{\mathbf{b}}_{2B} \end{bmatrix} \mathbf{q} + \begin{Bmatrix} \bar{\mathbf{P}}_{10} \\ \bar{\mathbf{P}}_{20} \end{Bmatrix} \text{ or } \bar{\mathbf{P}} = \bar{\mathbf{b}}_B \mathbf{q} + \bar{\mathbf{P}}_0 \tag{3.11}$$

where the vectors $\bar{\mathbf{P}}_{10}$ and $\bar{\mathbf{P}}_{20}$ contain the member end forces due to the effects of member loads and $\bar{\mathbf{b}}_B$ is the member equilibrium matrix for the simply supported beam model. It is observed that only the dependent forces are affected by the presence of member loads.

In the case of uniformly distributed member loads (Figure 3.4), the components of the member load vector $\bar{\mathbf{P}}_0$ are:

$$\bar{P}_{10} = -w_{x0}L; \quad \bar{P}_{20} = -\frac{w_{y0}L}{2}; \quad \bar{P}_{30} = 0; \quad \bar{P}_{40} = 0; \quad \bar{P}_{50} = -\frac{w_{y0}L}{2}; \text{ and } \bar{P}_{60} = 0 \tag{3.12}$$

### 3.2.1   REMARK

(i) It is noted that the choice of a basic force system is not unique, and there are other choices for the basic force system, for example: fixed-connected collar-roller beam system. The derivation of the matrix equilibrium equation for the fixed-connected collar-roller beam system is left to the reader as his/her exercise.

(ii) For other types of member loadings (e.g. linearly distributed member loads), the derivation of the member load vector $\bar{\mathbf{P}}_0$ can be carried out in a similar fashion and is left to the reader as his/her end-of-chapter exercise.

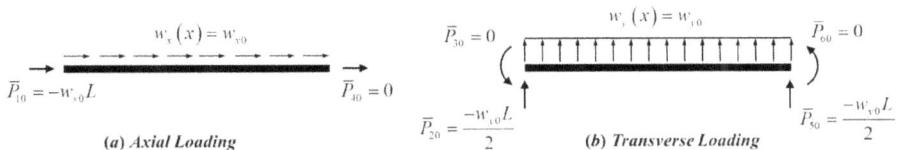

FIGURE 3.4   Uniformly Distributed Loads: Simply Supported Beam Model.

## 3.3   COMPONENT TRANSFORMATION OF A FORCE VECTOR

In the previous section, the matrix equilibrium Eq. (3.11) has been expressed with respect to the local reference system ($x$, $y$, and $z$). However, it is also essential to express the matrix equilibrium equation in the global reference system (X, Y, and Z). In other words, the interrelationship between the member end forces in the local and global reference systems must be provided. This can be achieved through the transformation of the vector components.

In Figure 3.5, the local reference axes are inclined at an angle $\phi$ with respect to the global reference axes. This implies that the angle $\phi$ is measured from the global to a local reference system. The sign of the angle $\phi$ follows the right-hand system. The vector $\vec{A}$ in Figure 3.5 can be decomposed with respect to both $x$-$y$ and X-Y components. It is noted that $\vec{A}$ can be thought of as a force vector acting either at node $i$ or $j$. It is now necessary to seek for the relationship between vector components defined with respect to the global and local reference systems.

From geometric consideration of the vector $\vec{A}$ in Figure 3.5, the relations for transforming vector components from the $x$-$y$ to X-Y reference systems are:

$$A_X = A_x \cos\phi - A_y \sin\phi \text{ and } A_Y = A_x \sin\phi + A_y \cos\phi \qquad (3.13)$$

or in matrix form as:

$$\begin{Bmatrix} A_X \\ A_Y \end{Bmatrix} = \begin{bmatrix} \cos\phi & -\sin\phi \\ \sin\phi & \cos\phi \end{bmatrix} \begin{Bmatrix} A_x \\ A_y \end{Bmatrix} \text{ or } \mathbf{A}_{XY} = \mathbf{T}_{ROT}^T \mathbf{A}_{xy} \qquad (3.14)$$

where $A_X$ and $A_Y$ are the X and Y components of vector $\vec{A}$, respectively; and $A_x$ and $A_y$ are the $x$ and $y$ components of vector $\vec{A}$, respectively. The matrix $\mathbf{T}_{ROT}$ is known as the "Rotational

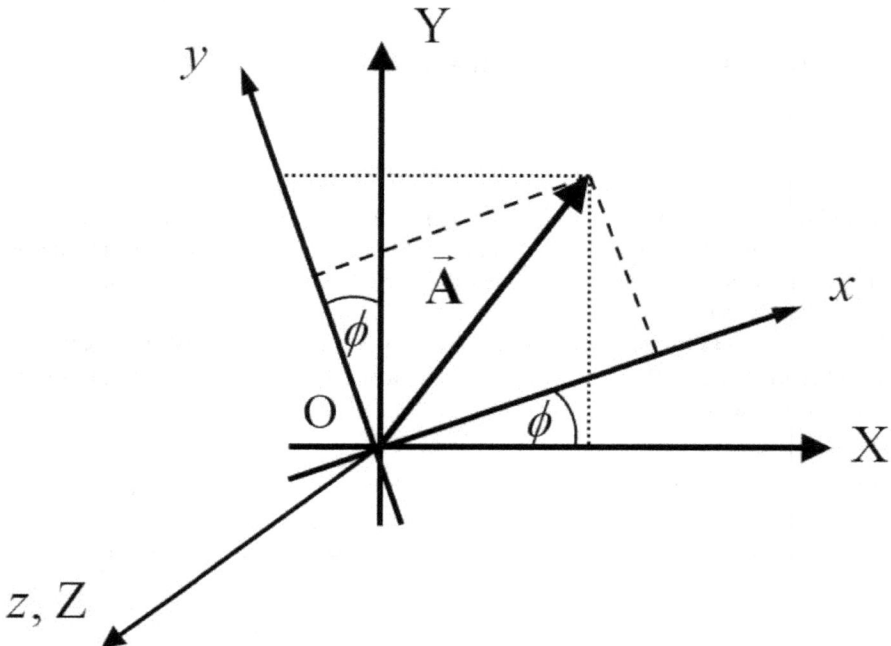

**FIGURE 3.5**   Component Transformation of a Plane Vector $\vec{A}$.

Matrix". Vice versa, the relations for transforming vector components from the X-Y to x-y reference systems are:

$$A_x = A_X \cos\phi + A_Y \sin\phi \text{ and } A_y = -A_X \sin\phi + A_Y \cos\phi \qquad (3.15)$$

or in matrix form as:

$$\left\{ \begin{array}{c} A_x \\ A_y \end{array} \right\} = \left[ \begin{array}{cc} \cos\phi & \sin\phi \\ -\sin\phi & \cos\phi \end{array} \right] \left\{ \begin{array}{c} A_X \\ A_Y \end{array} \right\} \text{ or } \mathbf{A}_{xy} = \mathbf{T}_{ROT}\mathbf{A}_{XY} \qquad (3.16)$$

From Eqs. (3.14) and (3.16), it can be concluded that $\mathbf{T}_{ROT}$ is an orthogonal matrix possessing a very useful property:

$$\mathbf{T}_{ROT}^T = \mathbf{T}_{ROT}^{-1} \text{ and } |\mathbf{T}_{ROT}| = 1 \qquad (3.17)$$

As mentioned earlier, it is convenient to keep the local z-axis coinciding with the global Z-axis for a plane structure. Therefore, if the vector $\vec{\mathbf{A}}$ has some projection along the z-axis, it is the same as along the Z-axis, or vice versa. The transformation relations of Eqs. (3.14) and (3.16) can be augmented as:

$$\left\{ \begin{array}{c} A_X \\ A_Y \\ A_Z \end{array} \right\} = \left[ \begin{array}{ccc} \cos\phi & -\sin\phi & 0 \\ \sin\phi & \cos\phi & 0 \\ 0 & 0 & 1 \end{array} \right] \left\{ \begin{array}{c} A_x \\ A_y \\ A_z \end{array} \right\} \text{ or } \mathbf{A}_{XYZ} = \mathbf{T}_{ROT}^T\mathbf{A}_{xyz} \qquad (3.18)$$

and

$$\left\{ \begin{array}{c} A_x \\ A_y \\ A_z \end{array} \right\} = \left[ \begin{array}{ccc} \cos\phi & \sin\phi & 0 \\ -\sin\phi & \cos\phi & 0 \\ 0 & 0 & 1 \end{array} \right] \left\{ \begin{array}{c} A_X \\ A_Y \\ A_Z \end{array} \right\} \text{ or } \mathbf{A}_{xyz} = \mathbf{T}_{ROT}\mathbf{A}_{XYZ} \qquad (3.19)$$

respectively. It is noted that the orthogonal property of the rotational matrix $\mathbf{T}_{ROT}$ is still preserved.

Since the present goal is to establish the transformation relations between the end force components acting in local and global coordinate systems, the force acting at node $i$ shown in Figure 3.6 is considered. Figure 3.6a shows the force components along the local reference system while Figure 3.6b along the global reference system. Following the transformation relations of Eq. (3.18), the relations between the force components shown in Figure 3.6a and b are:

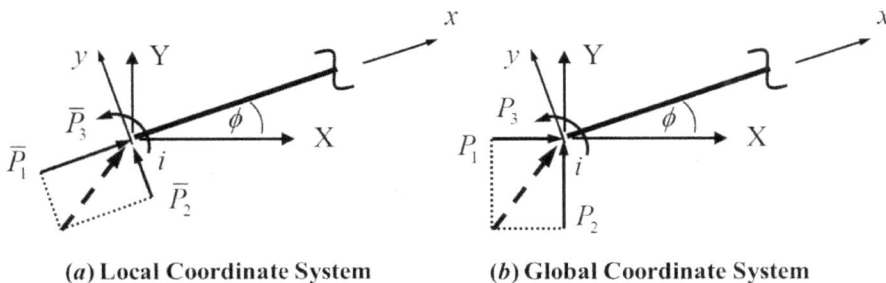

(a) Local Coordinate System        (b) Global Coordinate System

**FIGURE 3.6**   Transformation of Member End Forces at Node $i$.

$$\begin{Bmatrix} P_1 \\ P_2 \\ P_3 \end{Bmatrix} = \begin{bmatrix} \cos\phi & -\sin\phi & 0 \\ \sin\phi & \cos\phi & 0 \\ 0 & 0 & 1 \end{bmatrix} \begin{Bmatrix} \bar{P}_1 \\ \bar{P}_2 \\ \bar{P}_3 \end{Bmatrix} \text{or} \, \mathbf{P}_1 = \mathbf{T}_{ROT}^T \bar{\mathbf{P}}_1 \tag{3.20}$$

and vice versa,

$$\begin{Bmatrix} \bar{P}_1 \\ \bar{P}_2 \\ \bar{P}_3 \end{Bmatrix} = \begin{bmatrix} \cos\phi & \sin\phi & 0 \\ -\sin\phi & \cos\phi & 0 \\ 0 & 0 & 1 \end{bmatrix} \begin{Bmatrix} P_1 \\ P_2 \\ P_3 \end{Bmatrix} \text{or} \, \bar{\mathbf{P}}_1 = \mathbf{T}_{ROT} \mathbf{P}_1 \tag{3.21}$$

Similar consideration is applied to the force acting at node $j$ (Figure 3.7) and leads to the same transformation relations as:

$$\begin{Bmatrix} P_4 \\ P_5 \\ P_6 \end{Bmatrix} = \begin{bmatrix} \cos\phi & -\sin\phi & 0 \\ \sin\phi & \cos\phi & 0 \\ 0 & 0 & 1 \end{bmatrix} \begin{Bmatrix} \bar{P}_4 \\ \bar{P}_5 \\ \bar{P}_6 \end{Bmatrix} \text{or} \, \mathbf{P}_2 = \mathbf{T}_{ROT}^T \bar{\mathbf{P}}_2 \tag{3.22}$$

and vice versa,

$$\begin{Bmatrix} \bar{P}_4 \\ \bar{P}_5 \\ \bar{P}_6 \end{Bmatrix} = \begin{bmatrix} \cos\phi & \sin\phi & 0 \\ -\sin\phi & \cos\phi & 0 \\ 0 & 0 & 1 \end{bmatrix} \begin{Bmatrix} P_4 \\ P_5 \\ P_6 \end{Bmatrix} \text{or} \, \bar{\mathbf{P}}_2 = \mathbf{T}_{ROT} \mathbf{P}_2 \tag{3.23}$$

Combining Eqs. (3.20) and (3.22) leads to the transformation relation for a member-end force:

$$\begin{Bmatrix} P_1 \\ P_2 \\ P_3 \\ P_4 \\ P_5 \\ P_6 \end{Bmatrix} = \begin{bmatrix} \cos\phi & -\sin\phi & 0 & 0 & 0 & 0 \\ \sin\phi & \cos\phi & 0 & 0 & 0 & 0 \\ 0 & 0 & 1 & 0 & 0 & 0 \\ 0 & 0 & 0 & \cos\phi & -\sin\phi & 0 \\ 0 & 0 & 0 & \sin\phi & \cos\phi & 0 \\ 0 & 0 & 0 & 0 & 0 & 1 \end{bmatrix} \begin{Bmatrix} \bar{P}_1 \\ \bar{P}_2 \\ \bar{P}_3 \\ \bar{P}_4 \\ \bar{P}_5 \\ \bar{P}_6 \end{Bmatrix} \text{or} \, \mathbf{P} = \Gamma_{ROT}^T \bar{\mathbf{P}} \tag{3.24}$$

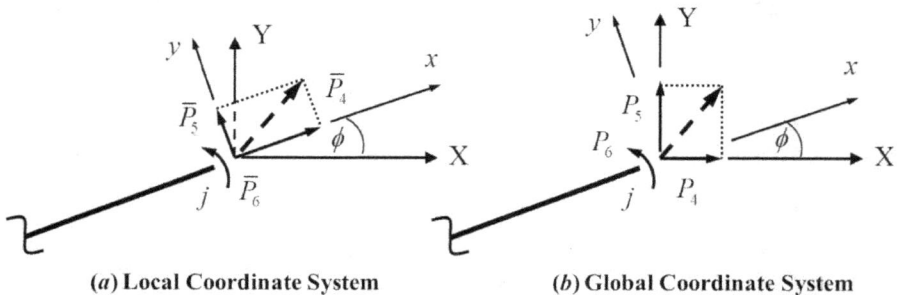

(a) Local Coordinate System           (b) Global Coordinate System

**FIGURE 3.7**    Transformation of Member End Forces at Node $j$.

or vice versa

$$\begin{Bmatrix} \overline{P}_1 \\ \overline{P}_2 \\ \overline{P}_3 \\ \overline{P}_4 \\ \overline{P}_5 \\ \overline{P}_6 \end{Bmatrix} = \begin{bmatrix} \cos\phi & \sin\phi & 0 & 0 & 0 & 0 \\ -\sin\phi & \cos\phi & 0 & 0 & 0 & 0 \\ 0 & 0 & 1 & 0 & 0 & 0 \\ 0 & 0 & 0 & \cos\phi & \sin\phi & 0 \\ 0 & 0 & 0 & -\sin\phi & \cos\phi & 0 \\ 0 & 0 & 0 & 0 & 0 & 1 \end{bmatrix} \begin{Bmatrix} P_1 \\ P_2 \\ P_3 \\ P_4 \\ P_5 \\ P_6 \end{Bmatrix} \text{ or } \overline{\mathbf{P}} = \Gamma_{ROT}\mathbf{P} \qquad (3.25)$$

where

$$\Gamma_{ROT} = \begin{bmatrix} \mathbf{T}_{ROT} & \mathbf{0} \\ \mathbf{0} & \mathbf{T}_{ROT} \end{bmatrix} \qquad (3.26)$$

It is observed that the off-diagonal terms in the matrix $\Gamma_{ROT}$ are zero. This implies that the components of end forces at nodes $i$ and $j$ are transformed independently. It is noted that the component arrangement of the vector $\mathbf{P}$ is similar to that of the vector $\overline{\mathbf{P}}$. Starting from node $i$, one assign first the degree of freedom (DOF) in the global X-axis (first component), then the DOF in the global Y-axis (second component), then the DOF about the global Z-axis (third component), and then proceed in this fashion to node $j$ of the member.

In computer implementation of the matrix analysis program, the value of angle $\phi$ of each member is not necessarily given since values of the direction cosines ($\cos\phi$ and $\sin\phi$) are not calculated directly. Instead, they are computed indirectly based on the coordinate data of nodes $i$ and $j$ (Figure 3.8) as:

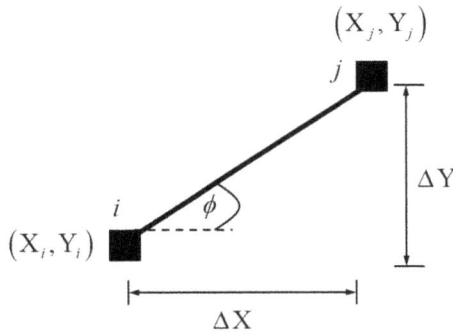

FIGURE 3.8   Indirect Calculation of Direction Cosines.

$$\Delta X = X_j - X_i; \quad \Delta Y = Y_j - Y_i; \text{ and } L = \sqrt{(\Delta X)^2 + (\Delta Y)^2} \tag{3.27}$$

$$\cos\phi = \Delta X / L \text{ and } \sin\phi = \Delta Y / L \tag{3.28}$$

## 3.4 EQUILIBRIUM IN THE GLOBAL REFERENCE SYSTEM: A FRAME MEMBER

Up to now, the matrix equilibrium equation for the simply supported beam system has been established with respect to the local reference axes. Also, the transformation relations (3.24) between the member end forces described in local and global coordinate systems were obtained. Based on those transformation relations in hand, the matrix equilibrium equation for the simply supported beam system can easily be expressed with respect to the global reference axes.

Substitution of Eqs. (3.10) into (3.24) yields a matrix equilibrium equation with respect to the global reference system as:

$$
\begin{Bmatrix} P_1 \\ P_2 \\ P_3 \\ P_4 \\ P_5 \\ P_6 \end{Bmatrix} =
\begin{bmatrix}
-\cos\phi & -\dfrac{\sin\phi}{L} & -\dfrac{\sin\phi}{L} \\
-\sin\phi & \dfrac{\cos\phi}{L} & \dfrac{\cos\phi}{L} \\
0 & 1 & 0 \\
\cos\phi & \dfrac{\sin\phi}{L} & \dfrac{\sin\phi}{L} \\
\sin\phi & -\dfrac{\cos\phi}{L} & -\dfrac{\cos\phi}{L} \\
0 & 0 & 1
\end{bmatrix}
\begin{Bmatrix} q_1 \\ q_2 \\ q_3 \end{Bmatrix} +
\begin{Bmatrix}
-\cos\phi\, N_0 - \sin\phi\left(\dfrac{M_0}{L} - V_0\right) \\
\cos\phi\left(\dfrac{M_0}{L} - V_0\right) - \sin\phi\, N_C \\
0 \\
\sin\phi\dfrac{M_0}{L} \\
-\cos\phi\dfrac{M_0}{L} \\
0
\end{Bmatrix} \tag{3.29}
$$

or in concise format as:

$$\begin{Bmatrix} \mathbf{P}_1 \\ \mathbf{P}_2 \end{Bmatrix} = \begin{bmatrix} \mathbf{b}_{1B} \\ \mathbf{b}_{2B} \end{bmatrix} \mathbf{q} + \begin{Bmatrix} \mathbf{P}_{10} \\ \mathbf{P}_{20} \end{Bmatrix} \text{ or } \mathbf{P} = \mathbf{b}_B \mathbf{q} + \mathbf{P}_0 \tag{3.30}$$

where $\mathbf{b}_B$ is the equilibrium matrix for the simply supported beam model in the global reference system. The transformation process of member forces from basic to complete systems in global reference axes is schematically presented in Figure 3.9.

## 3.5 EQUILIBRIUM: A FRAME STRUCTURE

In this section, the main objective is to establish the matrix equilibrium equation of a frame structure. This matrix equation relates the basic member forces to the externally applied loads on the structure. The external loads can be categorized as member loads and nodal applied loads. The member-load actions have already been included in the member equilibrium equations as discussed in the previous sections, thus leading to the equivalent nodal loads. The nodal applied loads can be split into two parts following the restrained conditions at degrees of freedom. The first ones are typically known forces associated with the unconstrained (free) degrees of freedom. The second ones are generally unknown forces (support reactions) and correspond to the constrained degrees of freedom. The matrix equilibrium equation of the structure can be obtained by considering the equilibrium conditions at all degrees of freedom in a collective manner.

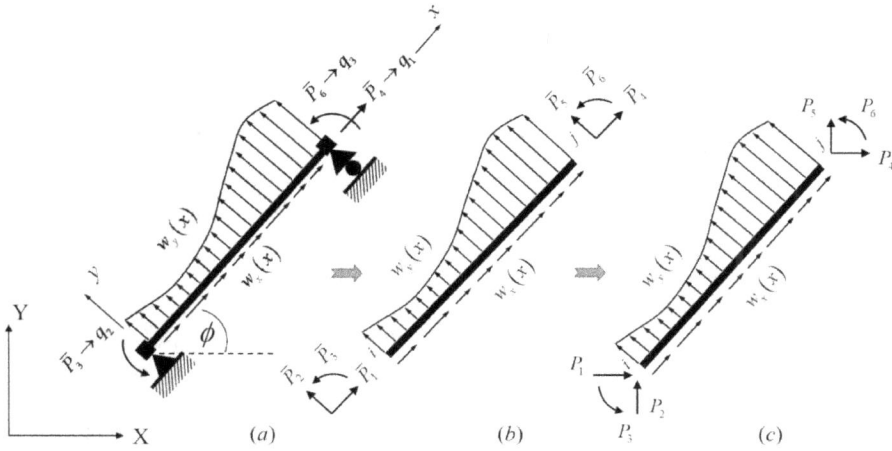

**FIGURE 3.9** Simply Supported Beam Model: (*a*) Basic System; (*b*) Complete System in Local Reference Axes; (*c*) Complete System in Global Reference Axes.

For the frame as a whole structure, an augmented vector $\mathbf{Q}$ contains the basic forces of all members and is defined as:

$$\mathbf{Q} = \lfloor \mathbf{Q}_1 \quad \mathbf{Q}_2 \quad \cdots \quad \mathbf{Q}_k \quad \cdots \quad \mathbf{Q}_m \rfloor_{3m \times 1}^T \tag{3.31}$$

where $m$ is the number of members and the sub-vector $\mathbf{Q}_k$ contains the basic forces of member $k$. Therefore, the size of $\mathbf{Q}$ is $3m \times 1$.

Similarly, the nodal loads are grouped into a vector $\mathbf{F}$ defined as:

$$\mathbf{F} = \lfloor \mathbf{F}_1 \quad \mathbf{F}_2 \quad \cdots \quad \mathbf{F}_k \quad \cdots \quad \mathbf{F}_n \rfloor_{3n \times 1}^T \tag{3.32}$$

where $n$ is the number of nodes and the sub-vector $\mathbf{F}_k$ contains the applied loads at node $k$. Therefore, the size of $\mathbf{F}$ is $3n \times 1$.

Once equilibrium conditions are collectively considered at all degrees of freedom, the matrix equilibrium equation is obtained as:

$$
\begin{bmatrix}
\mathbf{b}_1^{(1)} & \mathbf{b}_1^{(2)} & \cdots & \cdots \\
\mathbf{b}_2^{(1)} & \mathbf{b}_2^{(2)} & \cdots & \cdots \\
\vdots & \vdots & & \\
\vdots & \vdots & &
\end{bmatrix}_{3n \times 3m}
\begin{Bmatrix}
\mathbf{Q}_1 \\
\mathbf{Q}_2 \\
\vdots \\
\mathbf{Q}_m
\end{Bmatrix}_{3m \times 1}
+
\begin{Bmatrix}
\mathbf{F}_1^0 \\
\mathbf{F}_2^0 \\
\vdots \\
\mathbf{F}_n^0
\end{Bmatrix}_{3n \times 1}
=
\begin{Bmatrix}
\mathbf{F}_1 \\
\mathbf{F}_2 \\
\vdots \\
\mathbf{F}_n
\end{Bmatrix}_{3n \times 1}
\tag{3.33}
$$

or in a concise format as:

$$\mathbf{B}\mathbf{Q} + \mathbf{F}^0 = \mathbf{F} \text{ or } \mathbf{B}\mathbf{Q} = \mathbf{F} - \mathbf{F}^0 \tag{3.34}$$

where vector $\mathbf{F}^0$ contains the equivalent nodal loads due to the presence of member loads.

Furthermore, the matrix equilibrium equation of Eq. (3.34) can be partitioned following known and unknown nodal loads as:

$$\mathbf{F} = \left\{ \begin{array}{c} \mathbf{F}_{free} \\ \mathbf{F}_{constr} \end{array} \right\} = \left[ \begin{array}{c} \mathbf{B}_{free} \\ \mathbf{B}_{constr} \end{array} \right] \mathbf{Q} + \left\{ \begin{array}{c} \mathbf{F}^0_{free} \\ \mathbf{F}^0_{constr} \end{array} \right\} \text{ or } \mathbf{F}^* = \left\{ \begin{array}{c} \mathbf{F}_{free} \\ \mathbf{F}_{constr} \end{array} \right\} - \left\{ \begin{array}{c} \mathbf{F}^0_{free} \\ \mathbf{F}^0_{constr} \end{array} \right\} = \left[ \begin{array}{c} \mathbf{B}_{free} \\ \mathbf{B}_{constr} \end{array} \right] \mathbf{Q} \quad (3.35)$$

where $\mathbf{F}_{free}$ and $\mathbf{F}_{constr}$ contain the nodal loads acting at free and constrained degrees of freedom, respectively, $\mathbf{B}_{free}$ and $\mathbf{B}_{constr}$ are the equilibrium matrices associated with free and constrained degrees of freedom, respectively, and $\mathbf{F}^0_{free}$ and $\mathbf{F}^0_{constr}$ contain the member-equivalent nodal loads corresponding to free and constrained degrees of freedom, respectively. It is observed that by moving the member loads to the left hand side, the member-load actions are eventually converted to the equivalent applied nodal loads $\mathbf{F}^*$.

To illustrate the aforementioned concept of establishing the matrix equilibrium equation of a structure, a prototypical frame shown in Figure 3.10 is considered. Figure 3.11 shows its local reference axes and numbering systems of nodes, members, and degrees of freedom.

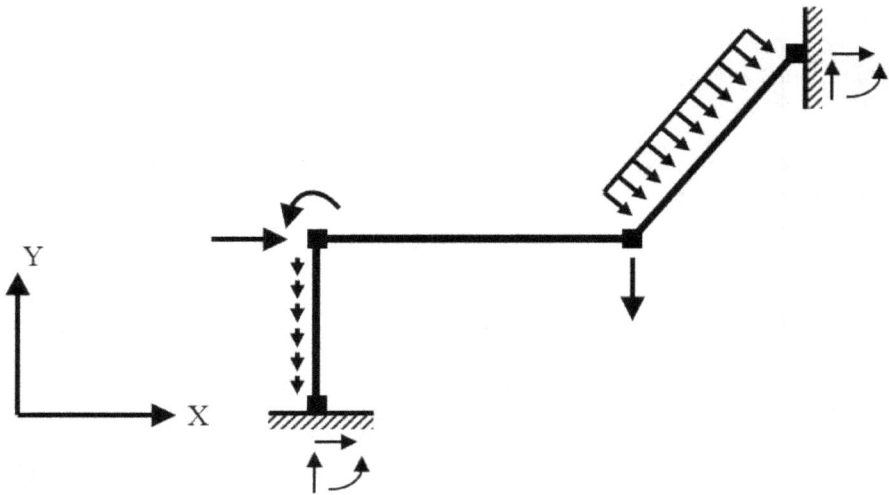

**FIGURE 3.10**  A Prototypical Frame: Nodal Loads, Member Loads, and Support Reactions.

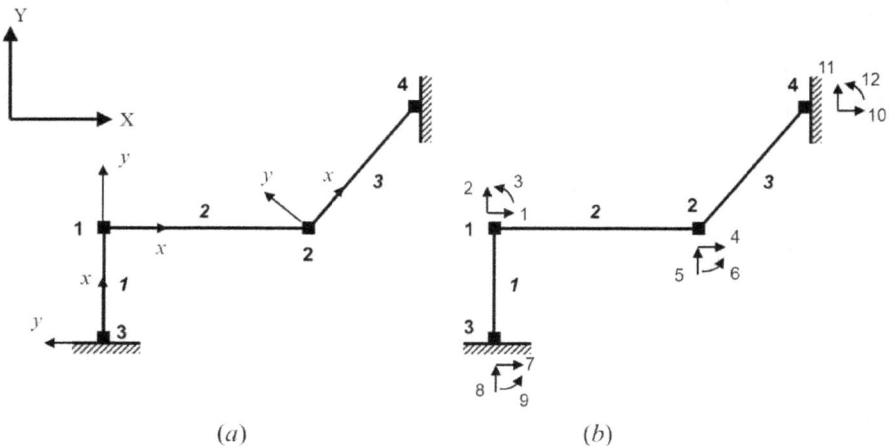

**FIGURE 3.11**  A Prototypical Frame: (*a*) Local Reference Axes; (*b*) Nodal, Member, and DOFs Numbering Systems.

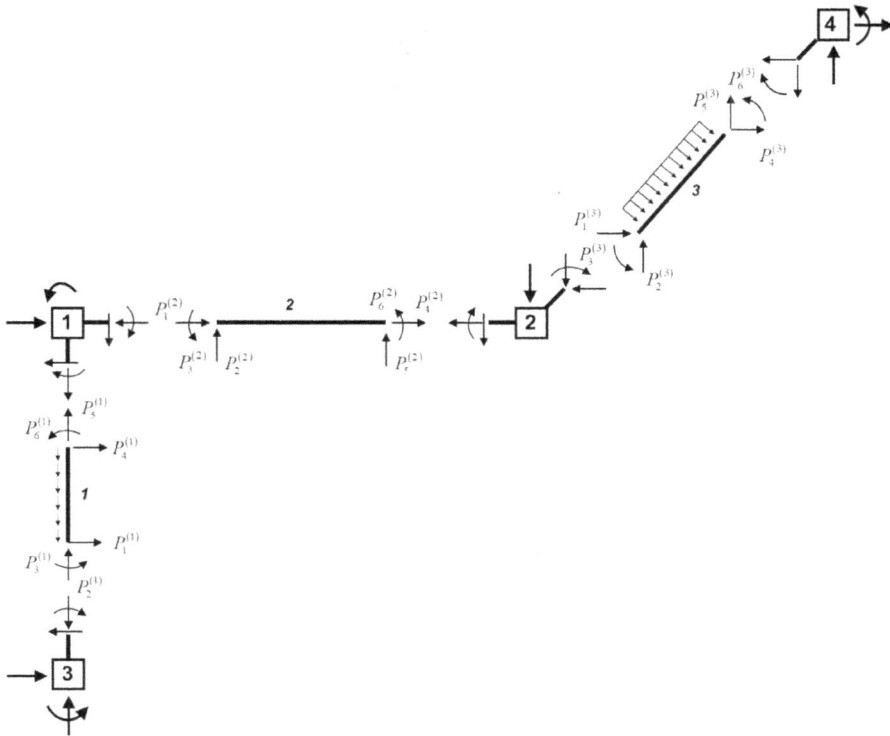

**FIGURE 3.12**   Nodal Equilibrium Conditions.

From the nodal equilibrium conditions shown in Figure 3.12, the equilibrium conditions at node 1 are:

$$\left\{\begin{array}{l}\sum F_X = 0 \\ \sum F_Y = 0 \\ \sum M_Z = 0\end{array}\right\} \Rightarrow \left\{\begin{array}{l}F_1 \\ F_2 \\ F_3\end{array}\right\} = \left\{\begin{array}{l}P_4^{(1)} \\ P_5^{(1)} \\ P_6^{(1)}\end{array}\right\} + \left\{\begin{array}{l}P_1^{(2)} \\ P_2^{(2)} \\ P_3^{(2)}\end{array}\right\} \text{ or } \mathbf{F}_1 = \mathbf{P}_2^{(1)} + \mathbf{P}_1^{(2)} \tag{3.36}$$

It is assumed that the simply supported beam model (Figure 3.9) is used to represent the basic member system. Consequently, the end forces $\mathbf{P}_2^{(1)}$ and $\mathbf{P}_1^{(2)}$ can be expressed in terms of basic member forces $\mathbf{Q}_1$ and $\mathbf{Q}_2$ as:

$$\mathbf{P}_2^{(1)} = \mathbf{b}_{2B}^{(1)}\mathbf{Q}_1 + \mathbf{P}_{20}^{(1)} \text{ and } \mathbf{P}_1^{(2)} = \mathbf{b}_{1B}^{(2)}\mathbf{Q}_2 + \mathbf{P}_{10}^{(2)} \tag{3.37}$$

The nodal equilibrium equations at node 1 can be expressed in terms of basic member forces by substituting Eqs. (3.37) into (3.36):

$$\mathbf{F}_1 = \mathbf{b}_{2B}^{(1)}\mathbf{Q}_1 + \mathbf{b}_{1B}^{(2)}\mathbf{Q}_2 + \mathbf{P}_{20}^{(1)} + \mathbf{P}_{10}^{(2)} \tag{3.38}$$

Similarly at

*Node 2:*

$$\mathbf{F}_2 = \mathbf{b}_{2B}^{(2)}\mathbf{Q}_2 + \mathbf{b}_{1B}^{(3)}\mathbf{Q}_3 + \mathbf{P}_{20}^{(2)} + \mathbf{P}_{10}^{(3)} \tag{3.39}$$

*Node 3:*

$$\mathbf{F}_3 = \mathbf{b}_{1B}^{(1)}\mathbf{Q}_1 + \mathbf{P}_{10}^{(1)} \tag{3.40}$$

*Node 4:*

$$\mathbf{F}_4 = \mathbf{b}_{2B}^{(3)}\mathbf{Q}_3 + \mathbf{P}_{20}^{(3)} \tag{3.41}$$

Equations (3.38), (3.39), (3.40), and (3.41) can be written in a collective manner, hence resulting in the matrix equilibrium equation of the structure as:

$$\begin{Bmatrix} \mathbf{F}_1 \\ \mathbf{F}_2 \\ \mathbf{F}_3 \\ \mathbf{F}_4 \end{Bmatrix}_{12\times1} = \begin{bmatrix} \mathbf{b}_{2B}^{(1)} & \mathbf{b}_{1B}^{(2)} & \mathbf{0} \\ \mathbf{0} & \mathbf{b}_{2B}^{(2)} & \mathbf{b}_{1B}^{(3)} \\ \mathbf{b}_{1B}^{(1)} & \mathbf{0} & \mathbf{0} \\ \mathbf{0} & \mathbf{0} & \mathbf{b}_{2B}^{(3)} \end{bmatrix}_{12\times9} \begin{Bmatrix} \mathbf{Q}_1 \\ \mathbf{Q}_2 \\ \mathbf{Q}_3 \end{Bmatrix}_{9\times1} + \begin{Bmatrix} \mathbf{P}_{20}^{(1)} + \mathbf{P}_{10}^{(2)} \\ \mathbf{P}_{20}^{(2)} + \mathbf{P}_{10}^{(3)} \\ \mathbf{P}_{10}^{(1)} \\ \mathbf{P}_{20}^{(3)} \end{Bmatrix}_{12\times1} \tag{3.42}$$

or in a concise format as:

$$\mathbf{F} = \mathbf{BQ} + \mathbf{F}^0 \text{ or } \mathbf{F} - \mathbf{F}^0 = \mathbf{BQ} \tag{3.43}$$

or in a partitioned format as:

$$\mathbf{F}^* = \begin{Bmatrix} \mathbf{F}_{free} \\ \hline \mathbf{F}_{constr} \end{Bmatrix} - \begin{Bmatrix} \mathbf{F}_{free}^0 \\ \hline \mathbf{F}_{constr}^0 \end{Bmatrix} = \begin{bmatrix} \mathbf{B}_{free} \\ \hline \mathbf{B}_{constr} \end{bmatrix} \mathbf{Q} \tag{3.44}$$

where

$$\mathbf{F}_{free} = \lfloor \mathbf{F}_1 \quad \mathbf{F}_2 \rfloor^T \text{ and } \mathbf{F}_{constr} = \lfloor \mathbf{F}_3 \quad \mathbf{F}_4 \rfloor^T \tag{3.45}$$

$$\mathbf{B}_{free} = \begin{bmatrix} \mathbf{b}_{2B}^{(1)} & \mathbf{b}_{1B}^{(2)} & \mathbf{0} \\ \mathbf{0} & \mathbf{b}_{2B}^{(2)} & \mathbf{b}_{1B}^{(3)} \end{bmatrix} \text{ and } \mathbf{B}_{constr} = \begin{bmatrix} \mathbf{b}_{1B}^{(1)} & \mathbf{0} & \mathbf{0} \\ \mathbf{0} & \mathbf{0} & \mathbf{b}_{2B}^{(3)} \end{bmatrix} \tag{3.46}$$

$$\mathbf{F}_{free}^0 = \lfloor \mathbf{P}_{20}^{(1)} + \mathbf{P}_{10}^{(2)} \quad \mathbf{P}_{20}^{(2)} + \mathbf{P}_{10}^{(3)} \rfloor^T \text{ and } \mathbf{F}_{constr}^0 = \lfloor \mathbf{P}_{10}^{(1)} \quad \mathbf{P}_{20}^{(3)} \rfloor^T \tag{3.47}$$

In this matrix equilibrium equation, there are 15 unknown forces, namely 9 basic member forces $\mathbf{Q}$ and 6 applied forces at constrained degrees of freedom (reaction forces), $\mathbf{F}_{constr}$ but only 12 equilibrium equations are available. Therefore, this structure is statically indeterminate, and there exists an infinite number of equilibrium solutions. The degree of statical indeterminacy of the system is 15–12 = 3.

It is also noted that the degree of statical indeterminacy can easily be determined from the size of equilibrium matrix $\mathbf{B}_{free}$. The matrix $\mathbf{B}_{free}$ in Eq. (3.46) has the size of 6×9. This implies that there are six equilibrium equations at free degrees of freedom, but there are nine basic member forces in the system. Therefore, the degree of statical indeterminacy of the system is 9–6 = 3. This is exactly the same as that determined when the entire structural equilibrium equations are considered. It is pointed out that equilibrium equations at constrained degrees of freedom do not help in determining the basic member forces. This is due to the fact that, at constrained degrees of freedom, the number of equilibrium equations is as many as that of the unknown forces (reactions).

Generalizing, the equilibrium matrix $\mathbf{B}_{free}$ has as many rows as equilibrium equations available at free degrees of freedom and as many columns as basic unknown forces of the system. Consequently, the degree of statical indeterminacy ($SI$) is the difference between the column $n_{column}$ and row $n_{row}$ numbers.

$$SI = n_{column} - n_{row} \tag{3.48}$$

Mathematically, the degree of statical indeterminacy is related to the rank of the equilibrium matrix $\mathbf{B}_{free}$. This is summarized in Table 3.1 for various scenarios.

## EXAMPLE 3.1

For the frame structure shown in Figure 3.13, the following tasks are assigned:

(a) Number the degrees of freedom (both constrained and unconstrained) and basic (independent) member forces using the simply supported beam model.
(b) Formulate the equilibrium equations at constrained and unconstrained degrees of freedom using the simply supported beam model.

**Solution:** Member, node, and DOFs numbering systems are shown in Figure 3.14a:

From the numbering system for DOF, the mapping vector of each member is:

## TABLE 3.1
## Statical Indeterminacy and Rank of Equilibrium Matrix $\mathbf{B}_{free}$

| Scenario | $n_{row} \times n_{column}$ of Matrix $\mathbf{B}_{free}$ | Rank $r$ of Matrix $\mathbf{B}_{free}$ | Structural Type |
|---|---|---|---|
| 1 | $n_{row} = n_{column}$ | $r = n_{row}$ | Statically Determinate and Stable |
| 2 | $n_{row} < n_{column}$ | $r = n_{row}$ | Statically Indeterminate and Stable |
| 3 | $n_{row} \leq n_{column}$ | $r < n_{row}$ | Either some parts of the structure form a mechanism or rigid body modes are not properly suppressed by supports |
| 4 | $n_{row} > n_{column}$ | $r \leq n_{column}$ | |

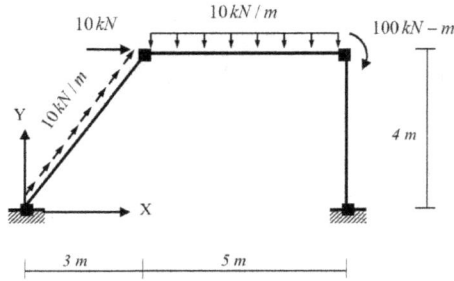

**FIGURE 3.13**  Example 3.1.

$$LM_1 = \begin{Bmatrix} 7 \\ 8 \\ 9 \\ 1 \\ 2 \\ 3 \end{Bmatrix}_1 ; LM_2 = \begin{Bmatrix} 1 \\ 2 \\ 3 \\ 4 \\ 5 \\ 6 \end{Bmatrix}_2 ; \text{and } LM_3 = \begin{Bmatrix} 10 \\ 11 \\ 12 \\ 4 \\ 5 \\ 6 \end{Bmatrix}_3$$

The basic (independent) member force numbering system is shown in Figure 3.14b.

Based on the matrix equilibrium equation for a simply supported beam model, considering the force system of each member:

Member 1 (Figure 3.15):

From Basic System to Local Reference System

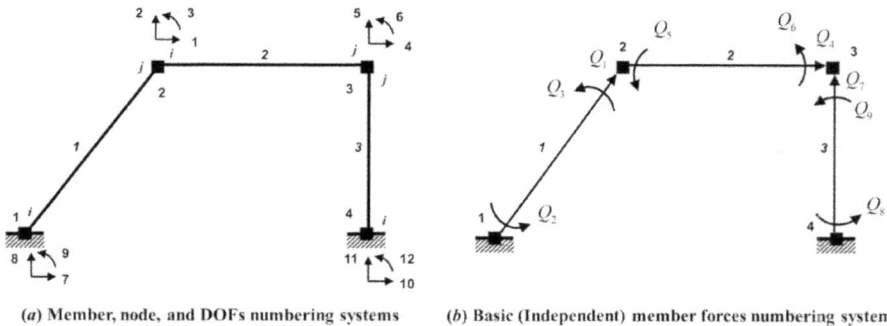

(a) Member, node, and DOFs numbering systems     (b) Basic (Independent) member forces numbering system

**FIGURE 3.14**  Example 3.1 (Continued).

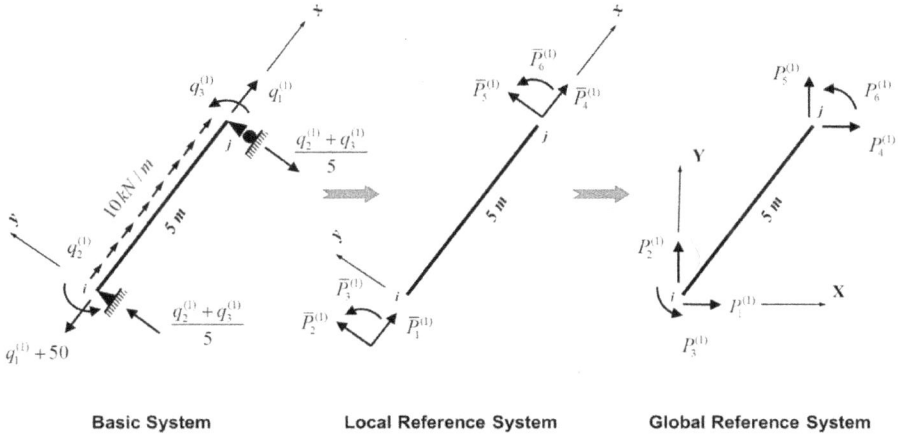

**FIGURE 3.15**   Example 3.1 (Continued).

$$
\begin{Bmatrix} \overline{P}_1^{(1)} \\ \overline{P}_2^{(1)} \\ \overline{P}_3^{(1)} \\ \overline{P}_4^{(1)} \\ \overline{P}_5^{(1)} \\ \overline{P}_6^{(1)} \end{Bmatrix} = \begin{bmatrix} -1 & 0 & 0 \\ 0 & \dfrac{1}{5} & \dfrac{1}{5} \\ 0 & 1 & 0 \\ 1 & 0 & 0 \\ 0 & -\dfrac{1}{5} & -\dfrac{1}{5} \\ 0 & 0 & 1 \end{bmatrix} \begin{Bmatrix} q_1^{(1)} \\ q_2^{(1)} \\ q_3^{(1)} \end{Bmatrix} + \begin{Bmatrix} -50 \\ 0 \\ 0 \\ 0 \\ 0 \\ 0 \end{Bmatrix}
$$

Assign $\left\lfloor q_1^{(1)} \quad q_2^{(1)} \quad q_3^{(1)} \right\rfloor^T$ as $\left\lfloor Q_1 \quad Q_2 \quad Q_3 \right\rfloor^T$, then

$$
\begin{Bmatrix} \overline{P}_1^{(1)} \\ \overline{P}_2^{(1)} \\ \overline{P}_3^{(1)} \\ \overline{P}_4^{(1)} \\ \overline{P}_5^{(1)} \\ \overline{P}_6^{(1)} \end{Bmatrix} = \begin{bmatrix} -1 & 0 & 0 \\ 0 & \dfrac{1}{5} & \dfrac{1}{5} \\ 0 & 1 & 0 \\ 1 & 0 & 0 \\ 0 & -\dfrac{1}{5} & -\dfrac{1}{5} \\ 0 & 0 & 1 \end{bmatrix} \begin{Bmatrix} Q_1 \\ Q_2 \\ Q_3 \end{Bmatrix} + \begin{Bmatrix} -50 \\ 0 \\ 0 \\ 0 \\ 0 \\ 0 \end{Bmatrix}
$$

From Local Reference System to Global Reference System

$$
\begin{Bmatrix} P_1^{(1)} \\ P_2^{(1)} \\ P_3^{(1)} \\ P_4^{(1)} \\ P_5^{(1)} \\ P_6^{(1)} \end{Bmatrix} = \begin{bmatrix} -0.6 & -0.16 & -0.16 \\ -0.8 & 0.12 & 0.12 \\ 0 & 1 & 0 \\ 0.6 & 0.16 & 0.16 \\ 0.8 & -0.12 & -0.12 \\ 0 & 0 & 1 \end{bmatrix} \begin{Bmatrix} Q_1 \\ Q_2 \\ Q_3 \end{Bmatrix} + \begin{Bmatrix} -30 \\ -40 \\ 0 \\ 0 \\ 0 \\ 0 \end{Bmatrix}
$$

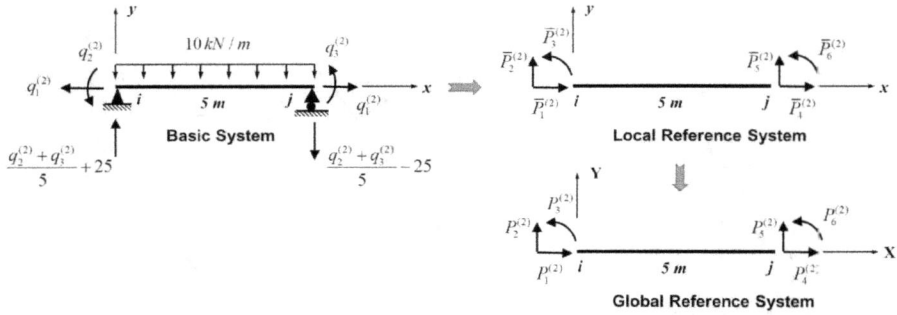

**FIGURE 3.16**  Example 3.1 (Continued).

Member 2 (Figure 3.16):

From Basic System to Local Reference System

$$
\begin{Bmatrix} \bar{P}_1^{(2)} \\ \bar{P}_2^{(2)} \\ \bar{P}_3^{(2)} \\ \bar{P}_4^{(2)} \\ \bar{P}_5^{(2)} \\ \bar{P}_6^{(2)} \end{Bmatrix} =
\begin{bmatrix} -1 & 0 & 0 \\ 0 & 0.2 & 0.2 \\ 0 & 1 & 0 \\ 1 & 0 & 0 \\ 0 & -0.2 & -0.2 \\ 0 & 0 & 1 \end{bmatrix}
\begin{Bmatrix} q_1^{(2)} \\ q_2^{(2)} \\ q_3^{(2)} \end{Bmatrix} +
\begin{Bmatrix} 0 \\ 25 \\ 0 \\ 0 \\ 25 \\ 0 \end{Bmatrix}
$$

Assign $\left\lfloor q_1^{(2)} \quad q_2^{(2)} \quad q_3^{(2)} \right\rfloor^T$ as $\left\lfloor Q_4 \quad Q_5 \quad Q_6 \right\rfloor^T$, then

$$
\begin{Bmatrix} \bar{P}_1^{(2)} \\ \bar{P}_2^{(2)} \\ \bar{P}_3^{(2)} \\ \bar{P}_4^{(2)} \\ \bar{P}_5^{(2)} \\ \bar{P}_6^{(2)} \end{Bmatrix} =
\begin{bmatrix} -1 & 0 & 0 \\ 0 & 0.2 & 0.2 \\ 0 & 1 & 0 \\ 1 & 0 & 0 \\ 0 & -0.2 & -0.2 \\ 0 & 0 & 1 \end{bmatrix}
\begin{Bmatrix} Q_4 \\ Q_5 \\ Q_6 \end{Bmatrix} +
\begin{Bmatrix} 0 \\ 25 \\ 0 \\ 0 \\ 25 \\ 0 \end{Bmatrix}
$$

From Local Reference System to Global Reference System

$$
\begin{Bmatrix} P_1^{(2)} \\ P_2^{(2)} \\ P_3^{(2)} \\ P_4^{(2)} \\ P_5^{(2)} \\ P_6^{(2)} \end{Bmatrix} =
\begin{bmatrix} -1 & 0 & 0 \\ 0 & 0.2 & 0.2 \\ 0 & 1 & 0 \\ 1 & 0 & 0 \\ 0 & -0.2 & -0.2 \\ 0 & 0 & 1 \end{bmatrix}
\begin{Bmatrix} Q_4 \\ Q_5 \\ Q_6 \end{Bmatrix} +
\begin{Bmatrix} 0 \\ 25 \\ 0 \\ 0 \\ 25 \\ 0 \end{Bmatrix}
$$

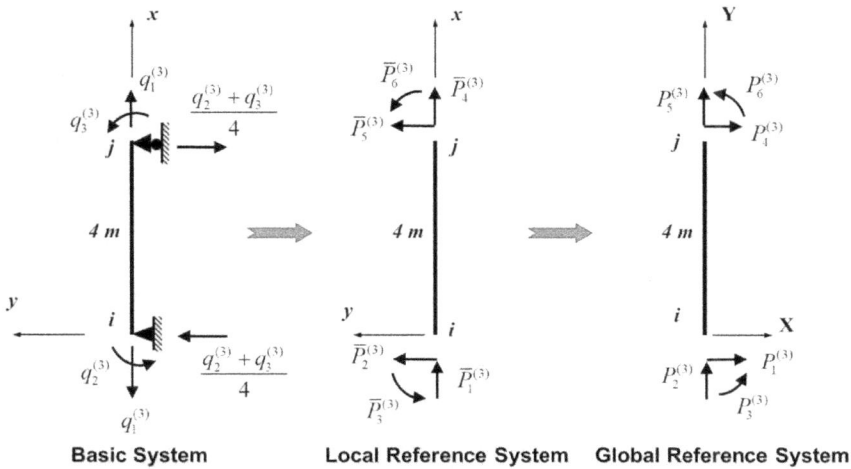

**Basic System**      **Local Reference System**      **Global Reference System**

**FIGURE 3.17**    Example 3.1 (Continued).

Member 3 (Figure 3.17):

From Basic System to Local Reference System

$$
\begin{Bmatrix} \overline{P}_1^{(3)} \\ \overline{P}_2^{(3)} \\ \overline{P}_3^{(3)} \\ \overline{P}_4^{(3)} \\ \overline{P}_5^{(3)} \\ \overline{P}_6^{(3)} \end{Bmatrix}
=
\begin{bmatrix} -1 & 0 & 0 \\ 0 & 0.25 & 0.25 \\ 0 & 1 & 0 \\ 1 & 0 & 0 \\ 0 & -0.25 & -0.25 \\ 0 & 0 & 1 \end{bmatrix}
\begin{Bmatrix} q_1^{(3)} \\ q_2^{(3)} \\ q_3^{(3)} \end{Bmatrix}
$$

Assign $\begin{bmatrix} q_1^{(3)} & q_2^{(3)} & q_3^{(3)} \end{bmatrix}^T$ as $\begin{bmatrix} Q_7 & Q_8 & Q_9 \end{bmatrix}^T$, then

$$
\begin{Bmatrix} \overline{P}_1^{(3)} \\ \overline{P}_2^{(3)} \\ \overline{P}_3^{(3)} \\ \overline{P}_4^{(3)} \\ \overline{P}_5^{(3)} \\ \overline{P}_6^{(3)} \end{Bmatrix}
=
\begin{bmatrix} -1 & 0 & 0 \\ 0 & 0.25 & 0.25 \\ 0 & 1 & 0 \\ 1 & 0 & 0 \\ 0 & -0.25 & -0.25 \\ 0 & 0 & 1 \end{bmatrix}
\begin{Bmatrix} Q_7 \\ Q_8 \\ Q_9 \end{Bmatrix}
$$

From Local Reference System to Global Reference System

$$
\begin{Bmatrix} P_1^{(3)} \\ P_2^{(3)} \\ P_3^{(3)} \\ P_4^{(3)} \\ P_5^{(3)} \\ P_6^{(3)} \end{Bmatrix}
=
\begin{bmatrix} 0 & -0.25 & -0.25 \\ -1 & 0 & 0 \\ 0 & 1 & 0 \\ 0 & 0.25 & 0.25 \\ 1 & 0 & 0 \\ 0 & 0 & 1 \end{bmatrix}
\begin{Bmatrix} Q_7 \\ Q_8 \\ Q_9 \end{Bmatrix}
$$

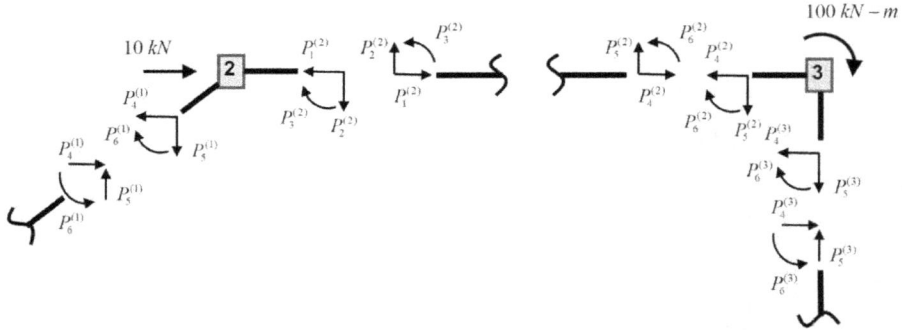

**FIGURE 3.18**  Example 3.1 (Continued).

Consider equilibrium conditions at free (unconstrained) nodes shown in Figure 3.18:

*Node 2:*

$$\sum F_X = 0; \ 10 - P_4^{(1)} - P_1^{(2)} = 0 \rightarrow = 0.6Q_1 + 0.16Q_2 + 0.16Q_3 - Q_4$$

$$\sum F_Y = 0; \ -P_5^{(1)} - P_2^{(2)} = 0 \rightarrow 0 = 0.8Q_1 - 0.12Q_2 - 0.2Q_3 + 0.2Q_4 + 0.2Q_5 + 25$$

$$\sum M_Z = 0; \ -P_6^{(1)} - P_3^{(2)} = 0 \rightarrow 0 = Q_3 + Q_5$$

*Node 3:*

$$\sum F_X = 0; \ -P_4^{(2)} - P_4^{(3)} = 0 \rightarrow 0 = Q_4 + 0.25Q_8 + 0.25Q_9$$

$$\sum F_Y = 0; \ -P_5^{(2)} - P_5^{(3)} = 0 \rightarrow 0 = -0.2Q_5 - 0.2Q_6 + Q_7 + 25$$

$$\sum M_Z = 0; \ -100 - P_6^{(2)} - P_6^{(3)} = 0 \rightarrow -100 = Q_6 + Q_9$$

Consider equilibrium conditions at constrained nodes shown in Figure 3.19:

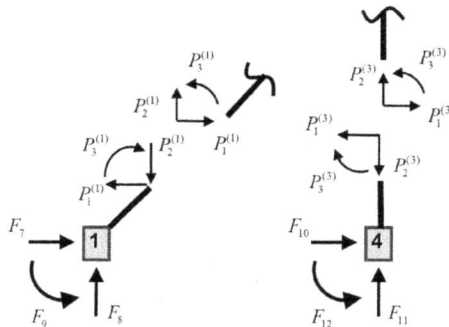

**FIGURE 3.19**  Example 3.1 (Continued).

*Node 1:*

$$\sum F_X = 0; \ F_7 - P_1^{(1)} = 0 \ \rightarrow \ F_7 = -0.6Q_1 - 0.16Q_2 - 0.16Q_3 - 30$$

$$\sum F_Y = 0; \ F_8 - P_2^{(1)} = 0 \ \rightarrow \ F_8 = -0.8Q_1 + 0.12Q_2 + 0.12Q_3 - 40$$

$$\sum M_Z = 0; \ F_9 - P_3^{(1)} = 0 \ \rightarrow \ F_9 = Q_2$$

*Node 4:*

$$\sum F_X = 0; \ F_{10} - P_1^{(3)} = 0 \ \rightarrow \ F_{10} = -0.25Q_8 - 0.25Q_9$$

$$\sum F_Y = 0; \ F_{11} - P_2^{(3)} = 0 \ \rightarrow \ F_{11} = -Q_7$$

$$\sum M_Z = 0; \ F_{12} - P_3^{(3)} = 0 \ \rightarrow \ F_{12} = Q_8$$

Collectively, all nodal equilibrium equations can be written in the matrix form as:

$$
\begin{Bmatrix} 10 \\ 0 \\ 0 \\ 0 \\ 0 \\ -100 \\ F_7 \\ F_8 \\ F_9 \\ F_{10} \\ F_{11} \\ F_{12} \end{Bmatrix} =
\begin{bmatrix}
0.6 & 0.16 & 0.16 & -1 & 0 & 0 & 0 & 0 & 0 \\
0.8 & -0.12 & -0.12 & 0 & 0.2 & 0.2 & 0 & 0 & 0 \\
0 & 0 & 1 & 0 & 1 & 0 & 0 & 0 & 0 \\
0 & 0 & 0 & 1 & 0 & 0 & 0 & 0.25 & 0.25 \\
0 & 0 & 0 & 0 & -0.2 & -0.2 & 1 & 0 & 0 \\
0 & 0 & 0 & 0 & 0 & 1 & 0 & 0 & 1 \\
-0.6 & -0.16 & -0.16 & 0 & 0 & 0 & 0 & 0 & 0 \\
-0.8 & 0.12 & 0.12 & 0 & 0 & 0 & 0 & 0 & 0 \\
0 & 1 & 0 & 0 & 0 & 0 & 0 & 0 & 0 \\
0 & 0 & 0 & 0 & 0 & 0 & 0 & -0.25 & -0.25 \\
0 & 0 & 0 & 0 & 0 & 0 & -1 & 0 & 0 \\
0 & 0 & 0 & 0 & 0 & 0 & 0 & 1 & 0
\end{bmatrix}
\begin{Bmatrix} Q_1 \\ Q_2 \\ Q_3 \\ Q_4 \\ Q_5 \\ Q_6 \\ Q_7 \\ Q_8 \\ Q_9 \end{Bmatrix} +
\begin{Bmatrix} 0 \\ 25 \\ 0 \\ 0 \\ 25 \\ 0 \\ -30 \\ -40 \\ 0 \\ 0 \\ 0 \\ 0 \end{Bmatrix}
$$

or in the compact format:

$$\mathbf{F} = \mathbf{B}\mathbf{Q} + \mathbf{F}^0 \ \text{ or } \begin{Bmatrix} \mathbf{F}_{free} \\ \mathbf{F}_{constr} \end{Bmatrix} = \begin{bmatrix} \mathbf{B}_{free} \\ \mathbf{B}_{constr} \end{bmatrix} \mathbf{Q} + \begin{Bmatrix} \mathbf{F}^0_{free} \\ \mathbf{F}^0_{constr} \end{Bmatrix}$$

Separately, equilibrium equations at free and constrained degrees of freedom are expressed as:

$$\mathbf{F}_{free} = \mathbf{B}_{free}\mathbf{Q} + \mathbf{F}^0_{free} \ \text{ and } \ \mathbf{F}_{constr} = \mathbf{B}_{constr}\mathbf{Q} + \mathbf{F}^0_{constr}$$

By moving the member loads to the left hand side, the member load effects are converted to equivalent applied loads as:

$$\mathbf{F}_{free} - \mathbf{F}^0_{free} = \mathbf{B}_{free}\mathbf{Q} \ \text{ and } \ \mathbf{F}_{constr} - \mathbf{F}^0_{constr} = \mathbf{B}_{constr}\mathbf{Q}$$

Equilibrium equations associated with the free (unconstrained) degrees of freedom are:

$$
\begin{Bmatrix} 10 \\ -25 \\ 0 \\ 0 \\ -25 \\ -100 \end{Bmatrix}
=
\begin{bmatrix}
0.6 & 0.16 & 0.16 & -1 & 0 & 0 & 0 & 0 & 0 \\
0.8 & -0.12 & -0.12 & 0 & 0.2 & 0.2 & 0 & 0 & 0 \\
0 & 0 & 1 & 0 & 1 & 0 & 0 & 0 & 0 \\
0 & 0 & 0 & 1 & 0 & 0 & 0 & 0.25 & 0.25 \\
0 & 0 & 0 & 0 & -0.2 & -0.2 & 1 & 0 & 0 \\
0 & 0 & 0 & 0 & 0 & 1 & 0 & 0 & 1
\end{bmatrix}
\begin{Bmatrix} Q_1 \\ Q_2 \\ Q_3 \\ Q_4 \\ Q_5 \\ Q_6 \\ Q_7 \\ Q_8 \\ Q_9 \end{Bmatrix}
$$

In this reduced matrix form of equilibrium equations, there are 6 equilibrium equations and 9 basic member force unknowns. Therefore, the degree of statical indeterminacy is 9–6 = 3.

If three hinges (moment releases) are inserted at member ends as shown in Figure 3.20, this structure becomes statically determinate.

The equilibrium equations of this modified structure can be obtained from the equilibrium equations of the original structure by setting $Q_2 = Q_8 = Q_9 = 0$. It is observed that these moment releases can easily be inserted in the model. This is due to the use of the simply supported beam model as the basic system.

$$
\begin{Bmatrix} 10 \\ -25 \\ 0 \\ 0 \\ -25 \\ -100 \end{Bmatrix}
=
\begin{bmatrix}
0.6 & 0.16 & 0.16 & -1 & 0 & 0 & 0 & 0 & 0 \\
0.8 & -0.12 & -0.12 & 0 & 0.2 & 0.2 & 0 & 0 & 0 \\
0 & 0 & 1 & 0 & 1 & 0 & 0 & 0 & 0 \\
0 & 0 & 0 & 1 & 0 & 0 & 0 & 0.25 & 0.25 \\
0 & 0 & 0 & 0 & -0.2 & -0.2 & 1 & 0 & 0 \\
0 & 0 & 0 & 0 & 0 & 1 & 0 & 0 & 1
\end{bmatrix}
\begin{Bmatrix} Q_1 \\ 0 \\ Q_3 \\ Q_4 \\ Q_5 \\ Q_6 \\ Q_7 \\ 0 \\ 0 \end{Bmatrix}
$$

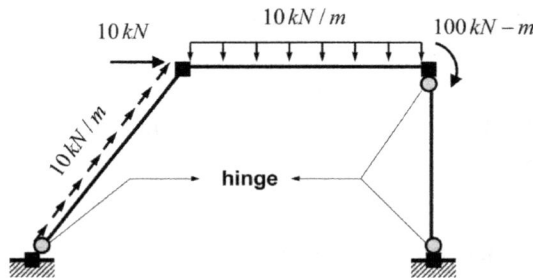

**FIGURE 3.20**   Example 3.1 (Continued).

$$\begin{Bmatrix} 10 \\ -25 \\ 0 \\ 0 \\ -25 \\ -100 \end{Bmatrix} = \begin{bmatrix} 0.6 & 0.16 & -1 & 0 & 0 & 0 \\ 0.8 & -0.12 & 0 & 0.2 & 0.2 & 0 \\ 0 & 1 & 0 & 1 & 0 & 0 \\ 0 & 0 & 1 & 0 & 0 & 0 \\ 0 & 0 & 0 & -0.2 & -0.2 & 1 \\ 0 & 0 & 0 & 0 & 1 & 0 \end{bmatrix} \begin{Bmatrix} Q_1 \\ Q_3 \\ Q_4 \\ Q_5 \\ Q_6 \\ Q_7 \end{Bmatrix}$$

Now, there are 6 basic unknown forces and 6 equilibrium equations. Thus, this statically determinate system can be solved and yields:

$Q_1 = 7.5\,kN;\ Q_3 = 34.375\,kN - m;\ Q_4 = 0;\ Q_5 = -34.375\,kN - m;\ Q_6 = -100\,kN - m;$ and $Q_7 = -51.875\,kN$

$$\mathbf{Q} = \lfloor Q_1 \quad Q_2 \quad Q_3 \quad Q_4 \quad Q_5 \quad Q_6 \quad Q_7 \quad Q_8 \quad Q_9 \rfloor^T$$
$$= \lfloor 7.5 \quad 0 \quad 34.375 \quad 0 \quad -34.375 \quad -100 \quad -51.875 \quad 0 \quad 0 \rfloor^T$$

The reaction forces shown in Figure 3.21 can easily be obtained:

$$\mathbf{F}_{constr} = \mathbf{B}_{constr}\mathbf{Q} + \mathbf{F}^0_{constr} \text{ or } \begin{Bmatrix} F_7 \\ F_8 \\ F_9 \\ F_{10} \\ F_{11} \\ F_{12} \end{Bmatrix} = \begin{bmatrix} -0.6 & -0.16 & -0.16 & 0 & 0 & 0 & 0 & 0 & 0 \\ -0.8 & 0.12 & 0.12 & 0 & 0 & 0 & 0 & 0 & 0 \\ 0 & 1 & 0 & 0 & 0 & 0 & 0 & 0 & 0 \\ 0 & 0 & 0 & 0 & 0 & 0 & 0 & -0.25 & -0.25 \\ 0 & 0 & 0 & 0 & 0 & 0 & -1 & 0 & 0 \\ 0 & 0 & 0 & 0 & 0 & 0 & 0 & 1 & 0 \end{bmatrix} \begin{Bmatrix} 7.5 \\ 0 \\ 34.375 \\ 0 \\ -34.375 \\ -100 \\ -51.875 \\ 0 \\ 0 \end{Bmatrix} + \begin{Bmatrix} -30 \\ -40 \\ 0 \\ 0 \\ 0 \\ 0 \end{Bmatrix}$$

$F_7 = -40\,kN;\ F_8 = -41.875\,kN;\ F_9 = 0;\ F_{10} = 0;\ F_{11} = 51.875kN;$ and $F_{12} = 0$

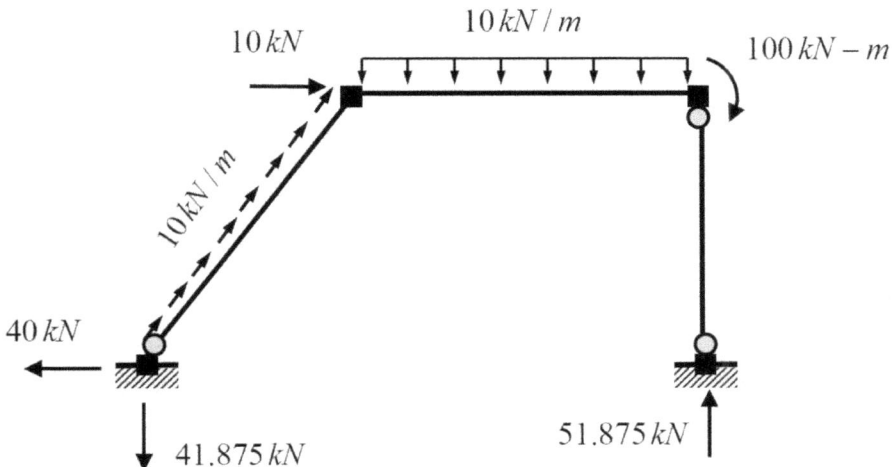

**FIGURE 3.21**  Example 3.1 (Continued).

The member end forces can be obtained by substituting their basic forces into the member equilibrium equations.

For example, member 1 (Figure 3.22):

$$
\begin{Bmatrix} P_1^{(1)} \\ P_2^{(1)} \\ P_3^{(1)} \\ P_4^{(1)} \\ P_5^{(1)} \\ P_6^{(1)} \end{Bmatrix} = \begin{bmatrix} -0.6 & -0.16 & -0.16 \\ -0.8 & 0.12 & 0.12 \\ 0 & 1 & 0 \\ 0.6 & 0.16 & 0.16 \\ 0.8 & -0.12 & -0.12 \\ 0 & 0 & 1 \end{bmatrix} \begin{Bmatrix} 7.5 \\ 0 \\ 34.375 \end{Bmatrix} + \begin{Bmatrix} -30 \\ -40 \\ 0 \\ 0 \\ 0 \\ 0 \end{Bmatrix} = \begin{Bmatrix} -40 \\ -41.875 \\ 0 \\ 10 \\ 1.875 \\ 34.375 \end{Bmatrix}
$$

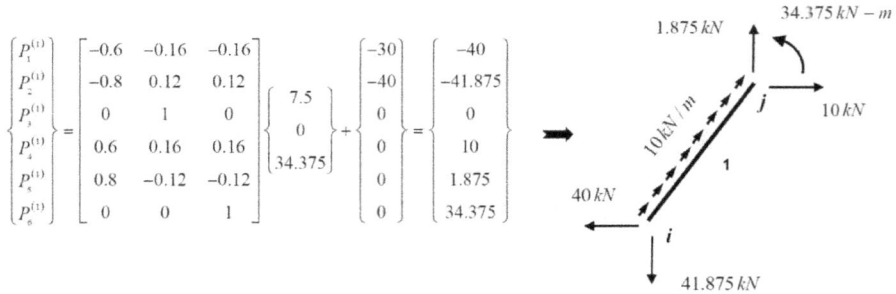

**FIGURE 3.22**   Example 3.1 (Continued).

## 3.6   EQUILIBRIUM: TRUSS (PIN-ENDED) MEMBER AND PIN-JOINTED STRUCTURE

So far, only structures consisting of frame members are considered. In this section, the attention is on pin-jointed structures commonly known as truss structures. The matrix equilibrium equations derived in the previous sections can easily be applied to a truss structure since the truss member is merely a special form of the frame member. This can be achieved by modifying the simply supported beam model of Figure 3.3 as follows. In the absence of member loads ($w_x(x) = w_y(x) = 0$), a truss member is obtained by releasing end moments of the basic system (Figure 3.3b). Physically, this is equivalent to the insertion of a hinge at each member end. Mathematically, this implies that:

$$
q_2 = 0 \text{ and } q_3 = 0 \tag{3.49}
$$

Substituting Eqs. (3.49) into (3.10) and removing end moments from the end force vector $\bar{\mathbf{P}}$ result in:

$$
\begin{Bmatrix} \bar{P}_1 \\ \bar{P}_2 \end{Bmatrix} = \begin{bmatrix} -1 \\ 1 \end{bmatrix} q_1 \text{ or } \bar{\mathbf{P}} = \bar{\mathbf{b}}^{truss} q_1 \tag{3.50}
$$

where $\bar{\mathbf{b}}^{truss}$ is the equilibrium matrix of the basic truss system in the local reference axes. A schematic representation of the whole modification process is shown in Figure 3.23.

The matrix equilibrium equation of the truss member in the global reference axes can be obtained in a similar fashion. Substituting Eqs. (3.49) into (3.29) and removing end moments from the end force vector $\mathbf{P}$ result in:

$$
\begin{Bmatrix} P_1 \\ P_2 \\ P_3 \\ P_4 \end{Bmatrix} = \begin{bmatrix} -\cos\phi \\ -\sin\phi \\ \cos\phi \\ \sin\phi \end{bmatrix} q_1 \text{ or } \mathbf{P} = \mathbf{b}^{truss} q_1 \tag{3.51}
$$

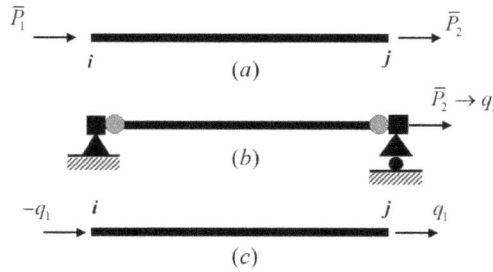

**FIGURE 3.23** Truss Member in Local Reference System: (a) Complete System, (b) Basic System, (c) Basic System with Reactive Forces.

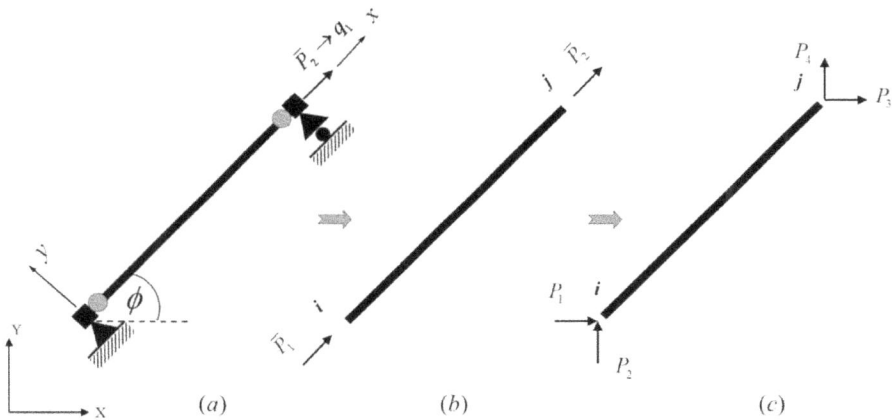

**FIGURE 3.24** Truss Member in Global Reference System: (a) Basic System, (b) Complete System in Local Reference Axes, (c) Complete System in Global Reference Axes.

where $\mathbf{b}^{truss}$ is the equilibrium matrix of the basic truss system in the global reference axes. A schematic representation of the whole modification process is shown in Figure 3.24.

## EXAMPLE 3.2

Analyze the truss shown on Figure 3.25 for internal forces and support reactions through the following steps:

(a) Number the degrees of freedom (both constrained and unconstrained) and basic (independent) member forces.
(b) Formulate the equilibrium equations at unconstrained degrees of freedom.
(c) Solve the basic member forces.

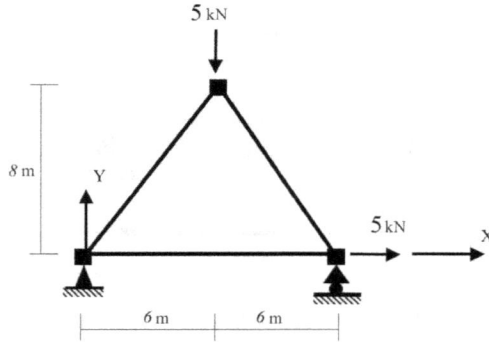

**FIGURE 3.25**   Example 3.2.

(d)   Employ the equilibrium equations associated with the constrained degrees of freedom to compute the support reactions.

**Solution:** Member, node, and DOFs numbering systems are shown in Figure 3.26a:
From the numbering system for DOF, the mapping vector of each member is:

$$LM_1 = \begin{Bmatrix} 5 \\ 6 \\ 1 \\ 2 \end{Bmatrix}_1 ; \; LM_2 = \begin{Bmatrix} 1 \\ 2 \\ 3 \\ 4 \end{Bmatrix}_2 ; \text{ and } LM_3 = \begin{Bmatrix} 5 \\ 6 \\ 3 \\ 4 \end{Bmatrix}_3$$

The basic (independent) member force numbering system is shown in Figure 3.26b.

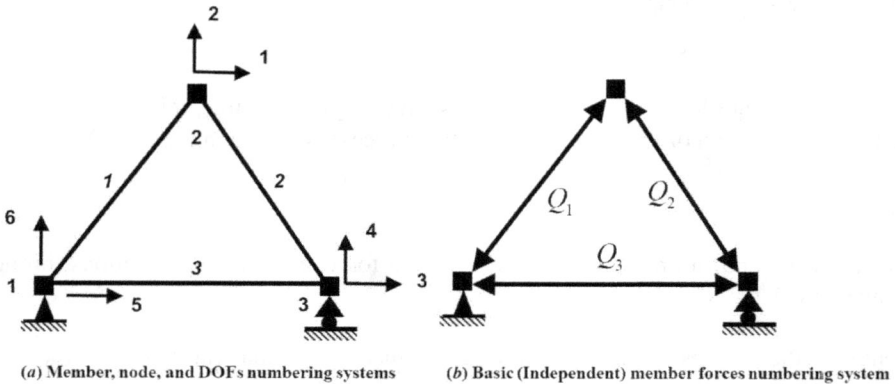

(a) Member, node, and DOFs numbering systems        (b) Basic (Independent) member forces numbering system

**FIGURE 3.26**   Example 3.2 (Continued).

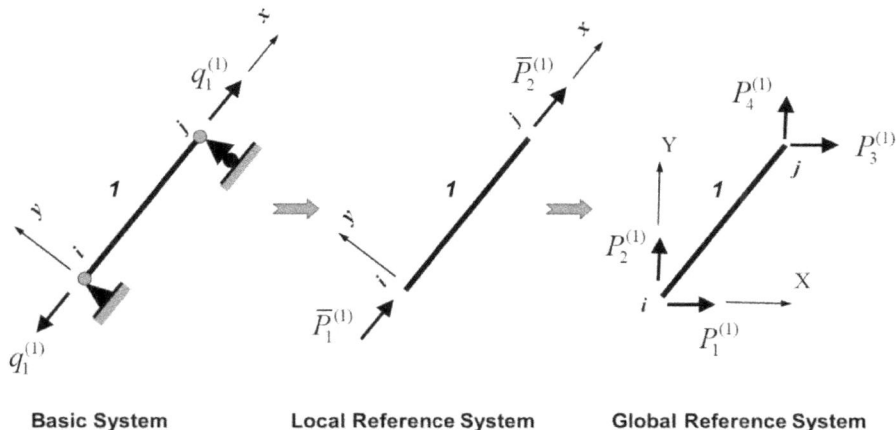

**Basic System**  **Local Reference System**  **Global Reference System**

**FIGURE 3.27** Example 3.2 (Continued).

Based on the matrix equilibrium equation for a truss member, considering the force system of each member:

Member 1 (Figure 3.27):

From Basic System to Local Reference System

$$\left\{ \begin{matrix} \bar{P}_1^{(1)} \\ \bar{P}_2^{(1)} \end{matrix} \right\} = \begin{bmatrix} -1 \\ 1 \end{bmatrix} q_1^{(1)} \rightarrow Q_1 = q_1^{(1)} \rightarrow \left\{ \begin{matrix} \bar{P}_1^{(1)} \\ \bar{P}_2^{(1)} \end{matrix} \right\} = \begin{bmatrix} -1 \\ 1 \end{bmatrix} Q_1$$

From Local Reference System to Global Reference System

$$\left\{ \begin{matrix} P_1^{(1)} \\ P_2^{(1)} \\ P_3^{(1)} \\ P_4^{(1)} \end{matrix} \right\} = \begin{bmatrix} -0.6 \\ -0.8 \\ 0.6 \\ 0.8 \end{bmatrix} Q_1$$

Member 2 (Figure 3.28):

From Basic System to Local Reference System

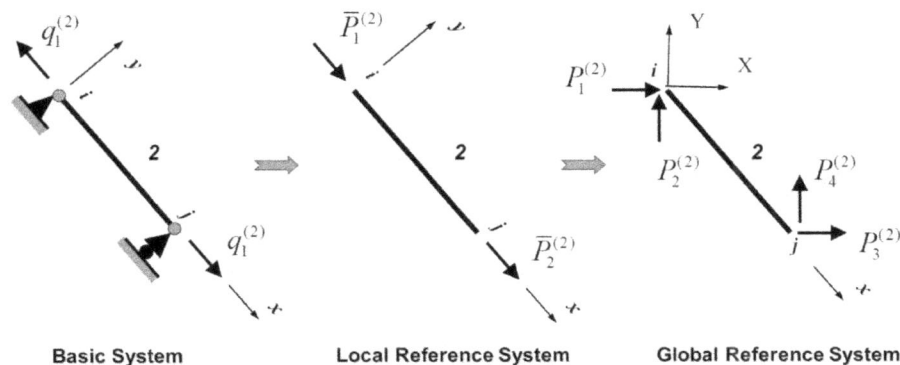

**Basic System**  **Local Reference System**  **Global Reference System**

**FIGURE 3.28** Example 3.2 (Continued).

$$\left\{ \begin{matrix} \overline{P}_1^{(2)} \\ \overline{P}_2^{(2)} \end{matrix} \right\} = \begin{bmatrix} -1 \\ 1 \end{bmatrix} q_1^{(2)} \rightarrow Q_2 = q_1^{(2)} \rightarrow \left\{ \begin{matrix} \overline{P}_1^{(2)} \\ \overline{P}_2^{(2)} \end{matrix} \right\} = \begin{bmatrix} -1 \\ 1 \end{bmatrix} Q_2$$

From Local Reference System to Global Reference System

$$\left\{ \begin{matrix} P_1^{(2)} \\ P_2^{(2)} \\ P_3^{(2)} \\ P_4^{(2)} \end{matrix} \right\} = \begin{bmatrix} -0.6 \\ 0.8 \\ 0.6 \\ -0.8 \end{bmatrix} Q_2$$

Member 3 (Figure 3.29):

From Basic System to Local Reference System

$$\left\{ \begin{matrix} \overline{P}_1^{(3)} \\ \overline{P}_2^{(3)} \end{matrix} \right\} = \begin{bmatrix} -1 \\ 1 \end{bmatrix} q_1^{(3)} \rightarrow Q_3 = q_1^{(3)} \rightarrow \left\{ \begin{matrix} \overline{P}_1^{(3)} \\ \overline{P}_2^{(3)} \end{matrix} \right\} = \begin{bmatrix} -1 \\ 1 \end{bmatrix} Q_3$$

From Local Reference System to Global Reference System

$$\left\{ \begin{matrix} P_1^{(3)} \\ P_2^{(3)} \\ P_3^{(3)} \\ P_4^{(3)} \end{matrix} \right\} = \begin{bmatrix} -1 \\ 0 \\ 1 \\ 0 \end{bmatrix} Q_3$$

Consider equilibrium conditions at all nodes as shown in Figure 3.30:

*Node 1:*

$$\sum F_X = 0; \quad F_5 - P_1^{(1)} - P_1^{(3)} = 0 \rightarrow F_5 = -0.6Q_1 - Q_3$$

$$\sum F_Y = 0; \quad F_6 - P_2^{(1)} - P_2^{(3)} = 0 \rightarrow F_6 = -0.8Q_1$$

*Node 2:*

$$\sum F_X = 0; \quad -P_3^{(1)} - P_1^{(2)} = 0 \rightarrow 0 = 0.6Q_1 - 0.6Q_2$$

$$\sum F_Y = 0; \quad -5 - P_4^{(1)} - P_2^{(2)} = 0 \rightarrow -5 = 0.8Q_1 + 0.8Q_2$$

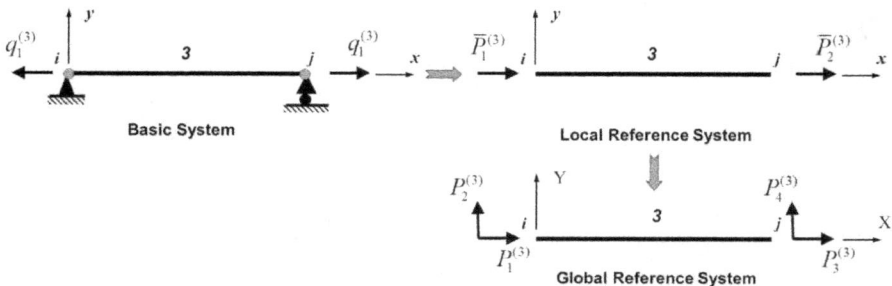

**FIGURE 3.29** Example 3.2 (Continued).

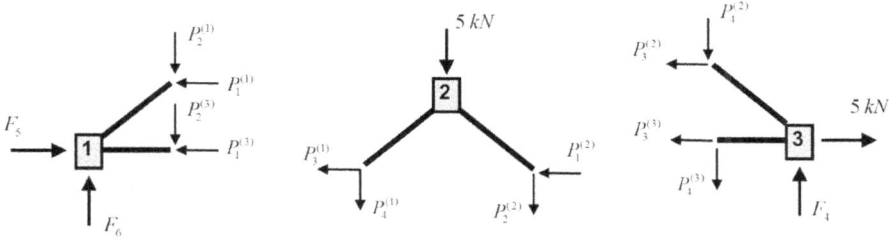

**FIGURE 3.30** Example 3.2 (Continued).

*Node 3:*

$$\sum F_X = 0;\ 5 - P_3^{(2)} - P_3^{(3)} = 0 \rightarrow 5 = 0.6Q_2 + Q_3$$

$$\sum F_Y = 0;\ F_4 - P_4^{(2)} - P_4^{(3)} = 0 \rightarrow F_4 = -0.8Q_2$$

Collectively, all nodal equilibrium equations can be written in a matrix form as:

$$\begin{Bmatrix} 0 \\ -5 \\ 5 \\ F_4 \\ F_5 \\ F_6 \end{Bmatrix} = \begin{bmatrix} 0.6 & -0.6 & 0 \\ 0.8 & 0.8 & 0 \\ 0 & 0.6 & 1 \\ 0 & -0.8 & 0 \\ -0.6 & 0 & -1 \\ -0.8 & 0 & 0 \end{bmatrix} \begin{Bmatrix} Q_1 \\ Q_2 \\ Q_3 \end{Bmatrix}$$

or in a compact form as:

$$\mathbf{F} = \mathbf{BQ} \text{ or } \left\{ \frac{\mathbf{F}_{free}}{\mathbf{F}_{constr}} \right\} = \left[ \frac{\mathbf{B}_{free}}{\mathbf{B}_{constr}} \right] \mathbf{Q}$$

Separately, equilibrium equations at free and constrained degrees of freedom are expressed as:

$$\mathbf{F}_{free} = \mathbf{B}_{free}\mathbf{Q} \text{ and } \mathbf{F}_{constr} = \mathbf{B}_{constr}\mathbf{Q}$$

In this matrix equilibrium equation, there are 6 unknown forces; namely 3 basic member forces $\mathbf{Q}$ and 3 applied forces at constrained degrees of freedom (reaction forces) $\mathbf{F}_{constr}$ and 6 equilibrium equations are available. Therefore, this truss is statically determinate and its degree of statical indeterminacy is 6–6 = 0.

Equilibrium equations associated with free (unconstrained) degrees of freedom are:

$$\mathbf{F}_{free} = \mathbf{B}_{free}\mathbf{Q} \text{ or } \begin{Bmatrix} 0 \\ -5 \\ 5 \end{Bmatrix} = \begin{bmatrix} 0.6 & -0.6 & 0 \\ 0.8 & 0.8 & 0 \\ 0 & 0.6 & 1 \end{bmatrix} \begin{Bmatrix} Q_1 \\ Q_2 \\ Q_3 \end{Bmatrix}$$

In this reduced matrix form of equilibrium equations, there are 3 equilibrium equations and 3 basic member force unknowns. Therefore, the degree of statical indeterminacy is 3–3 = 0. Thus, solving this reduced matrix equation yields the basic member forces as:

$$Q_1 = -3.125\ kN;\ Q_2 = -3.125\ kN;\ \text{and } Q_3 = 6.875\ kN$$

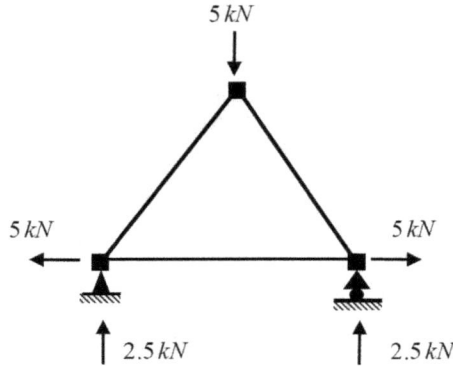

**FIGURE 3.31**   Example 3.2 (Continued).

**FIGURE 3.32**   Example 3.2 (Continued).

Following sign convention, a positive value of the basic member force represents tension while a negative value represents compression.

Substituting these basic member forces into the equilibrium equations associated with the constrained degrees of freedom, the support reactions are determined as shown in Figure 3.31:

$$\mathbf{F}_{constr} = \mathbf{B}_{constr}\mathbf{Q} \text{ or } \begin{Bmatrix} F_4 \\ F_5 \\ F_6 \end{Bmatrix} = \begin{bmatrix} 0 & -0.8 & 0 \\ -0.6 & 0 & -1 \\ -0.8 & 0 & 0 \end{bmatrix} \begin{bmatrix} -3.125 \\ -3.125 \\ 6.875 \end{bmatrix} = \begin{Bmatrix} 2.5 \\ -5 \\ 2.5 \end{Bmatrix} kN$$

The member end forces can be obtained by substituting their relevant basic forces into the member equations.

For example, member 1 (Figure 3.32):

## 3.7   AUTOMATED ASSEMBLY OF EQUILIBRIUM EQUATIONS

In the previous sections, the matrix equilibrium equation of a structure had been constructed in the following manner: one first expressed the local end forces $\overline{\mathbf{P}}$ of each member in terms of its basic (independent) member forces $\mathbf{q}$, then transformed the local member end forces $\overline{\mathbf{P}}$ to the global member end forces $\mathbf{P}$ by the rotational matrix, and finally considered equilibrium equations at all degrees of freedom in a collective fashion. The aforementioned process of setting up the equilibrium equations can be extremely tedious in the case of a large structural system.

In this section, a systematic approach to assembling equilibrium equations of a structure is presented. It is natural to extend this approach to computer-oriented assembly of the equilibrium equations.

To demonstrate the automated assembling approach, the frame structure of Example 3.1 is revisited. From the viewpoint of each member, each end force in the member degree of freedom is associated with only one structural degree of freedom. This one-to-one correspondence is represented by the mapping vector as discussed in Chapter 1. It is worth repeating that the mapping vectors of members 1, 2, and 3 are:

$$
\begin{Bmatrix} 7 \\ 8 \\ 9 \\ 1 \\ 2 \\ 3 \end{Bmatrix}_{structure} \rightarrow \begin{Bmatrix} 1 \\ 2 \\ 3 \\ 4 \\ 5 \\ 6 \end{Bmatrix}_{member\ 1} \qquad \begin{Bmatrix} 1 \\ 2 \\ 3 \\ 4 \\ 5 \\ 6 \end{Bmatrix}_{structure} \rightarrow \begin{Bmatrix} 1 \\ 2 \\ 3 \\ 4 \\ 5 \\ 6 \end{Bmatrix}_{member\ 2} \qquad \begin{Bmatrix} 10 \\ 11 \\ 12 \\ 4 \\ 5 \\ 6 \end{Bmatrix}_{structure} \rightarrow \begin{Bmatrix} 1 \\ 2 \\ 3 \\ 4 \\ 5 \\ 6 \end{Bmatrix}_{member\ 3} \tag{3.52}
$$

The global equilibrium equations at all degrees of freedom can be written in terms of member end forces as:

$$
\begin{Bmatrix} 10 \\ 0 \\ 0 \\ 0 \\ 0 \\ -100 \\ F_7 \\ F_8 \\ F_9 \\ F_{10} \\ F_{11} \\ F_{12} \end{Bmatrix} - \begin{Bmatrix} 0 \\ 25 \\ 0 \\ 0 \\ 25 \\ 0 \\ -30 \\ -40 \\ 0 \\ 0 \\ 0 \\ 0 \end{Bmatrix} = \begin{bmatrix} 0 & 0 & 0 & 1 & 0 & 0 \\ 0 & 0 & 0 & 0 & 1 & 0 \\ 0 & 0 & 0 & 0 & 0 & 1 \\ 0 & 0 & 0 & 0 & 0 & 0 \\ 0 & 0 & 0 & 0 & 0 & 0 \\ 0 & 0 & 0 & 0 & 0 & 0 \\ 1 & 0 & 0 & 0 & 0 & 0 \\ 0 & 1 & 0 & 0 & 0 & 0 \\ 0 & 0 & 1 & 0 & 0 & 0 \\ 0 & 0 & 0 & 0 & 0 & 0 \\ 0 & 0 & 0 & 0 & 0 & 0 \\ 0 & 0 & 0 & 0 & 0 & 0 \end{bmatrix} \begin{Bmatrix} P_1^{(1)} \\ P_2^{(1)} \\ P_3^{(1)} \\ P_4^{(1)} \\ P_5^{(1)} \\ P_6^{(1)} \end{Bmatrix} + \begin{bmatrix} 1 & 0 & 0 & 0 & 0 & 0 \\ 0 & 1 & 0 & 0 & 0 & 0 \\ 0 & 0 & 1 & 0 & 0 & 0 \\ 0 & 0 & 0 & 1 & 0 & 0 \\ 0 & 0 & 0 & 0 & 1 & 0 \\ 0 & 0 & 0 & 0 & 0 & 1 \\ 0 & 0 & 0 & 0 & 0 & 0 \\ 0 & 0 & 0 & 0 & 0 & 0 \\ 0 & 0 & 0 & 0 & 0 & 0 \\ 0 & 0 & 0 & 0 & 0 & 0 \\ 0 & 0 & 0 & 0 & 0 & 0 \\ 0 & 0 & 0 & 0 & 0 & 0 \end{bmatrix} \begin{Bmatrix} P_1^{(2)} \\ P_2^{(2)} \\ P_3^{(2)} \\ P_4^{(2)} \\ P_5^{(2)} \\ P_6^{(2)} \end{Bmatrix} + \begin{bmatrix} 0 & 0 & 0 & 0 & 0 & 0 \\ 0 & 0 & 0 & 0 & 0 & 0 \\ 0 & 0 & 0 & 0 & 0 & 0 \\ 0 & 0 & 0 & 1 & 0 & 0 \\ 0 & 0 & 0 & 0 & 1 & 0 \\ 0 & 0 & 0 & 0 & 0 & 1 \\ 0 & 0 & 0 & 0 & 0 & 0 \\ 0 & 0 & 0 & 0 & 0 & 0 \\ 0 & 0 & 0 & 0 & 0 & 0 \\ 1 & 0 & 0 & 0 & 0 & 0 \\ 0 & 1 & 0 & 0 & 0 & 0 \\ 0 & 0 & 1 & 0 & 0 & 0 \end{bmatrix} \begin{Bmatrix} P_1^{(3)} \\ P_2^{(3)} \\ P_3^{(3)} \\ P_4^{(3)} \\ P_5^{(3)} \\ P_6^{(3)} \end{Bmatrix}
$$

$$\tag{3.53}$$

or in a compact form as:

$$
\mathbf{F} - \mathbf{F}^0 = \mathbf{H}^{(1)}\mathbf{P}^{(1)} + \mathbf{H}^{(2)}\mathbf{P}^{(2)} + \mathbf{H}^{(3)}\mathbf{P}^{(3)} \tag{3.54}
$$

In the global equilibrium equation (3.54), matrices $\mathbf{H}$ pre-multiplying the member end forces are called the "mapping matrices" and contain only 1's and 0's. It is noted that a matrix containing only 1's and 0's is known as a "Boolean Matrix". The mapping vector can be used to construct the mapping matrix of each member.

For example, the 1's of the mapping matrix for member 1 are at the following positions:

$$(1, 4), (2, 5), (3, 6), (7, 1), (8, 2), \text{ and } (9, 3)$$

It is observed that the row index corresponds to the structural degree of freedom, while the column index corresponds to the member degree of freedom. This becomes clear when one observes the correspondence between the structural and member degrees of freedom in Eq. (3.52). The next step is to express the member end forces in terms of the basic member forces via Eq. (3.30) and substitute into Eq. (3.54).

In Eq. (3.53), the equilibrium equations appear to involve a pre-multiplication of the member end force vector by the member mapping matrix. However, in reality, this operational process is not necessary since it can be achieved by direct indexing.

For example, the first equation of Eq. (3.53) involves the end force $P_4^{(1)}$ of member 1. It is recalled from Example 3.1 that

$$P_4^{(1)} = b_{41}^{(1)} Q_1 + b_{42}^{(1)} Q_2 + b_{43}^{(1)} Q_3 = 0.6 Q_1 + 0.16 Q_2 + 0.16 Q_3 \tag{3.55}$$

Imagine, after conducting the multiplication of the mapping matrix with the member end forces expressed in terms of the basic member forces, the coefficients of $Q_1$, $Q_2$, and $Q_3$ will fill in the first row and columns 1, 2, and 3 of equilibrium matrix $\mathbf{B}$, respectively. Generally, the $i$th row of the $\mathbf{b}$ matrix of member 1 will fill in the row number of equilibrium matrix $\mathbf{B}$ corresponding to the structural degree of freedom that is located in the $i$th row of the mapping vector of member 1. The first three columns of the equilibrium matrix $\mathbf{b}$ are associated with $Q_1$, $Q_2$, and $Q_3$, respectively. For member 2, the scenario is the same as long as the row number is considered. The columns of the equilibrium matrix $\mathbf{b}$ of member 2 correspond to the basic forces $Q_4$, $Q_5$, and $Q_6$. Therefore, they fill in columns 4, 5, and 6 of equilibrium matrix $\mathbf{B}$, respectively. The schematic representation of setting up of the first row of the equilibrium matrix $\mathbf{B}$ is shown in Figure 3.33. The same process is applied to all other degrees of freedom (Filippou 2002).

### Example 3.3

For the braced frame shown in Figure 3.34, the following tasks are assigned:

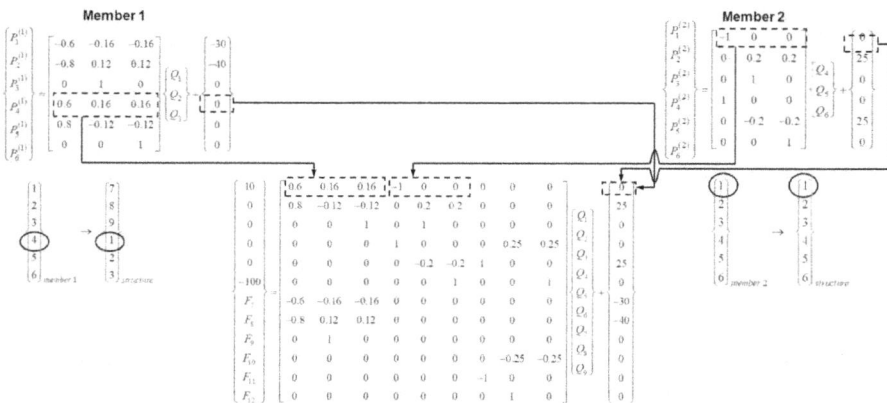

**FIGURE 3.33** Demonstration of Automated Assembly of Structural Equilibrium Matrix: The first Row.

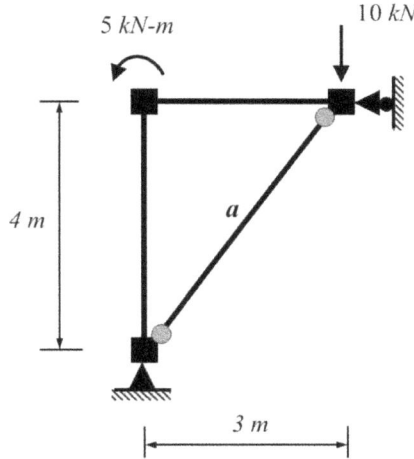

**FIGURE 3.34** Example 3.3.

(a) Number the degrees of freedom (both constrained and unconstrained) and basic (inde-pendent) member forces based on the simply supported beam model.
(b) Formulate the equilibrium equations at unconstrained degrees of freedom. There should be 1 equation less than basic unknown forces. Hence, the structure is statically indeterminate.
(c) Given the axial force in the bracing member $a$ with value of $-6.8$ $kN$. Use this given basic force to solve the other unknown forces and support reactions.

The member properties are: $IE = 20 \times 10^3$ $kN - m^2$ and $AE = 50 \times 10^3$ $kN$ for members 1 and 2, and $AE = 10 \times 10^3$ $kN$ for bracing member $a$.

**Solution:** Member, node, and DOFs numbering systems are shown in Figure 3.35a:

The basic (independent) member force numbering system based on the simply supported beam model is shown in Figure 3.35b:

The mapping vector and matrix equilibrium equation of each member are:

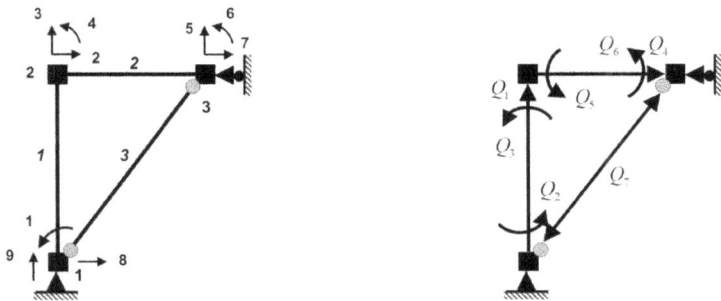

(a) Member, node, and DOFs numbering systems   (b) Basic (Independent) member forces numbering system

**FIGURE 3.35** Example 3.3 (Continued).

$$LM_1 = \begin{Bmatrix} 8 \\ 9 \\ 1 \\ 2 \\ 3 \\ 4 \end{Bmatrix}_1 ; \begin{Bmatrix} P_1^{(1)} \\ P_2^{(1)} \\ P_3^{(1)} \\ P_4^{(1)} \\ P_5^{(1)} \\ P_6^{(1)} \end{Bmatrix} = \begin{bmatrix} 0 & -0.25 & -0.25 \\ -1 & 0 & 0 \\ 0 & 1 & 0 \\ 0 & 0.25 & 0.25 \\ 1 & 0 & 0 \\ 0 & 0 & 1 \end{bmatrix} \begin{Bmatrix} Q_1 \\ Q_2 \\ Q_3 \end{Bmatrix} ;$$

$$LM_2 = \begin{Bmatrix} 2 \\ 3 \\ 4 \\ 7 \\ 5 \\ 6 \end{Bmatrix}_2 ; \begin{Bmatrix} P_1^{(2)} \\ P_2^{(2)} \\ P_3^{(2)} \\ P_4^{(2)} \\ P_5^{(2)} \\ P_6^{(2)} \end{Bmatrix} = \begin{bmatrix} -1 & 0 & 0 \\ 0 & 0.333 & 0.333 \\ 0 & 1 & 0 \\ 1 & 0 & 0 \\ 0 & -0.333 & -0.333 \\ 0 & 0 & 1 \end{bmatrix} \begin{Bmatrix} Q_4 \\ Q_5 \\ Q_6 \end{Bmatrix} ; \text{ and } LM_3 = \begin{Bmatrix} 8 \\ 9 \\ 7 \\ 5 \end{Bmatrix}_3 ; \begin{Bmatrix} P_1^{(3)} \\ P_2^{(3)} \\ P_3^{(3)} \\ P_4^{(3)} \end{Bmatrix} = \begin{bmatrix} -0.6 \\ -0.8 \\ 0.6 \\ 0.8 \end{bmatrix} Q_7.$$

The demonstration of automated assembly of structural equilibrium matrix is presented in Figure 3.36:

The compact form of the matrix equilibrium equation of the braced frame is:

$$F = BQ \text{ or } \begin{Bmatrix} F_{free} \\ \hline F_{constr} \end{Bmatrix} = \begin{bmatrix} B_{free} \\ \hline B_{constr} \end{bmatrix} Q$$

Separately, equilibrium equations at free and constrained degrees of freedom are expressed as:

$$F_{free} = B_{free} Q \text{ and } F_{constr} = B_{constr} Q$$

In this matrix equilibrium equation, there are 10 unknown forces, namely seven basic member forces $Q$ and three applied forces at constrained degrees of freedom (reaction forces), $F_{constr}$ and nine equilibrium equations are available. Therefore, this structure is statically indeterminate, and its degree of statical indeterminacy is $10-9 = 1$.

Equilibrium equations associated with free (unconstrained) degrees of freedom are:

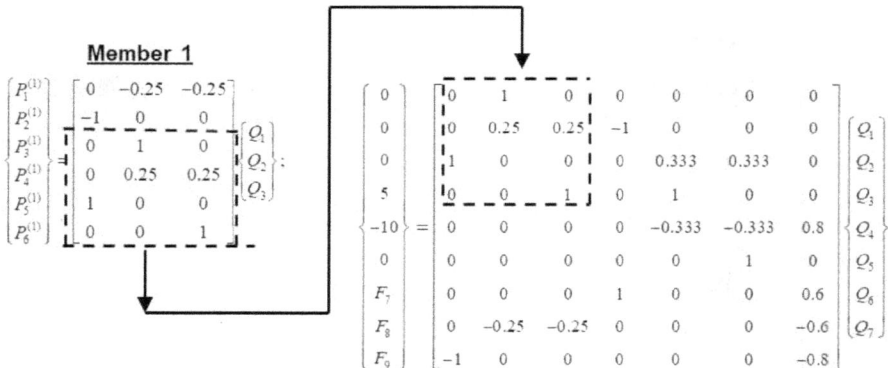

**FIGURE 3.36** Example 3.3 (Continued).

$$\mathbf{F}_{free} = \mathbf{B}_{free}\mathbf{Q} \text{ or } \begin{Bmatrix} 0 \\ 0 \\ 0 \\ 5 \\ -10 \\ 0 \end{Bmatrix} = \begin{bmatrix} 0 & 1 & 0 & 0 & 0 & 0 & 0 \\ 0 & 0.25 & 0.25 & -1 & 0 & 0 & 0 \\ 1 & 0 & 0 & 0 & 0.333 & 0.333 & 0 \\ 0 & 0 & 1 & 0 & 1 & 0 & 0 \\ 0 & 0 & 0 & 0 & -0.333 & -0.333 & 0.8 \\ 0 & 0 & 0 & 0 & 0 & 1 & 0 \end{bmatrix} \begin{Bmatrix} Q_1 \\ Q_2 \\ Q_3 \\ Q_4 \\ Q_5 \\ Q_6 \\ Q_7 \end{Bmatrix}$$

In this reduced matrix form of equilibrium equations, there are six equilibrium equations and seven basic member force unknowns. Therefore, the degree of statical indeterminacy is 7–6 = 1. If one basic member force, namely, $Q_7$ is known, this system becomes statically determinate.

Now, the equilibrium matrix $\mathbf{B}_{free}$ could be split into two sub-matrices, namely, $\mathbf{B}^1_{free}$ and $\mathbf{B}^2_{free}$ associated with unknown and known basic forces, respectively.

$$\mathbf{B}_{free} = \left[ \mathbf{B}^1_{free} \,\middle|\, \mathbf{B}^2_{free} \right]$$

where

$$\mathbf{B}^1_{free} = \begin{bmatrix} 0 & 1 & 0 & 0 & 0 & 0 \\ 0 & 0.25 & 0.25 & -1 & 0 & 0 \\ 1 & 0 & 0 & 0 & 0.333 & 0.333 \\ 0 & 0 & 1 & 0 & 1 & 0 \\ 0 & 0 & 0 & 0 & -0.333 & -0.333 \\ 0 & 0 & 0 & 0 & 0 & 1 \end{bmatrix} \text{ and } \mathbf{B}^2_{free} = \begin{bmatrix} 0 \\ 0 \\ 0 \\ 0 \\ 0.8 \\ 0 \end{bmatrix}$$

In a similar fashion, the basic force vector $\mathbf{Q}$ could be separated accordingly as:

$$\mathbf{Q} = \left\lfloor \mathbf{Q}^1 \quad \mathbf{Q}^2 \right\rfloor^T$$

where

$$\mathbf{Q}^1 = \left\lfloor Q_1 \quad Q_2 \quad Q_3 \quad Q_4 \quad Q_5 \quad Q_6 \right\rfloor^T \Rightarrow unknowns \text{ and } \mathbf{Q}^2 = \left\lfloor Q_7 \right\rfloor^T \Rightarrow known$$

The matrix equilibrium equations of free degrees of freedom can be rewritten as:

$$\mathbf{F}_{free} = \left[ \mathbf{B}^1_{free} \,\middle|\, \mathbf{B}^2_{free} \right] \begin{Bmatrix} \mathbf{Q}^1 \\ \mathbf{Q}^2 \end{Bmatrix} = \mathbf{B}^1_{free}\mathbf{Q}^1 + \mathbf{B}^2_{free}\mathbf{Q}^2 \Rightarrow \mathbf{F}_{free} - \mathbf{B}^2_{free}\mathbf{Q}^2 = \mathbf{B}^1_{free}\mathbf{Q}^1$$

The known basic force is given as:

$$Q_7 = -6.8 \, kN \text{ or } \mathbf{Q}^2 = \left\lfloor -6.8 \right\rfloor^T$$

It is observed that the role of term $\mathbf{B}^2_{free}\mathbf{Q}^2$ is similar to that of member loads in Example 3.1. Now, the basic unknown forces can be solved as:

$$\mathbf{Q}^1 = \left[ \mathbf{B}^1_{free} \right]^{-1} \left( \mathbf{F}_{free} - \mathbf{B}^2_{free}\mathbf{Q}^2 \right)$$

$$= \left\lfloor -4.56 \quad 0 \quad -8.69 \quad -2.17 \quad 13.69 \quad 0 \right\rfloor^T$$

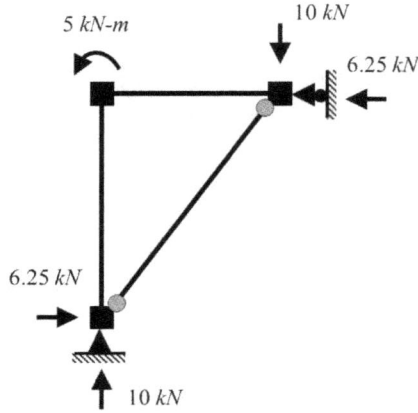

**FIGURE 3.37** Example 3.3 (Continued).

Substituting these basic member forces into the equilibrium equations associated with constrained degrees of freedom, the support reactions (Figure 3.37) are computed as:

$$
\mathbf{F}_{constr} = \mathbf{B}_{constr}\mathbf{Q} \text{ or } \begin{Bmatrix} F_7 \\ F_8 \\ F_9 \end{Bmatrix} = \begin{bmatrix} 0 & 0 & 0 & 1 & 0 & 0 & 0.6 \\ 0 & -0.25 & -0.25 & 0 & 0 & 0 & -0.6 \\ -1 & 0 & 0 & 0 & 0 & 0 & -0.8 \end{bmatrix} \begin{Bmatrix} -4.56 \\ 0 \\ -8.69 \\ -2.17 \\ 13.69 \\ 0 \\ -6.8 \end{Bmatrix} = \begin{Bmatrix} -6.25 \\ 6.25 \\ 10 \end{Bmatrix} kN
$$

The member end forces can be obtained by substituting their basic forces into the member equations.

For example, member 2 (Figure 3.38):

$$
\begin{Bmatrix} P_1^{(2)} \\ P_2^{(2)} \\ P_3^{(2)} \\ P_4^{(2)} \\ P_5^{(2)} \\ P_6^{(2)} \end{Bmatrix} = \begin{bmatrix} -1 & 0 & 0 \\ 0 & 0.333 & 0.333 \\ 0 & 1 & 0 \\ 1 & 0 & 0 \\ 0 & -0.333 & -0.333 \\ 0 & 0 & 1 \end{bmatrix} \begin{Bmatrix} -2.17\,kN \\ 13.69\,kN\text{-}m \\ 0\,kN\text{-}m \end{Bmatrix} = \begin{Bmatrix} 2.17\,kN \\ 4.56\,kN \\ 13.69\,kN\text{-}m \\ -2.17\,kN \\ -4.56\,kN \\ 0\,kN\text{-}m \end{Bmatrix}
$$

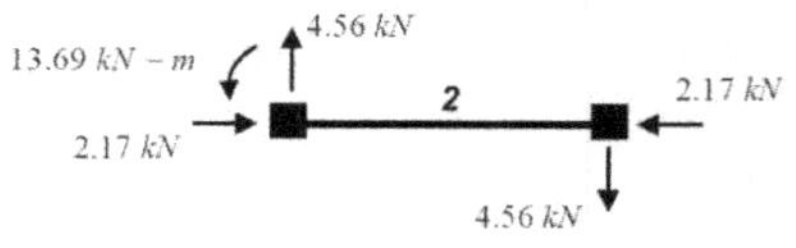

**FIGURE 3.38** Example 3.3 (Continued).

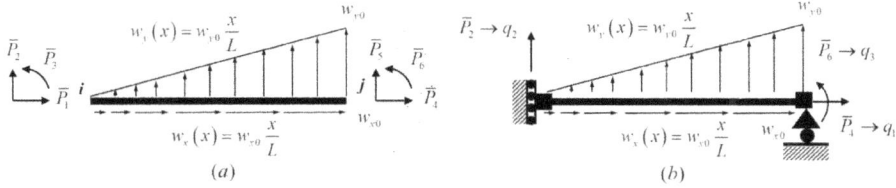

FIGURE 3.39    Problem 3.1.

## 3.8   EXERCISES

**Problem 3.1:** For a frame member under end forces and linearly distributed member loadings shown in Figure 3.39a, please establish its matrix equilibrium equation by selecting $\bar{P}_2$, $\bar{P}_4$, and $\bar{P}_6$ as the independent end forces. This choice of the independent end forces can be viewed as a fixed-connected collar-roller beam system shown in Figure 3.39b.

**Problem 3.2:** Analyze the truss shown in Figure 3.40 for internal forces and support reactions through the following steps:

(a)   Number the degrees of freedom (both constrained and unconstrained) and the basic (independent) member forces.
(b)   Formulate the equilibrium equations at unconstrained degrees of freedom.
(c)   Solve the basic member forces.
(d)   Employ the equilibrium equations associated with the constrained degrees of freedom to compute the support reactions.

**Problem 3.3:** For the queen post bridge structure shown in Figure 3.41, the following tasks are assigned:

(a)   Number the degrees of freedom (both constrained and unconstrained) and the basic (independent) member forces based on a simply supported beam model.

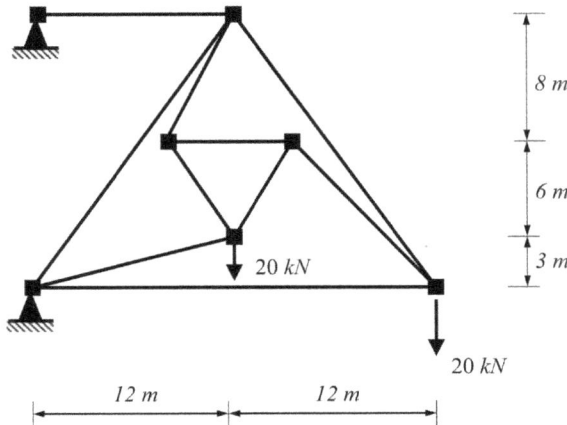

FIGURE 3.40    Problem 3.2.

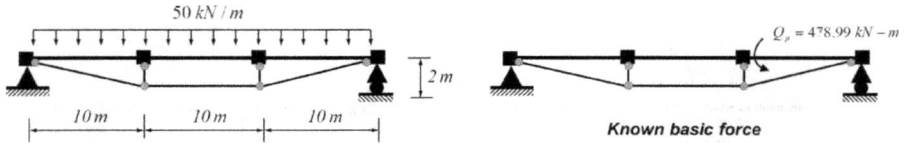

**FIGURE 3.41**   Problem 3.3.

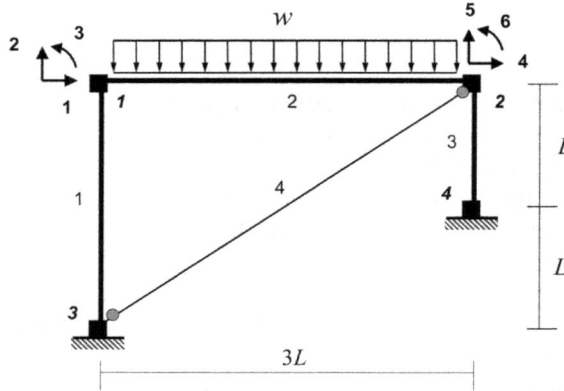

**FIGURE 3.42**   Problem 3.4.

(b)   Formulate the equilibrium equations at unconstrained degrees of freedom. There should be one equation less than the basic unknown forces. Hence, the structure is statically indeterminate.

Given basic force shown in Figure 3.41, use this known force ($Q_p = 478.99\ kN - m$) to solve the other unknown forces and support reactions. It is noted that the known force given acts at the member ends.

   **Problem 3.4:** For the braced structure shown in Figure 3.42, the following tasks are assigned:

(a)   Identify its degree of statical indeterminacy.
(b)   Systematically number the member forces **Q** using the simply supported beam model as basic system.
(c)   Write down the matrix equilibrium equations associated with free degrees of freedom.

## REFERENCES

Dawe, D.J. 1984. *Matrix and finite element displacement analysis of structures.* Clarendon Press.
Elias, Z.M. 1986. *Theory and methods of structural analysis.* John Wiley and Son Inc.
Filippou, F.C. 2002. *CE 220 course notes: Structural analysis: Theory and applications.* Department of Civil and Environmental Engineering, University of California.
Kanchi, M.B. 1993. *Matrix methods of structural analysis* (2nd Edition). Wiley Eastern Limited.
Limkatanyu, S., W. Prachasaree, G. Kaewkulchai, and E. Spacone. 2014. Unification of mixed Euler-Bernoulli-von Karman planar frame model and corotational approach. *Mechanics Based Design of Structures and Machines: An International Journal* 42(4): 419–441.

Livesley, R.K. 1974. *Matrix methods of structural analysis*. Pergamon Press.

Przemieniecki, J.S. 1985. *Theory of matrix structural analysis*. Dover Publications Inc.

Rajasekaran, S. and G. Sankarasubramanian. 2006. *Computational structural mechanics*. Prentice-Hall of India Private Limited.

Whittaker, A. 2001. *CIE 423 course notes: Structures III*. Department of Civil, Structural, and Environmental Engineering, State University of New York.

# 4 Geometric Compatibility

## 4.1 INTRODUCTION

As mentioned in Chapter 1, compatibility is one of the three fundamental ingredients required in any structural analysis approach. In this chapter, the main objective is on the different views of compatibility (kinematical) considerations faced in structural analysis. There are various forms of describing compatibility conditions. They can be expressed in terms of strains or deformations (resultant strains). The first form is often encountered in a course on Continuum Mechanics while the second one, discussed in Chapter 2, is found in basic courses on the Mechanics of Materials and Structural Analysis. Compatibility relations must hold for all levels of free bodies including a differential segment, a member, a part of the structure, and the whole structure.

In this textbook, only compatibility considerations of frame (skeletal) members and structures are of interest.

## 4.2 COMPATIBILITY IN A LOCAL REFERENCE SYSTEM: A FRAME MEMBER

A two-dimensional frame structure with its global reference axes (X-Y-Z) is shown in Figure 4.1. The Z-axis is normal to the structure plane and points toward the reader. Each assembled member is referred to by its local reference axes (x-y-z). Conventionally, node $i$ is chosen as the starting one while node $j$ is selected as the ending one. The positive local x-axis points toward node $j$ from node $i$. For plane structures, it is preferable to keep the local z-axis coinciding with the global Z-axis pointing toward the reader. Consequently, the local y-axis is uniquely defined. Structural members of the frame in Figure 4.1 cannot deform independently since they are connected to one another to form the structural framework, which as a whole displaces itself following member connectivity and constraints imposed by the supports. The member $a$ of the frame in Figure 4.1 is taken as a generic frame member. The kinematics of this generic member is described by its end displacements in the local reference system (Figure 4.2). The member-end displacements are grouped in the vector $\bar{\mathbf{U}}$ as:

$$\bar{\mathbf{U}} = \lfloor \bar{\mathbf{U}}_1 \quad \bar{\mathbf{U}}_2 \rfloor^T = \lfloor \bar{U}_1 \quad \bar{U}_2 \quad \bar{U}_3 \quad \bar{U}_4 \quad \bar{U}_5 \quad \bar{U}_6 \rfloor^T \tag{4.1}$$

The components of the vector $\bar{\mathbf{U}}$ are arranged in the following fashion: beginning from node $i$, one assigns first the degree of freedom (DOF) in the local x-axis (first component), then the DOF in the local y-axis (second component), then the DOF about the local z-axis (third component), and then proceed in this fashion to node $j$ of the member. These six components form the complete set of end displacements for a member's free body in the local reference system.

In this section, the main objective is to establish the geometric compatibility relations associating member-end displacements with "basic member deformations". The basic member deformations are the conjugated work pairs of the basic member forces discussed in Chapter 3. Consequently, the simply supported beam system shown in Figure 4.3 is intentionally selected to establish the geometric compatibility relations.

In this basic system, the corotational reference axes $(x'-y'-z')$ are rigidly bounded to the cross section at the end $i$ in the undeformed configuration and at the end $i'$ in the deformed

DOI: 10.1201/9781003595458-4

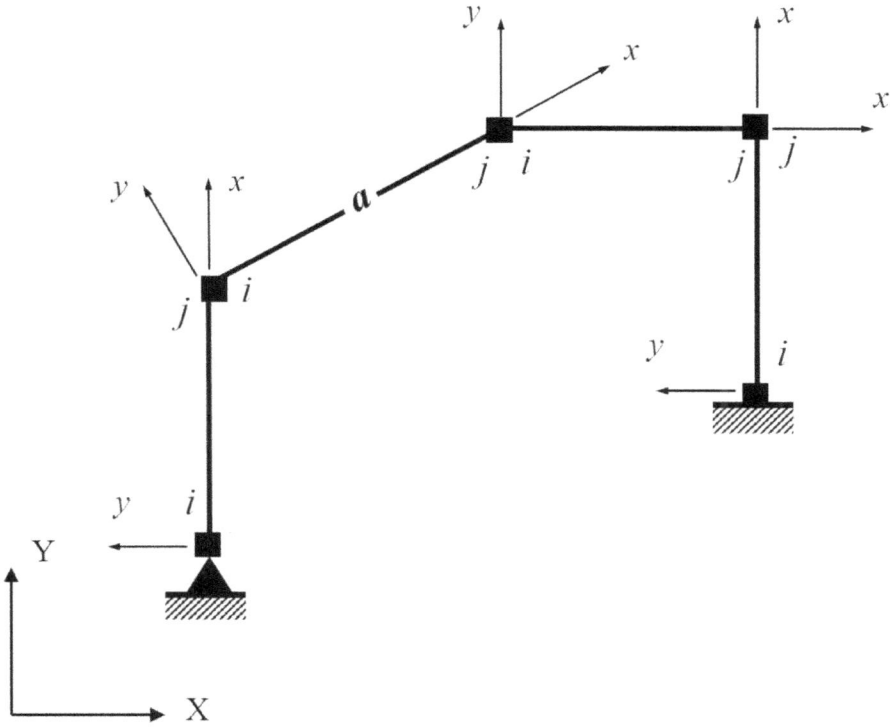

**FIGURE 4.1**  A Two-Dimensional Frame Structure: Global and Local Reference Axes.

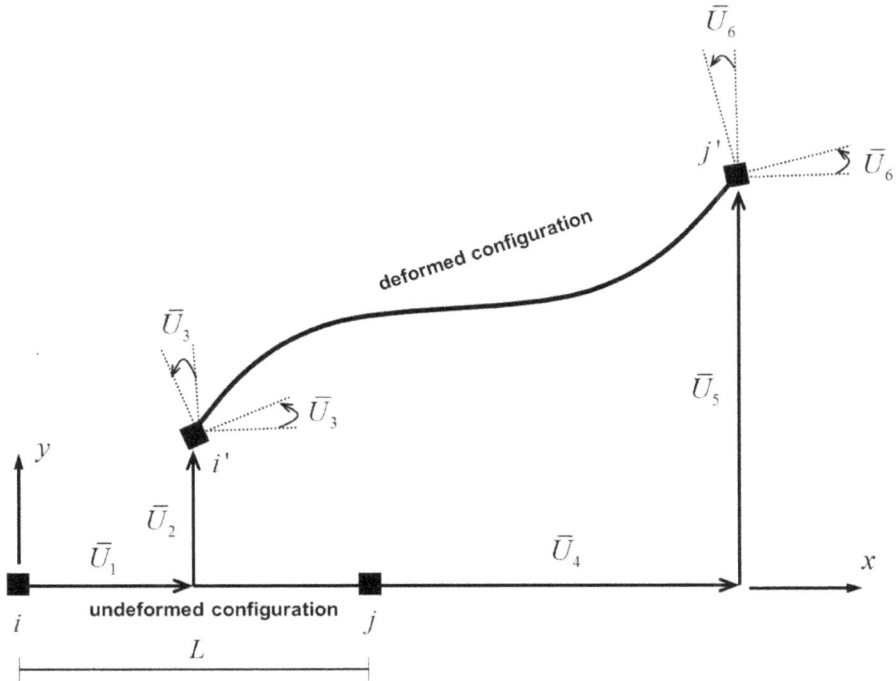

**FIGURE 4.2**  Member-End Displacements in Local Reference System.

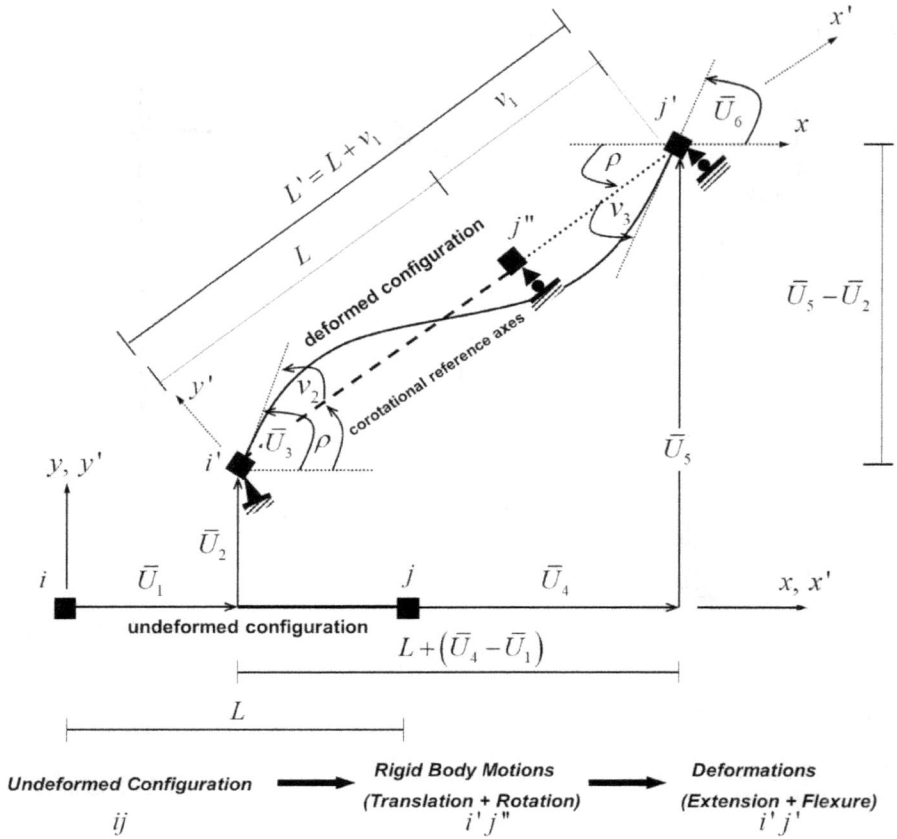

**FIGURE 4.3** Deformations and Displacements of Simply Supported Beam Model: Exact Kinematical Description.

configuration. In the undeformed configuration, the corotational reference axes $(x'- y'- z')$ coincide with the local reference axes $(x - y - z)$. If the member displaces as a rigid body, the corotational reference axes are carried with it and no member deformations are induced. As shown in Figure 4.3, the basic member deformations are defined as displacements relative to the corotational reference axes. Relative to this reference system, the member can be visualized as a beam simply supported with a pin at end $i'$ and roller at ends $j'$ and $j''$. This virtual supporting system simply helps the reader to suppress the rigid body modes. The basic member deformations of this system are defined as: the relative displacement $v_1$ at the end $j'$ along the chord connecting ends $i'$ and $j'$; the rotational deformation $v_2$ at end $i'$ relative to the chord $i'j'$; and the rotational deformation $v_3$ at end $j'$ relative to the chord $i'j'$. Consequently, these basic member deformations are grouped in vector form:

$$\mathbf{v} = \lfloor v_1 \quad v_2 \quad v_3 \rfloor^T \tag{4.2}$$

and defined as:

$$v_1 = L' - L; \quad v_2 = \bar{U}_3 - \rho; \text{and} \, v_3 = \bar{U}_6 - \rho \tag{4.3}$$

where $L$ is the member length in the undeformed configuration and $\overset{.}{L}$ the member length in the final position can be obtained by Pythagoras theorem as:

$$\overset{.}{L} = \sqrt{\left(L+\left(\bar{U}_4-\bar{U}_1\right)\right)^2+\left(\bar{U}_5-\bar{U}_2\right)^2} = \sqrt{\left(L+\Delta\bar{U}_x\right)^2+\left(\Delta\bar{U}_y\right)^2} \qquad (4.4)$$

$L$ is factored out and the binomial term is expanded. This leads to:

$$\overset{.}{L} = L\sqrt{\left(1+\frac{\Delta\bar{U}_x}{L}\right)^2+\left(\frac{\Delta\bar{U}_y}{L}\right)^2} \qquad (4.5)$$

$$\overset{.}{L} = L\sqrt{1+2\frac{\Delta\bar{U}_x}{L}+\left(\frac{\Delta\bar{U}_x}{L}\right)^2+\left(\frac{\Delta\bar{U}_y}{L}\right)^2} \qquad (4.6)$$

$$\overset{.}{L} = L\sqrt{1+2\frac{\bar{U}_4-\bar{U}_1}{L}+\left(\frac{\bar{U}_4-\bar{U}_1}{L}\right)^2+\left(\frac{\bar{U}_5-\bar{U}_2}{L}\right)^2} \qquad (4.7)$$

$\rho$ is the chord rotation angle and defined as:

$$\rho = \arctan\frac{\left(\bar{U}_5-\bar{U}_2\right)}{L+\left(\bar{U}_4-\bar{U}_1\right)} = \arctan\frac{\left(\dfrac{\left(\bar{U}_5-\bar{U}_2\right)}{L}\right)}{\left(1+\dfrac{\left(\bar{U}_4-\bar{U}_1\right)}{L}\right)} = \arctan\frac{\dfrac{\Delta\bar{U}_y}{L}}{1+\dfrac{\Delta\bar{U}_x}{L}} \qquad (4.8)$$

$$\cos\rho = \frac{L+\left(\bar{U}_4-\bar{U}_1\right)}{\overset{.}{L}} = \frac{L+\Delta\bar{U}_x}{\overset{.}{L}} \text{ and } \sin\rho = \frac{\left(\bar{U}_5-\bar{U}_2\right)}{\overset{.}{L}} = \frac{\Delta\bar{U}_y}{\overset{.}{L}} \qquad (4.9)$$

Equation (4.3) represents the "exact" geometric compatibility relations of the frame member in Figure 4.3 and is used by Limkatanyu et al. (2014) to develop the geometrically nonlinear frame model based on the unification of the Euler-Bernoulli-von Karman beam theory and corotational concept. These relations are labeled "exact" since the kinematics of rigid body motion is treated without any approximation. It is obvious that these geometric compatibility relations are non-linear functions. These are relevant to the geometrically nonlinear analysis of structures. More details of the analysis of geometrically nonlinear structures are presented in advanced textbooks on matrix structural analysis and finite element analysis (e.g. McGuire et al. 2000; de Borst et al. 2012).

With the following assumptions,

1.  The axial deformation $v_1$ is small compared to the member length $L$.
2.  $v_1/L$ is presumed to be of the same magnitude of square of $\Delta\bar{U}_y/L$.

The aforementioned geometric relations (4.3) can be approximated as follows:
First, the power series is employed to expand Eq. (4.6) as:

$$\overset{.}{L} = L\left(1+\frac{\Delta\bar{U}_x}{L}+\frac{1}{2}\left(\frac{\Delta\bar{U}_x}{L}\right)^2+\frac{1}{2}\left(\frac{\Delta\bar{U}_y}{L}\right)^2-\frac{1}{8}\left(\frac{2\Delta\bar{U}_x}{L}\right)^2+\text{higher order terms}\right) \qquad (4.10)$$

Noting that (Kreyszig 2011):

$$\sqrt{1+\alpha} = 1 + \frac{1}{2}\alpha - \frac{1}{8}\alpha^2 + \text{higher order terms} \tag{4.11}$$

Simplifying Eq. (4.6) and neglecting third- and higher-order terms lead to the first level approximation of the axial deformation as:

$$v_1 = \left(\bar{U}_4 - \bar{U}_1\right) + \frac{\left(\bar{U}_5 - \bar{U}_2\right)^2}{2L} \tag{4.12}$$

Second, the power series is used to expand the term $\left(1 + \frac{\Delta\bar{U}_x}{L}\right)^{-1}$ of Eq. (4.8) as:

$$\left(1 + \frac{\Delta\bar{U}_x}{L}\right)^{-1} = 1 - \frac{\Delta\bar{U}_x}{L} + \left(\frac{\Delta\bar{U}_x}{L}\right)^2 + \text{higher order terms} \tag{4.13}$$

Noting that (Kreyszig 2011):

$$\left(1+\alpha\right)^{-1} = 1 - \alpha + \alpha^2 + \text{higher order terms} \tag{4.14}$$

Substituting Eqs. (4.13) into (4.8) leads to:

$$\rho = \arctan\left[\frac{\Delta\bar{U}_y}{L}\left(1 - \frac{\Delta\bar{U}_x}{L} + \left(\frac{\Delta\bar{U}_x}{L}\right)^2 + \text{higher order terms}\right)\right] \tag{4.15}$$

Eq. (4.15) is expanded by Taylor's Series as:

$$\rho = \left(\frac{\Delta\bar{U}_y}{L} - \frac{\Delta\bar{U}_x}{L}\frac{\Delta\bar{U}_y}{L} + \left(\frac{\Delta\bar{U}_x}{L}\right)^2\right) + \text{higher order terms} \tag{4.16}$$

Noting that (Kreyszig 2011):

$$\arctan\alpha = \alpha - \frac{\alpha^3}{3} + \frac{\alpha^5}{5} + \text{higher order terms} \tag{4.17}$$

Since the term $\Delta\bar{U}_x$ in Eq. (4.16) is comparable to $v_1$ and small relative to $L$, the chord rotation angle can be simplified as:

$$\rho = \frac{\Delta\bar{U}_y}{L} = \frac{\bar{U}_5 - \bar{U}_2}{L} \tag{4.18}$$

Substituting Eqs. (4.18) into (4.3) leads to the first level approximation of the rotational deformations as:

$$v_2 = \bar{U}_3 - \frac{\left(\bar{U}_5 - \bar{U}_2\right)}{L} \text{ and } v_3 = \bar{U}_6 - \frac{\left(\bar{U}_5 - \bar{U}_2\right)}{L} \tag{4.19}$$

It is observed that the chord rotation angle $\rho$ in Eq. (4.18) is independent of member-end displacements in the $x$ direction. Therefore, it is implied that the member-end rotations relative to the chord $v_2$ and $v_3$ do not affect the change in member length. In other words, the axial deformation $v_1$ is not affected by the member-end rotational deformations $v_2$ and $v_3$.

Equations (4.12) and (4.19) are the geometric compatibility relations following the first-level approximation. It is worth expressing them together as:

$$v_1 = \left(\bar{U}_4 - \bar{U}_1\right) + \frac{\left(\bar{U}_5 - \bar{U}_2\right)^2}{2L}; \quad v_2 = \bar{U}_3 - \frac{\left(\bar{U}_5 - \bar{U}_2\right)}{L}; \text{ and } v_3 = \bar{U}_6 - \frac{\left(\bar{U}_5 - \bar{U}_2\right)}{L} \quad (4.20)$$

These geometric compatibility relations are relevant to the second-order structural analysis considering their $P - \Delta$ effects. It is observed that only the geometric relation for the axial deformation is nonlinear while those for the rotational deformation are linear. However, the interest of this textbook is limited only to geometrically linear structures. This implies that basic member deformations $\mathbf{v}$ (as well as resultant strains) are expressed as linear combinations of member-end displacements $\bar{\mathbf{U}}$. Neglecting all nonlinear terms in Eq. (4.20) leads to the linear geometric compatibility relations of the frame shown in Figure 4.3 as:

$$v_1 = \left(\bar{U}_4 - \bar{U}_1\right); \quad v_2 = \bar{U}_3 - \frac{\left(\bar{U}_5 - \bar{U}_2\right)}{L}; \text{ and } v_3 = \bar{U}_6 - \frac{\left(\bar{U}_5 - \bar{U}_2\right)}{L} \quad (4.21)$$

in the matrix form as:

$$\begin{Bmatrix} v_1 \\ v_2 \\ v_3 \end{Bmatrix} = \begin{bmatrix} -1 & 0 & 0 & 1 & 0 & 0 \\ 0 & \dfrac{1}{L} & 1 & 0 & -\dfrac{1}{L} & 0 \\ 0 & \dfrac{1}{L} & 0 & 0 & -\dfrac{1}{L} & 1 \end{bmatrix} \begin{Bmatrix} \bar{U}_1 \\ \bar{U}_2 \\ \bar{U}_3 \\ \bar{U}_4 \\ \bar{U}_5 \\ \bar{U}_6 \end{Bmatrix} \quad (4.22)$$

or in the concise form as:

$$\mathbf{v} = \begin{bmatrix} \bar{\mathbf{a}}_{1B} & \bar{\mathbf{a}}_{2B} \end{bmatrix} \begin{Bmatrix} \bar{\mathbf{U}}_1 \\ \bar{\mathbf{U}}_2 \end{Bmatrix} \text{ or } \mathbf{v} = \bar{\mathbf{a}}_B \bar{\mathbf{U}} \quad (4.23)$$

where matrix $\bar{\mathbf{a}}_B$ is called the compatibility matrix for the simply supported beam system with respect to local reference axes. From the compatibility matrix $\bar{\mathbf{a}}_B$, it is observed that the axial deformation $v_1$ depends only on member-end displacements ($\bar{U}_1$ and $\bar{U}_4$) along the $x$-axis while the member-end rotational $v_2$ and $v_3$ deformations depend on member-end displacements ($\bar{U}_2$ and $\bar{U}_5$) along the $y$-axis and member-end rotations ($\bar{U}_3$ and $\bar{U}_6$). This lack of coupling between axial displacement components with respect to the local reference system on one hand and transverse displacement components and rotations on the other hand is due to the linearization of geometric compatibility relations.

Furthermore, it is observed that the matrix $\bar{\mathbf{a}}_B$ is equal to the transpose of the equilibrium matrix $\bar{\mathbf{b}}_B$ (3/3.11) for the simply supported beam system defined in Chapter 3.

$$\bar{\mathbf{P}} - \bar{\mathbf{P}}_0 = \bar{\mathbf{b}}_B \mathbf{q} \quad (3/3.11)$$

It is interesting to point out that linearization of member compatibility relations is associated with consideration of member equilibrium conditions under its undeformed configuration.

One may simply state that the statical (3/3.11) and kinematical (4.23) transformations are *contragradient*.

It can be concluded that:

1. The basic member deformations **v** of Eq. (4.23) and basic member forces **q** of Eq. (3/3.11) are conjugate-work pairs. In other words, the scalar product of these two quantities produces consistent work.
2. The member-end displacements $\bar{\mathbf{U}}$ of Eq. (4.23) and member-end forces $\bar{\mathbf{P}} - \bar{\mathbf{P}}_0$ of Eq. (3/3.11) are conjugate-work pairs. Otherwise stated, the scalar product of these two quantities produces consistent work.

This form of relationship between equilibrium and compatibility transformation matrices is common, and its general proof is given by the virtual work principle as will be discussed in the following chapter.

Summarily, there are three basic member deformations for the frame member, namely the axial deformation $v_1$, end-$i$ rotational deformation $v_2$, and end-$j$ rotational deformation $v_3$. Since there are six end displacements, there exist three rigid-body displaced modes producing zero deformations. Each of the rigid-body displaced modes can be obtained by imposing zero basic member deformations in Eq. (4.22):

$$\begin{Bmatrix} v_1 \\ v_2 \\ v_3 \end{Bmatrix} = \begin{Bmatrix} 0 \\ 0 \\ 0 \end{Bmatrix} = \begin{bmatrix} -1 & 0 & 0 & 1 & 0 & 0 \\ 0 & \dfrac{1}{L} & 1 & 0 & -\dfrac{1}{L} & 0 \\ 0 & \dfrac{1}{L} & 0 & 0 & -\dfrac{1}{L} & 1 \end{bmatrix} \begin{Bmatrix} \bar{U}_1 \\ \bar{U}_2 \\ \bar{U}_3 \\ \bar{U}_4 \\ \bar{U}_5 \\ \bar{U}_6 \end{Bmatrix} \tag{4.24}$$

From Eq. (4.24), rigid body displaced modes are:

$$\begin{aligned} v_1 = 0 &\Rightarrow -\bar{U}_1 + \bar{U}_4 = 0 &\Rightarrow \bar{U}_4 = \bar{U}_1 \\ v_2 = 0 &\Rightarrow \frac{\bar{U}_2}{L} + \bar{U}_3 - \frac{\bar{U}_5}{L} = 0 &\Rightarrow \bar{U}_3 = -\frac{\bar{U}_2}{L} + \frac{\bar{U}_5}{L} \\ v_3 = 0 &\Rightarrow \frac{\bar{U}_2}{L} - \frac{\bar{U}_5}{L} + \bar{U}_6 = 0 &\Rightarrow \bar{U}_6 = -\frac{\bar{U}_2}{L} + \frac{\bar{U}_5}{L} \end{aligned} \tag{4.25}$$

and are depicted in Figure 4.4.

From the rigid-body displacements in Eq. (4.25), one selects $\bar{U}_1$, $\bar{U}_2$, and $\bar{U}_5$ as displacements restraining the rigid body modes and express them in the vector form as:

$$\bar{\mathbf{U}}_w = \lfloor \bar{U}_1 \quad \bar{U}_2 \quad \bar{U}_5 \rfloor^T \tag{4.26}$$

$\bar{U}_3$, $\bar{U}_4$, and $\bar{U}_6$ are selected as remaining displacements and are expressed in vector form as:

$$\bar{\mathbf{U}}_r = \lfloor \bar{U}_3 \quad \bar{U}_4 \quad \bar{U}_6 \rfloor^T \tag{4.27}$$

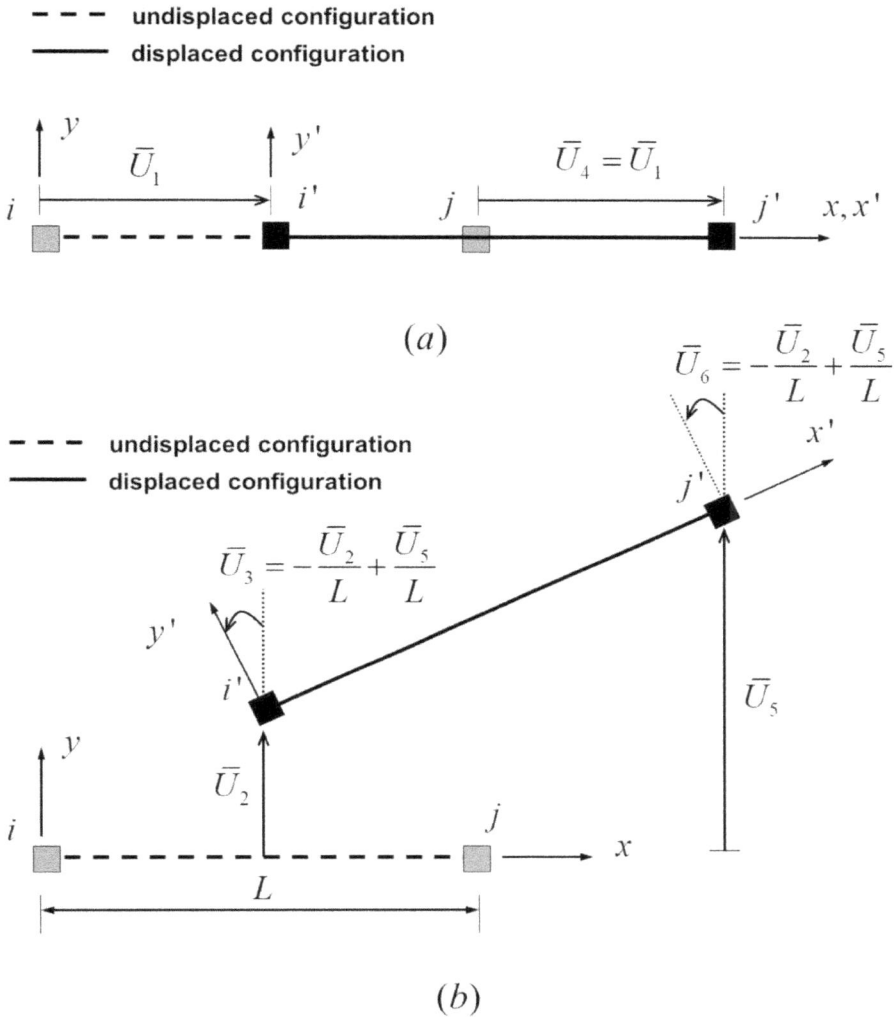

FIGURE 4.4 Rigid Body Displaced Modes: Simply Supported Beam Model.

The member-end displacements $\bar{\mathbf{U}}$ can be defined as:

$$\bar{\mathbf{U}} = \lfloor \bar{\mathbf{U}}_w \quad \bar{\mathbf{U}}_r \rfloor^T \tag{4.28}$$

Likewise, the member-end forces $\bar{\mathbf{P}}$ can be partitioned so that:

$$\bar{\mathbf{P}} = \lfloor \bar{\mathbf{P}}_w \quad \bar{\mathbf{P}}_r \rfloor^T \tag{4.29}$$

where $\bar{\mathbf{P}}_w = \lfloor \bar{P}_1 \quad \bar{P}_2 \quad \bar{P}_5 \rfloor^T$ and $\bar{\mathbf{P}}_r = \lfloor \bar{P}_3 \quad \bar{P}_4 \quad \bar{P}_6 \rfloor^T$ are the end forces associated with $\bar{\mathbf{U}}_w$ and $\bar{\mathbf{U}}_r$, respectively.

Under the condition of rigid body motion ($\mathbf{v} = \mathbf{0}$), $\bar{\mathbf{U}}_r$ and $\bar{\mathbf{U}}_w$ are related through the following transformation expression:

$$\begin{Bmatrix} \bar{U}_3 \\ \bar{U}_4 \\ \bar{U}_6 \end{Bmatrix} = \begin{bmatrix} 0 & -\dfrac{1}{L} & \dfrac{1}{L} \\ 1 & 0 & 0 \\ 0 & -\dfrac{1}{L} & \dfrac{1}{L} \end{bmatrix} \begin{Bmatrix} \bar{U}_1 \\ \bar{U}_2 \\ \bar{U}_5 \end{Bmatrix} \text{ or } \bar{\mathbf{U}}_r = \bar{\mathbf{T}}_B \bar{\mathbf{U}}_w \tag{4.30}$$

With the existence of member deformations ($\mathbf{v} \neq \mathbf{0}$), the kinematical relationships of the frame are:

$$\begin{Bmatrix} \bar{U}_3 \\ \bar{U}_4 \\ \bar{U}_6 \end{Bmatrix} = \begin{Bmatrix} v_1 \\ v_2 \\ v_3 \end{Bmatrix} + \begin{bmatrix} 0 & -\dfrac{1}{L} & \dfrac{1}{L} \\ 1 & 0 & 0 \\ 0 & -\dfrac{1}{L} & \dfrac{1}{L} \end{bmatrix} \begin{Bmatrix} \bar{U}_1 \\ \bar{U}_2 \\ \bar{U}_5 \end{Bmatrix} \text{ or } \bar{\mathbf{U}}_r = \mathbf{v} + \bar{\mathbf{T}}_B \bar{\mathbf{U}}_w \tag{4.31}$$

The complete set of the linearized member kinematics (deformation + rigid body motion) associated with Eq. (4.31) is schematically presented in Figure 4.5.

The member equilibrium conditions can be obtained from the kinematical relations (4.31) through the virtual displacement principle as follows:

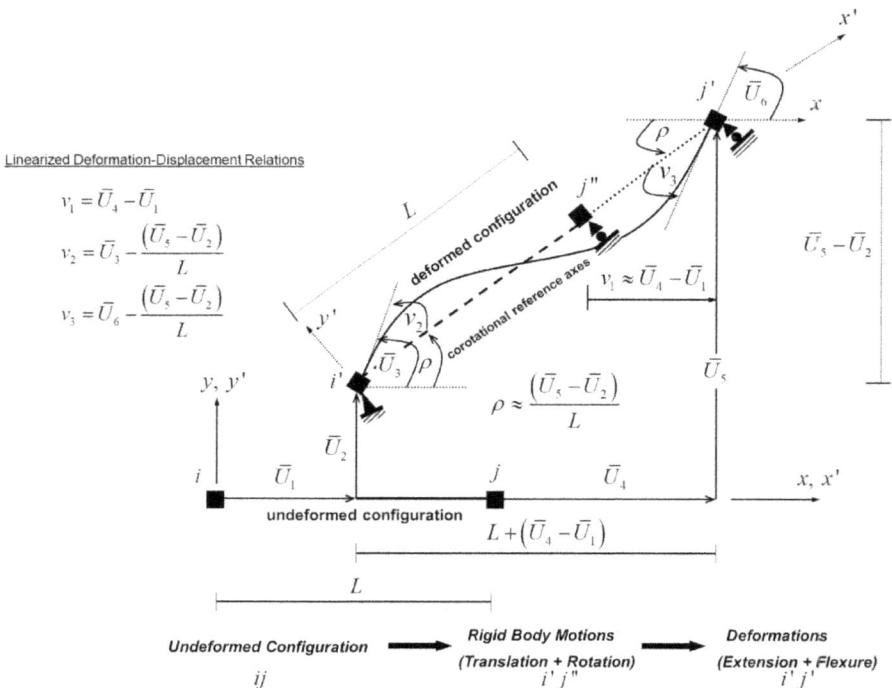

**FIGURE 4.5** Deformations and Displacements of a Simply Supported Beam Model: Linearized Kinematical Description.

Applying the arbitrary virtual displacements $\delta\bar{\mathbf{U}}_w$, then from (4.31)

$$\delta\bar{\mathbf{U}}_r = \bar{\mathbf{T}}_B \delta\bar{\mathbf{U}}_w \tag{4.32}$$

and the external virtual work

$$\delta W_{ext} = \delta\bar{\mathbf{U}}^T \bar{\mathbf{P}} = \delta\bar{\mathbf{U}}_w{}^T \bar{\mathbf{P}}_w + \delta\bar{\mathbf{U}}_r{}^T \bar{\mathbf{P}}_r = 0 \tag{4.33}$$

Substituting Eqs. (4.32) into (4.33) results in:

$$\delta W_{ext} = \delta\bar{\mathbf{U}}_w{}^T \left( \bar{\mathbf{P}}_w + \bar{\mathbf{T}}_B{}^T \bar{\mathbf{P}}_r \right) = 0 \tag{4.34}$$

Due to the arbitrariness of $\delta\bar{\mathbf{U}}_w$, one can conclude that:

$$\bar{\mathbf{P}}_w + \bar{\mathbf{T}}_B{}^T \bar{\mathbf{P}}_r = \mathbf{0} \tag{4.35}$$

Explicitly,

$$\begin{Bmatrix} \bar{P}_1 \\ \bar{P}_2 \\ \bar{P}_5 \end{Bmatrix} + \begin{bmatrix} 0 & 1 & 0 \\ -\dfrac{1}{L} & 0 & -\dfrac{1}{L} \\ \dfrac{1}{L} & 0 & \dfrac{1}{L} \end{bmatrix} \begin{Bmatrix} \bar{P}_3 \\ \bar{P}_4 \\ \bar{P}_6 \end{Bmatrix} = \begin{Bmatrix} 0 \\ 0 \\ 0 \end{Bmatrix} \tag{4.36}$$

Equation (4.36) represents the whole (rigid-body) member equilibrium conditions, namely:

$$\bar{P}_4 + \bar{P}_1 = 0 \rightarrow \text{Horizontal Equilibrium} \tag{4.37}$$

$$\bar{P}_2 - \frac{\bar{P}_3}{L} - \frac{\bar{P}_6}{L} = 0 \rightarrow \text{Moment Equilibrium at End } j \tag{4.38}$$

$$\bar{P}_5 + \frac{\bar{P}_3}{L} + \frac{\bar{P}_6}{L} = 0 \rightarrow \text{Moment Equilibrium at End } i \tag{4.39}$$

***Remark:***
(i) It is noted that the choice of a basic deformation system is not unique, and there are other choices for the basic deformation system, for example: Cantilever Beam System. The derivation of the matrix compatibility equation for the cantilever beam system is left to the reader as his/her end-of-chapter exercise.

## 4.3   COMPONENT TRANSFORMATIONS OF A DISPLACEMENT VECTOR

In the previous section, the matrix compatibility of Eq. (4.22) was expressed with respect to the local reference system ($x$, $y$, and $z$). However, it is also essential to express that matrix compatibility equation in the global reference system (X, Y, and Z). In other words, the interrelationship

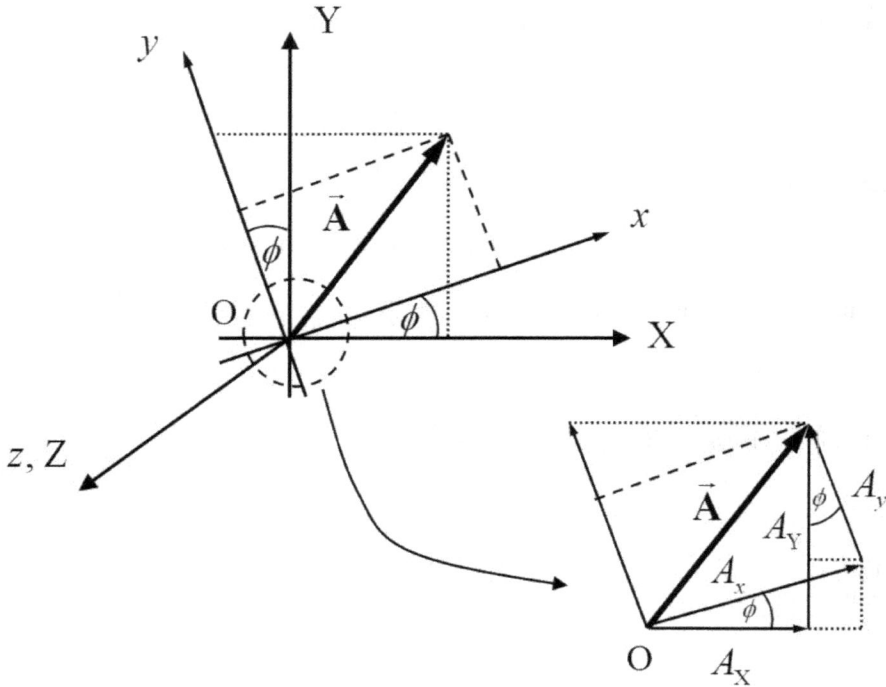

**FIGURE 4.6**  Component Transformation of a Plane Vector.

between the member-end displacements in local and global reference systems must be provided. This could be achieved through the transformation of vector components.

In Figure 4.6, the local reference axes are inclined at an angle $\phi$ with respect to the global reference axes. This implies that the angle $\phi$ is measured from the global reference to the local reference system. The sign of the angle $\phi$ follows the right-hand system. The vector $\vec{A}$ in Figure 4.6 can be decomposed with respect to both $x$-$y$ and X-Y components. It is noted that $\vec{A}$ can be thought of as a displacement vector at either node $i$ or $j$. In Chapter 3, one had also considered the vector $\vec{A}$ as a force vector at either node $i$ or $j$. Therefore, the relations between vector components defined with respect to the global and local reference systems given in Eqs. (3/3.18) and (3/3.19) can be used and are worth repeating them here.

$$
\begin{Bmatrix} A_X \\ A_Y \\ A_Z \end{Bmatrix} = \begin{bmatrix} \cos\phi & -\sin\phi & 0 \\ \sin\phi & \cos\phi & 0 \\ 0 & 0 & 1 \end{bmatrix} \begin{Bmatrix} A_x \\ A_y \\ A_z \end{Bmatrix} \text{ or } \mathbf{A}_{XYZ} = \mathbf{T}_{ROT}^T \mathbf{A}_{xyz} \tag{4.40}
$$

$$
\begin{Bmatrix} A_x \\ A_y \\ A_z \end{Bmatrix} = \begin{bmatrix} \cos\phi & \sin\phi & 0 \\ -\sin\phi & \cos\phi & 0 \\ 0 & 0 & 1 \end{bmatrix} \begin{Bmatrix} A_X \\ A_Y \\ A_Z \end{Bmatrix} \text{ or } \mathbf{A}_{xyz} = \mathbf{T}_{ROT} \mathbf{A}_{XYZ} \tag{4.41}
$$

As the present goal is to establish the transformation relations between the end-displacement components with respect to local and global coordinate systems, the displacement vector $\vec{U}_i$ at node $i$ shown in Figure 4.7 is considered. Figure 4.7a shows the displacement components along

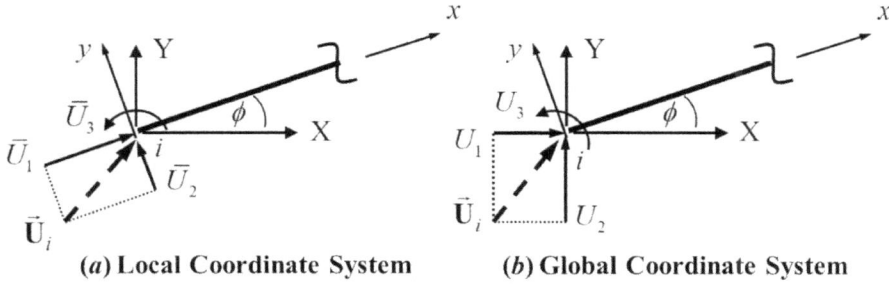

(a) Local Coordinate System          (b) Global Coordinate System

**FIGURE 4.7**   Transformation of Member-End Displacement Vector at Node $i$.

the local reference axes, while Figure 4.7$b$ shows the displacement components along the global reference axes. Following the transformation relations of Eq. (4.40), the relations between the displacement components shown in Figure 4.7$a$ and $b$ are:

$$\begin{Bmatrix} U_1 \\ U_2 \\ U_3 \end{Bmatrix} = \begin{bmatrix} \cos\phi & -\sin\phi & 0 \\ \sin\phi & \cos\phi & 0 \\ 0 & 0 & 1 \end{bmatrix} \begin{Bmatrix} \bar{U}_1 \\ \bar{U}_2 \\ \bar{U}_3 \end{Bmatrix} \text{ or } \mathbf{U}_1 = \mathbf{T}_{ROT}^T \bar{\mathbf{U}}_1 \tag{4.42}$$

and vice versa,

$$\begin{Bmatrix} \bar{U}_1 \\ \bar{U}_2 \\ \bar{U}_3 \end{Bmatrix} = \begin{bmatrix} \cos\phi & \sin\phi & 0 \\ -\sin\phi & \cos\phi & 0 \\ 0 & 0 & 1 \end{bmatrix} \begin{Bmatrix} U_1 \\ U_2 \\ U_3 \end{Bmatrix} \text{ or } \bar{\mathbf{U}}_1 = \mathbf{T}_{ROT} \mathbf{U}_1 \tag{4.43}$$

A similar consideration is applied to the displacement vector $\vec{\mathbf{U}}_j$ at node $j$ (Figure 4.8) and leads to the same transformation relations as:

$$\begin{Bmatrix} U_4 \\ U_5 \\ U_6 \end{Bmatrix} = \begin{bmatrix} \cos\phi & -\sin\phi & 0 \\ \sin\phi & \cos\phi & 0 \\ 0 & 0 & 1 \end{bmatrix} \begin{Bmatrix} \bar{U}_4 \\ \bar{U}_5 \\ \bar{U}_6 \end{Bmatrix} \text{ or } \mathbf{U}_2 = \mathbf{T}_{ROT}^T \bar{\mathbf{U}}_2 \tag{4.44}$$

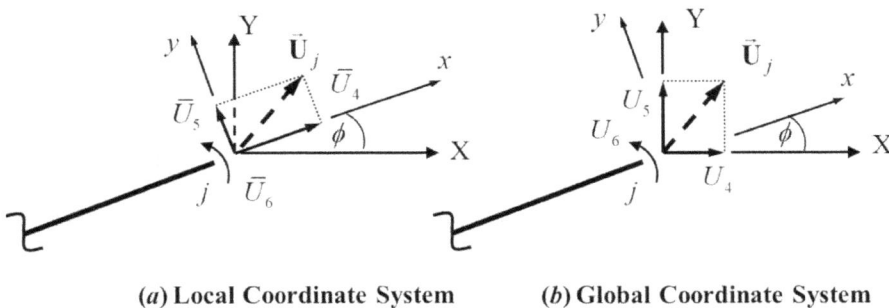

(a) Local Coordinate System          (b) Global Coordinate System

**FIGURE 4.8**   Transformation of Member-End Displacement Vector at Node $j$.

and vice versa

$$\begin{Bmatrix} \bar{U}_4 \\ \bar{U}_5 \\ \bar{U}_6 \end{Bmatrix} = \begin{bmatrix} \cos\phi & \sin\phi & 0 \\ -\sin\phi & \cos\phi & 0 \\ 0 & 0 & 1 \end{bmatrix} \begin{Bmatrix} U_4 \\ U_5 \\ U_6 \end{Bmatrix} \text{or} \, \bar{U}_2 = T_{ROT} U_2 \tag{4.45}$$

Combining Eqs. (4.42) and (4.44) leads to the transformation relations of member-end displacements:

$$\begin{Bmatrix} U_1 \\ U_2 \\ U_3 \\ U_4 \\ U_5 \\ U_6 \end{Bmatrix} = \begin{bmatrix} \cos\phi & -\sin\phi & 0 & 0 & 0 & 0 \\ \sin\phi & \cos\phi & 0 & 0 & 0 & 0 \\ 0 & 0 & 1 & 0 & 0 & 0 \\ 0 & 0 & 0 & \cos\phi & -\sin\phi & 0 \\ 0 & 0 & 0 & \sin\phi & \cos\phi & 0 \\ 0 & 0 & 0 & 0 & 0 & 1 \end{bmatrix} \begin{Bmatrix} \bar{U}_1 \\ \bar{U}_2 \\ \bar{U}_3 \\ \bar{U}_4 \\ \bar{U}_5 \\ \bar{U}_6 \end{Bmatrix} \text{or} \, U = \Gamma_{ROT}^T \bar{U} \tag{4.46}$$

or vice versa

$$\begin{Bmatrix} \bar{U}_1 \\ \bar{U}_2 \\ \bar{U}_3 \\ \bar{U}_4 \\ \bar{U}_5 \\ \bar{U}_6 \end{Bmatrix} = \begin{bmatrix} \cos\phi & \sin\phi & 0 & 0 & 0 & 0 \\ -\sin\phi & \cos\phi & 0 & 0 & 0 & 0 \\ 0 & 0 & 1 & 0 & 0 & 0 \\ 0 & 0 & 0 & \cos\phi & \sin\phi & 0 \\ 0 & 0 & 0 & -\sin\phi & \cos\phi & 0 \\ 0 & 0 & 0 & 0 & 0 & 1 \end{bmatrix} \begin{Bmatrix} U_1 \\ U_2 \\ U_3 \\ U_4 \\ U_5 \\ U_6 \end{Bmatrix} \text{or} \, \bar{U} = \Gamma_{ROT} U \tag{4.47}$$

where

$$\Gamma_{ROT} = \begin{bmatrix} T_{ROT} & 0 \\ 0 & T_{ROT} \end{bmatrix} \tag{4.48}$$

The component arrangement of the vector $U$ is similar to that of the vector $\bar{U}$. Starting from node $i$, one assigns first the degree of freedom (DOF) in global X-axis (first component), then the DOF in global Y-axis (second component), then the DOF about the global Z-axis (third component), and then proceed in this fashion to node $j$ of the member.

In computer implementation of a matrix analysis program, the value of the angle $\phi$ of each member is not necessarily given since values of direction cosines ($\cos\phi$ and $\sin\phi$) are not computed directly. Instead, they are computed indirectly based on the coordinate data of nodes $i$ and $j$ (Figure 4.9) as:

$$\Delta X = X_j - X_i; \quad \Delta Y = Y_j - Y_i; \text{and} \, L = \sqrt{(\Delta X)^2 + (\Delta Y)^2} \tag{4.49}$$

$$\cos\phi = \frac{\Delta X}{L} \text{and} \sin\phi = \frac{\Delta Y}{L} \tag{4.50}$$

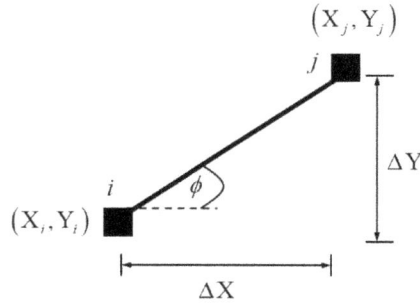

**FIGURE 4.9** Indirect Calculation of Direction Cosines.

## 4.4 COMPATIBILITY IN THE GLOBAL REFERENCE SYSTEM: A FRAME MEMBER

Up to now, the matrix compatibility equation of Eq. (4.22) has been established with respect to the local reference axes. Also, the transformation relations (4.47) between the member-end displacements described in local and global coordinate systems were obtained. Based on those transformation relations in hand, the matrix compatibility equation can easily be expressed with respect to the global reference axes.

Substitution of Eqs. (4.47) into (4.22) yields the matrix compatibility equation with respect to the global reference system as:

$$
\left\{ \begin{array}{c} v_1 \\ v_2 \\ v_3 \end{array} \right\} = \begin{bmatrix} -\cos\phi & -\sin\phi & 0 & \cos\phi & \sin\phi & 0 \\ \dfrac{\sin\phi}{L} & \dfrac{\cos\phi}{L} & 1 & \dfrac{\sin\phi}{L} & -\dfrac{\cos\phi}{L} & 0 \\ -\dfrac{\sin\phi}{L} & \dfrac{\cos\phi}{L} & 0 & \dfrac{\sin\phi}{L} & -\dfrac{\cos\phi}{L} & 1 \end{bmatrix} \left\{ \begin{array}{c} U_1 \\ U_2 \\ U_3 \\ U_4 \\ U_5 \\ U_6 \end{array} \right\} \tag{4.51}
$$

or in the concise format as:

$$
\mathbf{v} = \begin{bmatrix} \mathbf{a}_{1B} & \mathbf{a}_{2B} \end{bmatrix} \left\{ \begin{array}{c} \mathbf{U}_1 \\ \mathbf{U}_2 \end{array} \right\} \text{ or } \mathbf{v} = \mathbf{a}_B \mathbf{U} \tag{4.52}
$$

where $\mathbf{a}_B$ is the compatibility matrix for the simply supported beam model in the global reference system. Furthermore, it is observed that the matrix $\mathbf{a}_B$ is equal to the transpose of the equilibrium matrix $\mathbf{b}_B$ (3/3.30) for the simply supported beam system defined in Chapter 3.

$$
\mathbf{P} - \mathbf{P}_0 = \mathbf{b}_B \mathbf{q} \tag{3/3.30}
$$

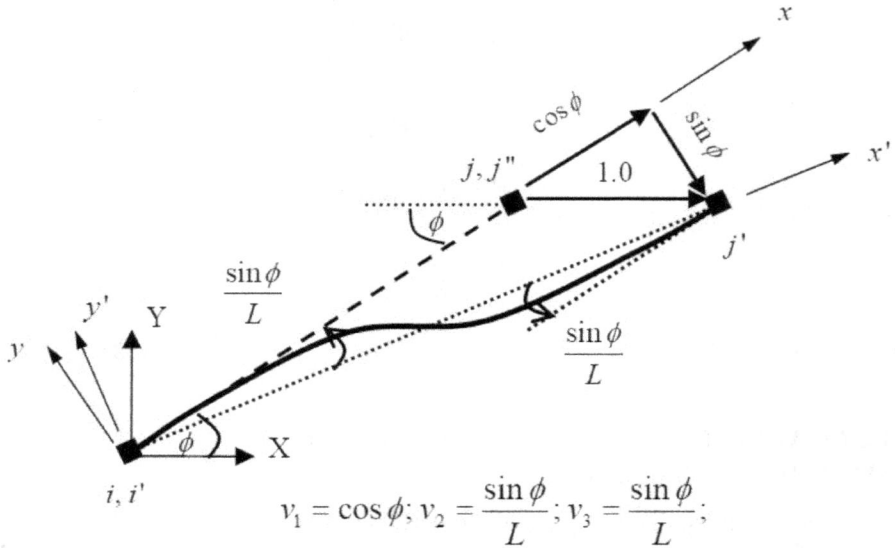

**FIGURE 4.10** Independent Deformed Mode of a Simply Supported Beam Model Associated with Imposed Unit Displacement $U_4 = 1.0$.

One may simply state that the statical (3/3.30) and kinematical (4.52) transformations are contragradient.

It can be concluded that:

1. The basic member deformations $\mathbf{v}$ of Eq. (4.52) and basic member forces $\mathbf{q}$ of Eq. (3/3.30) are conjugate-work pairs. In other words, the scalar product of these two quantities produces consistent work.

2. The member-end displacements $\mathbf{U}$ of Eq. (4.52) and member-end forces $\mathbf{P} - \mathbf{P}_0$ of Eq. (3/3.30) are conjugate-work pairs. Otherwise stated, the scalar product of these two quantities produces consistent work.

Furthermore, it is noted that each column of the compatibility matrix represents the member deformation mode for a unit displacement imposed at the corresponding displacement degree of freedom in the global reference system.

For example, the geometric interpretation of the fourth column of the compatibility matrix $\mathbf{a}_B$ is shown in Figure 4.10.

## 4.5  COMPATIBILITY: A FRAME STRUCTURE

In this section, the main objective is to establish the matrix compatibility equation of a frame structure. This matrix equation relates the basic member deformations to the nodal displacements at the global degrees of freedom. The matrix compatibility equation of the structure can be obtained by considering the compatibility equations of all members in a collective manner.

For the frame structure as a whole, an augmented vector $\mathbf{V}$ is used to collect the basic deformations of each member and is defined as:

$$\mathbf{V} = \lfloor \mathbf{V}_1 \quad \mathbf{V}_2 \quad \cdots \quad \mathbf{V}_k \quad \cdots \quad \mathbf{V}_m \rfloor^T_{3m \times 1} \tag{4.53}$$

where $m$ is the number of members and the sub-vector $\mathbf{V}_k$ contains the basic deformations of member $k$. Therefore, the size of the vector $\mathbf{V}$ is $3m \times 1$.

Similarly, the nodal displacements are grouped into an augmented vector $\mathbf{U}$ defined as:

$$\mathbf{U} = \lfloor \mathbf{U}_1 \quad \mathbf{U}_2 \quad \cdots \quad \mathbf{U}_k \quad \cdots \quad \mathbf{U}_n \rfloor^T_{3n \times 1} \tag{4.54}$$

where $n$ is the number of nodes and the sub-vector $\mathbf{U}_k$ contains the displacements of node $k$. Therefore, the size of the vector $\mathbf{U}$ is $3n \times 1$.

Once compatibility conditions are collectively considered for all members, the matrix compatibility equation is obtained as:

$$
\begin{Bmatrix} \mathbf{V}_1 \\ \mathbf{V}_2 \\ \vdots \\ \mathbf{V}_n \end{Bmatrix}_{3m \times 1}
=
\begin{bmatrix} \mathbf{a}_1^{(1)} & \mathbf{a}_2^{(1)} & \cdots & \cdots \\ \mathbf{a}_1^{(2)} & \mathbf{a}_2^{(2)} & \cdots & \cdots \\ \vdots & \vdots & & \\ \vdots & \vdots & & \end{bmatrix}_{3m \times 3n}
\begin{Bmatrix} \mathbf{U}_1 \\ \mathbf{U}_2 \\ \vdots \\ \mathbf{U}_n \end{Bmatrix}_{3n \times 1}
\tag{4.55}
$$

or in the concise format as:

$$\mathbf{V} = \mathbf{A}\mathbf{U} \tag{4.56}$$

where $\mathbf{A}$ is the structural compatibility matrix. The matrix $\mathbf{A}$ is formulated in terms of matrices $\mathbf{a}$ of members through the member compatibility equation of Eq. (4.52). The member node numbers are replaced by the structural node numbers through the member connectivity.

For example, let a member $a$ of Figure 4.11 have its ends $i$ and $j$ incident at structural nodes $I$ and $J$, respectively. The matrix compatibility equation of member $a$ is written as:

$$\mathbf{V}_a = \mathbf{a}_1^{(a)} \mathbf{U}_1^{(a)} + \mathbf{a}_2^{(a)} \mathbf{U}_2^{(a)} \tag{4.57}$$

Due to the connectivity of member $a$:

$$\mathbf{U}_I = \mathbf{U}_1^{(a)} \text{ and } \mathbf{U}_J = \mathbf{U}_2^{(a)} \tag{4.58}$$

The compatibility equation of member $a$ is modified as:

$$\mathbf{V}_a = \mathbf{a}_1^{(a)} \mathbf{U}_I + \mathbf{a}_2^{(a)} \mathbf{U}_J \tag{4.59}$$

This process is repeated for all members in succession. Eventually, the structural compatibility equation (4.56) is obtained. It is also noted that from Eq. (4.56), the member deformations $\mathbf{V}$ can always be determined from given displacement degrees of freedom $\mathbf{U}$.

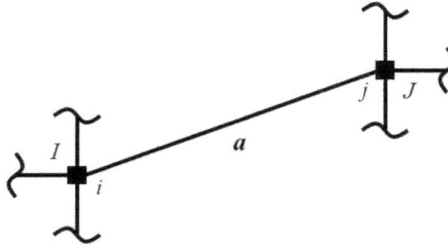

**FIGURE 4.11**    A Generic Frame Member $a$ with Its Ends $i$ and $j$ Connecting to Structural Nodes $I$ and $J$.

The matrix compatibility of Eq. (4.56) can be partitioned following known and unknown nodal displacements as:

$$V = \begin{bmatrix} A_{free} & \vdots & A_{constr} \end{bmatrix} \left\{ \dfrac{U_{free}}{U_{constr}} \right\} \tag{4.60}$$

where $U_{free}$ and $U_{constr}$ contain the nodal displacements associated with free and constrained degrees of freedom, respectively; $A_{free}$ and $A_{constr}$ are the compatibility matrices associated with free and constrained degrees of freedom, respectively. $U_{constr}$ are known and typically associated with the support conditions. In the case where there are no support movements ($U_{constr} = 0$), Eq. (4.60) is reduced to:

$$V = V_{free} = A_{free} U_{free} \tag{4.61}$$

It is noted that the number of rows of the compatibility matrix $A_{free}$ corresponds to the number of basic deformations, which is equal to the number of basic forces as discussed in Chapter 3. The number of columns of the compatibility matrix $A_{free}$ is associated with the number of free degrees of freedom. Consequently, the number of rows exceeds the number of columns by the degree of statical indeterminacy ($SI$). In other words, the degree of statical indeterminacy ($SI$) can be determined from the dimension of the matrix $A_{free}$ as:

$$SI = n_{row} - n_{column} \tag{4.62}$$

To illustrate the aforementioned concept of establishing the matrix compatibility equation of a structure, a prototypical frame shown in Figure 4.12 is considered. It is assumed that the simply supported beam model (Figure 4.3) is used to represent the basic member system. The basic deformation system of the frame is shown in Figure 4.13.

Consider the compatibility equation of member 1:

Based on the kinematical relations of Eq. (4.52), the member compatibility is:

$$v^{(1)} = a_{1B}^{(1)} U_1^{(1)} + a_{2B}^{(1)} U_2^{(1)} \tag{4.63}$$

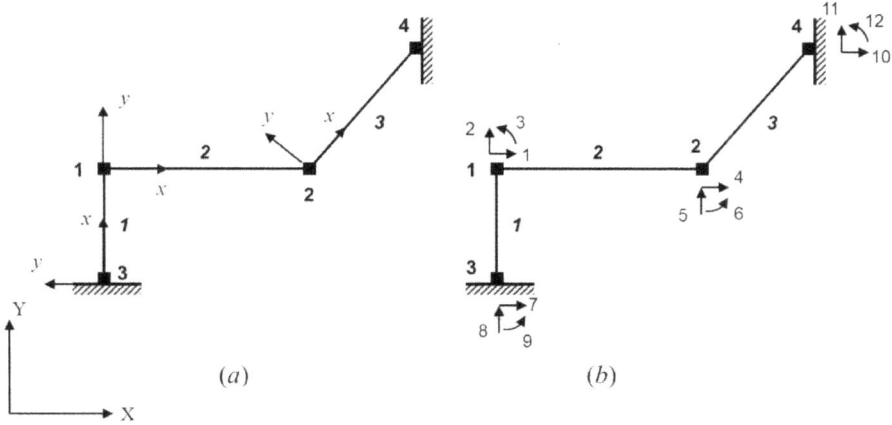

**FIGURE 4.12** A Prototypical Frame: (a) Local Reference Axes; (b) Nodal, Member, and DOFs Numbering Systems.

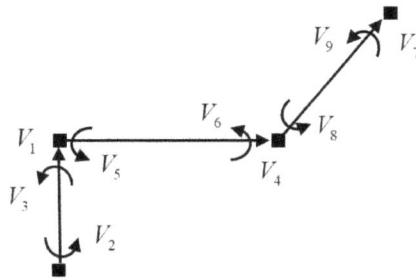

**FIGURE 4.13** A Prototypical Frame: Basic Deformation System.

Applying the member connectivity and deformation identification:

$$\mathbf{U}_1^{(1)} \Rightarrow \mathbf{U}_3 \text{ and } \mathbf{U}_2^{(1)} \Rightarrow \mathbf{U}_1 \tag{4.64}$$

$$\mathbf{v}^{(1)} \Rightarrow \mathbf{V}_1 \tag{4.65}$$

Eq. (4.63) can be written in terms of structural variables as:

$$\mathbf{V}_1 = \mathbf{a}_{1B}^{(1)}\mathbf{U}_3 + \mathbf{a}_{2B}^{(1)}\mathbf{U}_1 \tag{4.66}$$

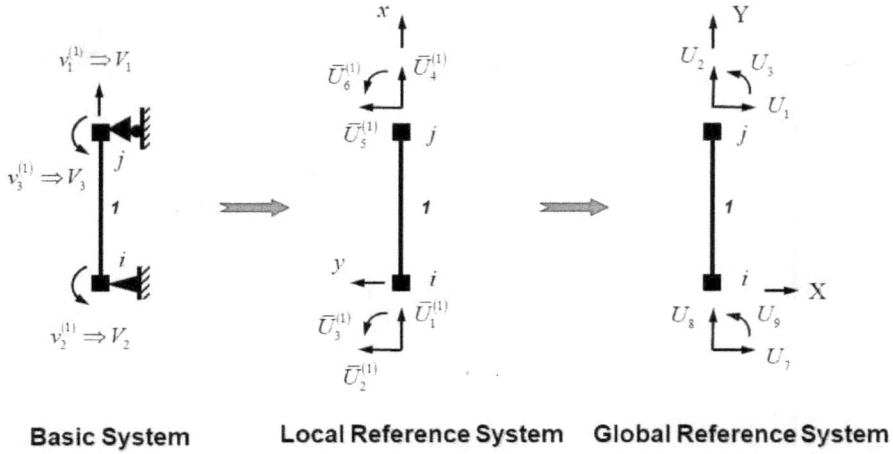

**Basic System         Local Reference System     Global Reference System**

**FIGURE 4.14** Deformation–Displacement Transformations of Member 1: Basic System to Global Reference System.

and is placed in the matrix compatibility equation of the structure as:

$$\begin{Bmatrix} \mathbf{V}_1 \\ \bullet \\ \bullet \end{Bmatrix} = \begin{bmatrix} \mathbf{a}_{2B}^{(1)} & \mathbf{0} & \mathbf{a}_{1B}^{(1)} & \mathbf{0} \\ \bullet & \bullet & \bullet & \bullet \\ \bullet & \bullet & \bullet & \bullet \end{bmatrix} \begin{Bmatrix} \mathbf{U}_1 \\ \mathbf{U}_2 \\ \mathbf{U}_3 \\ \mathbf{U}_4 \end{Bmatrix}$$

(4.67)

The whole process of establishing the compatibility relation of member 1 is schematically represented in Figure 4.14.

The same process is also applied to members 2 and 3. Consequently, their member compatibility relations expressed in the structural variables are:

$$\mathbf{V}_2 = \mathbf{a}_{1B}^{(2)}\mathbf{U}_1 + \mathbf{a}_{2B}^{(2)}\mathbf{U}_2$$

(4.68)

$$\mathbf{V}_3 = \mathbf{a}_{1B}^{(3)}\mathbf{U}_2 + \mathbf{a}_{2B}^{(3)}\mathbf{U}_4$$

(4.69)

These compatibility relations are placed in the structural compatibility relation as:

$$\begin{Bmatrix} \mathbf{V}_1 \\ \mathbf{V}_2 \\ \mathbf{V}_3 \end{Bmatrix}_{9\times1} = \begin{bmatrix} \mathbf{a}_{2B}^{(1)} & \mathbf{0} & \mathbf{a}_{1B}^{(1)} & \mathbf{0} \\ \mathbf{a}_{1B}^{(2)} & \mathbf{a}_{2B}^{(2)} & \mathbf{0} & \mathbf{0} \\ \mathbf{0} & \mathbf{a}_{1B}^{(3)} & \mathbf{0} & \mathbf{a}_{2B}^{(3)} \end{bmatrix}_{9\times12} \begin{Bmatrix} \mathbf{U}_1 \\ \mathbf{U}_2 \\ \mathbf{U}_3 \\ \mathbf{U}_4 \end{Bmatrix}_{12\times1}$$

(4.70)

This is the matrix compatibility equation of the whole structure and can be expressed in the compact format as:

$$\mathbf{V} = \mathbf{AU}$$

(4.71)

It is observed that this compatibility matrix is the transpose of the equilibrium matrix (3/3.42) of the same structure. One concludes that:

$$\mathbf{A} = \mathbf{B}^T \tag{4.72}$$

where

$$\mathbf{B} = \begin{bmatrix} \mathbf{b}_{2B}^{(1)} & \mathbf{b}_{1B}^{(2)} & \mathbf{0} \\ \mathbf{0} & \mathbf{b}_{2B}^{(2)} & \mathbf{b}_{1B}^{(3)} \\ \mathbf{b}_{1B}^{(1)} & \mathbf{0} & \mathbf{0} \\ \mathbf{0} & \mathbf{0} & \mathbf{b}_{2B}^{(3)} \end{bmatrix}_{12\times9} \tag{3/3.42}$$

$$\mathbf{a}_{1B} = \mathbf{b}_{1B}^T \text{ and } \mathbf{a}_{2B} = \mathbf{b}_{2B}^T \tag{4.73}$$

This type of relationship is due to the contragradient nature between the statical and kinematical relations and can be proved by the virtual work principle in the following chapter.

The boundary conditions of this prototypical frame are:

$$\mathbf{U}_{constr} = \lfloor \mathbf{U}_3 \quad \mathbf{U}_4 \rfloor^T = \mathbf{0} \quad \text{(no support settlements)} \tag{4.74}$$

Substitution of Eqs. (4.74) into (4.70) results in the compatibility relations associated with free degrees of freedom as:

$$\begin{Bmatrix} \mathbf{V}_1 \\ \mathbf{V}_2 \\ \mathbf{V}_3 \end{Bmatrix}_{9\times1} = \begin{bmatrix} \mathbf{a}_{2B}^{(1)} & \mathbf{0} \\ \mathbf{a}_{1B}^{(2)} & \mathbf{a}_{2B}^{(2)} \\ \mathbf{0} & \mathbf{a}_{1B}^{(3)} \end{bmatrix}_{9\times6} \begin{Bmatrix} \mathbf{U}_1 \\ \mathbf{U}_2 \end{Bmatrix}_{6\times1} \quad \text{or } \mathbf{V} = \mathbf{V}_{free} = \mathbf{A}_{free}\mathbf{U}_{free} \tag{4.75}$$

It is also interesting to note that each column of the matrix $\mathbf{A}_{free}$ represents the member deformations associated with the imposed unit displacement at each degree of freedom

For example, the member deformations corresponding to the imposed unit displacement $U_5 = 1.0$ are presented in Figure 4.15.

## Example 4.1

For the frame structure of Example 3.1, formulate the matrix compatibility equation relating global free degrees of freedom to member deformations.

**Solution:** Employing numbering systems of Example 3.1 and the basic member deformation numbering system of Figure 4.16, the establishing process of the geometric compatibility equation of member 1 is schematically presented in Figure 4.17.

From Basic System to Local Reference System:

$$\begin{Bmatrix} v_1^{(1)} \\ v_2^{(1)} \\ v_3^{(1)} \end{Bmatrix} = \begin{bmatrix} -1 & 0 & 0 & 1 & 0 & 0 \\ 0 & 0.2 & 1 & 0 & -0.2 & 0 \\ 0 & 0.2 & 0 & 0 & -0.2 & 1 \end{bmatrix} \begin{Bmatrix} \bar{U}_1^{(1)} \\ \bar{U}_2^{(1)} \\ \bar{U}_3^{(1)} \\ \bar{U}_4^{(1)} \\ \bar{U}_5^{(1)} \\ \bar{U}_6^{(1)} \end{Bmatrix}$$

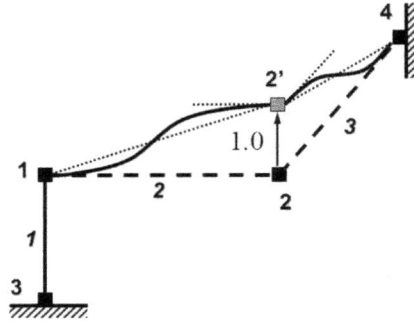

**FIGURE 4.15**   Member Deformations due to Imposed Unit Displacement $U_5 = 1.0$.

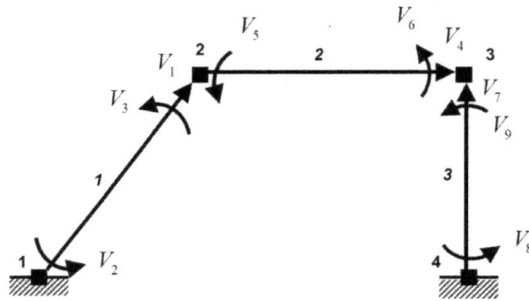

**FIGURE 4.16**   Example 4.1.

Member connectivity and deformation identification:

$$LM_1 = \begin{Bmatrix} 7 \\ 8 \\ 9 \\ 1 \\ 2 \\ 3 \end{Bmatrix}_1 ; \begin{Bmatrix} U_1^{(1)} \\ U_2^{(1)} \\ U_3^{(1)} \\ U_4^{(1)} \\ U_5^{(1)} \\ U_6^{(1)} \end{Bmatrix} \Rightarrow \begin{Bmatrix} U_7 \\ U_8 \\ U_9 \\ U_1 \\ U_2 \\ U_3 \end{Bmatrix} ; \text{ and } \begin{Bmatrix} v_1^{(1)} \\ v_2^{(1)} \\ v_3^{(1)} \end{Bmatrix} = \begin{Bmatrix} V_1 \\ V_2 \\ V_3 \end{Bmatrix}$$

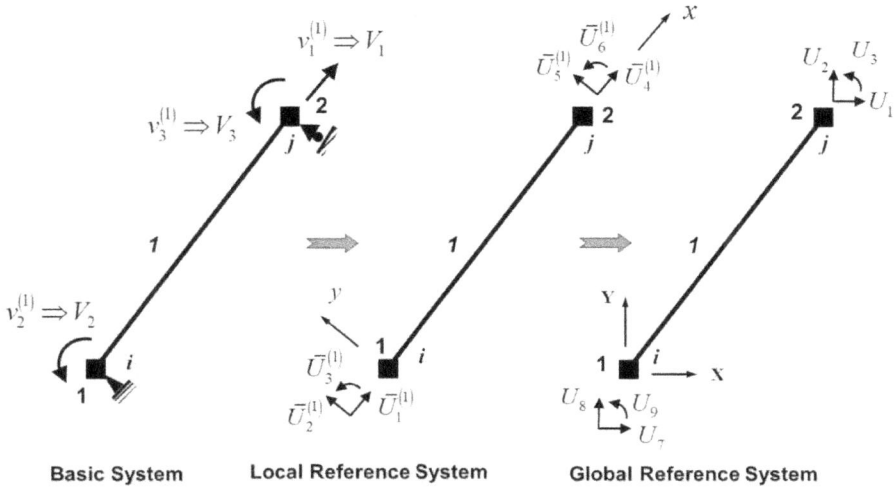

**FIGURE 4.17**   Example 4.1 (Continued).

From Local Reference System to Global Reference System:

$$
\begin{Bmatrix} V_1 \\ V_2 \\ V_3 \end{Bmatrix} = \begin{bmatrix} -0.6 & -0.8 & 0 & 0.6 & 0.8 & 0 \\ -0.16 & 0.12 & 1 & 0.16 & -0.12 & 0 \\ -0.16 & 0.12 & 0 & 0.16 & -0.12 & 1 \end{bmatrix} \begin{Bmatrix} U_7 \\ U_8 \\ U_9 \\ U_1 \\ U_2 \\ U_3 \end{Bmatrix}
$$

The matrix compatibility equation of member 1 is placed in the matrix compatibility equation of the structure as:

$$
\begin{Bmatrix} V_1 \\ V_2 \\ V_3 \\ \bullet \\ \bullet \\ \bullet \\ \bullet \\ \bullet \\ \bullet \end{Bmatrix} = \begin{bmatrix} 0.6 & 0.8 & 0 & 0 & 0 & 0 & -0.6 & -0.8 & 0 & 0 & 0 & 0 \\ 0.16 & -0.12 & 0 & 0 & 0 & 0 & -0.16 & 0.12 & 1 & 0 & 0 & 0 \\ 0.16 & -0.12 & 1 & 0 & 0 & 0 & -0.16 & 0.12 & 0 & 0 & 0 & 0 \\ \bullet & \bullet & \bullet & \bullet & \bullet & \bullet & \bullet & \bullet & \bullet & \bullet & \bullet & \bullet \\ \bullet & \bullet & \bullet & \bullet & \bullet & \bullet & \bullet & \bullet & \bullet & \bullet & \bullet & \bullet \\ \bullet & \bullet & \bullet & \bullet & \bullet & \bullet & \bullet & \bullet & \bullet & \bullet & \bullet & \bullet \\ \bullet & \bullet & \bullet & \bullet & \bullet & \bullet & \bullet & \bullet & \bullet & \bullet & \bullet & \bullet \\ \bullet & \bullet & \bullet & \bullet & \bullet & \bullet & \bullet & \bullet & \bullet & \bullet & \bullet & \bullet \\ \bullet & \bullet & \bullet & \bullet & \bullet & \bullet & \bullet & \bullet & \bullet & \bullet & \bullet & \bullet \end{bmatrix} \begin{Bmatrix} U_1 \\ U_2 \\ U_3 \\ U_4 \\ U_5 \\ U_6 \\ U_7 \\ U_8 \\ U_9 \\ U_{10} \\ U_{11} \\ U_{12} \end{Bmatrix}
$$

Applying the same procedure to members 2 and 3, the matrix compatibility equation of the whole structure is obtained as:

$$
\begin{Bmatrix} V_1 \\ V_2 \\ V_3 \\ V_4 \\ V_5 \\ V_6 \\ V_7 \\ V_8 \\ V_9 \end{Bmatrix} =
\begin{bmatrix}
0.6 & 0.8 & 0 & 0 & 0 & 0 & -0.6 & -0.8 & 0 & 0 & 0 & 0 \\
0.16 & -0.12 & 0 & 0 & 0 & 0 & -0.16 & 0.12 & 1 & 0 & 0 & 0 \\
0.16 & -0.12 & 1 & 0 & 0 & 0 & -0.16 & 0.12 & 0 & 0 & 0 & 0 \\
-1 & 0 & 0 & 1 & 0 & 0 & 0 & 0 & 0 & 0 & 0 & 0 \\
0 & 0.2 & 1 & 0 & -0.2 & 0 & 0 & 0 & 0 & 0 & 0 & 0 \\
0 & 0.2 & 0 & 0 & -0.2 & 1 & 0 & 0 & 0 & 0 & 0 & 0 \\
0 & 0 & 0 & 0 & 1 & 0 & 0 & 0 & 0 & 0 & -1 & 0 \\
0 & 0 & 0 & 0.25 & 0 & 0 & 0 & 0 & 0 & -0.25 & 0 & 1 \\
0 & 0 & 0 & 0.25 & 0 & 1 & 0 & 0 & 0 & -0.25 & 0 & 0
\end{bmatrix}
\begin{Bmatrix} U_1 \\ U_2 \\ U_3 \\ U_4 \\ U_5 \\ U_6 \\ U_7 \\ U_8 \\ U_9 \\ U_{10} \\ U_{11} \\ U_{12} \end{Bmatrix}
$$

or in the compact format as:

$$
\mathbf{V} = \mathbf{A}\mathbf{U} \text{ or } \mathbf{V} = \begin{bmatrix} \mathbf{A}_{free} : \mathbf{A}_{constr} \end{bmatrix} \begin{Bmatrix} \mathbf{U}_{free} \\ \hline \mathbf{U}_{constr} \end{Bmatrix}
$$

Matrix $\mathbf{A}$ is the structural compatibility matrix relating the member deformations $\mathbf{V}$ to the nodal displacements $\mathbf{U}$. It is observed that this compatibility matrix is the transpose of the equilibrium matrix $\mathbf{B}$ of the same structure (Example 3.1).

Basic member deformations can be separated as:

$$
\mathbf{V} = \mathbf{V}_{free} + \mathbf{V}_{constr}
$$

where $\mathbf{V}_{free} = \mathbf{A}_{free}\mathbf{U}_{free}$ are basic deformations associated with free displacement degrees of freedom, and $\mathbf{V}_{constr} = \mathbf{A}_{constr}\mathbf{U}_{constr}$ are basic deformations associated with constrained displacement degrees of freedom.

In this example, there are no support settlements ($\mathbf{U}_{constr} = \mathbf{0}$). Therefore, the matrix compatibility equation is reduced to:

$$
\mathbf{V} = \mathbf{V}_{free} = \mathbf{A}_{free}\mathbf{U}_{free} \text{ or }
\begin{Bmatrix} V_1 \\ V_2 \\ V_3 \\ V_4 \\ V_5 \\ V_6 \\ V_7 \\ V_8 \\ V_9 \end{Bmatrix} =
\begin{bmatrix}
0.6 & 0.8 & 0 & 0 & 0 & 0 \\
0.16 & -0.12 & 0 & 0 & 0 & 0 \\
0.16 & -0.12 & 1 & 0 & 0 & 0 \\
-1 & 0 & 0 & 1 & 0 & 0 \\
0 & 0.2 & 1 & 0 & -0.2 & 0 \\
0 & 0.2 & 0 & 0 & -0.2 & 1 \\
0 & 0 & 0 & 0 & 1 & 0 \\
0 & 0 & 0 & 0.25 & 0 & 0 \\
0 & 0 & 0 & 0.25 & 0 & 1
\end{bmatrix}
\begin{Bmatrix} U_1 \\ U_2 \\ U_3 \\ U_4 \\ U_5 \\ U_6 \end{Bmatrix}
$$

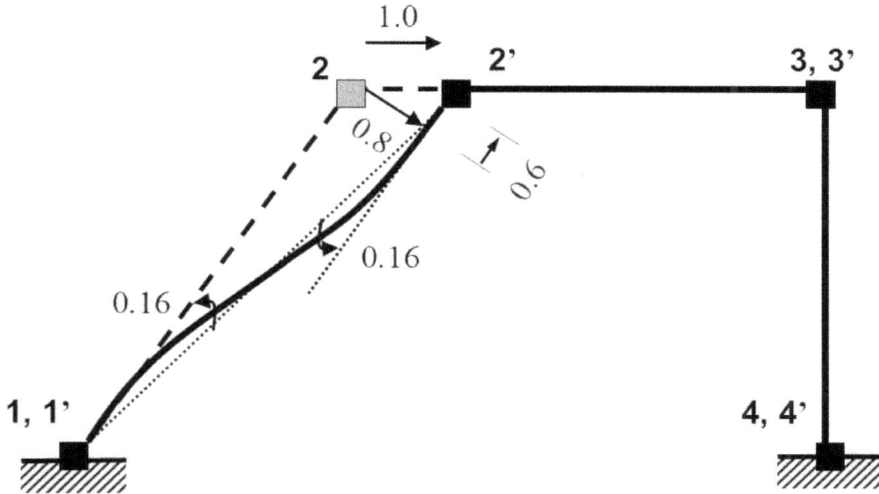

**FIGURE 4.18** Example 4.1 (Continued).

As well, it is observed that the compatibility matrix $\mathbf{A}_{free}$ is the transpose of the equilibrium matrix $\mathbf{B}_{free}$ of the same structure (Example 3.1).

It is also interesting to note that each column of the matrix $\mathbf{A}_{free}$ represents the member deformations associated with imposed unit displacement at each degree of freedom.

For example, the first column of the matrix $\mathbf{A}_{free}$ is the member deformations associated with imposed unit displacement at the first degree of freedom ($U_1 = 1.0$) as shown in Figure 4.18.

## 4.6  COMPATIBILITY: TRUSS (PIN-ENDED) MEMBER AND PIN-JOINTED STRUCTURE

Up to now, only a structure consisting of frame members has been considered. In this section, attention will be on the pin-jointed structure commonly known as a truss structure. The matrix compatibility equations derived in the previous sections can easily be extended to the case of a truss structure since the truss member is simply a special form of the frame member. This can be achieved by observing that the axial deformation is uncoupled from the transverse and rotational deformations due to the linearization of geometric compatibility relations. This independence of the axial deformation mode is obvious in the member compatibility Eq. (4.22) as shown in Figure 4.19.

Consequently, the geometric compatibility relations for a truss member with respect to the local reference axes are established as:

$$v_1 = \begin{bmatrix} -1 & 1 \end{bmatrix} \begin{Bmatrix} \bar{U}_1 \\ \bar{U}_2 \end{Bmatrix} \text{ or } v_1 = \bar{\mathbf{a}}^{truss}\bar{\mathbf{U}} \tag{4.76}$$

where $\bar{\mathbf{a}}^{truss}$ is the compatibility matrix of the basic truss system (Figure 4.20) in the local reference axes.

The matrix compatibility equation of the truss member in the global reference axes can be obtained by transforming the member-end displacements from local to global reference systems.

$$
\left\{ \begin{array}{c} v_1 \\ v_2 \\ v_3 \end{array} \right\} = \left[ \begin{array}{ccc:ccc} -1 & 0 & 0 & 1 & 0 & 0 \\ \hdashline 0 & \dfrac{1}{L} & 1 & 0 & -\dfrac{1}{L} & 0 \\ 0 & \dfrac{1}{L} & 0 & 0 & -\dfrac{1}{L} & 1 \end{array} \right] \left\{ \begin{array}{c} \bar{U}_1 \\ \bar{U}_2 \\ \bar{U}_3 \\ \bar{U}_4 \\ \bar{U}_5 \\ \bar{U}_6 \end{array} \right\}
$$

**FIGURE 4.19** Uncoupling Response between Axial and Transverse Deformations in a Frame Member due to Small Displacement Assumption.

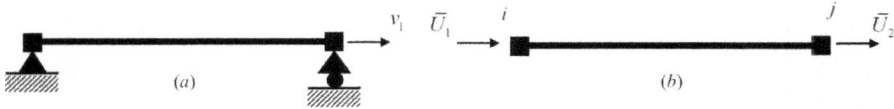

**FIGURE 4.20** The Truss Member in the Local Reference System: (a) Basic System; (b) Complete System.

This leads to the geometric compatibility relation for a truss member with respect to the global reference axes as:

$$
v_1 = \begin{bmatrix} -\cos\phi & -\sin\phi & \cos\phi & \sin\phi \end{bmatrix} \left\{ \begin{array}{c} U_1 \\ U_2 \\ U_3 \\ U_4 \end{array} \right\} \text{ or } v_1 = \mathbf{a}^{truss}\mathbf{U} \tag{4.77}
$$

where $\mathbf{a}^{truss}$ is the compatibility matrix of the basic truss system in the global reference axes. The deformation-displacement relation of the truss member is schematically presented in Figure 4.21.

## EXAMPLE 4.2

For the truss structure of Example 3.2, formulate the matrix compatibility equation relating global free degrees of freedom to member deformations.

**Solution:** Employing numbering systems of Example 3.2 and the basic member deformation numbering system of Figure 4.22, the establishing process of the geometric compatibility equation of member 1 is schematically presented in Figure 4.23.

From Basic System to Local Reference System:

$$
v_1^{(1)} = \begin{bmatrix} -1 & 1 \end{bmatrix} \left\{ \begin{array}{c} \bar{U}_1^{(1)} \\ \bar{U}_2^{(1)} \end{array} \right\}
$$

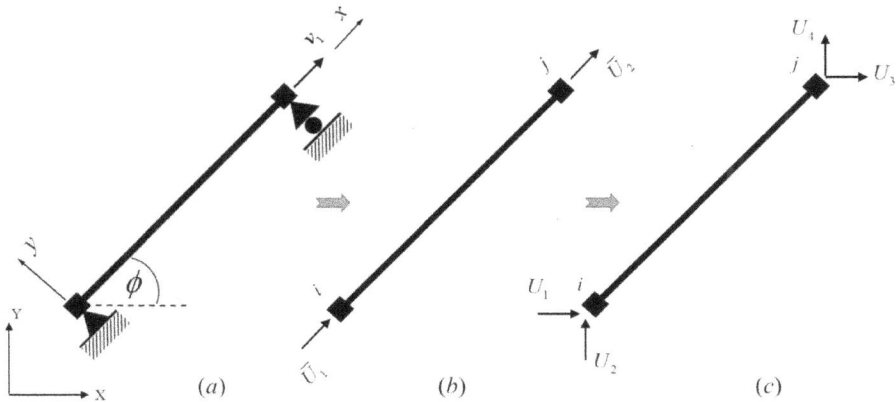

**FIGURE 4.21**    Truss Member in the Global Reference System: (*a*) Basic System; (*b*) Complete System in Local Reference Axes; (*c*) Complete System in Global Reference Axes.

Member connectivity and deformation identification:

$$LM_1 = \begin{Bmatrix} 5 \\ 6 \\ 1 \\ 2 \end{Bmatrix}_1 \; ; \; \begin{Bmatrix} U_1^{(1)} \\ U_2^{(1)} \\ U_3^{(1)} \\ U_4^{(1)} \end{Bmatrix} \Rightarrow \begin{Bmatrix} U_5 \\ U_6 \\ U_1 \\ U_2 \end{Bmatrix} \; ; \text{ and } v_1^{(1)} = V_1$$

From Local Reference System to Global Reference System:

$$V_1 = \begin{bmatrix} -0.6 & -0.8 & 0.6 & 0.8 \end{bmatrix} \begin{Bmatrix} U_5 \\ U_6 \\ U_1 \\ U_2 \end{Bmatrix}$$

The matrix compatibility equation of member 1 is placed in the matrix compatibility equation of the structure as:

$$\begin{Bmatrix} V_1 \\ \bullet \\ \bullet \end{Bmatrix} = \begin{bmatrix} 0.6 & 0.8 & 0 & 0 & -0.6 & -0.8 \\ \bullet & \bullet & \bullet & \bullet & \bullet & \bullet \\ \bullet & \bullet & \bullet & \bullet & \bullet & \bullet \end{bmatrix} \begin{Bmatrix} U_1 \\ U_2 \\ U_3 \\ U_4 \\ U_5 \\ U_6 \end{Bmatrix}$$

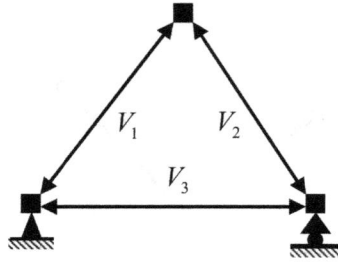

**FIGURE 4.22**   Example 4.2.

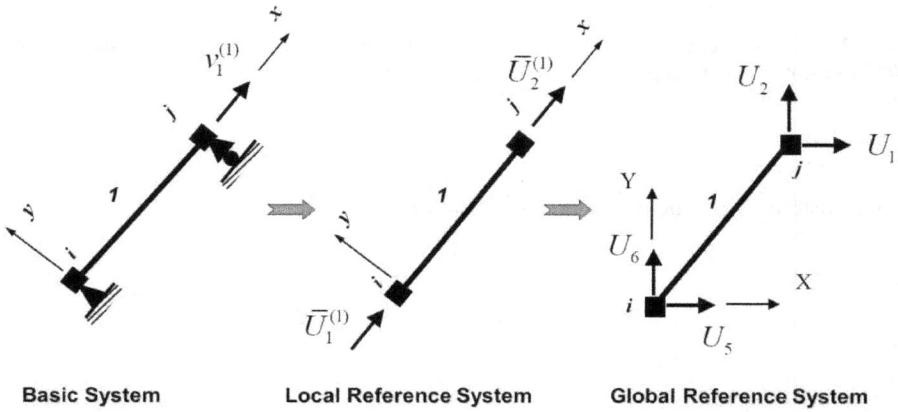

**FIGURE 4.23**   Example 4.2 (Continued).

Applying the same procedure to members 2 and 3, the matrix compatibility equation of the whole truss is obtained as:

$$
\begin{Bmatrix} V_1 \\ V_2 \\ V_3 \end{Bmatrix} = \begin{bmatrix} 0.6 & 0.8 & 0 & 0 & -0.6 & -0.8 \\ -0.6 & 0.8 & 0.6 & -0.8 & 0 & 0 \\ 0 & 0 & 1 & 0 & -1 & 0 \end{bmatrix} \begin{Bmatrix} U_1 \\ U_2 \\ U_3 \\ U_4 \\ U_5 \\ U_6 \end{Bmatrix}
$$

or in the compact format as:

$$\mathbf{V} = \mathbf{AU} \text{ or } \mathbf{V} = \left[ \mathbf{A}_{free} : \mathbf{A}_{constr} \right] \left\{ \frac{\mathbf{U}_{free}}{\mathbf{U}_{constr}} \right\}$$

The matrix $\mathbf{A}$ is the structural compatibility matrix relating the member deformations $\mathbf{V}$ to the nodal displacements $\mathbf{U}$. It is observed that this compatibility matrix is the transpose of the equilibrium matrix $\mathbf{B}$ of the same structure (Example 3.2).

In this example, there are no support settlements ($\mathbf{U}_{constr} = \mathbf{0}$). Therefore, the matrix compatibility equation is reduced to:

$$\mathbf{V} = \mathbf{V}_{free} = \mathbf{A}_{free} \mathbf{U}_{free} \text{ or } \left\{ \begin{matrix} V_1 \\ V_2 \\ V_3 \end{matrix} \right\} = \begin{bmatrix} 0.6 & 0.8 & 0 \\ -0.6 & 0.8 & 0.6 \\ 0 & 0 & 1 \end{bmatrix} \left\{ \begin{matrix} U_1 \\ U_2 \\ U_3 \end{matrix} \right\}$$

## 4.7   AUTOMATED ASSEMBLY OF COMPATIBILITY EQUATIONS

In the previous sections, the matrix compatibility equation of a structure has been constructed in the following manner: the basic deformations $\mathbf{v}$ of each member are first expressed in terms of the local member-end displacements $\overline{\mathbf{U}}$, then transformed the local member-end displacements $\overline{\mathbf{U}}$ to the global member-end displacements $\mathbf{U}$ by the rotational matrix, and finally considered the connectivity of all members in the collective fashion. The aforementioned process of setting up the compatibility equations can be extremely tedious in the case of large structural systems.

In this section, the systematic approach of assembling compatibility equations of the structure is presented. It is natural to extend this approach to the computer-oriented assembly of the compatibility equations.

To demonstrate the automated assembling approach, the frame structure of Example 4.1 is revisited. From the viewpoint of each member, each end displacement in the member degree of freedom is associated with only one structural degree of freedom. This one-to-one correspondence is represented by the mapping vector as discussed in Chapter 1. It is worth repeating that the mapping vectors of members 1, 2, and 3 are:

$$\left\{ \begin{matrix} 7 \\ 8 \\ 9 \\ 1 \\ 2 \\ 3 \end{matrix} \right\}_{structure} \rightarrow \left\{ \begin{matrix} 1 \\ 2 \\ 3 \\ 4 \\ 5 \\ 6 \end{matrix} \right\}_{member\ 1} \quad \left\{ \begin{matrix} 1 \\ 2 \\ 3 \\ 4 \\ 5 \\ 6 \end{matrix} \right\}_{structure} \rightarrow \left\{ \begin{matrix} 1 \\ 2 \\ 3 \\ 4 \\ 5 \\ 6 \end{matrix} \right\}_{member\ 2} \quad \left\{ \begin{matrix} 10 \\ 11 \\ 12 \\ 4 \\ 5 \\ 6 \end{matrix} \right\}_{structure} \rightarrow \left\{ \begin{matrix} 1 \\ 2 \\ 3 \\ 4 \\ 5 \\ 6 \end{matrix} \right\}_{member\ 3}$$

$$(4.78)$$

From the mapping vector of each member, its mapping matrix is constructed as:
Member 1:

$$
\begin{Bmatrix} U_1^{(1)} \\ U_2^{(1)} \\ U_3^{(1)} \\ U_4^{(1)} \\ U_5^{(1)} \\ U_6^{(1)} \end{Bmatrix} =
\begin{bmatrix}
0 & 0 & 0 & 0 & 0 & 0 & 1 & 0 & 0 & 0 & 0 & 0 \\
0 & 0 & 0 & 0 & 0 & 0 & 0 & 1 & 0 & 0 & 0 & 0 \\
0 & 0 & 0 & 0 & 0 & 0 & 0 & 0 & 1 & 0 & 0 & 0 \\
1 & 0 & 0 & 0 & 0 & 0 & 0 & 0 & 0 & 0 & 0 & 0 \\
0 & 1 & 0 & 0 & 0 & 0 & 0 & 0 & 0 & 0 & 0 & 0 \\
0 & 0 & 1 & 0 & 0 & 0 & 0 & 0 & 0 & 0 & 0 & 0
\end{bmatrix}
\begin{Bmatrix} U_1 \\ U_2 \\ U_3 \\ U_4 \\ U_5 \\ U_6 \\ U_7 \\ U_8 \\ U_9 \\ U_{10} \\ U_{11} \\ U_{12} \end{Bmatrix} \text{ or } \mathbf{U}^{(1)} = \mathbf{G}^{(1)}\mathbf{U} \qquad (4.79)
$$

Member 2:

$$
\begin{Bmatrix} U_1^{(2)} \\ U_2^{(2)} \\ U_3^{(2)} \\ U_4^{(2)} \\ U_5^{(2)} \\ U_6^{(2)} \end{Bmatrix} =
\begin{bmatrix}
1 & 0 & 0 & 0 & 0 & 0 & 0 & 0 & 0 & 0 & 0 & 0 \\
0 & 1 & 0 & 0 & 0 & 0 & 0 & 0 & 0 & 0 & 0 & 0 \\
0 & 0 & 1 & 0 & 0 & 0 & 0 & 0 & 0 & 0 & 0 & 0 \\
0 & 0 & 0 & 1 & 0 & 0 & 0 & 0 & 0 & 0 & 0 & 0 \\
0 & 0 & 0 & 0 & 1 & 0 & 0 & 0 & 0 & 0 & 0 & 0 \\
0 & 0 & 0 & 0 & 0 & 1 & 0 & 0 & 0 & 0 & 0 & 0
\end{bmatrix}
\begin{Bmatrix} U_1 \\ U_2 \\ U_3 \\ U_4 \\ U_5 \\ U_6 \\ U_7 \\ U_8 \\ U_9 \\ U_{10} \\ U_{11} \\ U_{12} \end{Bmatrix} \text{ or } \mathbf{U}^{(2)} = \mathbf{G}^{(2)}\mathbf{U} \qquad (4.80)
$$

Member 3:

$$
\begin{Bmatrix} U_1^{(3)} \\ U_2^{(3)} \\ U_3^{(3)} \\ U_4^{(3)} \\ U_5^{(3)} \\ U_6^{(3)} \end{Bmatrix} = \begin{bmatrix} 0 & 0 & 0 & 0 & 0 & 0 & 0 & 0 & 0 & 1 & 0 & 0 \\ 0 & 0 & 0 & 0 & 0 & 0 & 0 & 0 & 0 & 0 & 1 & 0 \\ 0 & 0 & 0 & 0 & 0 & 0 & 0 & 0 & 0 & 0 & 0 & 1 \\ 0 & 0 & 0 & 1 & 0 & 0 & 0 & 0 & 0 & 0 & 0 & 0 \\ 0 & 0 & 0 & 0 & 1 & 0 & 0 & 0 & 0 & 0 & 0 & 0 \\ 0 & 0 & 0 & 0 & 0 & 1 & 0 & 0 & 0 & 0 & 0 & 0 \end{bmatrix} \begin{Bmatrix} U_1 \\ U_2 \\ U_3 \\ U_4 \\ U_5 \\ U_6 \\ U_7 \\ U_8 \\ U_9 \\ U_{10} \\ U_{11} \\ U_{12} \end{Bmatrix} \text{ or } \mathbf{U}^{(3)} = \mathbf{G}^{(3)} \mathbf{U} \qquad (4.81)
$$

Referred to Eq. (3/3.54), it is observed that the compatibility mapping matrix $\mathbf{G}^{(i)}$ is the transpose of the equilibrium mapping matrix $\mathbf{H}^{(i)}$ discussed in Chapter 3.

$$
\mathbf{G}^{(1)} = \mathbf{H}^{(1)^T} ; \mathbf{G}^{(2)} = \mathbf{H}^{(2)^T} ; \text{and} \, \mathbf{G}^{(3)} = \mathbf{H}^{(3)^T} \qquad (4.82)
$$

where

$$
\mathbf{H}^{(1)} = \begin{bmatrix} 0 & 0 & 0 & 1 & 0 & 0 \\ 0 & 0 & 0 & 0 & 1 & 0 \\ 0 & 0 & 0 & 0 & 0 & 1 \\ 0 & 0 & 0 & 0 & 0 & 0 \\ 0 & 0 & 0 & 0 & 0 & 0 \\ 0 & 0 & 0 & 0 & 0 & 0 \\ 1 & 0 & 0 & 0 & 0 & 0 \\ 0 & 1 & 0 & 0 & 0 & 0 \\ 0 & 0 & 1 & 0 & 0 & 0 \\ 0 & 0 & 0 & 0 & 0 & 0 \\ 0 & 0 & 0 & 0 & 0 & 0 \\ 0 & 0 & 0 & 0 & 0 & 0 \end{bmatrix} ; \mathbf{H}^{(2)} = \begin{bmatrix} 1 & 0 & 0 & 0 & 0 & 0 \\ 0 & 1 & 0 & 0 & 0 & 0 \\ 0 & 0 & 1 & 0 & 0 & 0 \\ 0 & 0 & 0 & 1 & 0 & 0 \\ 0 & 0 & 0 & 0 & 1 & 0 \\ 0 & 0 & 0 & 0 & 0 & 1 \\ 0 & 0 & 0 & 0 & 0 & 0 \\ 0 & 0 & 0 & 0 & 0 & 0 \\ 0 & 0 & 0 & 0 & 0 & 0 \\ 0 & 0 & 0 & 0 & 0 & 0 \\ 0 & 0 & 0 & 0 & 0 & 0 \\ 0 & 0 & 0 & 0 & 0 & 0 \end{bmatrix} ; \text{ and } \mathbf{H}^{(3)} = \begin{bmatrix} 0 & 0 & 0 & 0 & 0 & 0 \\ 0 & 0 & 0 & 0 & 0 & 0 \\ 0 & 0 & 0 & 0 & 0 & 0 \\ 0 & 0 & 0 & 1 & 0 & 0 \\ 0 & 0 & 0 & 0 & 1 & 0 \\ 0 & 0 & 0 & 0 & 0 & 1 \\ 0 & 0 & 0 & 0 & 0 & 0 \\ 0 & 0 & 0 & 0 & 0 & 0 \\ 0 & 0 & 0 & 0 & 0 & 0 \\ 1 & 0 & 0 & 0 & 0 & 0 \\ 0 & 1 & 0 & 0 & 0 & 0 \\ 0 & 0 & 1 & 0 & 0 & 0 \end{bmatrix}
$$

$$(4.83)$$

From the member compatibility Eq. (4.52), the basic member deformations are written in terms of the member-end displacements in the global reference system as:

$$
\mathbf{v}^{(1)} = \mathbf{a}_B^{(1)} \mathbf{U}^{(1)} : \text{Member 1} \qquad (4.84)
$$

$$\mathbf{v}^{(2)} = \mathbf{a}_B^{(2)}\mathbf{U}^{(2)}: \text{Member 2} \qquad\qquad (4.85)$$

$$\mathbf{v}^{(3)} = \mathbf{a}_B^{(3)}\mathbf{U}^{(3)}: \text{Member 3} \qquad\qquad (4.86)$$

It is noted that these compatibility relations express the basic member deformations ($\mathbf{v}^{(1)}$, $\mathbf{v}^{(2)}$, and $\mathbf{v}^{(3)}$) in terms of member variables ($\mathbf{U}^{(1)}$, $\mathbf{U}^{(2)}$, and $\mathbf{U}^{(3)}$).

In order to express the basic member deformations in terms of the structural variable $\mathbf{U}$, one substitutes Eqs. (4.79), (4.80), and (4.81) into (4.84), (4.85), and (4.86), respectively. This leads to the following expressions:

$$\mathbf{v}^{(1)} = \mathbf{a}_B^{(1)}\mathbf{G}^{(1)}\mathbf{U} \qquad\qquad (4.87)$$

$$\mathbf{v}^{(2)} = \mathbf{a}_B^{(2)}\mathbf{G}^{(2)}\mathbf{U} \qquad\qquad (4.88)$$

$$\mathbf{v}^{(3)} = \mathbf{a}_B^{(3)}\mathbf{G}^{(3)}\mathbf{U} \qquad\qquad (4.89)$$

The compatibility relation of the structure is obtained by collecting the basic deformations of all members in a single vector $\mathbf{V}$ as:

$$\begin{Bmatrix} \mathbf{v}^{(1)} \\ \mathbf{v}^{(2)} \\ \mathbf{v}^{(3)} \end{Bmatrix} = \begin{bmatrix} \mathbf{a}_B^{(1)}\mathbf{G}^{(1)} \\ \mathbf{a}_B^{(2)}\mathbf{G}^{(2)} \\ \mathbf{a}_B^{(3)}\mathbf{G}^{(3)} \end{bmatrix} \mathbf{U} \text{ or } \mathbf{V} = \mathbf{AU} \qquad\qquad (4.90)$$

Since the compatibility mapping matrix is the Boolean matrix, there is no need to execute the product of $\mathbf{a}_B\mathbf{G}$ explicitly. Alternatively, the terms in matrix $\mathbf{a}_B$ can be inserted into matrix $\mathbf{A}$ following the column indices. It becomes obvious by multiplying matrix $\mathbf{a}_B^{(1)}$ with the first column of $\mathbf{G}^{(1)}$ matrix. It is noted that the result of this operation is a part of the first column of matrix $\mathbf{A}$. This implies that the fourth column of matrix $\mathbf{a}_B^{(1)}$ will fill into the first column of matrix $\mathbf{A}$ as shown in Figure 4.24. The other columns of matrix $\mathbf{a}_B^{(1)}$ are mapped into matrix $\mathbf{A}$ in a similar fashion; the fifth column of matrix $\mathbf{a}_B^{(1)}$ to the second column of matrix $\mathbf{A}$; the sixth column of matrix $\mathbf{a}_B^{(1)}$ to the third column of matrix $\mathbf{A}$, etc. After applying this mapping process to member 1, the first three rows of matrix $\mathbf{A}$ are obtained. Next, this process is repeated for the rest of the members (Filippou 2002).

It is clear that the aforementioned process of establishing the structural compatibility matrix is well suited to computer implementation.

## EXAMPLE 4.3

For the frame structure of Figure 4.25, formulate the compatibility relation between all displacement degrees of freedom and structural basic deformations.

$$\begin{bmatrix} 0 & 0 & 0 & 1 & 0 & 0 \end{bmatrix}$$

$$\begin{bmatrix} -0.6 & -0.8 & 0 & 0.6 & 0.8 & 0 \\ -0.16 & 0.12 & 1 & 0.16 & -0.12 & 0 \\ -0.16 & 0.12 & 0 & 0.16 & -0.12 & 1 \end{bmatrix} \underbrace{\qquad\qquad\qquad\qquad}_{\mathbf{a}_B^{(1)}} \underbrace{\begin{bmatrix} 0 & 0 & 0 \\ 0 & 0 & 0 \\ 0 & 0 & 0 \\ 1 & 0 & 0 \\ 0 & 1 & 0 \\ 0 & 0 & 1 \end{bmatrix}}_{\mathbf{G}^{(1)}} = \underbrace{\begin{bmatrix} 0.6 \\ 0.16 \\ 0.16 \end{bmatrix}}_{\mathbf{A}}$$

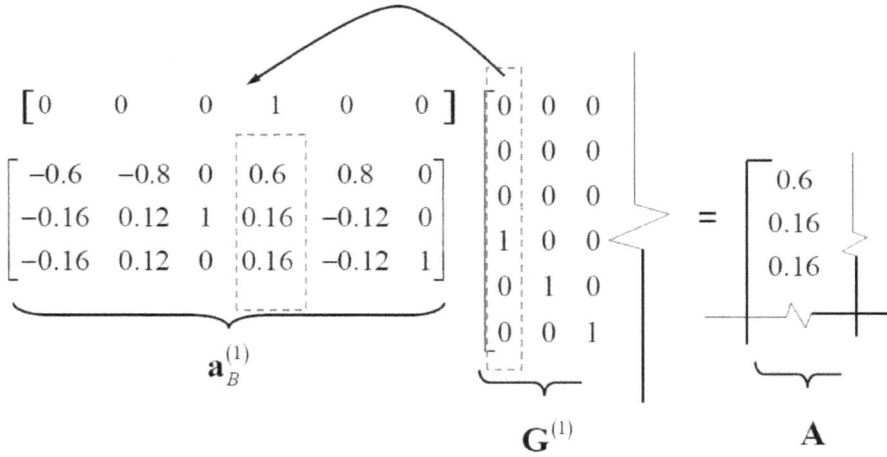

**FIGURE 4.24**   Schematic Presentation of the Product of $\mathbf{a}_B^{(1)}\mathbf{G}^{(1)}$.

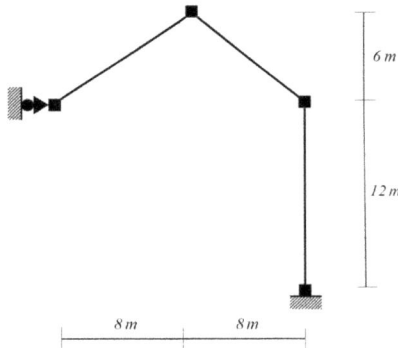

**FIGURE 4.25**   Example 4.3.

**Solution:** Member, node, and DOF numbering systems are shown in Figure 4.26a: From the numbering system for DOF, the mapping vector of each member is:

$$LM_1 = \begin{Bmatrix} 9 \\ 7 \\ 8 \\ 4 \\ 5 \\ 6 \end{Bmatrix}_1 ; \ LM_2 = \begin{Bmatrix} 4 \\ 5 \\ 6 \\ 1 \\ 2 \\ 3 \end{Bmatrix}_2 ; \ \text{and} \ LM_3 = \begin{Bmatrix} 1 \\ 2 \\ 3 \\ 10 \\ 11 \\ 12 \end{Bmatrix}_3$$

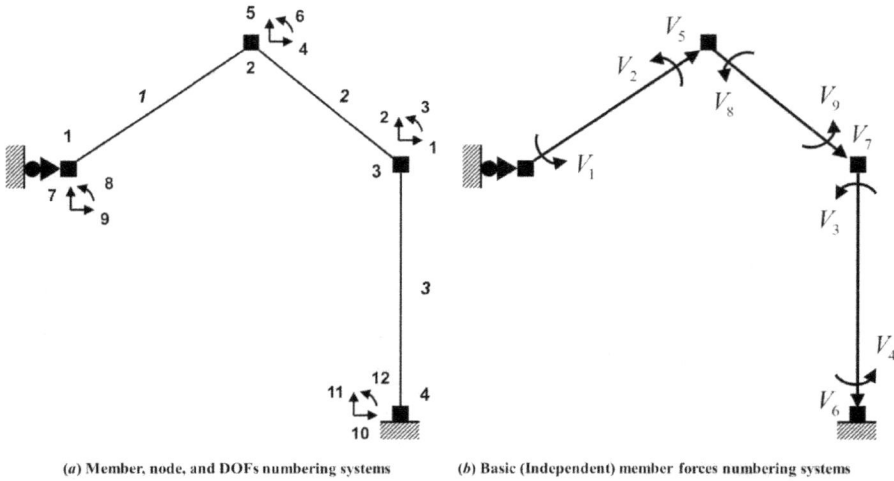

(a) Member, node, and DOFs numbering systems          (b) Basic (Independent) member forces numbering systems

**FIGURE 4.26**   Example 4.3 (Continued).

The basic member deformation numbering system is shown in Figure 4.26b. It is noted that the basic member deformations are numbered in such a way that the inextensible and inflexible constraints can easily be imposed as will be shown in Example 4.6.

The compatibility relations of each member are:

$$
\begin{Bmatrix} V_5 \\ V_1 \\ V_2 \end{Bmatrix} = \begin{bmatrix} -0.8 & -0.6 & 0 & 0.8 & 0.6 & 0 \\ -0.06 & 0.08 & 1 & 0.06 & -0.08 & 0 \\ -0.06 & 0.08 & 0 & 0.06 & -0.08 & 1 \end{bmatrix} \begin{Bmatrix} U_9 \\ U_7 \\ U_8 \\ U_4 \\ U_5 \\ U_6 \end{Bmatrix} ; \quad \begin{Bmatrix} V_7 \\ V_8 \\ V_9 \end{Bmatrix} = \begin{bmatrix} -0.8 & 0.6 & 0 & 0.8 & -0.6 & 0 \\ 0.06 & 0.08 & 1 & -0.06 & -0.08 & 0 \\ 0.06 & 0.08 & 0 & -0.06 & -0.08 & 1 \end{bmatrix} \begin{Bmatrix} U_4 \\ U_5 \\ U_6 \\ U_1 \\ U_2 \\ U_3 \end{Bmatrix} ; \text{ and}
$$

$$
\begin{Bmatrix} V_6 \\ V_3 \\ V_4 \end{Bmatrix} = \begin{bmatrix} 0 & 1 & 0 & 0 & -1 & 0 \\ 0.0833 & 0 & 1 & -0.0833 & 0 & 0 \\ 0.0833 & 0 & 0 & -0.0833 & 0 & 1 \end{bmatrix} \begin{Bmatrix} U_1 \\ U_2 \\ U_3 \\ U_{10} \\ U_{11} \\ U_{12} \end{Bmatrix}.
$$

The demonstration of the automated assembly of the structural compatibility matrix is presented in Figure 4.27.

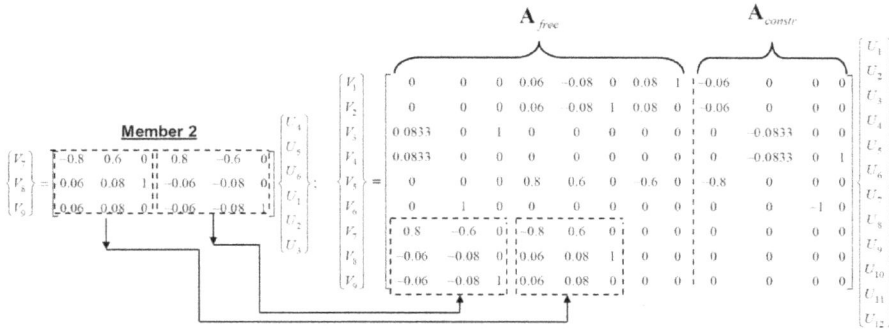

**FIGURE 4.27** Example 4.3 (Continued).

## 4.8 MULTI-FREEDOM CONSTRAINTS: MASTER-SLAVE APPROACH

So far, displacement constraints imposed on a single degree of freedom have been considered. For example,

- The displacement constraints due to support conditions of the frame in Figure 4.12.

$$U_7 = 0; \quad U_8 = 0; \quad U_9 = 0; \quad U_{10} = 0; \quad U_{11} = 0; \text{ and } U_{12} = 0 \qquad (4.91)$$

- The unit displacement imposed on each free degree of freedom to obtain the deformation mode corresponding to each column of the structural compatibility matrix $\mathbf{A}$ (Example 4.1).

$$U_1 = 1.0; \quad \Rightarrow \quad V_1 = 0.6; \ V_2 = 0.16; \ V_3 = 0.16; \ V_4 = -1; \text{ and } V_5 = V_6 = V_7 = V_8 = V_9 = 0 \qquad (4.92)$$

These are known as "single-freedom constraints" and can be imposed with ease. More specifically, the constraints in Eq. (4.91) are homogeneous single-freedom constraints, while that in Eq. (4.92) is a non-homogeneous single-freedom constraint. The next step forward in sophistication is the so-called "multi-freedom constraint". This type of constraint involves more than one independent degree of freedom. Multi-freedom constraints are frequently used in analyses of frame structures to take into account the effects of inclined supports, inextensibility, inflexibility of frame members, etc.

The general form of the multi-freedom constraint can be expressed as:

$$\Psi \text{ (Nodal Displacements)} = \text{Prescribed Value} \qquad (4.93)$$

This general form is called a "single-point" or "single-node" constraint if all nodal displacements in the function argument are at the same node. Otherwise, it is known as a "multi-point" or "multi-node" constraint. The constraint is called homogeneous if, after moving all displacement-dependent terms to the left-hand side, the term "Prescribed Value" on the left-hand side of Eq. (4.93) vanishes. Otherwise, it is known as nonhomogeneous. Furthermore, if the constrained function $\Psi$ of Eq. (4.93) is linear in nodal displacements, the constraint is called linear. Otherwise, it is known as nonlinear.

## Example 4.4

The following are examples of multi-freedom constraints:

1. $v + u \tan \alpha = 0$
2. $U_1 - U_6 = 0.2$
3. $(X_8 - X_3)(U_{16} - U_4) + (Y_8 - Y_3)(U_{17} - U_5) = 0$
4. $U_6 - U_3 = 0$ and $(U_4 - U_1) + 4U_3 = 0$
5. $(X_8 + U_{16} - X_3 - U_4)^2 + (Y_{17} + U_5 - Y_3 - U_5)^2 = 0$

The first one is a single-point, linear, and homogeneous constraint and corresponds to the constraint imposed by the inclined roller support of Figure 1.6a.

The second one is a multi-point, linear, and nonhomogeneous constraint. This constraint implies that the gap between $U_1$ and $U_6$ is 0.2.

The third one is a multi-point, linear, and homogeneous constraint. This constraint corresponds to the inextensibility of the frame member under the small-displacement assumption and will be discussed in the next section.

The fourth ones are multi-point, linear, and homogeneous constraints. These constraints correspond to the inflexibility of the frame member under the small-displacement assumption and will be discussed in the next section.

The last one is a multi-point, nonlinear, and homogeneous constraint. This constraint corresponds to the inextensibility of the frame member under the large-displacement assumption. This kind of constraint appears in geometrically nonlinear analyses of structures.

Generally, there are three approaches for treating the multi-freedom constraints, namely (Felippa 1998):

1. *Master-Slave Elimination*: This approach is the most straightforward one. All displacement degrees of freedom expressed in each constraint equation are categorized as Master and Slave. The slave degrees of freedom are explicitly eliminated by expressing them as the linear function of the master degrees of freedom. Consequently, the modified equation contains only the master degrees of freedom.
2. *Penalty Augmentation or Penalty Function*: In this approach, each multi-freedom constraint is represented in the approximate fashion by a fictitious elastic link device. This fictitious link connects together all displacement degrees of freedom appearing in the constraint equation. The accuracy of the constrained condition is controlled by the numerical weight of the link device. The "exact" constrained condition is gained once the numerical weight approaches infinity, resulting in an infinitely rigid link device.
3. *Lagrange Multiplier Adjunction*: In this approach, each multi-freedom constraint is represented by an additional unknown cast into the stiffness matrix equation. This unknown is called *the Lagrange Multiplier* $\lambda$. The physical interpretation of the Lagrange Multiplier is that it is the constrained force that enforces the constrained condition exactly should they be imposed on the unconstrained system. The rigorous proof of this approach is done by the calculus of variation (Felippa 1998).

The pros and cons of these three approaches to treating multi-freedom constraints can be found elsewhere (Felippa 1998). However, in this textbook, only the master-slave approach is considered due to its simplicity and straightforwardness.

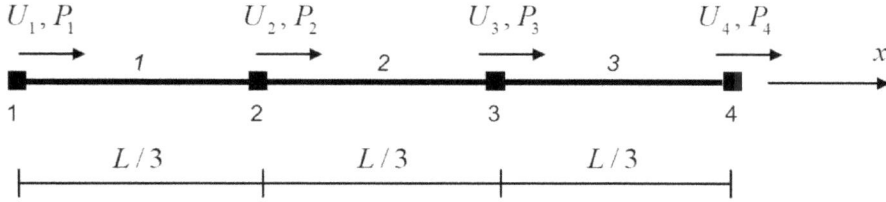

**FIGURE 4.28** Series Arrangement of Truss Members.

### 4.8.1 MASTER-SLAVE APPROACH

In the master-slave approach, one displacement degree of freedom is chosen as *slave* for each multi-freedom constraint. The rest of the displacement degrees of freedom in that constraint are named *master*. In fact, slave and master imply dependency and independency, respectively. Consequently, the new set of nodal displacements $\tilde{\mathbf{U}}$ is obtained by eliminating all slave displacement degrees of freedom from $\mathbf{U}$. A matrix transformation equation relating $\mathbf{U}$ to $\tilde{\mathbf{U}}$ is written as:

$$\mathbf{U} = \mathbf{A}_c \tilde{\mathbf{U}} \tag{4.94}$$

where $\mathbf{A}_c$ is the transformation matrix accounting for the multi-freedom constraints.

To demonstrate the concept of master-slave elimination, the truss structure shown in Figure 4.28 is considered. This truss comprises three members jointed by four nodes that can translate only along the $x$-axis. There are four displacement degrees of freedom for this structure, namely:

$$\mathbf{U} = \lfloor U_1 \quad U_2 \quad U_3 \quad U_4 \rfloor^T \tag{4.95}$$

Their associated nodal forces are:

$$\mathbf{P} = \lfloor P_1 \quad P_2 \quad P_3 \quad P_4 \rfloor^T \tag{4.96}$$

Let this truss subjected to the multi-freedom constraint stating that nodes 2 and 4 translate by the same amount:

$$U_2 - U_4 = 0 \rightarrow U_4 = U_2 \tag{4.97}$$

This constrained condition can be visualized as nodes 2 and 4 being connected by a rigid link since they are forced to move together.

From Eq. (4.97), $U_4$ and $U_2$ are selected as slave and master, respectively. Consequently, the new set of nodal displacements $\tilde{\mathbf{U}}$ is:

$$\tilde{\mathbf{U}} = \lfloor \tilde{U}_1 \quad \tilde{U}_2 \quad \tilde{U}_3 \rfloor^T = \lfloor U_1 \quad U_2 \quad U_3 \rfloor^T \tag{4.98}$$

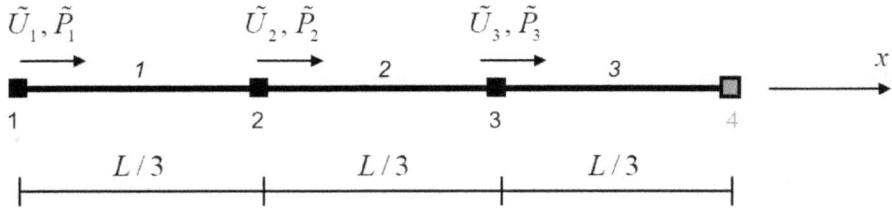

**FIGURE 4.29** Reduced Structure due to the Multi-Freedom Constraint of Eq. (4.97).

A matrix transformation equation relating $\mathbf{U}$ to $\tilde{\mathbf{U}}$ is written as:

$$
\begin{Bmatrix} U_1 \\ U_2 \\ U_3 \\ U_4 \end{Bmatrix} = \begin{bmatrix} 1 & 0 & 0 \\ 0 & 1 & 0 \\ 0 & 0 & 1 \\ 0 & 1 & 0 \end{bmatrix} \begin{Bmatrix} \tilde{U}_1 \\ \tilde{U}_2 \\ \tilde{U}_3 \end{Bmatrix} \text{ or } \mathbf{U} = \mathbf{A}_c \tilde{\mathbf{U}}
\tag{4.99}
$$

Now, the number of displacement degrees of freedom is reduced to 3 as shown in Figure 4.29.

From the contragradient relation between statical and kinematical systems, the matrix transformation equation relating $\tilde{\mathbf{P}}$ to $\mathbf{P}$ is:

$$
\begin{Bmatrix} \tilde{P}_1 \\ \tilde{P}_2 \\ \tilde{P}_3 \end{Bmatrix} = \begin{bmatrix} 1 & 0 & 0 & 0 \\ 0 & 1 & 0 & 1 \\ 0 & 0 & 1 & 0 \end{bmatrix} \begin{Bmatrix} P_1 \\ P_2 \\ P_3 \\ P_4 \end{Bmatrix} \text{ or } \tilde{\mathbf{P}} = \mathbf{A}_c^T \mathbf{P} = \mathbf{B}_c \mathbf{P}
\tag{4.100}
$$

Let the truss of Figure 4.28 subjected to the following multi-freedom constraint:

$$
U_2 = \frac{2}{3} U_1 + \frac{1}{3} U_4
\tag{4.101}
$$

$$
U_3 = \frac{1}{3} U_1 + \frac{2}{3} U_4
\tag{4.102}
$$

These constrained conditions can be visualized as all structural nodes are forced to translate linearly.

From Eq. (4.101), $U_2$ is selected as the slave while $U_1$ and $U_4$ are selected as masters. Similarly, from Eq. (4.102), $U_3$ is selected as the slave while $U_1$ and $U_4$ are selected as masters.

Consequently, the new set of nodal displacements $\tilde{\mathbf{U}}$ is:

$$
\tilde{\mathbf{U}} = \lfloor \tilde{U}_1 \quad \tilde{U}_2 \rfloor^T = \lfloor U_1 \quad U_4 \rfloor^T
\tag{4.103}
$$

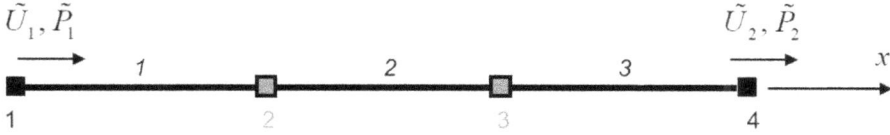

**FIGURE 4.30** Reduced Structure due to the Multi-Freedom Constraints of Eqs. (4.101) and (4.102).

A matrix transformation equation relating $\mathbf{U}$ to $\tilde{\mathbf{U}}$ is written as:

$$\begin{Bmatrix} U_1 \\ U_2 \\ U_3 \\ U_4 \end{Bmatrix} = \begin{bmatrix} 1 & 0 \\ \dfrac{2}{3} & \dfrac{1}{3} \\ \dfrac{1}{3} & \dfrac{2}{3} \\ 0 & 1 \end{bmatrix} \begin{Bmatrix} \tilde{U}_1 \\ \tilde{U}_2 \end{Bmatrix} \text{ or } \mathbf{U} = \mathbf{A}_c \tilde{\mathbf{U}} \tag{4.104}$$

Now, the number of displacement degrees of freedom is reduced to 2 as shown in Figure 4.30.

From the contragradient relations between statical and kinematical systems, the matrix transformation equation relating $\tilde{\mathbf{P}}$ to $\mathbf{P}$ is:

$$\begin{Bmatrix} \tilde{P}_1 \\ \tilde{P}_2 \end{Bmatrix} = \begin{bmatrix} 1 & \dfrac{2}{3} & \dfrac{1}{3} & 0 \\ 0 & \dfrac{1}{3} & \dfrac{2}{3} & 1 \end{bmatrix} \begin{Bmatrix} P_1 \\ P_2 \\ P_3 \\ P_4 \end{Bmatrix} \text{ or } \tilde{\mathbf{P}} = \mathbf{A}_c^T \mathbf{P} = \mathbf{B}_c \mathbf{P} \tag{4.105}$$

## EXAMPLE 4.5

The truss structure of Example 4.2 is subjected to the inclined roller support as shown in Figure 4.31. You are asked to establish the matrix compatibility equation accounting for the constraint by the inclined roller support.

**Solution:** The unconstrained DOFs numbering system is shown in Figure 4.32$a$.

The constraint due to the inclined roller support is:

$$U_4 = \frac{1}{\sqrt{3}} U_3$$

Selecting $U_4$ as the slave and $U_3$ as master, the constrained DOFs numbering system is shown in Figure 4.32$b$.

A matrix transformation equation relating $\mathbf{U}$ (Figure 4.32$a$) to $\tilde{\mathbf{U}}$ (Figure 4.32$b$) is written as:

$$\begin{Bmatrix} U_1 \\ U_2 \\ U_3 \\ U_4 \\ U_5 \\ U_6 \end{Bmatrix} = \begin{bmatrix} 1 & 0 & 0 & 0 & 0 \\ 0 & 1 & 0 & 0 & 0 \\ 0 & 0 & 1 & 0 & 0 \\ 0 & 0 & 1/\sqrt{3} & 0 & 0 \\ 0 & 0 & 0 & 1 & 0 \\ 0 & 0 & 0 & 0 & 1 \end{bmatrix} \begin{Bmatrix} \tilde{U}_1 \\ \tilde{U}_2 \\ \tilde{U}_3 \\ \tilde{U}_4 \\ \tilde{U}_5 \end{Bmatrix} \text{ or } \mathbf{U} = \mathbf{A}_c \tilde{\mathbf{U}} \tag{a}$$

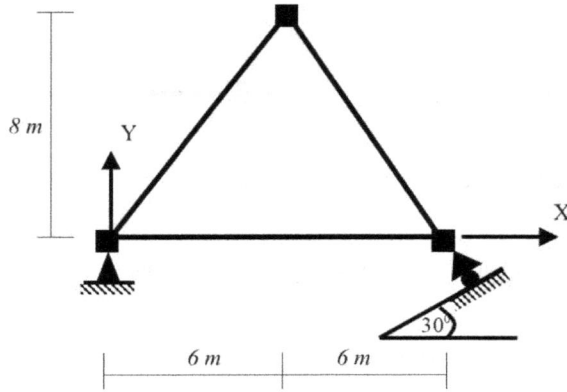

**FIGURE 4.31**   Example 4.5.

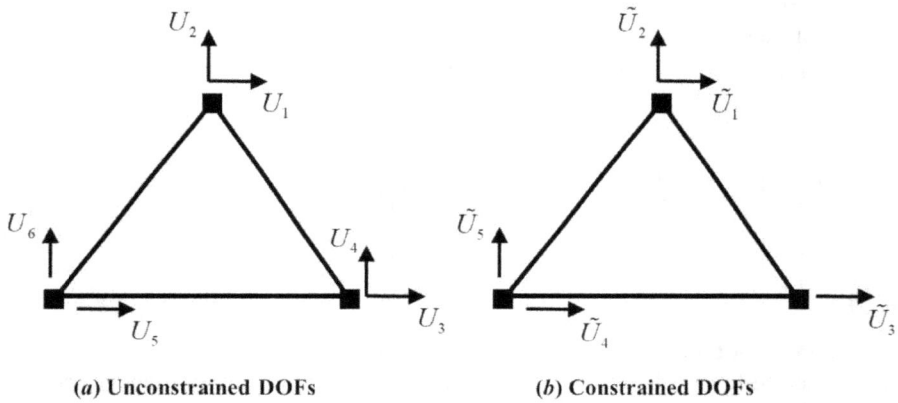

(*a*) Unconstrained  DOFs                    (*b*) Constrained  DOFs

**FIGURE 4.32**   Example 4.5 (Continued).

It is recalled from Example 4.2 that the structural compatibility equation without any constrained condition is:

$$
\begin{Bmatrix} V_1 \\ V_2 \\ V_3 \end{Bmatrix} = \begin{bmatrix} 0.6 & 0.8 & 0 & 0 & -0.6 & -0.8 \\ -0.6 & 0.8 & 0.6 & -0.8 & 0 & 0 \\ 0 & 0 & 1 & 0 & -1 & 0 \end{bmatrix} \begin{Bmatrix} U_1 \\ U_2 \\ U_3 \\ U_4 \\ U_5 \\ U_6 \end{Bmatrix} \text{ or } \mathbf{V} = \mathbf{AU} \qquad (b)
$$

Substitution of Eqs. (*a*) into (*b*) leads to the matrix compatibility equation of the truss structure with inclined roller support:

$$\begin{Bmatrix} V_1 \\ V_2 \\ V_3 \end{Bmatrix} = \begin{bmatrix} 0.6 & 0.8 & 0 & -0.6 & -0.8 \\ -0.6 & 0.8 & 0.1381 & 0 & 0 \\ 0 & 0 & 1 & -1 & 0 \end{bmatrix} \begin{Bmatrix} \tilde{U}_1 \\ \tilde{U}_2 \\ \tilde{U}_3 \\ \tilde{U}_4 \\ \tilde{U}_5 \end{Bmatrix} \text{ or } \mathbf{V} = \mathbf{A}\mathbf{A}_c\tilde{\mathbf{U}} = \tilde{\mathbf{A}}\tilde{\mathbf{U}}$$

The support conditions are:

$$\tilde{U}_4 = 0 \text{ and } \tilde{U}_5 = 0$$

Therefore, the matrix compatibility equation is reduced to:

$$\begin{Bmatrix} V_1 \\ V_2 \\ V_3 \end{Bmatrix} = \begin{bmatrix} 0.6 & 0.8 & 0 \\ -0.6 & 0.8 & 0.1381 \\ 0 & 0 & 1 \end{bmatrix} \begin{Bmatrix} \tilde{U}_1 \\ \tilde{U}_2 \\ \tilde{U}_3 \end{Bmatrix} \text{ or } \mathbf{V}_{free} = \tilde{\mathbf{A}}_{free}\tilde{\mathbf{U}}$$

## 4.9 STRUCTURAL COMPATIBILITY EQUATION: INEXTENSIBLE AND INFLEXIBLE CONSTRAINTS

With the absence of constraints, it is relatively easy to identify the number of independent degrees of freedom of a frame structure. This may not be true with the presence of constraints between independent displacement degrees of freedom. In analyses of frame structures, the common constraints imposed on frame models may result from:

- Inclined support conditions
- Inextensibility of members
- Inflexibility of members
- Rigid floor diaphragms in 3D structural models
- Rigid-end zones

In the previous section, the constrained problem due to inclined support conditions through the master-slave approach was tackled. In this section, the master-slave approach to the multi-freedom constraints corresponding to the inextensibility and inflexibility of structural members is discussed. The constrained conditions due to rigid-end zones are to be considered in the later chapter.

In analysis of typical frame structures, the number of independent degrees of freedom can be reduced by assuming that the axial stiffnesses of frame members are relatively large compared to their flexural stiffnesses. As a result, one can consider frame members as axially rigid, hence ignoring their axial deformations. Generally speaking, each axially rigid member provides one independent constraint between displacement degrees of freedom. In other words, the number of independent structural degrees of freedom is reduced by the number of inextensible members present in the structural model.

To understand the concept of the inextensible constraint, the rectangular frame of Figure 4.33 is considered. There are six free independent displacement degrees of freedom (Figure 4.33a) when axial deformations are included. If all members of the frame are considered as inextensible, the resulting constraints are imposed:

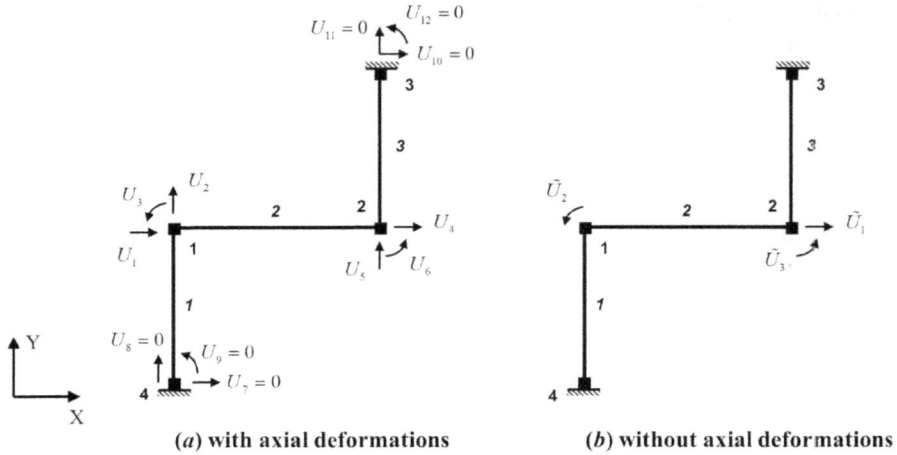

(a) with axial deformations                    (b) without axial deformations

**FIGURE 4.33**  Displacement Degrees of Freedom for a Rectangular Frame: (a) with Axial Deformations; (b) without Axial Deformations.

$$U_2 - \underset{0}{\underline{U_8}} = 0 \rightarrow U_2 = U_8 = 0 \qquad (4.106)$$

$$U_5 - \underset{0}{\underline{U_{11}}} = 0 \rightarrow U_5 = U_{11} = 0 \qquad (4.107)$$

$$U_4 - U_1 = 0 \rightarrow U_1 = U_4 \qquad (4.108)$$

It is noted that there are no master freedoms in Eqs. (4.106) and (4.107). From Eq. (4.108), $U_1$ is selected as the slave while $U_4$ is selected as the master. Consequently, the new set of nodal displacements $\tilde{\mathbf{U}}$ is defined as:

$$\tilde{\mathbf{U}} = \lfloor \tilde{U}_1 \quad \tilde{U}_2 \quad \tilde{U}_3 \rfloor^T = \lfloor U_4 \quad U_3 \quad U_6 \rfloor^T \qquad (4.109)$$

Therefore, the number of displacement degrees of freedom is reduced to $6-3 = 3$ as shown in Figure 4.33b. A matrix transformation equation relating $\mathbf{U}$ to $\tilde{\mathbf{U}}$ is written as:

$$\begin{Bmatrix} U_1 \\ U_2 \\ U_3 \\ U_4 \\ U_5 \\ U_6 \end{Bmatrix} = \begin{bmatrix} 1 & 0 & 0 \\ 0 & 0 & 0 \\ 0 & 1 & 0 \\ 1 & 0 & 0 \\ 0 & 0 & 0 \\ 0 & 0 & 1 \end{bmatrix} \begin{Bmatrix} \tilde{U}_1 \\ \tilde{U}_2 \\ \tilde{U}_3 \end{Bmatrix} \text{ or } \mathbf{U} = \mathbf{A}_c \tilde{\mathbf{U}} \qquad (4.110)$$

From the contragredient relations between statical and kinematical systems, the matrix transformation equation relating the nodal forces $\tilde{\mathbf{P}}$ to the nodal forces $\mathbf{P}$ is:

$$\begin{Bmatrix} \tilde{P}_1 \\ \tilde{P}_2 \\ \tilde{P}_3 \end{Bmatrix} = \begin{bmatrix} 1 & 0 & 0 & 1 & 0 & 0 \\ 0 & 0 & 1 & 0 & 0 & 0 \\ 0 & 0 & 0 & 0 & 0 & 1 \end{bmatrix} \begin{Bmatrix} P_1 \\ P_2 \\ P_3 \\ P_4 \\ P_5 \\ P_6 \end{Bmatrix} \text{ or } \tilde{\mathbf{P}} = \mathbf{B}_c \mathbf{P} \qquad (4.111)$$

The first relation of Eq. (4.111) is explicitly written as:

$$\tilde{P}_1 = P_1 + P_4 \qquad (4.112)$$

Equation (4.112) implies that the equilibrium equations along the X reference axis of nodes 1 and 2 cannot be considered independently. This corresponds to the inextensible constraint imposed on member 2 forcing it to translate as a rigid body along the X reference axis.

It is rather straightforward to impose the inextensible constraints on the frame structure of Figure 4.33. This is based on the fact that all of its constituent members are oriented either along the global X or Y reference axes. For the frame with the presence of inclined members, this may not be true.

The frame structure shown in Figure 4.34 is of interest. With axial deformations, there are 12 independent free degrees of freedom: 8 translations and 4 rotations. With the assumption that all frames are inextensible, there are five independent constraints; each one from each inextensible member. Consequently, without axial deformations, the number of independent free degrees of freedom is reduced to 12–5 = 7. As already mentioned, it is relatively easy to impose the

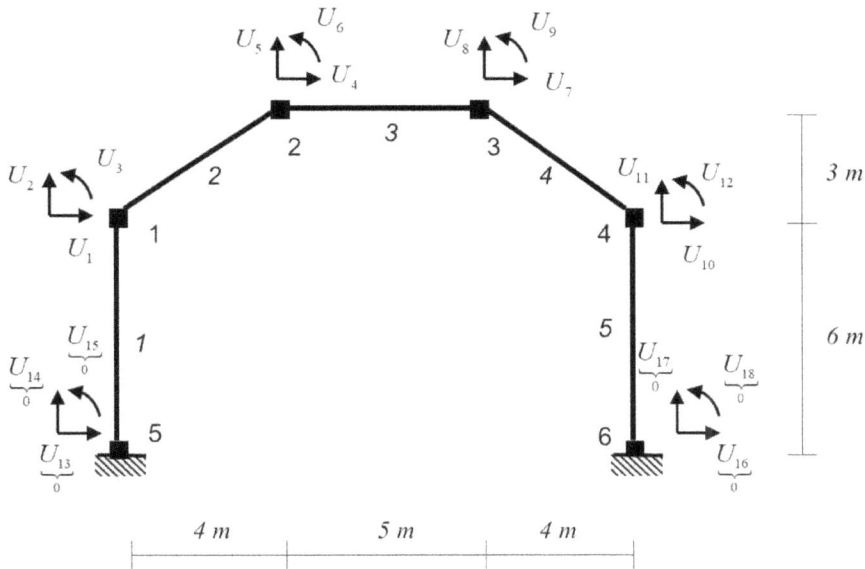

**FIGURE 4.34** Frame with Inclined Members Including Axial Deformations.

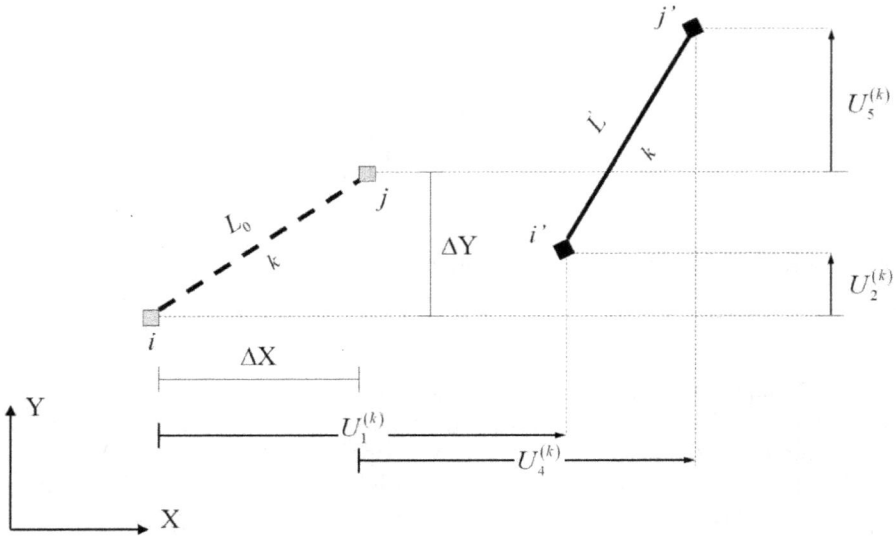

**FIGURE 4.35**  Kinematical Description of a Generic Frame Member $k$.

inextensible constraints on members (members 1, 3, and 5) oriented along the global X or Y reference axes. For inclined members (2 and 4), more effort is required.

To derive the general expression of the inextensible constraint for a frame member (horizontal, vertical, or inclined), member 2 of the frame in Figure 4.34 is taken as a generic frame shown in Figure 4.35. The undeformed $L_0$ and deformed $L'$ lengths of a generic frame $k$ in Figure 4.35 are defined as:

$$L_0 = \sqrt{(\Delta X)^2 + (\Delta Y)^2} \tag{4.113}$$

$$L' = \sqrt{\left(\Delta X + U_4^{(k)} - U_1^{(k)}\right)^2 + \left(\Delta Y + U_5^{(k)} - U_2^{(k)}\right)^2} \tag{4.114}$$

Since the frame is considered as inextensible, its length is supposed to remain constant under the member-end displacements $\mathbf{U}^{(k)}$. Therefore, the inextensible constraint can be defined as:

$$L' = \sqrt{\left(\Delta X + U_4^{(k)} - U_1^{(k)}\right)^2 + \left(\Delta Y + U_5^{(k)} - U_2^{(k)}\right)^2} = L_0 = \sqrt{(\Delta X)^2 + (\Delta Y)^2} \tag{4.115}$$

Equation (4.115) is simplified as:

$$\sqrt{1 + 2\left(\frac{\Delta X}{L_0}\right)\left(\frac{U_4^{(k)} - U_1^{(k)}}{L_0}\right) + \left(\frac{U_4^{(k)} - U_1^{(k)}}{L_0}\right)^2 + 2\left(\frac{\Delta Y}{L_0}\right)\left(\frac{U_5^{(k)} - U_2^{(k)}}{L_0}\right) + \left(\frac{U_5^{(k)} - U_2^{(k)}}{L_0}\right)^2} = 1 \tag{4.116}$$

Equation (4.116) represents the "exact" inextensible constraint for a frame member. This nonlinear multi-freedom constraint is labeled "exact" since the kinematics of the frame is represented without any approximation. Therefore, it can be applied to the problem of large displacements. However, this is beyond the scope of this textbook since only the linearized inextensible constraint for a frame member is of interest. This can be achieved by expanding the expression under the square root of Eq. (4.116) in a power series and neglecting all nonlinear terms, thus leading to:

$$\Delta X \left( U_4^{(k)} - U_1^{(k)} \right) + \Delta Y \left( U_5^{(k)} - U_2^{(k)} \right) = 0 \tag{4.117}$$

Equation (4.117) represents the linearized multi-freedom constraint due to the inextensibility of a frame member. It is noted that the inextensible constraint affects only the translational degrees of freedom.

Now, based on Eq. (4.117), inextensible constraints imposed on the frame structure of Figure 4.34 are expressed as:

Member 1:
$$6 \left( U_2 - \underbrace{U_{14}}_{0} \right) = 0 \quad \Rightarrow \quad U_2 = 0 \tag{4.118}$$

Member 5:
$$6 \left( U_{11} - \underbrace{U_{17}}_{0} \right) = 0 \quad \Rightarrow \quad U_{11} = 0 \tag{4.119}$$

Member 2:
$$4 \left( U_4 - U_1 \right) + 3 \left( U_5 - \underbrace{U_2}_{0} \right) = 0 \quad \Rightarrow \quad U_5 = \frac{4}{3} U_1 - \frac{4}{3} U_4 \tag{4.120}$$

Member 3:
$$5 \left( U_7 - U_4 \right) + (0) \left( U_8 - U_5 \right) = 0 \quad \Rightarrow \quad U_7 = U_4 \tag{4.121}$$

Member 4:
$$4 \left( U_{10} - U_7 \right) + (-3) \left( \underbrace{U_{11}}_{0} - U_8 \right) = 0 \quad \Rightarrow \quad U_8 = \frac{4}{3} U_7 - \frac{4}{3} U_{10} \tag{4.122}$$
$$\Rightarrow \quad U_8 = \frac{4}{3} U_4 - \frac{4}{3} U_{10}$$

It is noted that there are no master freedoms in Eqs. (4.118) and (4.119). From Eqs. (4.120), (4.121), and (4.122), $U_1$, $U_4$, and $U_{10}$ are selected as masters while $U_5$, $U_7$, and $U_8$ are selected as slaves. Consequently, the new set of independent free nodal displacements $\tilde{\mathbf{U}}$ is defined as:

$$\tilde{\mathbf{U}} = \lfloor \tilde{U}_1 \quad \tilde{U}_2 \quad \tilde{U}_3 \quad \tilde{U}_4 \quad \tilde{U}_5 \quad \tilde{U}_6 \quad \tilde{U}_7 \rfloor^T = \lfloor U_1 \quad U_4 \quad U_{10} \quad U_3 \quad U_6 \quad U_9 \quad U_{12} \rfloor^T \tag{4.123}$$

and is shown in Figure 4.36.

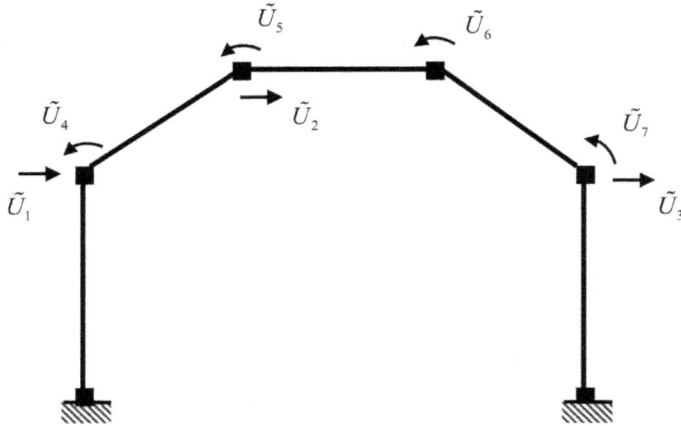

**FIGURE 4.36** Frame with Inclined Members Excluding Axial Deformations.

A matrix transformation equation relating $\mathbf{U}$ to $\tilde{\mathbf{U}}$ is written as:

$$
\begin{Bmatrix} U_1 \\ U_2 \\ U_3 \\ U_4 \\ U_5 \\ U_6 \\ U_7 \\ U_8 \\ U_9 \\ U_{10} \\ U_{11} \\ U_{12} \end{Bmatrix} = \begin{bmatrix} 1 & 0 & 0 & 0 & 0 & 0 & 0 \\ 0 & 0 & 0 & 0 & 0 & 0 & 0 \\ 0 & 0 & 0 & 1 & 0 & 0 & 0 \\ 0 & 1 & 0 & 0 & 0 & 0 & 0 \\ \dfrac{4}{3} & -\dfrac{4}{3} & 0 & 0 & 0 & 0 & 0 \\ 0 & 0 & 0 & 0 & 1 & 0 & 0 \\ 0 & 1 & 0 & 0 & 0 & 0 & 0 \\ 0 & \dfrac{4}{3} & -\dfrac{4}{3} & 0 & 0 & 0 & 0 \\ 0 & 0 & 0 & 0 & 0 & 1 & 0 \\ 0 & 0 & 1 & 0 & 0 & 0 & 0 \\ 0 & 0 & 0 & 0 & 0 & 0 & 0 \\ 0 & 0 & 0 & 0 & 0 & 0 & 1 \end{bmatrix} \begin{Bmatrix} \tilde{U}_1 \\ \tilde{U}_2 \\ \tilde{U}_3 \\ \tilde{U}_4 \\ \tilde{U}_5 \\ \tilde{U}_6 \\ \tilde{U}_7 \end{Bmatrix} \quad \text{or } \mathbf{U} = \mathbf{A}_c\tilde{\mathbf{U}} \qquad (4.124)
$$

It is noted that each column of the constraint matrix $\mathbf{A}_c$ represents the displaced shape of the frame corresponding to the imposed unit master displacement freedoms

For example, the displaced shape of the frame associated with the imposed unit master displacement freedom $\tilde{U}_3 = 1.0$ is shown in Figure 4.37.

Furthermore, it is clear from the three linear constraints of Eqs. (4.120), (4.121), and (4.122) that the choices of master freedoms are not unique. Alternative choices of master freedoms exist and are shown in Figure 4.38.

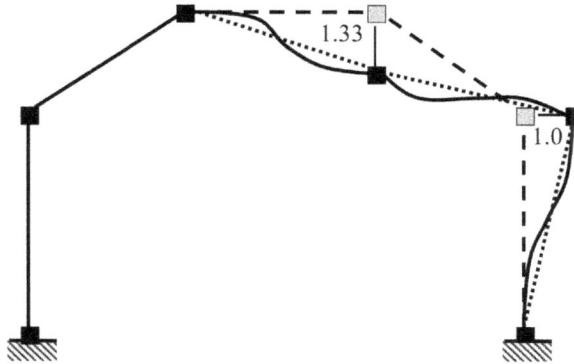

**FIGURE 4.37** Displaced Shapes of the Frame due to Imposed Unit Master Displacement Freedom $\tilde{U}_3 = 1.0$.

Another interesting constrained condition is due to the presence of rigid frame members in structural models. The frame members regarded as rigid are the ones possessing very high axial (*EA*) and flexural (*IE*) rigidities relative to those of other members in the model. These constraints involve both translational and rotational degrees of freedom and can be derived by considering the kinematical description of the rigid frame member shown in Figure 4.39.

The constraints due to the rigid frame are:

$$U_4^{(k)} - U_1^{(k)} = -\left(L_k \cos\phi - L_k \cos\left(\phi + U_3^{(k)}\right)\right)$$
$$U_5^{(k)} - U_2^{(k)} = \left(-L_k \sin\phi + L_k \sin\left(\phi + U_3^{(k)}\right)\right) \quad (4.125)$$
$$U_6^{(k)} = U_3^{(k)}$$

Equation (4.125) represents the "exact" constraint for a rigid frame member. This nonlinear multi-freedom constraint is labeled "exact" since the kinematics of the frame is represented without any approximation. Therefore, it can be applied to the problem of large displacements. However, this is beyond the scope of this textbook since only the linearized constraint for a rigid frame member is of interest. This could be achieved by expanding all trigonometric functions in Eq. (4.125) and using the following linearization:

$$\cos U_3^{(k)} \approx 1 \text{ and } \sin U_3^{(k)} \approx U_3^{(k)} \quad (4.126)$$

Substituting Eqs. (4.126) in (4.125), one obtains:

$$U_4^{(k)} - U_1^{(k)} = -\left(L_k \sin\phi\right)U_3^{(k)}; \quad U_5^{(k)} - U_2^{(k)} = \left(L_k \cos\phi\right)U_3^{(k)}; \text{ and } U_6^{(k)} = U_3^{(k)} \quad (4.127)$$

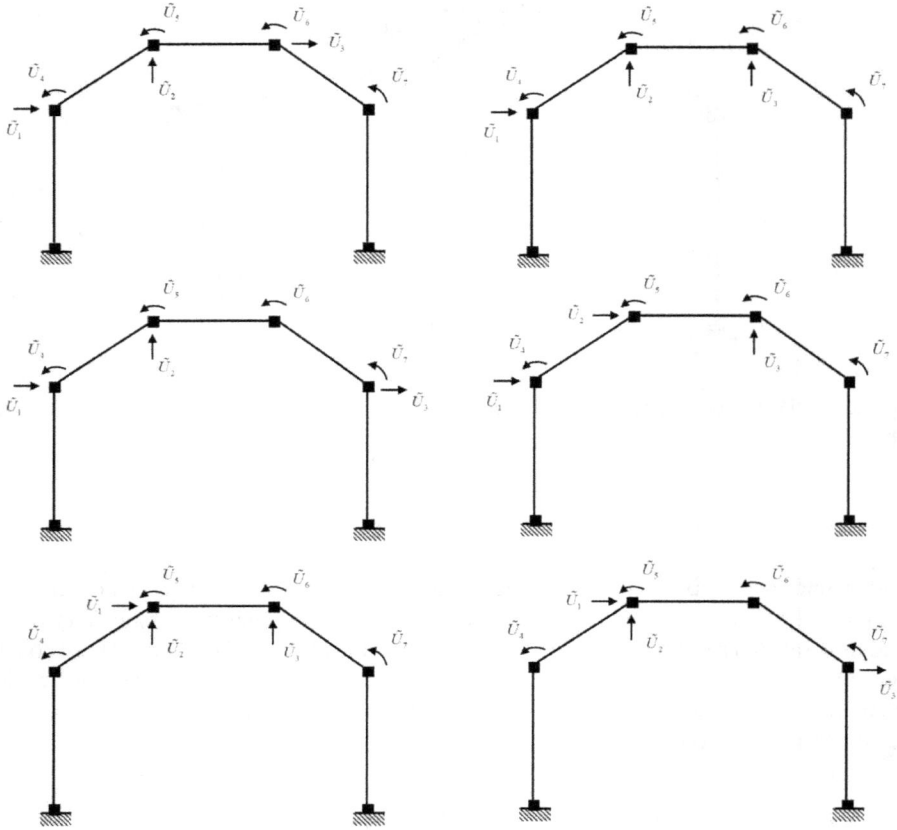

**FIGURE 4.38**    Alternative Choices of Master Freedoms.

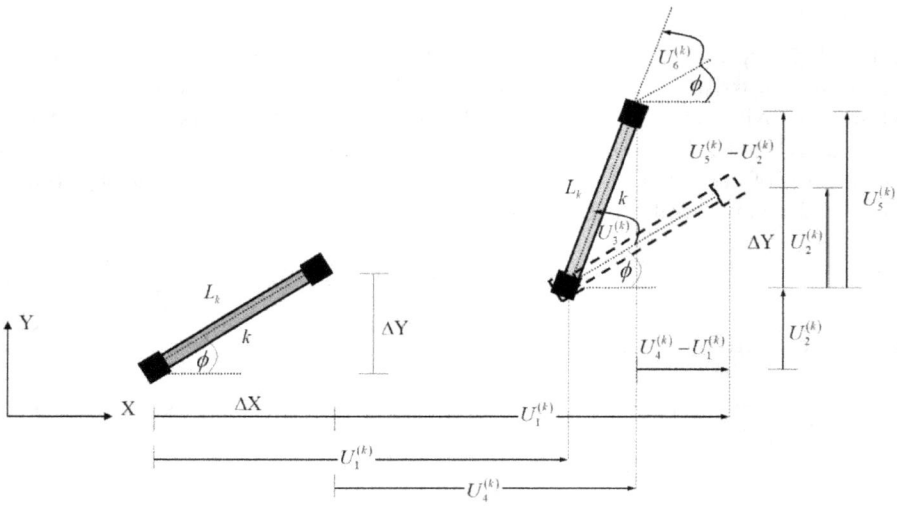

**FIGURE 4.39**    Kinematical Description of the Rigid Frame Member.

These are three linear multi-freedom constraints imposed on displacement freedoms of a rigid frame member. Generally speaking, each rigid frame member provides three independent constraints between six displacement degrees of freedom. Consequently, each rigid frame member present in the structural model reduces the degree-of-freedom number of a structure by three. It is interesting to point out that the inextensible constraint of Eq. (4.117) can be derived by combining the first and second constraints of Eq. (4.127) as:

$$\frac{\left(U_4^{(k)}-U_1^{(k)}\right)}{\left(U_5^{(k)}-U_2^{(k)}\right)}=-\frac{\Delta Y}{\Delta X}\Rightarrow\Delta X\left(U_4^{(k)}-U_1^{(k)}\right)+\Delta Y\left(U_5^{(k)}-U_2^{(k)}\right)=0 \qquad (4.128)$$

where

$$\frac{\Delta X}{L_k}=\cos\phi\,\text{and}\,\frac{\Delta Y}{L_k}=\sin\phi \qquad (4.129)$$

A single-story frame with inclined rigid roof shown in Figure 4.40 is considered. Without multi-freedom constraints, there are six degrees of freedom shown in Figure 4.41a. The constraint conditions due to the inclined rigid roof are:

$$U_4=U_1-4U_3;\ U_5=U_2+8U_3;\text{and}\,U_6=U_3 \qquad (4.130)$$

$U_1, U_2$, and $U_3$ are selected as masters while $U_4, U_5$, and $U_6$ are selected as slaves. Consequently, the new set of independent free nodal displacements $\tilde{\mathbf{U}}$ is defined as:

$$\tilde{\mathbf{U}}=\lfloor \tilde{U}_1\ \ \tilde{U}_2\ \ \tilde{U}_3\rfloor^T=\lfloor U_1\ \ U_2\ \ U_3\rfloor^T \qquad (4.131)$$

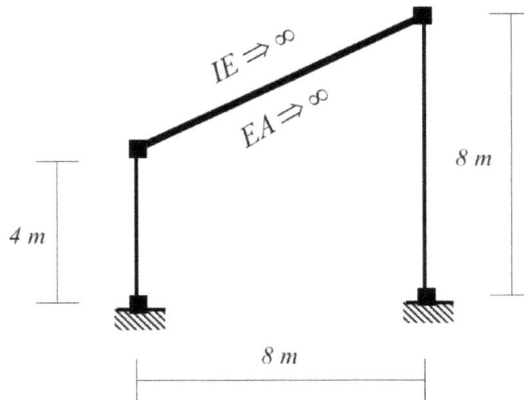

**FIGURE 4.40** A Single-Story Frame with Inclined Rigid Roof.

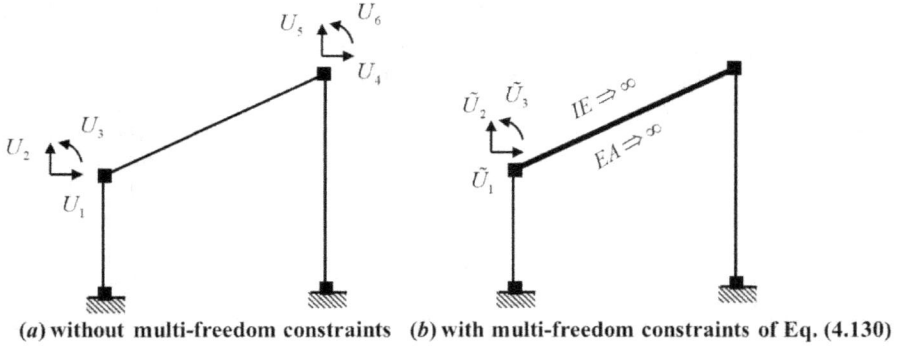

(a) without multi-freedom constraints   (b) with multi-freedom constraints of Eq. (4.130)

FIGURE 4.41   Degree-of-Freedom Systems: with and without Multi-Freedom Constraints.

and is shown in Figure 4.41b.

A matrix transformation equation relating $\mathbf{U}$ to $\tilde{\mathbf{U}}$ is written as:

$$\begin{Bmatrix} U_1 \\ U_2 \\ U_3 \\ U_4 \\ U_5 \\ U_6 \end{Bmatrix} = \begin{bmatrix} 1 & 0 & 0 \\ 0 & 1 & 0 \\ 0 & 0 & 1 \\ 1 & 0 & -4 \\ 0 & 1 & 8 \\ 0 & 0 & 1 \end{bmatrix} \begin{Bmatrix} \tilde{U}_1 \\ \tilde{U}_2 \\ \tilde{U}_3 \end{Bmatrix} \text{ or } \mathbf{U} = \mathbf{A}_c \tilde{\mathbf{U}} \tag{4.132}$$

From the contragradient relation between statical and kinematical systems, the matrix transformation equation relating the nodal forces $\tilde{\mathbf{P}}$ to the nodal forces $\mathbf{P}$ is:

$$\begin{Bmatrix} \tilde{P}_1 \\ \tilde{P}_2 \\ \tilde{P}_3 \end{Bmatrix} = \begin{bmatrix} 1 & 0 & 0 & 1 & 0 & 0 \\ 0 & 1 & 0 & 0 & 1 & 0 \\ 0 & 0 & 1 & -4 & 8 & 1 \end{bmatrix} \begin{Bmatrix} P_1 \\ P_2 \\ P_3 \\ P_4 \\ P_5 \\ P_6 \end{Bmatrix} \text{ or } \tilde{\mathbf{P}} = \mathbf{A}_c^T \mathbf{P} = \mathbf{B}_c \mathbf{P} \tag{4.133}$$

Explicitly,

$$\tilde{P}_1 = P_1 + P_4; \quad \tilde{P}_2 = P_2 + P_5; \quad \text{and } \tilde{P}_3 = P_3 - 4P_4 + 8P_5 + P_6 \tag{4.134}$$

It is interesting to observe from Eq. (4.134) that the nodal forces $\tilde{\mathbf{P}}$ are simply the equivalent forces of the nodal forces $\mathbf{P}$ subjected to the statical constraints. Consequently, matrix $\mathbf{B}_c$ is known as "the statical constraint matrix".

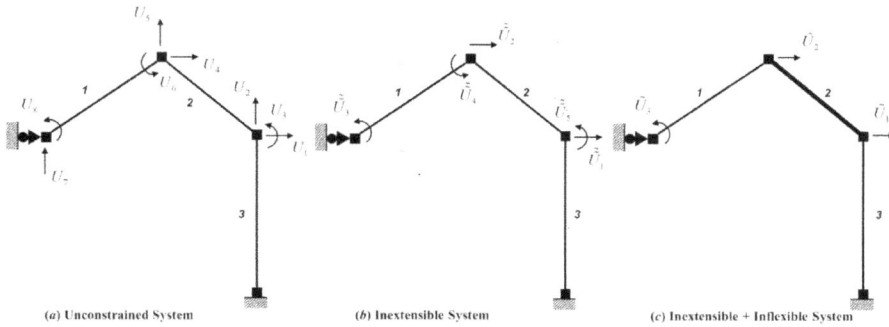

**FIGURE 4.42** Example 4.6.

## EXAMPLE 4.6

For the frame structure of Example 4.3 shown in Figure 4.42a, you are asked to establish the geometric compatibility relation between basic member deformations $V_{free}$ and independent free displacement degrees of freedom under the following condition.

(a) All members are inextensible (Figure 4.42b).

(b) Members 1 and 3 are inextensible while member 2 is considered as a rigid member (Figure 4.42c).

**Solution:** It is recalled from Example 4.3 that the matrix compatibility equation associated with free displacement degrees of freedom is:

$$
\mathbf{V}_{free} = \mathbf{A}_{free}\mathbf{U}_{free} \text{ or } \begin{Bmatrix} V_1 \\ V_2 \\ V_3 \\ V_4 \\ V_5 \\ V_6 \\ V_7 \\ V_8 \\ V_9 \end{Bmatrix} = \begin{bmatrix} 0 & 0 & 0 & 0.06 & -0.08 & 0 & 0.08 & 1 \\ 0 & 0 & 0 & 0.06 & -0.08 & 1 & 0.08 & 0 \\ 0.0833 & 0 & 1 & 0 & 0 & 0 & 0 & 0 \\ 0.0833 & 0 & 0 & 0 & 0 & 0 & 0 & 0 \\ 0 & 0 & 0 & 0.8 & 0.6 & 0 & -0.6 & 0 \\ 0 & 1 & 0 & 0 & 0 & 0 & 0 & 0 \\ 0.8 & -0.6 & 0 & -0.8 & 0.6 & 0 & 0 & 0 \\ -0.06 & -0.08 & 0 & 0.06 & 0.08 & 1 & 0 & 0 \\ -0.06 & -0.08 & 1 & 0.06 & 0.08 & 0 & 0 & 0 \end{bmatrix} \begin{Bmatrix} U_1 \\ U_2 \\ U_3 \\ U_4 \\ U_5 \\ U_6 \\ U_7 \\ U_8 \end{Bmatrix} \quad (a)
$$

**Condition (a): All members are inextensible**

The inextensible constraint for each member is considered as shown in Figure 4.43:

Following inextensible constraint conditions of members 1, 2, and 3, $U_1$, $U_4$, $U_8$, $U_6$, and $U_3$ are selected as master freedoms:

$$U_1 \Rightarrow \tilde{U}_1; U_4 \Rightarrow \tilde{U}_2; U_8 \Rightarrow \tilde{U}_3; U_6 \Rightarrow \tilde{U}_4; \text{ and } U_3 \Rightarrow \tilde{U}_5$$

and $U_2$, $U_5$, and $U_7$ as slave freedoms. The resulting constrained DOF numbering system is shown in Figure 4.42b.

(a) Member 1

(b) Member 3

(c) Member 2

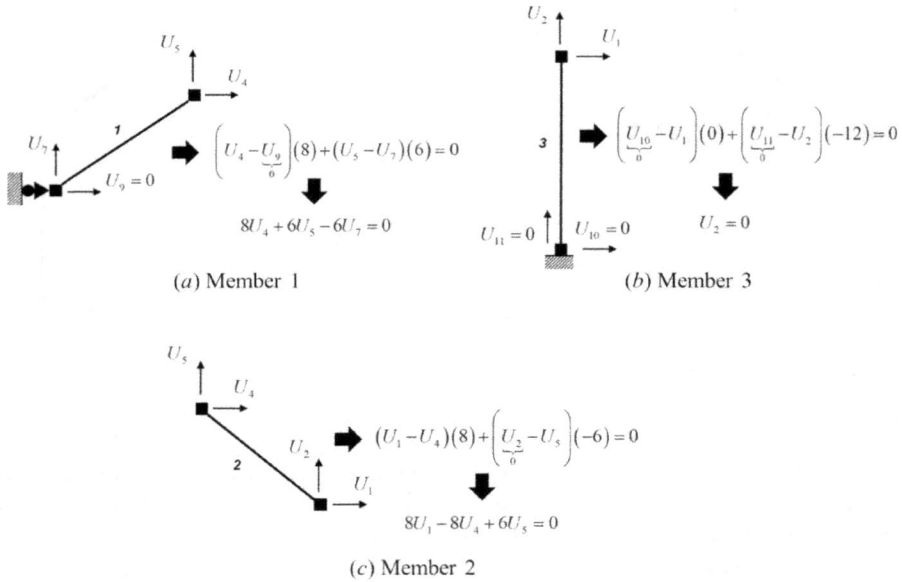

**FIGURE 4.43** Example 4.6 (Continued).

The kinematical relations corresponding to the inextensible constraints are written in matrix form as:

$$
\mathbf{U}_{free} = \mathbf{A}_c \tilde{\tilde{\mathbf{U}}} \text{ or }
\begin{Bmatrix} U_1 \\ U_2 \\ U_3 \\ U_4 \\ U_5 \\ U_6 \\ U_7 \\ U_8 \end{Bmatrix} =
\begin{bmatrix}
1 & 0 & 0 & 0 & 0 \\
0 & 0 & 0 & 0 & 0 \\
0 & 0 & 0 & 0 & 1 \\
0 & 1 & 0 & 0 & 0 \\
-4/3 & 4/3 & 0 & 0 & 0 \\
0 & 0 & 0 & 1 & 0 \\
-4/3 & 8/3 & 0 & 0 & 0 \\
0 & 0 & 1 & 0 & 0
\end{bmatrix}
\begin{Bmatrix} \tilde{\tilde{U}}_1 \\ \tilde{\tilde{U}}_2 \\ \tilde{\tilde{U}}_3 \\ \tilde{\tilde{U}}_4 \\ \tilde{\tilde{U}}_5 \end{Bmatrix}
\tag{b}
$$

Substituting the kinematical relations of Eq. (b) into the compatibility relations of Eq. (a) leads to:

$$
\mathbf{V}_{free} = \mathbf{A}_{free} \mathbf{A}_c \tilde{\tilde{\mathbf{U}}} = \tilde{\mathbf{A}}_{free} \tilde{\tilde{\mathbf{U}}} \text{ or }
\begin{Bmatrix} V_1 \\ V_2 \\ V_3 \\ V_4 \\ V_5 \\ V_6 \\ V_7 \\ V_8 \\ V_9 \end{Bmatrix} =
\begin{bmatrix}
0 & 0.167 & 1 & 0 & 0 \\
0 & 0.167 & 0 & 1 & 0 \\
0.0833 & 0 & 0 & 0 & 1 \\
0.0833 & 0 & 0 & 0 & 0 \\
0 & 0 & 0 & 0 & 0 \\
0 & 0 & 0 & 0 & 0 \\
0 & 0 & 0 & 0 & 0 \\
-0.167 & 0.167 & 0 & 1 & 0 \\
-0.167 & 0.167 & 0 & 0 & 1
\end{bmatrix}
\begin{Bmatrix} \tilde{\tilde{U}}_1 \\ \tilde{\tilde{U}}_2 \\ \tilde{\tilde{U}}_3 \\ \tilde{\tilde{U}}_4 \\ \tilde{\tilde{U}}_5 \end{Bmatrix}
\tag{c}
$$

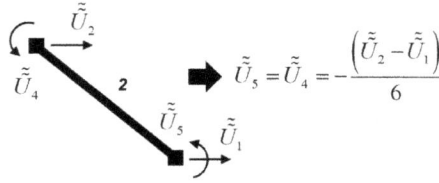

**FIGURE 4.44**   Example 4.6 (Continued).

Clearly, the basic member deformations $V_5$, $V_6$, and $V_7$ are zeros due to the inextensible constraints.

   **Condition (b): Members 1 and 3 are inextensible while member 2 is considered as a rigid member**

Inflexibility of member 2 is considered as shown in Figure 4.44.

$\tilde{U}_1$, $\tilde{U}_2$, and $\tilde{U}_3$ are selected as master freedoms:

$$\tilde{\tilde{U}}_1 \Rightarrow \tilde{U}_1; \; \tilde{\tilde{U}}_2 \Rightarrow \tilde{U}_2; \text{ and } \tilde{\tilde{U}}_3 \Rightarrow \tilde{U}_3$$

and $\tilde{\tilde{U}}_4$ and $\tilde{\tilde{U}}_5$ are selected as slave freedoms. The resulting constrained DOF numbering system is shown in Figure 4.42c.

   Consequently, the resulting kinematical relations are:

$$\tilde{\tilde{\mathbf{U}}} = \tilde{\mathbf{A}}_c \tilde{\mathbf{U}} \quad \text{or} \quad \begin{Bmatrix} \tilde{\tilde{U}}_1 \\ \tilde{\tilde{U}}_2 \\ \tilde{\tilde{U}}_3 \\ \tilde{\tilde{U}}_4 \\ \tilde{\tilde{U}}_5 \end{Bmatrix} = \begin{bmatrix} 1 & 0 & 0 \\ 0 & 1 & 0 \\ 0 & 0 & 1 \\ 1/6 & -1/6 & 0 \\ 1/6 & -1/6 & 0 \end{bmatrix} \begin{Bmatrix} \tilde{U}_1 \\ \tilde{U}_2 \\ \tilde{U}_3 \end{Bmatrix} \tag{d}$$

Consequently, the kinematical relations corresponding to the inextensible (members 1 and 2) and inflexible (member 3) constraints are written in matrix form as:

$$\mathbf{U}_{free} = \tilde{\mathbf{A}}_c \tilde{\tilde{\mathbf{A}}}_c \tilde{\mathbf{U}} = \mathbf{A}_c \tilde{\mathbf{U}} \quad \text{or} \quad \begin{Bmatrix} U_1 \\ U_2 \\ U_3 \\ U_4 \\ U_5 \\ U_6 \\ U_7 \\ U_8 \end{Bmatrix} = \begin{bmatrix} 1 & 0 & 0 \\ 0 & 0 & 0 \\ 1/6 & -1/6 & 0 \\ 0 & 1 & 0 \\ -4/3 & 4/3 & 0 \\ 1/6 & -1/6 & 0 \\ -4/3 & 8/3 & 0 \\ 0 & 0 & 1 \end{bmatrix} \begin{Bmatrix} \tilde{U}_1 \\ \tilde{U}_2 \\ \tilde{U}_3 \end{Bmatrix} \tag{e}$$

Substituting the kinematical relation of Eq. (*e*) into the compatibility relation of Eq. (*a*) leads to:

$$
\mathbf{V}_{free} = \mathbf{A}_{free}\mathbf{A}_c\tilde{\mathbf{U}} = \tilde{\mathbf{A}}_{free}\tilde{\mathbf{U}} \quad \text{or} \quad
\begin{Bmatrix} V_1 \\ V_2 \\ V_3 \\ V_4 \\ V_5 \\ V_6 \\ V_7 \\ V_8 \\ V_9 \end{Bmatrix} =
\begin{bmatrix}
0 & 0.167 & 1 \\
0.167 & 0 & 0 \\
0.25 & -0.167 & 0 \\
0.0833 & 0 & 0 \\
0 & 0 & 0 \\
0 & 0 & 0 \\
0 & 0 & 0 \\
0 & 0 & 0 \\
0 & 0 & 0
\end{bmatrix}
\begin{Bmatrix} \tilde{U}_1 \\ \tilde{U}_2 \\ \tilde{U}_3 \end{Bmatrix}
\tag{f}
$$

It is observed that the basic member deformations associated with axial deformations of all members and flexural deformation of member 2 vanish. This corresponds to the inextensible and inflexible constraints imposed on this frame.

## 4.10   EXERCISES

**Problem 4.1:** Consider the cantilever beam system shown in Figure 4.45.

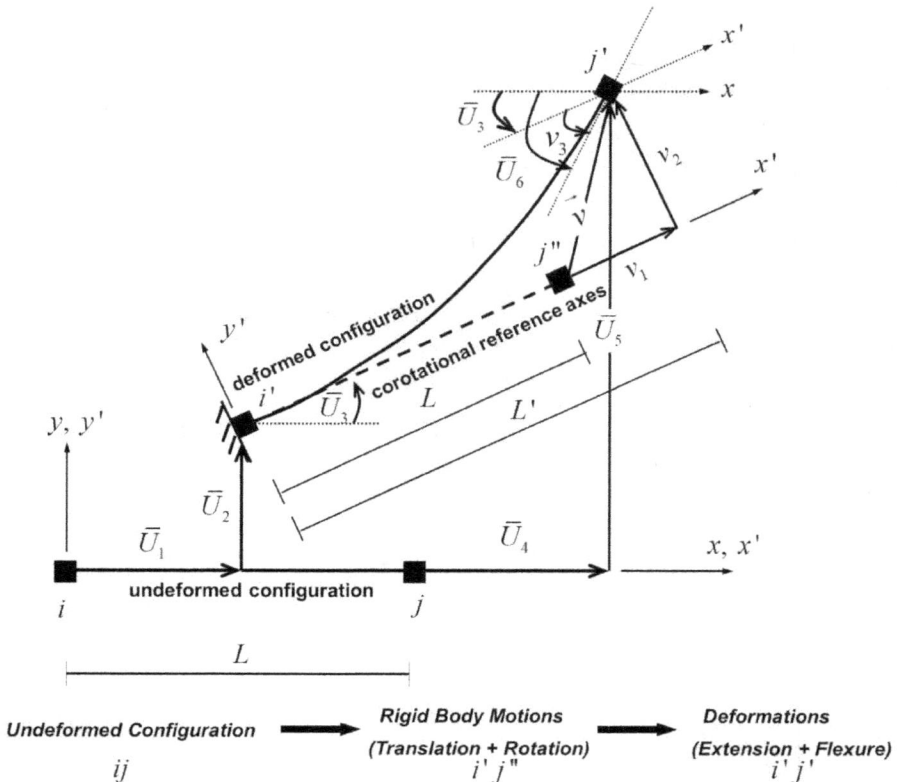

**FIGURE 4.45**   Problem 4.1.

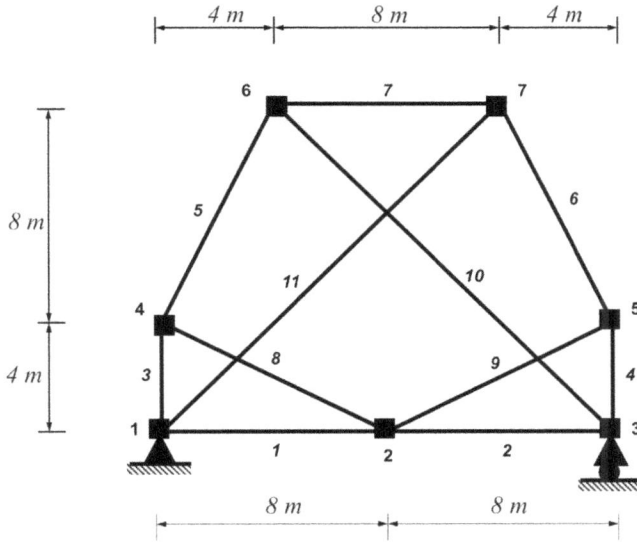

**FIGURE 4.46**  Problem 4.2.

Following the same procedure discussed for the simply supported beam system, derive the geometric compatibility relations associated with exact, second-order, and linear theories of structural analyses.

**Problem 4.2:** Consider the complex truss shown in Figure 4.46.

(a) Derive the linear geometric compatibility matrix $\mathbf{A}$ and $\mathbf{A}_{free}$ of the structure.
(b) Prove that the structural equilibrium matrices ($\mathbf{B}$ and $\mathbf{B}_{free}$) are the transpose of the structural compatibility matrices ($\mathbf{A}$ and $\mathbf{A}_{free}$).
(c) Determine the nodal displacements of the truss when members 7, 10, and 11 are heated up to 120°C. Given that the coefficient of thermal expansion of all members is $7.5 \times 10^{-6} / C^0$.

**Problem 4.3:** Consider the queen post bridge structure shown in Figure 4.47.

(a) Number the degrees of freedom (both constrained and unconstrained) and the basic member deformations based on the simply supported beam model.

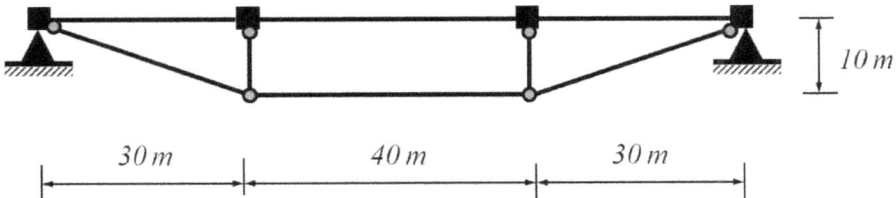

**FIGURE 4.47**  Problem 4.3.

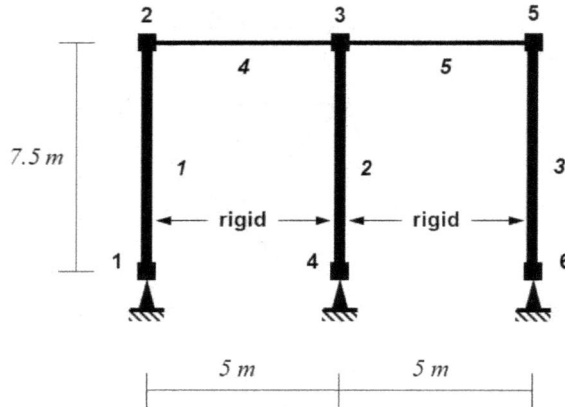

**FIGURE 4.48**   Problem 4.4.

(b) Formulate the compatibility equations at unconstrained degrees of freedom
($\mathbf{V}_{free} = \mathbf{A}_{free}\mathbf{U}_{free}$).

**Problem 4.4:** Consider the frame structure shown in Figure 4.48.

(a) Number the degrees of freedom (both constrained and unconstrained) and the basic
member deformations based on the simply supported beam model.
(b) Formulate the compatibility equations at free degrees of freedom ($\mathbf{V}_{free} = \mathbf{A}_{free}\mathbf{U}_{free}$). It
is noted that all members can be considered as inextensible. Furthermore, the flexural
stiffness of members 1, 2, and 3 is so large that they can be treated as infinitely rigid in
flexure.

## REFERENCES

de Borst, R., M.A. Crisfield, J.J.C. Remmers, and C.V. Verhoosel. 2012. *Nonlinear finite element analysis
of solids and structures.* John Wiley & Sons Inc.
Dawe, D.J. 1984. *Matrix and finite element displacement analysis of structures.* Clarendon Press.
Elias, Z.M. 1986. *Theory and methods of structural analysis.* John Wiley and Son Inc.
Felippa, C.A. 1998. *ASEN 5007 course notes: Introduction to finite element methods.* Department of
Aerospace Engineering, University of Colorado.
Filippou, F.C. 2002. *CE 220 course notes: Structural analysis: Theory and applications.* Department of
Civil and Environmental Engineering, University of California.
Kanchi, M.B. 1993. *Matrix methods of structural analysis* (2nd Edition). Wiley Eastern Limited.
Kreyszig, E. 2011. *Advanced engineering mathematics* (10th Edition). John Wiley & Sons Inc.
Limkatanyu, S., W. Prachasaree, G. Kaewkulchai, and E. Spacone. 2014. Unification of mixed Euler-
Bernoulli-von Karman planar frame model and corotational approach. *Mechanics Based Design of
Structures and Machines: An International Journal* 42(4): 419–441.
Livesley, R.K. 1974. *Matrix methods of structural analysis.* Pergamon Press.
McGuire, W., R.H. Gallagher, and R.D. Ziemian. 2000. *Matrix structural analysis* (2nd Edition). John
Wiley & Sons Inc.

Przemieniecki, J.S. 1985. *Theory of matrix structural analysis.* Dover Publications Inc.

Rajasekaran, S. and G. Sankarasubramanian. 2006. *Computational structural mechanics.* Prentice-Hall of India Private Limited.

Whittaker, A. 2001. *CIE 423 course notes: Structures III.* Department of Civil, Structural, and Environmental Engineering, State University of New York.

# 5 Application of Virtual Work Principles

## 5.1 INTRODUCTION

As mentioned in Chapter 1, there are three basic requirements that need to be fulfilled in analyzing any structure, namely, equilibrium, compatibility, and constitutive relations. In Chapter 3, equilibrium equations were discussed exclusively to ensure that the structure and each of its constituent members are in equilibrium and that the force (natural) boundaries are also satisfied. In Chapter 4, detailed discussions on the geometric compatibility equations were provided to ensure that the deformations are continuous throughout the structure and that the geometric (essential) boundaries are also satisfied. The discussions on the constitutive relations (stiffness and flexibility relations) will be postponed to later chapters with the help of the principles of virtual work, which is the main topic of the present chapter.

In this chapter, the basic treatments of virtual work principles will be provided. Mathematical derivations of these principles can be found in textbooks on calculus of variations (e.g. Washizu, 1982). As observed in Chapters 3 and 4, there are "*mysterious*" relations between the equilibrium and compatibility equations of members and structures. No formal derivations of these contragradient relations have been provided so far. By using the virtual work principles, these relations can be proved in a rational manner.

Generally, virtual work principles can be classified into two distinct categories, namely, the virtual displacement and virtual force principles. The former calls for equilibrium conditions in establishing system relations, while the latter invokes compatibility conditions in constituting system equations. In either principle, all basic requirements need to be satisfied, but at different stages. Any student who has taken basic courses on structural analysis must have encountered the virtual work principles in one form or another. It is recalled from basic courses on structural analysis that the slope-deflection method, the moment-distribution method, the dummy (unit) displacement method, and Castigliano's 1st theorem stem from the virtual displacement principle. On the other hand, the method of consistent displacements, the method of least work, the dummy (unit) load method, the three-moment method, and Castigliano's 2nd theorem all originate from the virtual force principle.

The invention of the virtual work principles emanated from the field of classical mechanics. In general, there are two paths of development in classical mechanics. One path known as "*Vectorial Mechanics*" follows Newton's law of motion ($\vec{F} = m\vec{a}$). The other path is known as "*the Principle of Virtual Work*" or "*Scalar Mechanics*". The origin of this principle probably dates back to Aristotle. In a fundamental form, this principle was known to da Vinci. This principle was noticed by Galileo as a general law that could be applied to simple machines (e.g. levers, pulleys). However, it was Johann Bernoulli who gave the principle of virtual work a general formulation that can apply to almost all mechanical systems. He gave birth to the principle of virtual displacement in his famous letter (January 26th 1717) to Varignon. The main contributor

DOI: 10.1201/9781003595458-5

to the mathematical formulation of the virtual work principle was Lagrange in his famous treatise *Me'canique analttique* (Lagrange, 1788). In the preface of this treatise, Lagrange stated that:

> ***"There are no figures in his book because the methods which he uses do not require geometrical or mechanical considerations, but only algebraic operations, which have to follow a prescribed order."***

With the works of Lagrange, d'Alembert, and Hamilton, the principle of virtual work was extended to kinetics, hence, leading to the famous Lagrange-Hamilton equation.

## 5.2 VIRTUAL WORK PRINCIPLE: PARTICLES AND RIGID BODIES

The derivation of the virtual displacement principle is first started with a single rigid particle. The particle $A$ of Figure 5.1 is subjected to a system of forces ($P_1, P_2, ..., P_i, ..., P_n$), which are in equilibrium. Therefore, the force equilibrium equation can be expressed as:

$$\vec{\mathbf{P}} = \sum_{i=1}^{n} \vec{\mathbf{P}}_i = \mathbf{0} \tag{5.1}$$

Suppose that the equilibrium state of the particle $A$ is disturbed by an infinitesimal and arbitrary displacement vector $\delta\vec{\mathbf{u}}$ in a certain direction $O'O$ as shown in Figure 5.2, that is, it is virtually moved from undisplaced to displaced configurations.

Such an imaginary displacement is termed a *"virtual"* displacement. It should be noticed that the virtual displacement $\delta\vec{\mathbf{u}}$ is totally fictitious and has nothing to deal with the real displacement of the particle may take place under the action of real forces. The requirement that the virtual displacement be infinitesimal is imposed so that the direction of the applied forces during the virtual movement is assumed to be unchanged. In other words, the equilibrium equation is taken with respect to the undisplaced configuration. If the virtual displacement is finite, an appropriate account must be considered in the description of virtual work. This circumstance is encountered in the problem of large displacements (e.g. Limkatanyu et al. 2014).

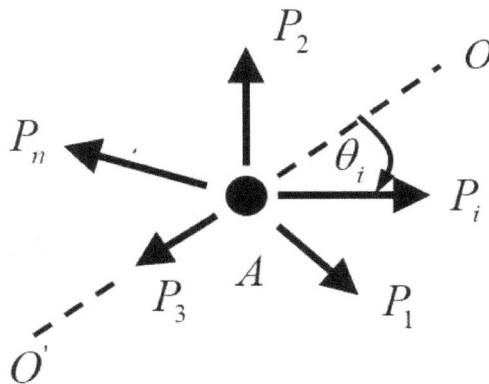

**FIGURE 5.1** A Particle Subjected to a System of Forces.

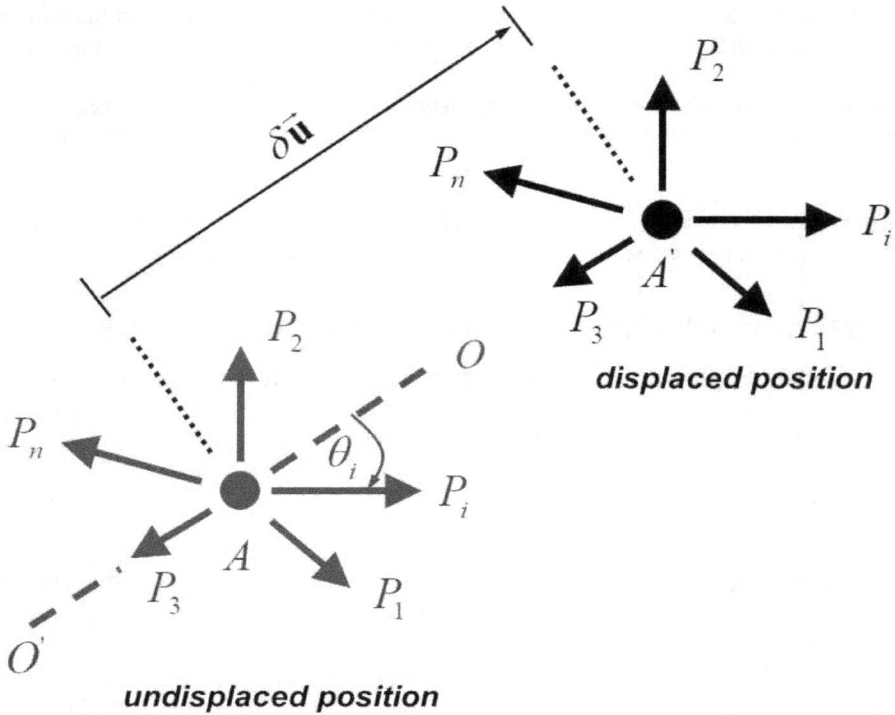

**FIGURE 5.2** Virtual Work Definition.

The work done by the real forces on the virtual displacement is known as virtual work $\delta W$ or, more precisely, as external virtual work $\delta W_{ext}$. The external virtual work is defined as a summation of the scalar products of the virtual displacement $\delta \vec{u}$ and all applied forces $\vec{P}$:

$$\delta W_{ext} = \delta \vec{u} \cdot \left( \sum_{i=1}^{n} \vec{P}_i \right) = \delta u_{O'O} P_1 \cos\theta_1 + \ldots + \delta u_{O'O} P_n \cos\theta_n = \delta u_{O'O} \left( \sum_{i=1}^{n} P_i \cos\theta_i \right) \quad (5.2)$$

where $\theta_i$ is denoted for the angle between the line of action of the force $\vec{P}_i$ and the direction of the virtual displacement $(O'O)$. It should be pointed out why there is no factor $1/2$ present in the virtual work expression. This is due to the assumption that during the virtual displacement, the force $\vec{P}_i$ acts at its constant value. This becomes clear when the load-displacement curve of the force $P_i$ is depicted as shown in Figure 5.3. From the response curve, as the force $P_i$ is gradually increased from the initial state (Point $O$), the corresponding displacement $u_i$ also increases until the equilibrium state is reached at Point $E$. The total virtual work resulting from disturbing this equilibrium state by the virtual displacement $\delta u_i$ is defined as:

$$\begin{aligned} \Delta W_{ext} &= \delta W_{ext} + \delta^2 W_{ext} + \ldots + higher\ order\ terms \\ &= P_i \delta u_i + \delta^2 W_{ext} + \ldots + higher\ order\ terms \end{aligned} \quad (5.3)$$

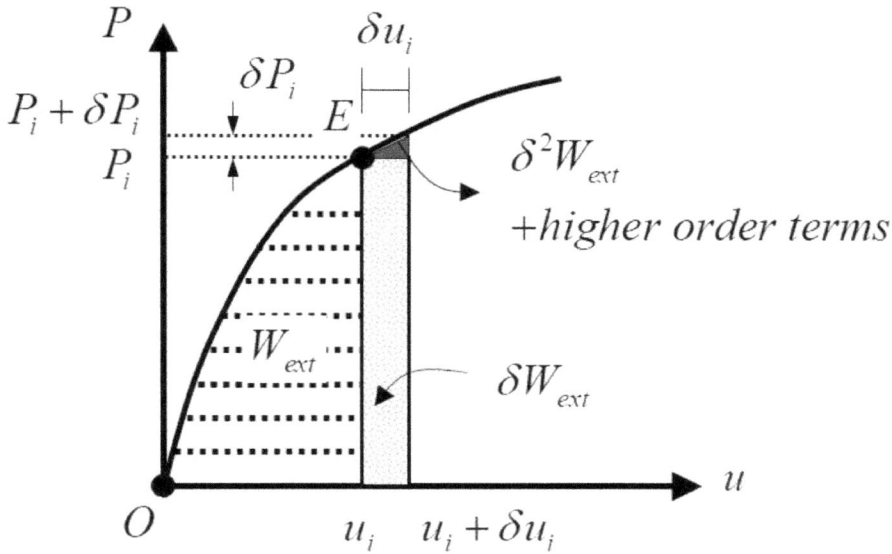

**FIGURE 5.3** Definition of External Virtual Work $\delta W_{ext}$.

Mathematically, $\delta W_{ext}$ is known as the first variation of the function (or in general functional) $W_{ext}$; $\delta^2 W_{ext}$ as the second variation of the function $W_{ext}$; and so on. Assuming that $\delta P_i$ and $\delta u_i$ are infinitesimal, the total virtual work can be sufficiently approximated as:

$$\Delta W_{ext} \approx \delta W_{ext} = P_i \delta u_i \tag{5.4}$$

However, for the nonlinear systems (e.g. nonlinear geometric structures), other than the first variation term must be kept in considering the total virtual work $\Delta W_{ext}$.

On the right hand side of Eq. (5.2), the bracket term vanishes due to the condition of force equilibrium (5.1). Consequently, it follows that:

$$\delta W = \delta W_{ext} = 0 \tag{5.5}$$

This is the equation for the virtual displacement principle and can be expressed in a verbal manner as:

*If a particle is in equilibrium under a system of real forces, the total work done by the real forces during any arbitrary virtual displacement vanishes.*

It is interesting to explore whether the converse of the aforementioned statement is valid. If valid, it would be very helpful in formulating the equilibrium equations of a particle. As shown in Figure 5.4, the virtual displacement $\delta \vec{u}$ can be expressed with respect to the $x$ and $y$ reference system:

$$\delta \vec{u} = \delta u_x \vec{i} + \delta u_y \vec{j} \tag{5.6}$$

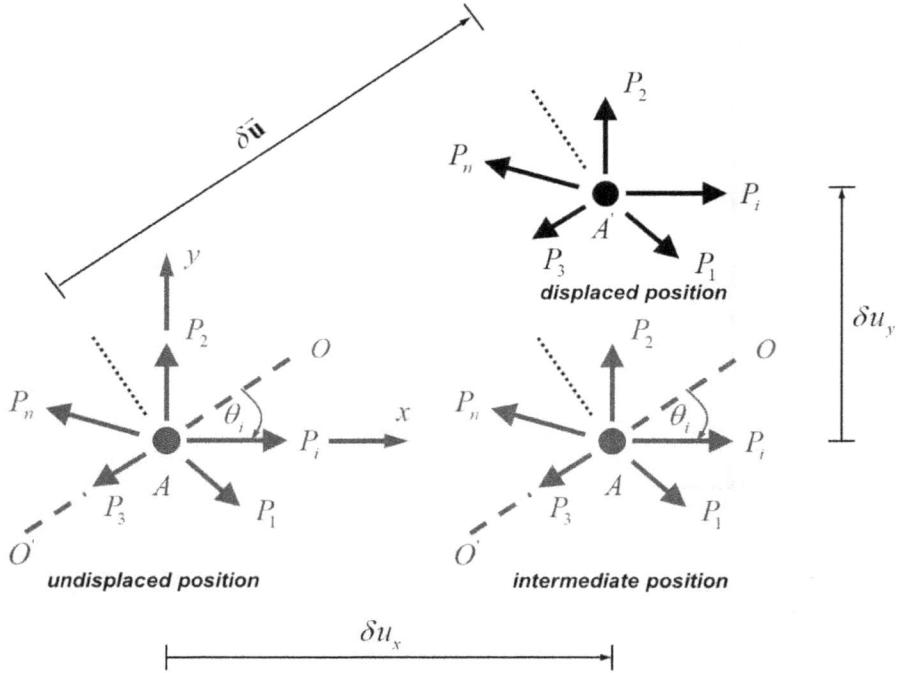

**FIGURE 5.4**   Decomposition of the Virtual Displacement Vector $\delta \bar{\mathbf{u}}$.

where $\vec{\mathbf{i}}$ and $\vec{\mathbf{j}}$ are the unit vectors along the $x$ and $y$ axes, respectively; and $\delta u_x$ and $\delta u_y$ are independent and arbitrary virtual displacement components along the $x$ and $y$ directions, respectively.

Similarly, the system of forces acting on the particle $A$ can be decomposed as:

$$\sum_{i=1}^{n} \vec{\mathbf{P}}_i = \left( \sum_{i=1}^{n} P_{xi} \right) \vec{\mathbf{i}} + \left( \sum_{i=1}^{n} P_{yi} \right) \vec{\mathbf{j}} \tag{5.7}$$

where $\sum_{i=1}^{n} P_{xi}$ is the resultant force along the $x$ axis while $\sum_{i=1}^{n} P_{yi}$ is the resultant force along the $y$ axis.

The virtual work is the scalar product of the vectors $\sum_{i=1}^{n} \vec{\mathbf{P}}_i$ and $\delta \bar{\mathbf{u}}$ and is defined as:

$$\delta W = \delta W_{ext} = \left( \sum_{i=1}^{n} P_{xi} \right) \delta u_x + \left( \sum_{i=1}^{n} P_{yi} \right) \delta u_y \tag{5.8}$$

Due to the arbitrariness of $\delta u_x$ and $\delta u_y$, the coefficients of $\delta u_x$ and $\delta u_y$ in Eq. (5.8) must be zero in order for $\delta W$ to vanish.

Therefore, one can conclude that:

$$\sum_{i=1}^{n} P_{xi} = 0 \tag{5.9}$$

$$\sum_{i=1}^{n} P_{yi} = 0 \tag{5.10}$$

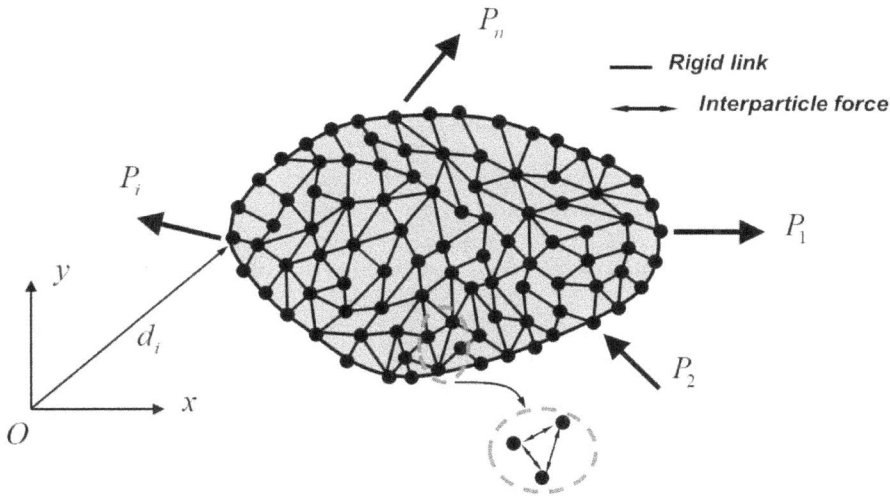

**FIGURE 5.5** A Rigid Body as an Assemblage of Rigid Particles and Rigid Links.

It is clear that Eqs. (5.9) and (5.10) are nothing else but the equilibrium equations of a particle along the $x$ and $y$ axes, respectively. The converse of the aforementioned virtual displacement principle can be stated as:

*A particle is in equilibrium under a system of real forces if the total work done by the real forces during every arbitrary and independent virtual displacement vanishes.*

The above statement yields the very crucial conclusion that the virtual displacement principle is just an alternative way to express the equilibrium equations. This conclusion seems trivial at this moment but it yields a very essential insight as presented in later applications.

Now, it is natural to extend the virtual work principle derived so far for a rigid particle to a rigid body since a rigid body can be simply envisaged as an assemblage of rigid particles connected together by rigid links as shown in Figure 5.5.

The associated force system comprises forces acting externally on the particle set and interparticle forces in rigid links. The rigid body (particle set plus rigid links) is at first presumed to be in equilibrium under the action of the force system. In other words, the particles are in overall equilibrium under the action of the external forces and each particle is in equilibrium under the action of external and interparticle forces. Consequently, the equilibrium equations for a rigid body can be written as:

**FORCE EQUILIBRIUM**

$$\sum_{i=1}^{n} \vec{P}_i = 0 \tag{5.11}$$

## MOMENT EQUILIBRIUM WITH RESPECT TO THE REFERENCE POINT $O$

$$\sum_{i=1}^{n} \vec{\mathbf{P}}_i \times \vec{\mathbf{d}}_i = 0 \tag{5.12}$$

where $\vec{\mathbf{d}}_i$ is the vector positioning the loading point of the force $\vec{\mathbf{P}}_i$ with respect to the reference point $O$. The equilibrated state is now disturbed by an arbitrary virtual movement. This virtual movement comprises the virtual translation $\delta \vec{\mathbf{u}}$ and the virtual rotation $\delta \vec{\omega}$. The resulting virtual work is defined as the summation of scalar products of the resultant force with the virtual translation vector and the resultant moment with the virtual rotation vector:

$$\delta W = \delta W_{ext} = \delta \vec{\mathbf{u}} \cdot \left( \sum_{i=1}^{n} \vec{\mathbf{P}}_i \right) + \delta \vec{\omega} \left( \sum_{i=1}^{n} \vec{\mathbf{P}}_i \times \vec{\mathbf{d}}_i \right) \tag{5.13}$$

After carrying out the scalar products in Eq. (5.13), one has:

$$\delta W = \delta W_{ext} = \left( \sum_{i=1}^{n} P_i \right) \delta u \cos \theta_i + \left( \sum_{i=1}^{n} P_i \times d_i \right) \delta \omega \cos \psi_i \tag{5.14}$$

where $\theta_i$ is the angle between the resultant force and the virtual translation vector while $\psi_i$ is the angle between the resultant moment and the virtual rotation vector. Clearly, the virtual work will vanish for any combination of $\delta \vec{\mathbf{u}}$ and $\delta \vec{\omega}$ if and only if the parenthesis terms in Eq. (5.14) are zero, that is, the equilibrium equations of a rigid body is obtained. Therefore, a general statement of the virtual displacement principle for a rigid body can be expressed as:

*A necessary and sufficient condition for a rigid body to be in an equilibrated state is that the virtual work of the real forces under an arbitrary virtual displacement field is equal to zero.*

It is important to point out that the virtual work done by each pair of interparticle forces in a rigid link under virtual displacement cancel out each other. This is the reason why there is no internal virtual work in the rigid body problem.

It is noted that visualizing the rigid body as an assemblage of rigid particles and rigid links, it is quite natural to extend the virtual work principle derived so far for a rigid body to the deformable one. This is done by replacing rigid links with deformable ones.

For example, the frame structure is made up of rigid joints (particles) connected together by deformable frame members. The continuum structure (e.g. dam) is made up of rigid joints connected together by deformable area or volume elements. In these cases, there are contributions of the internal virtual work to the total virtual work. this issue is to be pursued in the following sections.

## EXAMPLE 5.1

By the principle of virtual displacement, determine the force in member $A$ of the truss structure shown in Figure 5.6.

**Solution:** The force in member $A$ is first released. This is done by removing member $A$ and replacing it by the equal and opposite force $F_A$. Then, the virtual displaced state of the truss is visualized by assuming that the constraining force in member $A$ is absent. Finally, the displaced configuration of the truss is depicted in Figure 5.7.

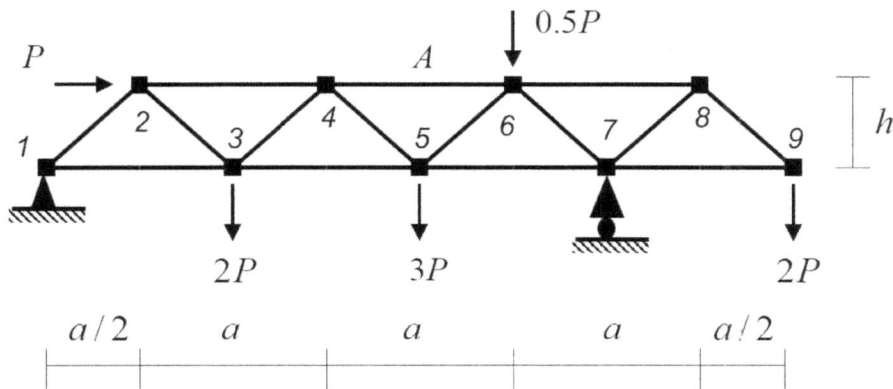

**FIGURE 5.6** Example 5.1.

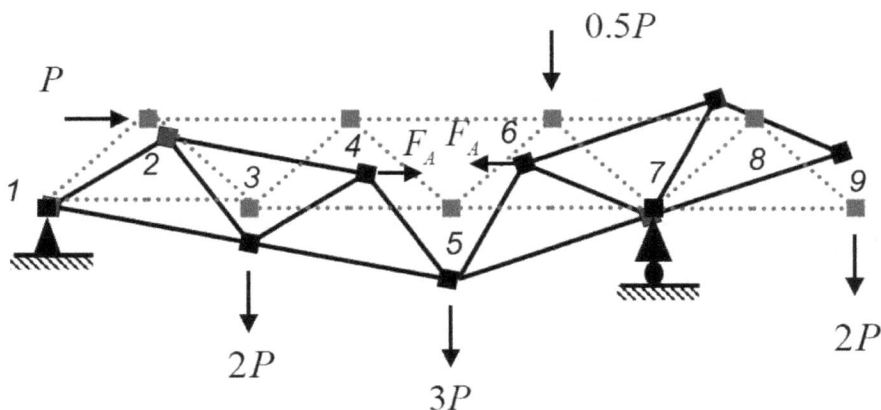

**FIGURE 5.7** Example 5.1 (Continued).

It should be noted that this truss is statically determinate. Therefore, introducing just one internal force release (member $A$) can form a mechanism.

The virtual work associated with the above virtual displacements is defined as:

$$\delta W = P(\delta U_{2X}) + 2P(\delta U_{3Y}) + 3P(\delta U_{5Y}) + 0.5P(\delta U_{6Y}) - 2P(\delta U_{9Y}) + F_A(\delta U_{4X}) + F_A(\delta U_{6X}) = 0$$

Since the virtual displaced state should represent properly the rigid-body mechanism, the following relations are obatined:

$$\delta U_{2X} = \frac{h}{2a}\delta U_{5Y}; \quad \delta U_{3Y} = \frac{1}{2}\delta U_{5Y}; \quad \delta U_{6Y} = \frac{1}{2}\delta U_{5Y}; \quad \delta U_{9Y} = \delta U_{5Y}; \quad \delta U_{4X} = \frac{h}{2a}\delta U_{5Y}; \quad \text{and}$$

$$\delta U_{6X} = \frac{h}{a}\delta U_{5Y}$$

These relations are kinematically admissible.

Substituting these relations into the virtual work expression leads to:

$$\delta W = \left( P\frac{h}{2a} + P + 3P + 0.25P - 2P + F_A\left(\frac{h}{2a} + \frac{h}{a}\right)\right)\delta U_{5Y} = 0$$

Due to the arbitrariness of $\delta U_{5Y}$, the force in member $A$ can be determined as:

$$F_A = -\left(\frac{1}{3} + \frac{3a}{2h}\right)P$$

It is left to the reader to verify this answer either by the method of joints or the method of sections.

## 5.3   VIRTUAL WORK PRINCIPLE: DEFORMABLE BODIES

In the previous section, the virtual work principle or, more specifically, the virtual displacement principle was derived and applied to a rigid particle and a rigid body. As mentioned before, the virtual displacement principle is just an alternative way to describe the equilibrium conditions. With this principle in hand, the support reactions and internal forces can only be determined for statically determinate structures. However, the most valuable application of matrix structural analysis is to analyze statically indeterminate structures. By simply accounting for the deformability of structures, statically indeterminate structures can be analyzed through the matrix virtual displacement principle.

To account for the deformable nature of solids and structures, there is a need to clarify the external and internal virtual work. The series arrangement of two deformable links shown in Figure 5.8 $a$ is used to convey this idea. The structure shown in Figure 5.8 $a$ consists of three rigid particles connected together by two springs. The free-body diagrams of these rigid particles and springs are shown in Figure 5.8 $b$. Giving the particles the virtual displacements $\delta u_a$, $\delta u_b$, and $\delta u_c$ at joints $a$, $b$, and $c$, respectively, the virtual work for these particles is:

$$\delta W = \left(F_a + N'_{ab}\right)\delta u_a + \left(F_b + N'_{ba} + N'_{bc}\right)\delta u_b + \left(F_c + N'_{cb}\right)\delta u_c \tag{5.15}$$

Since these three particles are rigid and are assumed to be in equilibrium state, the virtual work expression of Eq. (5.15) must vanish:

$$\delta W = \left(F_a + N'_{ab}\right)\delta u_a + \left(F_b + N'_{ba} + N'_{bc}\right)\delta u_b + \left(F_c + N'_{cb}\right)\delta u_c = 0 \tag{5.16}$$

Due to the arbitrariness and independence of $\delta u_a$, $\delta u_b$, and $\delta u_c$, the following equilibrium relations are obatined:

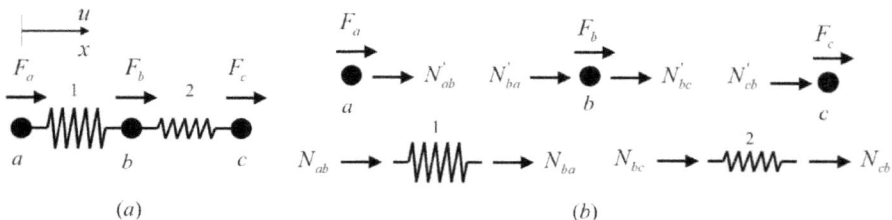

**FIGURE 5.8**   Series Arrangement of Two Deformable Links and Three Rigid Particles.

$$F_a + N'_{ab} = 0 \qquad \Rightarrow \quad F_a = -N'_{ab}$$
$$F_b + N'_{ba} + N'_{bc} = 0 \quad \Rightarrow \quad F_b = -N'_{ba} - N'_{bc} \qquad (5.17)$$
$$F_c + N'_{cb} = 0 \qquad \Rightarrow \quad F_c = -N'_{cb}$$

From Newton's 3rd law of motion, one has:

$$N'_{ab} = -N_{ab}; \ N'_{ba} = -N_{ba}; \ N'_{bc} = -N_{bc}; \text{and } N'_{cb} = -N_{cb} \qquad (5.18)$$

Substituting Eqs. (5.18) into (5.16), one has:

$$\delta W = \left( F_a - N_{ab} \right)\delta u_a + \left( F_b - N_{ba} - N_{bc} \right)\delta u_b + \left( F_c - N_{cb} \right)\delta u_c = 0 \qquad (5.19)$$

$$\delta W = \left( F_a \delta u_a + F_b \delta u_b + F_c \delta u_c \right) - \left( N_{ab}\delta u_a + N_{ba}\delta u_b + N_{bc}\delta u_b + N_{cb}\delta u_c \right) = 0 \quad (5.20)$$

Considering the equilibrium of each spring, one has:

$$N_1 = N_{ba} = -N_{ab} \text{ and } N_2 = N_{cb} = -N_{bc} \qquad (5.21)$$

Substituting Eqs. (5.21) into (5.20), one has:

$$\delta W = \left( F_a \delta u_a + F_b \delta u_b + F_c \delta u_c \right) + \left( -\left( N_1 \left( \delta u_b - \delta u_a \right) + N_2 \left( \delta u_c - \delta u_b \right) \right) \right) = 0 \quad (5.22)$$

The first term on the right-hand side of Eq. (5.22) represents the external virtual work denoted by $\delta W_{ext}$. The second term on the right-hand side is the internal virtual work denoted by $\delta W_{int}$. From the basic principle, the spring deformation can be defined in terms of its end displacements as:

$$\Delta_i = u_2^i - u_1^i \qquad (5.23)$$

Therefore, the virtual deformations compatible with the virtual end displacements are written as:

$$\delta\Delta_1 = \delta u_b - \delta u_a \text{ for spring 1} \qquad (5.24)$$

$$\delta\Delta_2 = \delta u_c - \delta u_b \text{ for spring 2} \qquad (5.25)$$

The internal virtual work is written as:

$$\delta W_{int} = -\left( N_1 \delta\Delta_1 + N_2 \delta\Delta_2 \right) \qquad (5.26)$$

The minus sign in Eq. (5.26) indicates that the internal virtual work is done *on* rather than *by* the internal forces. It is noted that the internal virtual work is equal in magnitude but opposite in sign to the so-called "*virtual strain energy*" $\delta U_{int}$.

$$\delta W_{int} = -\delta U_{int} = -\left( N_1 \delta\Delta_1 + N_2 \delta\Delta_2 \right) = -\sum_{i=1}^{2} N_i \delta\Delta_i \qquad (5.27)$$

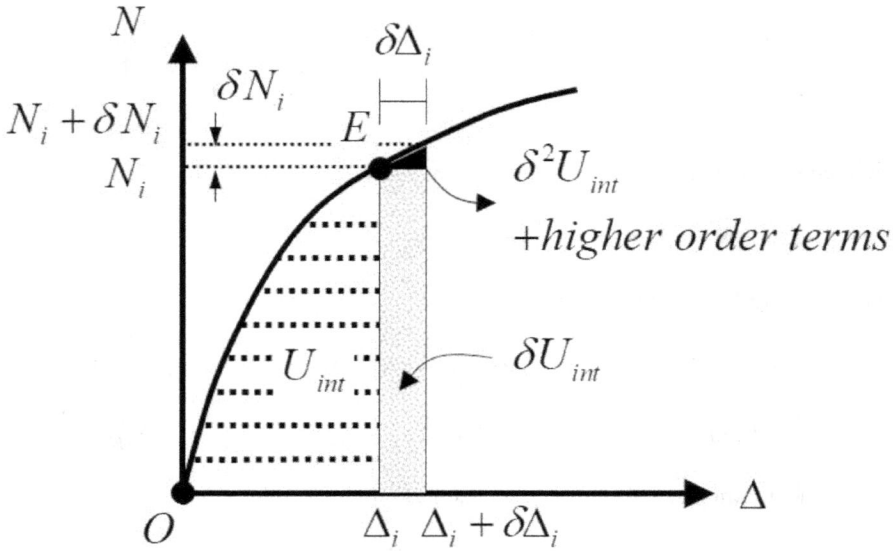

**FIGURE 5.9**  Definition of Virtual Strain Energy $\delta U_{int}$.

The reason why there is no factor 1/2 present in the internal work and virtual strain energy expressions should be pointed out. This is due to the assumption that during virtual deformation, the internal force $N_i$ acts at its constant value. This becomes clear when the force-deformation curve of the spring is depicted as shown in Figure 5.9. From the response curve, as the force $N_i$ is gradually increased from the initial state (Point $O$), the corresponding deformation $\Delta_i$ also increases until the equilibrium state is reached at Point $E$. The total virtual strain energy resulting from disturbing this equilibrium state by the virtual deformation $\delta\Delta_i$ compatible with the virtual displacement $\delta u_i$ is defined as:

$$\begin{aligned}
\Delta U_{int} &= \delta U_{int} + \delta^2 U_{int} + \ldots + higher\ order\ terms \\
&= N_i \delta\Delta_i + \delta^2 U_{int} + \ldots + higher\ order\ terms
\end{aligned} \tag{5.28}$$

Mathematically, $\delta U_{int}$ is known as the first variation of the function (or in general functional) $U_{int}$; $\delta^2 U_{int}$ as the second variation of the function $U_{int}$; and so on. Assuming that $\delta N_i$ and $\delta\Delta_i$ are infinitesimal, the total virtual strain energy can be sufficiently approximated as:

$$\Delta U_{int} \approx \delta U_{int} = N_i \delta\Delta_i \tag{5.29}$$

It is noted that for the nonlinear systems (e.g. large deformation structures), other than the first variation term must be kept in expressing the total virtual strain energy $\Delta U_{int}$.

Therefore, the algebraic statement of the virtual work principle is:

$$\delta W = \delta W_{ext} + \delta W_{int} = 0 \tag{5.30}$$

or

$$\delta W = \delta W_{ext} - \delta U_{int} = 0 \tag{5.31}$$

The verbal statement of the virtual work principle is:

> *A deformable body is in equilibrated state if the virtual work done by all external forces plus the virtual work done by all internal forces during any kinematically admissible virtual displacement is zero.*

In the above statement, the virtual displacement is restricted by the phrase *"kinematically admissible"*. This infers that the virtual displacement field is not completely arbitrary. In general, the virtual displacement field must be such that it maintains continuity between all parts of the system and introduces no tears or rips to the system.

Physical insights into the restriction on virtual displacement fields can be enriched by considering the tapered bar subjected to the uniformly distributed load as shown in Figure 5.10.

The candidates for the virtual displacement field are shown in Figure 5.11. The ones shown in Figure 5.11 *a*, *b*, and *c* are all kinematically admissible, but give different attributes. The virtual displacement field in Figure 5.11 *a* simply represents the rigid body motion along the *x*-axis. Furthermore, it violates the geometrical boundary at point *a*. By using this candidate, one can only compute the support reaction at point *a*. The virtual displacement field in Figure 5.11 *b* represents only the deformation mode and satisfies the geometrical boundary at point *a*. By using this candidate, one can compute the *"approximate"* solution to this problem. This aspect will explored in the following sections. The virtual displacement field in Figure 5.11 *c* represents both the rigid body motion and deformation mode. It violates the geometrical boundary at point *a* as well. By using this candidate, one can compute the *"approximate"* solution to this problem as well as the support reaction at point *a*. The virtual displacement field in Figure 5.11 *d* is absolutely not kinematically admissible since it introduces the discontinuity in the bar where the deformation can not uniquely be defined at point *c*.

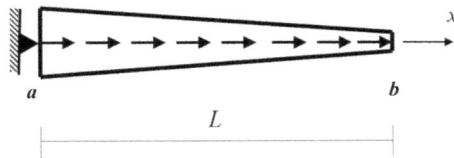

**FIGURE 5.10**   Tapered Bar Subjected to the Uniformly Distributed Load.

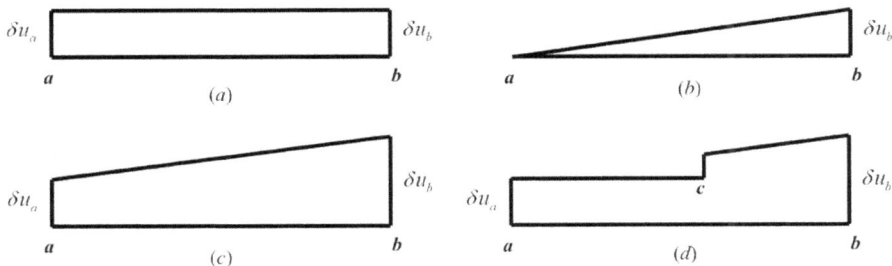

**FIGURE 5.11**   Candidates for the Virtual Displacement Field.

It should be noted that so far the constitutive relations between stress (or resultant stress) and strain (or resultant strain) have not been mentioned in deriving the virtual work principle. This implies that the virtual work principle can apply to both conservative (elastic) and unconservative (inelastic) systems. However, the main focus is limited to the conservative one.

Up to now, the virtual displacement principle has been developed only for a deformable body consisting of rigid particles and spring links. Hence, this principle can be applied to axially deformable structures (e.g. truss structures). Following the above procedure, one can extend the virtual work principle to any other type of deformable structures. Therefore, the remaining task in pursuing the virtual work principle is to derive the internal virtual work (virtual strain energy) expressions for specific modes of structural actions, namely axial, shear, and flexural forces. These will be accomplished in the next section.

## 5.4   INTERNAL VIRTUAL WORK AND VIRTUAL STRAIN ENERGY EXPRESSIONS: A FRAME MEMBER

Since frame structures are of main interest in this textbook, the primary modes of internal actions are axial, shear, and flexural responses. In the following subsections, the virtual strain energy expression will be derived for each type of structural actions.

### 5.4.1   Axial Response: Virtual Displacement Principle

Consider the infinitesimal segment $dx$ of an axially loaded member of length $L$ as shown in Figure 5.12. This segment is in equilibrium under actions of internal forces $N(x)$ and $N(x)+\dfrac{\partial N(x)}{\partial x}dx$ and of the external distributed load $w_x(x)$. It is noted that the displacement and deformation associated with these forces are qualified as real displacement and real deformation, respectively. The whole infinitesimal segment is considered as a free body and is disturbed by the virtual displacements $\delta u_0(x)$ at the left end and by $\delta u_0(x)+\dfrac{\partial \delta u_0(x)}{\partial x}dx$ at the right end. Consequently, the resulting virtual work is:

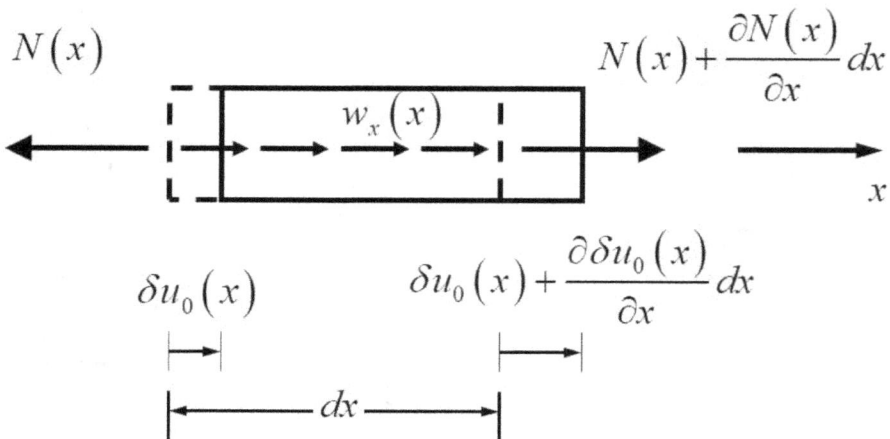

**FIGURE 5.12**   Virtual Strain Energy $\delta \bar{U}_{int}$ of an Infinitesimal Segment: Axial Response.

$$\delta W_{ext} = \left(w_x(x)dx\right)\delta u_0(x) - N(x)\delta u_0(x) + \left(N(x) + \frac{\partial N(x)}{\partial x}dx\right)\left(\delta u_0(x) + \frac{\partial \delta u_0(x)}{\partial x}dx\right) \quad (5.32)$$

$$\delta W_{ext} = \left(w_x(x)dx\right)\delta u_0(x) - N(x)\delta u_0(x) + N(x)\delta u_0(x) \; +$$

$$N(x)\frac{\partial \delta u_0(x)}{\partial x}dx + \underbrace{\left(\frac{\partial N(x)}{\partial x}dx\right)}_{-w_x(x)}\delta u_0(x) + \underbrace{\left(\frac{\partial N(x)}{\partial x}dx\right)\left(\frac{\partial \delta u_0(x)}{\partial x}dx\right)}_{2^{nd} \; order \; term} \quad (5.33)$$

By imposing the differential equilibrium equation (2/2.1) of a bar segment and neglecting the 2nd order term, Eq. (5.33) is written as:

$$\delta W_{ext} = N(x)\frac{\partial \delta u_0(x)}{\partial x}dx \quad (5.34)$$

Based on Eq. (5.31), one can conclude that:

$$\delta \bar{U}_{int} = N(x)\frac{\partial \delta u_0(x)}{\partial x}dx \quad (5.35)$$

where $\delta \bar{U}_{int}$ is denoted for the virtual strain energy of an infinitesimal segment $dx$. In Chapter 2, the axial strain is defined as the rate of change in the axial displacement with respect to the axial coordinate.

$$\varepsilon_0(x) = \frac{\partial u_0(x)}{\partial x} \quad (5.36)$$

Similarly, the virtual axial strain compatible with the virtual displacement $\delta u_0(x)$ is defined as:

$$\delta \varepsilon_0(x) = \frac{\partial \delta u_0(x)}{\partial x} \quad (5.37)$$

Therefore, Eq. (5.35) is written as:

$$\delta \bar{U}_{int} = N(x)\delta \varepsilon_0(x)dx \quad (5.38)$$

For the whole length of an axially loaded member, the virtual strain energy $\delta U_{int}$ is:

$$\delta U_{int} = \int_L \delta \bar{U}_{int} = \int_L N(x)\delta \varepsilon_0(x)dx \quad (5.39)$$

If only linearly elastic material is considered, the following section stiffness relation

$$N(x) = EA(x)\varepsilon_0(x) \quad (5.40)$$

is substituted into Eq. (5.39) and the virtual strain energy is entirely expressed in terms of strain (deformation) as:

$$\delta U_{int} = \int_L \varepsilon_0(x) EA(x) \delta\varepsilon_0(x) dx \tag{5.41}$$

It is noted that $\varepsilon_0(x)$ is the real axial strain associated with the real axial force $N(x)$ through the constitutive relation. By recalling the axial strain-displacement relations of Eqs. (5.36) and (5.37), the virtual strain energy is entirely expressed in terms of axial displacements as:

$$\delta U_{int} = \int_L \frac{\partial u_0(x)}{\partial x} EA(x) \frac{\partial \delta u_0(x)}{\partial x} dx \tag{5.42}$$

Based on Eq. (5.30), the internal virtual work is written as:

$$\delta W_{int} = -\int_L N(x) \delta\varepsilon_0(x) dx \tag{5.43}$$

Based on the compatibility expression relating the deformation (resultant strain) to the displacement field (5.36), it becomes clear that the "*kinematically admissible*" virtual displacement field has to satisfy the following requirements:

1. The virtual displacement field must be continuous and must possess the continuous first derivative at every point inside the member. In other words, it has $C^0$ continuity.
2. The magnitude of the virtual displacement must be in the range of validity of the small displacement theory.

If the virtual displacement field satisfies the above requirements as well as geometric (essential) boundaries, it becomes a "*geometrically admissible*" virtual displacement field. The "*kinematically admissible*" and "*geometrically admissible*" virtual strain fields are defined as ones which are related to the "*kinematically admissible*" and "*geometrically admissible*" virtual displacement fields, respectively through compatibility relation (5.36).

## 5.4.2  SHEAR RESPONSE: VIRTUAL DISPLACEMENT PRINCIPLE

Consider the infinitesimal segment $dx$ of a vertically loaded member of length $L$ as shown in Figure 5.13. This segment is in equilibrium under actions of internal forces $V(x)$ and $V(x) + \dfrac{\partial V(x)}{\partial x} dx$ and of the external distributed load $w_y(x)$. It is noted that the displacement and deformation associated with these forces are qualified as real displacement and real deformation, respectively. The whole infinitesimal segment is considered as a free body and is disturbed by the virtual displacement $\delta v_s(x)$ at the left end and by $\delta v_s(x) + \dfrac{\partial \delta v_s(x)}{\partial x} dx$ at the right end. Consequently, the resulting virtual work is:

$$\delta W_{ext} = \left(w_y(x) dx\right) \delta v_s(x) - V(x) \delta v_s(x) + \left(V(x) + \frac{\partial V(x)}{\partial x} dx\right)\left(\delta v_s(x) + \frac{\partial \delta v_s(x)}{\partial x} dx\right) \tag{5.44}$$

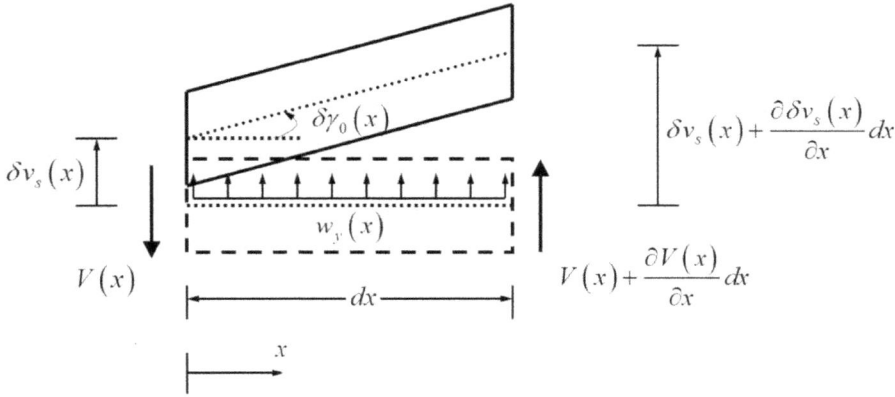

**FIGURE 5.13**   Virtual Strain Energy $\delta\bar{U}_{int}$ of an Infinitesimal Segment: Shear Response.

$$\delta W_{ext} = \left(w_y(x)dx\right)\delta v_s(x) - V(x)\delta v_s(x) + V(x)\delta v_s(x) \quad +$$

$$V(x)\frac{\partial \delta v_s(x)}{\partial x}dx + \left(\underbrace{\frac{\partial V(x)}{\partial x}dx}_{-w_y(x)}\right)\delta v_s(x) + \underbrace{\left(\frac{\partial V(x)}{\partial x}dx\right)\left(\frac{\partial \delta v_s(x)}{\partial x}dx\right)}_{2^{nd}\ order\ term} \qquad (5.45)$$

By imposing the vertical equilibrium equation (2/2.13) of a beam segment and neglecting the 2nd order term, Eq. (5.45) is written as:

$$\delta W_{ext} = V(x)\frac{\partial \delta v_s(x)}{\partial x}dx \qquad (5.46)$$

Based on Eq. (5.31), one can conclude that:

$$\delta\bar{U}_{int} = V(x)\frac{\partial \delta v_s(x)}{\partial x}dx \qquad (5.47)$$

where $\delta\bar{U}_{int}$ is denoted for the virtual strain energy of an infinitesimal segment $dx$. In Chapter 2, the shear strain is defined as the rate of change in the vertical displacement with respect to the axial coordinate.

$$\gamma_0(x) = \frac{\partial v_s(x)}{\partial x} \qquad (5.48)$$

Similarly, the virtual shear strain compatible with the virtual displacement $\delta v_s(x)$ is defined as:

$$\delta\gamma_0(x) = \frac{\partial \delta v_s(x)}{\partial x} \qquad (5.49)$$

Therefore, Eq. (5.47) is written as:

$$\delta\bar{U}_{int} = V(x)\delta\gamma_0(x)dx \qquad (5.50)$$

For the whole length of a vertically loaded member, the virtual strain energy $\delta U_{int}$ is:

$$\delta U_{int} = \int_L \delta \bar{U}_{int} = \int_L V(x)\delta\gamma_0(x)dx \qquad (5.51)$$

If only linearly elastic material is considered, the following section stiffness relation

$$V(x) = GA_s(x)\gamma_0(x) \qquad (5.52)$$

is substituted into Eq. (5.51) and the virtual strain energy is entirely expressed in terms of strain (deformation) as:

$$\delta U_{int} = \int_L \delta \bar{U}_{int} = \int_L \gamma_0(x)GA_s(x)\delta\gamma_0(x)dx \qquad (5.53)$$

It is noted that $\gamma_0(x)$ is the real shear strain associated with the real shear force $V(x)$ through the constitutive relation. By recalling the shear strain-displacement relations of Eqs. (5.48) and (5.49), the virtual strain energy is entirely expressed in terms of vertical displacements as:

$$\delta U_{int} = \int_L \frac{\partial v_s(x)}{\partial x}GA_s(x)\frac{\partial \delta v_s(x)}{\partial x}dx \qquad (5.54)$$

Based on Eq. (5.30), the internal virtual work is written as:

$$\delta W_{int} = -\int_L V(x)\delta\gamma_0(x)dx \qquad (5.55)$$

Based on the compatibility expression relating the deformation (resultant strain) to the displacement field (5.48), it becomes clear that the *"kinematically admissible"* virtual displacement field has to satisfy the following requirements:

1. The virtual displacement field must be continuous and must possess the continuous first derivative at every point inside the member. In other words, it has $C^0$ continuity.
2. The magnitude of the virtual displacement must be in the range of validity of the small displacement theory.

If the virtual displacement field satisfies the above requirements as well as geometric (essential) boundaries, it becomes a *"geometrically admissible"* virtual displacement field. The *"kinematically admissible"* and *"geometrically admissible"* virtual strain fields are defined as ones which are related to the *"kinematically admissible"* and *"geometrically admissible"* virtual displacement fields, respectively through compatibility relation (5.48).

### 5.4.3   FLEXURAL RESPONSE: VIRTUAL DISPLACEMENT PRINCIPLE

Consider the infinitesimal segment $dx$ of a vertically loaded member of length $L$ as shown in Figure 5.14. This segment is in equilibrium under actions of internal forces $(V(x), V(x)+\dfrac{\partial V(x)}{\partial x}dx$, $M(x)$, and $M(x)+\dfrac{\partial M(x)}{\partial x}dx)$ and of the external distributed load $w_y(x)$. It is noted that the displacements and deformations associated with these forces are qualified as real displacements and real deformations, respectively. The whole infinitesimal segment is considered as a free body

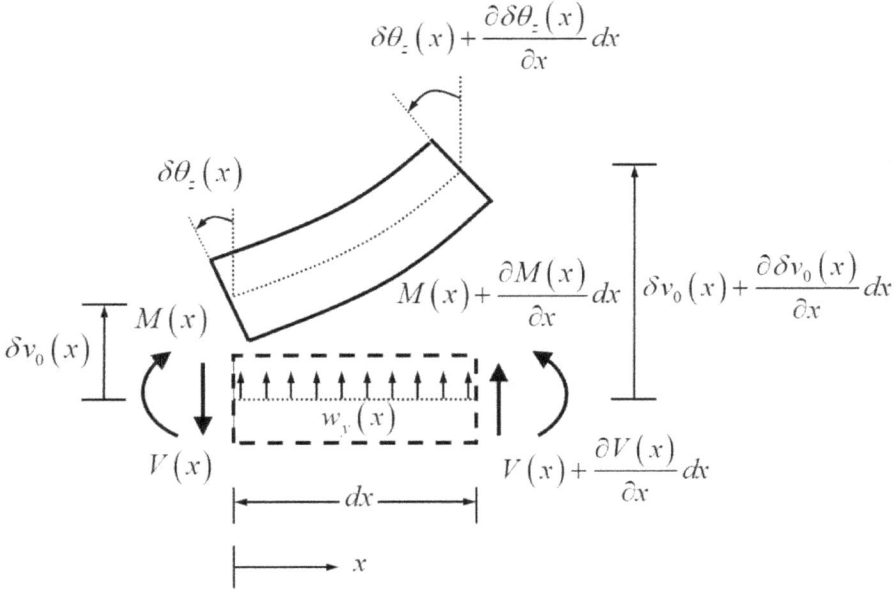

**FIGURE 5.14** Virtual Strain Energy $\delta \bar{U}_{int}$ of an Infinitesimal Segment: Flexural Response.

and is disturbed by the virtual displacement $\delta v_0(x)$ and virtual rotation $\delta \theta_z(x)$ at the left end and by $\delta v_0(x) + \dfrac{\partial \delta v_0(x)}{\partial x} dx$ and $\delta \theta_z(x) + \dfrac{\partial \delta \theta_z(x)}{\partial x} dx$ at the right end. Consequently, the resulting virtual work is:

$$\delta W_{ext} = \left( w_y(x) dx \right) \delta v_0(x) - V(x) \delta v_0(x) +$$
$$\left( V(x) + \frac{\partial V(x)}{\partial x} dx \right) \left( \delta v_0(x) + \frac{\partial \delta v_0(x)}{\partial x} dx \right) + \tag{5.56}$$
$$\left( M(x) + \frac{\partial M(x)}{\partial x} dx \right) \left( \delta \theta_z(x) + \frac{\partial \delta \theta_z(x)}{\partial x} dx \right)$$

$$\delta W_{ext} = \begin{pmatrix} \left( w_y(x) dx \right) \delta v_0(x) - V(x) \delta v_0(x) - M(x) \delta \theta_z(x) + V(x) \delta v_0(x) \quad + \\[2mm] V(x) \underbrace{\dfrac{\partial \delta v_0(x)}{\partial x} dx}_{\delta \theta_z(x)} + \left( \underbrace{\dfrac{\partial V(x)}{\partial x} dx}_{-w_y(x)} \right) \delta v_0(x) + \underbrace{\left( \dfrac{\partial V(x)}{\partial x} dx \right) \left( \dfrac{\partial \delta v_0(x)}{\partial x} dx \right)}_{2^{nd}\ order\ term} + \\[2mm] M(x) \delta \theta_z(x) + M(x) \dfrac{\partial \delta \theta_z(x)}{\partial x} dx + \left( \underbrace{\dfrac{\partial M(x)}{\partial x} dx}_{-V(x)} \right) \delta \theta_z(x) \quad + \\[2mm] \underbrace{\left( \dfrac{\partial M(x)}{\partial x} dx \right) \left( \dfrac{\partial \delta \theta_z(x)}{\partial x} dx \right)}_{2^{nd}\ order\ term} \end{pmatrix} \tag{5.57}$$

By imposing the differential equilibrium equations (2/2.13-14) of a beam segment, defining $\delta\theta_z(x) = \dfrac{\partial\delta v_0(x)}{\partial x}$, and neglecting the 2nd order terms, Eq. (5.57) is written as:

$$\delta W_{ext} = M(x)\frac{\partial\delta\theta_z(x)}{\partial x}dx \tag{5.58}$$

Based on Eq. (5.31), it can be concluded that:

$$\delta\bar{U}_{int} = M(x)\frac{\partial\delta\theta_z(x)}{\partial x}dx \tag{5.59}$$

where $\delta\bar{U}_{int}$ is denoted for the virtual strain energy of an infinitesimal segment $dx$. In Chapter 2, the flexural strain (curvature) is defined as the rate of change in the rotation with respect to the axial coordinate.

$$\kappa_0(x) = \frac{\partial\theta_z(x)}{\partial x} \tag{5.60}$$

Since the rotation is defined as the first derivative of the vertical displacement $(\theta_z(x) = \dfrac{\partial v_0(x)}{\partial x})$, the flexural curvature is defined in term of the vertical displacement as:

$$\kappa_0(x) = \frac{\partial^2 v_0(x)}{\partial x^2} \tag{5.61}$$

Similarly, the virtual flexural strain compatible with the virtual vertical displacement $\delta v_0(x)$ is defined as:

$$\delta\kappa_0(x) = \frac{\partial^2\delta v_0(x)}{\partial x^2} \tag{5.62}$$

Therefore, Eq. (5.59) is written as:

$$\delta\bar{U}_{int} = M(x)\delta\kappa_0(x)dx \tag{5.63}$$

For the whole length of a vertically loaded beam, the virtual strain energy $\delta U_{int}$ is:

$$\delta U_{int} = \int_L \delta\bar{U}_{int} = \int_L M(x)\delta\kappa_0(x)dx \tag{5.64}$$

If only linearly elastic material is considered, the following section stiffness relation

$$M(x) = IE(x)\kappa_0(x) \tag{5.65}$$

is substituted into Eq. (5.64) and the virtual strain energy is entirely expressed in terms of strain (deformation) as:

$$\delta U_{int} = \int_L \delta\bar{U}_{int} = \int_L \kappa_0(x)IE(x)\delta\kappa_0(x)dx \tag{5.66}$$

It is noted that $\kappa_0(x)$ is the real flexural strain associated with the real moment $M(x)$ through a constitutive relation. By recalling the flexural strain-displacement relations of Eqs. (5.61) and (5.62), the virtual strain energy is entirely expressed in terms of vertical displacements as:

$$\delta U_{int} = \int_L \delta \bar{U}_{int} = \int_L \frac{\partial^2 v_0(x)}{\partial x^2} IE(x) \frac{\partial^2 \delta v_0(x)}{\partial x^2} dx \qquad (5.67)$$

Based on Eq. (5.30), the internal virtual work is written as:

$$\delta W_{int} = -\int_L M(x) \delta \kappa_0(x) dx \qquad (5.68)$$

Based on the compatibility expression relating the deformation (resultant strain) to the displacement field (5.61), it becomes clear that the *"kinematically admissible"* virtual displacement field has to satisfy the following requirements:

1. The virtual displacement field and its first derivative must be continuous and must possess the continuous second derivative at every point inside the member. In other words, it has $C^1$ continuity.
2. The magnitude of the virtual displacement must be in the range of validity of the small displacement theory.

If the virtual displacement field satisfies the above requirements as well as geometric (essential) boundaries, it becomes a *"geometrically admissible"* virtual displacement field. The *"kinematically admissible"* and *"geometrically admissible"* virtual strain fields as are defined ones which are related to the *"kinematically admissible"* and *"geometrically admissible"* virtual displacement fields, respectively through compatibility relation (5.61).

Eqs. (5.41), (5.51), and (5.64) form the virtual strain energy expression for a plane Timoshenko frame member and can be written in the matrix form as:

$$\delta U_{int} = \int_L \delta \varepsilon_{TB}(x)^T \sigma_{TB}(x) dx \qquad (5.69)$$

where the vector $\delta \varepsilon_{TB}(x)$ contains the virtual resultant strains and is defined as:

$$\delta \varepsilon_{TB}(x) = \lfloor \delta \varepsilon_0(x) \quad \delta \kappa_0(x) \quad \delta \gamma_0(x) \rfloor^T \qquad (5.70)$$

and the vector $\sigma_{TB}(x)$ contains the real resultant stresses and is defined as:

$$\sigma_{TB}(x) = \lfloor N(x) \quad M(x) \quad V(x) \rfloor^T \qquad (5.71)$$

If only linearly elastic material is considered, the following section stiffness matrix

$$\begin{Bmatrix} N(x) \\ M(x) \\ V(x) \end{Bmatrix} = \begin{bmatrix} EA(x) & 0 & 0 \\ 0 & EI(x) & 0 \\ 0 & 0 & GA_s(x) \end{bmatrix} \begin{Bmatrix} \varepsilon_0(x) \\ \kappa_0(x) \\ \gamma_0(x) \end{Bmatrix} \text{ or } \sigma_{TB}(x) = \mathbf{E}_{TB}(x) \varepsilon_{TB}(x) \qquad (5.72)$$

is substituted into Eq. (5.69) and the virtual strain energy is entirely expressed in terms of strains (resultant strains) as:

$$\delta U_{int} = \int_L \begin{Bmatrix} \delta\varepsilon_0(x) \\ \delta\kappa_0(x) \\ \delta\gamma_0(x) \end{Bmatrix}^T \begin{bmatrix} EA(x) & 0 & 0 \\ 0 & EI(x) & 0 \\ 0 & 0 & GA_s(x) \end{bmatrix} \begin{Bmatrix} \varepsilon_0(x) \\ \kappa_0(x) \\ \gamma_0(x) \end{Bmatrix} dx \qquad (5.73)$$

or in the compact form as:

$$\delta U_{int} = \int_L \delta\varepsilon_{TB}(x)^T \mathbf{E}_{TB}(x)\varepsilon_{TB}(x)dx \qquad (5.74)$$

Eqs. (2/2.33-35) relate the sectional deformations (resultant strains) to the sectional displacements and can be expressed in the matrix fashion as:

$$\begin{Bmatrix} \varepsilon_0(x) \\ \kappa_0(x) \\ \gamma_0(x) \end{Bmatrix} = \begin{bmatrix} \dfrac{\partial}{\partial x} & 0 & 0 \\ 0 & 0 & \dfrac{\partial}{\partial x} \\ 0 & \dfrac{\partial}{\partial x} & -1 \end{bmatrix} \begin{Bmatrix} u_0(x) \\ v_0(x) \\ \alpha_0(x) \end{Bmatrix} \text{ or } \varepsilon_{TB}(x) = \mathbf{L}_{TB}\mathbf{u}_{TB}(x) \qquad (5.75)$$

In similar fashion, the matrix expression relating the virtual sectional deformations (resultant strains) to the virtual sectional displacements is

$$\begin{Bmatrix} \delta\varepsilon_0(x) \\ \delta\kappa_0(x) \\ \delta\gamma_0(x) \end{Bmatrix} = \begin{bmatrix} \dfrac{\partial}{\partial x} & 0 & 0 \\ 0 & 0 & \dfrac{\partial}{\partial x} \\ 0 & \dfrac{\partial}{\partial x} & -1 \end{bmatrix} \begin{Bmatrix} \delta u_0(x) \\ \delta v_0(x) \\ \delta\alpha_0(x) \end{Bmatrix} \text{ or } \delta\varepsilon_{TB}(x) = \mathbf{L}_{TB}\delta\mathbf{u}_{TB}(x) \qquad (5.76)$$

Substituting Eqs. (5.75) and (5.76) into (5.74), the virtual strain energy is entirely expressed in terms of displacements as:

$$\delta U_{int} = \int_L \begin{Bmatrix} \delta u_0(x) \\ \delta v_0(x) \\ \delta\alpha_0(x) \end{Bmatrix}^T \begin{bmatrix} \dfrac{\partial}{\partial x} & 0 & 0 \\ 0 & 0 & \dfrac{\partial}{\partial x} \\ 0 & \dfrac{\partial}{\partial x} & -1 \end{bmatrix}^T \begin{bmatrix} EA(x) & 0 & 0 \\ 0 & EI(x) & 0 \\ 0 & 0 & GA_s(x) \end{bmatrix} \begin{bmatrix} \dfrac{\partial}{\partial x} & 0 & 0 \\ 0 & 0 & \dfrac{\partial}{\partial x} \\ 0 & \dfrac{\partial}{\partial x} & -1 \end{bmatrix} \begin{Bmatrix} u_0(x) \\ v_0(x) \\ \alpha_0(x) \end{Bmatrix} dx$$

$$(5.77)$$

or in the compact form as:

$$\delta U_{int} = \int_L \delta\mathbf{u}_{TB}(x)^T \mathbf{L}_{TB}^T \mathbf{E}_{TB}(x)\mathbf{L}_{TB}\mathbf{u}_{TB}(x)dx \qquad (5.78)$$

For an Euler-Bernoulli frame member, the shear effects are ignored. Therefore, its virtual strain energy is written as:

$$\delta U_{int} = \int_L \delta \varepsilon_{EB}(x)^T \sigma_{EB}(x) dx = \int_L \begin{Bmatrix} \delta \varepsilon_0(x) \\ \delta \kappa_0(x) \end{Bmatrix}^T \begin{Bmatrix} N(x) \\ M(x) \end{Bmatrix} dx \qquad (5.79)$$

where the vector $\delta \varepsilon_{EB}(x)$ contains the virtual resultant strains and is defined as:

$$\delta \varepsilon_{EB}(x) = \lfloor \delta \varepsilon_0(x) \quad \delta \kappa_0(x) \rfloor^T \qquad (5.80)$$

and the vector $\sigma_{EB}(x)$ contains the real resultant stresses and is defined as:

$$\sigma_{EB}(x) = \lfloor N(x) \quad M(x) \rfloor^T \qquad (5.81)$$

If only linearly elastic material is considered, the following section stiffness matrix

$$\begin{Bmatrix} N(x) \\ M(x) \end{Bmatrix} = \begin{bmatrix} EA(x) & 0 \\ 0 & EI(x) \end{bmatrix} \begin{Bmatrix} \varepsilon_0(x) \\ \kappa_0(x) \end{Bmatrix} \text{ or } \sigma_{EB}(x) = \mathbf{E}_{EB}(x) \varepsilon_{EB}(x) \qquad (5.82)$$

is substituted into Eq. (5.79) and the virtual strain energy is expressed entirely in terms of strains (resultant strains) as:

$$\delta U_{int} = \int_L \delta \varepsilon_{EB}(x)^T \mathbf{E}_{EB}(x) \varepsilon_{EB}(x) dx = \int_L \begin{Bmatrix} \delta \varepsilon_0(x) \\ \delta \kappa_0(x) \end{Bmatrix}^T \begin{bmatrix} EA(x) & 0 \\ 0 & EI(x) \end{bmatrix} \begin{Bmatrix} \varepsilon_0(x) \\ \kappa_0(x) \end{Bmatrix} dx \qquad (5.83)$$

Eqs. (2/2.3) and (2/2.17) relate the sectional deformations (resultant strains) to the sectional displacements and can be expressed in the matrix fashion as:

$$\begin{Bmatrix} \varepsilon_0(x) \\ \kappa_0(x) \end{Bmatrix} = \begin{bmatrix} \dfrac{\partial}{\partial x} & 0 \\ 0 & \dfrac{\partial^2}{\partial x^2} \end{bmatrix} \begin{Bmatrix} u_0(x) \\ v_0(x) \end{Bmatrix} \text{ or } \varepsilon_{EB}(x) = \mathbf{L}_{EB} \mathbf{u}_{EB}(x) \qquad (5.84)$$

In similar fashion, the matrix expression relating the virtual sectional deformations (resultant strains) to the virtual sectional displacements is

$$\begin{Bmatrix} \delta \varepsilon_0(x) \\ \delta \kappa_0(x) \end{Bmatrix} = \begin{bmatrix} \dfrac{\partial}{\partial x} & 0 \\ 0 & \dfrac{\partial^2}{\partial x^2} \end{bmatrix} \begin{Bmatrix} \delta u_0(x) \\ \delta v_0(x) \end{Bmatrix} \text{ or } \delta \varepsilon_{EB}(x) = \mathbf{L}_{EB} \delta \mathbf{u}_{EB}(x) \qquad (5.85)$$

Substituting Eqs. (5.84) and (5.85) into (5.83), the virtual strain energy is expressed entirely in terms of displacements as:

$$\delta U_{int} = \int_L \begin{Bmatrix} \delta u_0(x) \\ \delta v_0(x) \end{Bmatrix}^T \begin{bmatrix} \dfrac{\partial}{\partial x} & 0 \\ 0 & \dfrac{\partial^2}{\partial x^2} \end{bmatrix}^T \begin{bmatrix} EA(x) & 0 \\ 0 & EI(x) \end{bmatrix} \begin{bmatrix} \dfrac{\partial}{\partial x} & 0 \\ 0 & \dfrac{\partial^2}{\partial x^2} \end{bmatrix} \begin{Bmatrix} u_0(x) \\ v_0(x) \end{Bmatrix} dx \qquad (5.86)$$

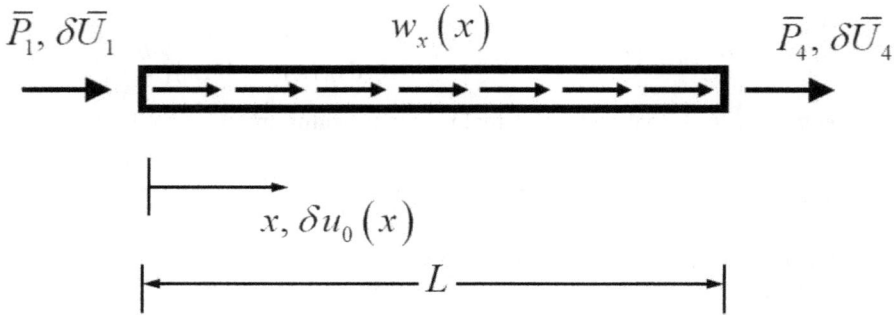

**FIGURE 5.15**  External Virtual Work: Bar.

or in the compact form as:

$$\delta U_{int} = \int_L \delta \mathbf{u}_{EB}(x)^T \mathbf{L}_{EB}^T \mathbf{E}_{EB}(x) \mathbf{L}_{EB} \mathbf{u}_{EB}(x) dx \tag{5.87}$$

## 5.5  EXTERNAL VIRTUAL WORK: A FRAME MEMBER

In the previous section, the expressions of the virtual strain energy for several structural actions were derived. In this section, the main objective is to derive the expressions of the external virtual work due to several external loadings.

Consider the axially loaded member (bar) of Figure 5.15. This member is in equilibrium under the system of real forces: external end forces ($\bar{P}_1$ and $\bar{P}_4$) and member load $w_x(x)$. The external virtual work done by the system of real forces on the virtual end displacements ($\delta \bar{U}_1$ and $\delta \bar{U}_4$) and virtual displacement field $\delta u_0(x)$ is:

$$\delta W_{ext} = \delta \bar{U}_1 \bar{P}_1 + \delta \bar{U}_4 \bar{P}_4 + \int_L \delta u_0(x) w_x(x) dx \tag{5.88}$$

Consider the vertically loaded member (beam) of Figure 5.16. This member is in equilibrium under the system of real forces: external end forces ($\bar{P}_2$ and $\bar{P}_5$), external end moments ($\bar{P}_3$ and

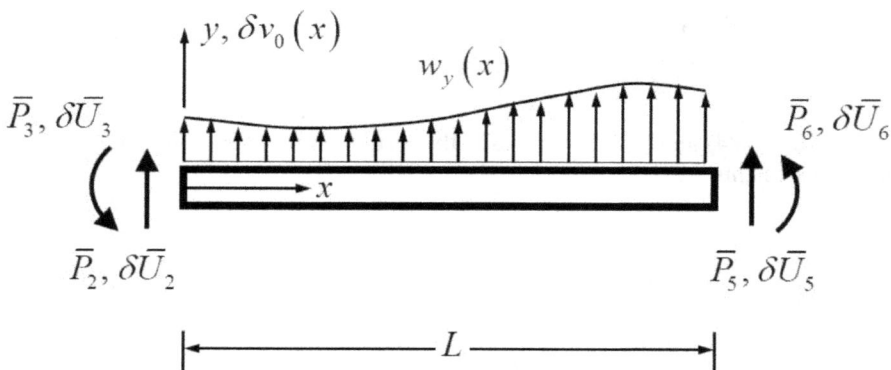

**FIGURE 5.16**  External Virtual Work: Beam.

$\overline{P}_6$), and member load $w_y(x)$. The external virtual work done by the system of real forces on the virtual end displacements ($\delta\overline{U}_2$ and $\delta\overline{U}_5$), virtual end rotations ($\delta\overline{U}_3$ and $\delta\overline{U}_6$), and virtual displacement field $\delta v_0(x)$ is:

$$\delta W_{ext} = \delta\overline{U}_2\overline{P}_2 + \delta\overline{U}_3\overline{P}_3 + \delta\overline{U}_5\overline{P}_5 + \delta\overline{U}_6\overline{P}_6 + \int_L \delta v_0(x)w_y(x)dx \qquad (5.89)$$

Eqs. (5.88) and (5.89) form the virtual work expression for a plane frame member and can be written together as: (5.90)

$$\delta W_{ext} = \delta\overline{U}_1\overline{P}_1 + \delta\overline{U}_2\overline{P}_2 + \delta\overline{U}_3\overline{P}_3 + \delta\overline{U}_4\overline{P}_4 + \delta\overline{U}_5\overline{P}_5 + \delta\overline{U}_6\overline{P}_6 +$$
$$\int_L \delta u_0(x)w_x(x)dx + \int_L \delta v_0(x)w_y(x)dx \qquad (5.90)$$

or in the matrix form as:

$$\delta W_{ext} = \delta\overline{\mathbf{U}}^T\overline{\mathbf{P}} + \int_L \delta\mathbf{u}(x)^T\mathbf{w}(x)dx \qquad (5.91)$$

where the member end displacement vector $\overline{\mathbf{U}}$ is defined as $\lfloor \overline{U}_1 \quad \overline{U}_2 \quad \overline{U}_3 \quad \overline{U}_4 \quad \overline{U}_5 \quad \overline{U}_6 \rfloor^T$; the member end force vector $\overline{\mathbf{P}}$ is defined as $\lfloor \overline{P}_1 \quad \overline{P}_2 \quad \overline{P}_3 \quad \overline{P}_4 \quad \overline{P}_5 \quad \overline{P}_6 \rfloor^T$; the member displacement field vector $\mathbf{u}(x)$ is defined as $\lfloor u_0(x) \quad v_0(x) \rfloor^T$; and the member load vector $\mathbf{w}(x)$ is defined as $\lfloor w_x(x) \quad w_y(x) \rfloor^T$.

## 5.6   VIRTUAL DISPLACEMENT PRINCIPLE: A FRAME MEMBER

In the previous two sections, the expressions for the virtual strain energy and the external virtual work of a frame member under several loading actions were derived. In this section, the virtual displacement equations for several types of frame members are set up. It should be noted that the virtual displacement equation is an alternative way to express the equilibrium equation.

### BAR MEMBER:

From Eqs. (5.88) and (5.39), the total virtual work expression (5.31) for a bar member can be written as:

$$\delta W = \underbrace{\delta\overline{U}_1\overline{P}_1 + \delta\overline{U}_4\overline{P}_4 + \int_L \delta u_0(x)w_x(x)dx}_{\delta W_{ext}} - \underbrace{\int_L N(x)\delta\varepsilon_0(x)dx}_{\delta U_{int}} = 0 \qquad (5.92)$$

### TIMOSHENKO BEAM MEMBER:

From Eqs. (5.89), (5.63), and (5.51), the total virtual work expression (5.31) for a Timoshenko beam member can be written as:

$$\delta W = \begin{bmatrix} \underbrace{\delta \bar{U}_2 \bar{P}_2 + \delta \bar{U}_3 \bar{P}_3 + \delta \bar{U}_5 \bar{P}_5 + \delta \bar{U}_6 \bar{P}_6 + \int_L \delta v_0(x) w_y(x) dx}_{\delta W_{ext}} \\ \underbrace{-\left( \int_L V(x) \delta \gamma_0(x) dx + \int_L M(x) \delta \kappa_0(x) dx \right)}_{\delta U_{int}} \end{bmatrix} = 0 \qquad (5.93)$$

### EULER-BERNOULLI BEAM MEMBER:

From Eqs. (5.89) and (5.63), the total virtual work expression (5.31) for an Euler-Bernoulli beam member can be written as:

$$\delta W = \underbrace{\delta \bar{U}_2 \bar{P}_2 + \delta \bar{U}_3 \bar{P}_3 + \delta \bar{U}_5 \bar{P}_5 + \delta \bar{U}_6 \bar{P}_6 + \int_L \delta v_0(x) w_y(x) dx}_{\delta W_{ext}} - \underbrace{\left( \int_L M(x) \delta \kappa_0(x) dx \right)}_{\delta U_{int}} = 0 \qquad (5.94)$$

### TIMOSHENKO FRAME MEMBER:

From Eqs. (5.90), (5.39), (5.51), and (5.63), the total virtual work expression (5.31) for a Timoshenko frame member can be written as:

$$\delta W = \begin{bmatrix} \underbrace{\begin{bmatrix} \delta \bar{U}_1 \bar{P}_1 + \delta \bar{U}_2 \bar{P}_2 + \delta \bar{U}_3 \bar{P}_3 + \delta \bar{U}_4 \bar{P}_4 + \delta \bar{U}_5 \bar{P}_5 + \delta \bar{U}_6 \bar{P}_6 \\ + \int_L \delta u_0(x) w_x(x) dx + \int_L \delta v_0(x) w_y(x) dx \end{bmatrix}}_{\delta W_{ext}} \\ \underbrace{-\left[ \int_L N(x) \delta \varepsilon_0(x) dx + \int_L V(x) \delta \gamma_0(x) dx + \int_L M(x) \delta \kappa_0(x) dx \right]}_{\delta U_{int}} \end{bmatrix} = 0 \qquad (5.95)$$

in the matrix form as:

$$\delta W = \underbrace{\delta \bar{\mathbf{U}}^T \bar{\mathbf{P}} + \int_L \delta \mathbf{u}(x)^T \mathbf{w}(x) dx}_{\delta W_{ext}} - \underbrace{\int_L \delta \varepsilon_{TB}(x)^T \sigma_{TB}(x) dx}_{\delta U_{int}} = 0 \qquad (5.96)$$

### EULER-BERNOULLI FRAME MEMBER:

From Eqs. (5.90), (5.39), and (5.63), the total virtual work expression (5.31) for an Euler-Bernoulli frame member can be written as:

$$\delta W = \begin{bmatrix} \underbrace{\begin{bmatrix} \delta \bar{U}_1 \bar{P}_1 + \delta \bar{U}_2 \bar{P}_2 + \delta \bar{U}_3 \bar{P}_3 + \delta \bar{U}_4 \bar{P}_4 + \delta \bar{U}_5 \bar{P}_5 + \delta \bar{U}_6 \bar{P}_6 \\ + \int_L \delta u_0(x) w_x(x) dx + \int_L \delta v_0(x) w_y(x) dx \end{bmatrix}}_{\delta W_{ext}} \\ \underbrace{-\left[ \int_L N(x) \delta \varepsilon_0(x) dx + \int_L M(x) \delta \kappa_0(x) dx \right]}_{\delta U_{int}} \end{bmatrix} = 0 \qquad (5.97)$$

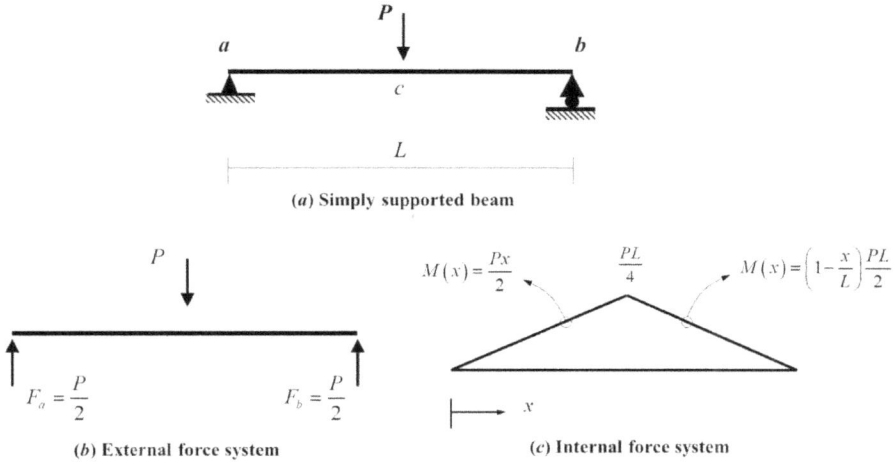

(a) Simply supported beam

(b) External force system                     (c) Internal force system

**FIGURE 5.17**   Example 5.2.

in the matrix form as:

$$\delta W = \delta \overline{\mathbf{U}}^T \overline{\mathbf{P}} + \underbrace{\int_L \delta \mathbf{u}(x)^T \mathbf{w}(x)dx}_{\delta W_{ext}} - \underbrace{\int_L \delta \varepsilon_{EB}(x)^T \sigma_{EB}(x)dx}_{\delta U_{int}} = 0 \qquad (5.98)$$

## EXAMPLE 5.2

Consider the simply-supported beam subjected to the concentrated load $P$ at mid-span as shown in Figure 5.17 $a$. The equilibrated external and internal force systems are given in Figure 5.17 $b$ and $c$, respectively.

Verify that the concentrated point load $P$ can be determined from the virtual displacement principle based on the following virtual displacements:

(a) $\delta v_0(x) = -\dfrac{16}{5L^4}\left(L^3 x - 2Lx^3 + x^4\right)$; (b) $\delta v_0(x) = -\sin\left(\dfrac{\pi x}{L}\right)$; and

(c) $\delta v_0(x) = -\left(\dfrac{1}{2} - \cos\left(\dfrac{\pi x}{2L}\right)\right)$

## SOLUTION:

(a) $\delta v_0(x) = -\dfrac{16}{5L^4}\left(L^3 x - 2Lx^3 + x^4\right)$

This given virtual displacement field is kinematically admissible since it satisfies the following requirements:

1. *The virtual displacement field must be continuous and must possess the continuous first derivative at every point inside the member.*
2. *The magnitude of the virtual displacement must be in the range of validity of the small displacement theory.*

It also satisfies the geometric (essential) boundaries:

$$\delta v_0(0) = 0 \text{ and } \delta v_0(L) = 0 \rightarrow \delta v_a = 0 \text{ and } \delta v_b = 0$$

Therefore, this virtual displacement field is geometrically admissible.

The virtual flexural curvature (resultant strain) compatible with the virtual displacement is defined as:

$$\delta\kappa_0(x) = \frac{\partial^2 \delta v_0(x)}{\partial x^2} = -16\frac{(-12Lx+12x^2)}{5L^4}$$

The external virtual work is defined as:

$$\delta W_{ext} = F_c \underbrace{\delta v_c}_{-1} + F_a \underbrace{\delta v_a}_{0} + F_b \underbrace{\delta v_b}_{0}$$

The internal virtual work is defined as:

$$\delta W_{int} = -\int_L M(x)\delta\kappa_0(x)\,dx$$

$$\delta W_{int} = -\delta U_{int} = -\left[\int_0^{\frac{L}{2}} \frac{Px}{2}\left(-16\frac{(-12Lx+12x^2)}{5L^4}\right)dx + \int_{\frac{L}{2}}^L \left(1-\frac{x}{L}\right)\frac{PL}{2}\left(-16\frac{(-12Lx+12x^2)}{5L^4}\right)dx\right] = -P$$

The virtual displacement equation:

$$\delta W = \delta W_{ext} + \delta W_{int} = -F_c - P = 0 \rightarrow F_c = -P = P \quad \downarrow$$

(b) $\delta v_0(x) = -\sin\left(\frac{\pi x}{L}\right)$

This given virtual displacement field is kinematically admissible since it satisfies the following requirements:

1.  *The virtual displacement field must be continuous and must possess the continuous first derivative at every point inside the member.*
2.  *The magnitude of the virtual displacement must be in the range of validity of the small displacement theory.*

It also satisfies geometric (essential) boundaries:

$$\delta v_0(0) = 0 \text{ and } \delta v_0(L) = 0 \rightarrow \delta v_a = 0 \text{ and } \delta v_b = 0$$

Therefore, this virtual displacement field is geometrically admissible.

The virtual flexural curvature (resultant strain) compatible with the virtual displacement is defined as:

$$\delta\kappa_0(x) = \frac{\partial^2 \delta v_0(x)}{\partial x^2} = \frac{\pi^2}{L^2}\sin\frac{\pi x}{L}$$

The external virtual work is defined as:

$$\delta W_{ext} = F_c \underbrace{\delta v_c}_{-1} + F_a \underbrace{\delta v_a}_{0} + F_b \underbrace{\delta v_b}_{0}$$

The internal virtual work is defined as:

$$\delta W_{int} = -\int_L M(x)\delta\kappa_0(x)dx$$

$$\delta W_{int} = -\delta U_{int} = -\left[\int_0^{\frac{L}{2}}\frac{Px}{2}\left(\frac{\pi^2}{L^2}\sin\frac{\pi x}{L}\right)dx + \int_{\frac{L}{2}}^L\left(1-\frac{x}{L}\right)\frac{PL}{2}\left(\frac{\pi^2}{L^2}\sin\frac{\pi x}{L}\right)dx\right] = -P$$

The virtual displacement equation:

$$\delta W = \delta W_{ext} + \delta W_{int} = -F_c - P = 0 \;\rightarrow\; F_c = -P = P \;\downarrow$$

**Discussion of (a) and (b):** To obtain the *"exact"* value of the mid-span load $P$, it is simply required to use the *"kinematically admissible"* virtual displacement field. This is true as long as the real force systems (both internal and external) satisfy the governing differential equation:

$$\frac{\partial^2 M(x)}{\partial x^2} - w_y(x) = 0$$

as well as the force boundary conditions. It should be kept in mind that the virtual displacement equation is an alternative way to express the equilibrium equation.

It is noted that these two choices of virtual displacement field are preferable since they eliminate the external virtual work contributed by the support reactions. Therefore, there is no need to determine the support reactions beforehand.

(c) $\delta v_0(x) = -\left(\dfrac{1}{2} - \cos\left(\dfrac{\pi x}{2L}\right)\right)$

This given virtual displacement field is kinematically admissible since it satisfies the following requirements:

1. *The virtual displacement field must be continuous and must possess the continuous first derivative at every point inside the member.*
2. *The magnitude of the virtual displacement must be in the range of validity of the small displacement theory.*

However, it violates geometric (essential) boundaries:

$$\delta v_0(0) = 0.5 \text{ and } \delta v_0(L) = -0.5 \;\rightarrow\; \delta v_a = 0.5 \text{ and } \delta v_b = -0.5$$

The virtual flexural curvature (resultant strain) compatible with the virtual displacement is defined as:

$$\delta\kappa_0(x) = \frac{\partial^2\delta v_0(x)}{\partial x^2} = -\frac{\pi^2\cos\left(\dfrac{\pi x}{2L}\right)}{4L^2}$$

The external virtual work is defined as:

$$\delta W_{ext} = F_c\delta v_c + F_a\delta v_a + F_b\delta v_b$$

$$\delta v_c = 0.207107$$

$$\delta W_{ext} = (0.207107)\, F_c + (0.5)\frac{P}{2} + (-0.5)\frac{P}{2} = 0.207107 F_c$$

The internal virtual work is defined as:

$$\delta W_{int} = -\int_L M(x)\,\delta\kappa_0(x)\,dx$$

$$\delta W_{int} = -\delta U_{int} = -\left[ \int_0^{\frac{L}{2}} \frac{Px}{2}\left( -\frac{\pi^2 \cos\left(\dfrac{\pi x}{2L}\right)}{4L^2} \right)dx + \int_{\frac{L}{2}}^L \left(1-\frac{x}{L}\right)\frac{PL}{2}\left( -\frac{\pi^2 \cos\left(\dfrac{\pi x}{2L}\right)}{4L^2} \right)dx \right] = 0.207107P$$

The virtual displacement equation:

$$\delta W = \delta W_{ext} + \delta W_{int} = 0.207107 F_c + 0.207107P = 0 \quad \Rightarrow \quad F_c = -P = P \;\;\downarrow$$

**Discussion of (c):** In this case, even though the virtual displacement field violates geometric boundaries, the "*exact*" value of the mid-span load $P$ can still be recovered as long as the real force systems (both internal and external) satisfy the governing differential equation:

$$\frac{\partial^2 M(x)}{\partial x^2} - w_y(x) = 0$$

as well as the force boundary conditions. However, this choice might not be a wise one since it requires that the support reactions must be given.

## 5.7   COMPLEMENTARY VIRTUAL WORK PRINCIPLE: PARTICLES AND RIGID BODIES

In the virtual work principle, the changes in external and internal work accounting for a disturbance of a virtual displaced state about a real state of forces (resultant stresses) are considered. The virtual work is defined as the work done by the real forces on the virtual displacement. It is noticed that the virtual work equation is an alternative way to express the equilibrium condition. Since forces (resultant stresses) and deformations (resultant strains) are the dual quantities of structural analysis, it is quite natural to expect that the conjugate pair of the virtual work principle must exist. This is certainly true and the dual principle is known as "*the complementary virtual work principle*" or "*the virtual force principle*". In this principle, the virtual work is defined as the work done by an imaginary virtual force system in moving through real displacements.

The derivation of the virtual force principle for a particle under a system of virtual forces $\delta \vec{\mathbf{P}}$ starts with the assumption that the particle displaces through the real displacement $\bar{\mathbf{u}}$ ($O'O$). The resulting virtual work is written as:

$$\delta W_{ext}^* = \bar{\mathbf{u}}.\left( \sum_{i=1}^n \delta \vec{\mathbf{P}}_i \right) \tag{5.99}$$

After carrying out the scalar products in Eq. (5.99), one has:

$$\delta W_{ext}^* = u_{O'O}\, \delta P_1 \cos\theta_1 + \ldots + u_{O'O}\, \delta P_n \cos\theta_n = u_{O'O}\left(\sum_{i=1}^{n} \delta P_i \cos\theta_i\right) \qquad (5.100)$$

If the virtual force system is selected such that they are in equilibrium:

$$\sum_{i=1}^{n} \delta P_i \cos\theta_i = 0 \qquad (5.101)$$

Consequently, the virtual work of these equilibrated virtual forces under the real displacement vanishes:

$$\delta W_{ext}^* = 0 \qquad (5.102)$$

The understanding of the virtual force principle can be enriched by considering the rigid block of Figure 5.18 subjected to an external force $F$ and a magnetic force $F_M$. The magnetic force is induced by a magnet located at a distance $L$ from the rigid block. The magnitude of $F_M$ is assumed to be proportional to the inverse of the third-degree of the distance $x$ from the block to the magnet.

$$F_M = \frac{k}{x^3} \text{ or } x = \sqrt[3]{\frac{k}{F_M}} \qquad (5.103)$$

where $k$ is the constant for the magnetic field.

Let the block reach its equilibrated state after moving through the real displacement $\Delta$. That is,

$$F - F_M = 0 \qquad (5.104)$$

This equilibrated state is disturbed by imposing the virtual force system, which satisfies the equilibrium equation:

$$\delta F = \delta F_M \qquad (5.105)$$

The virtual work done by this virtual force system through the real displacement $\Delta$ is

$$\delta W^* = -\delta F_M \Delta + \delta F \Delta \qquad (5.106)$$

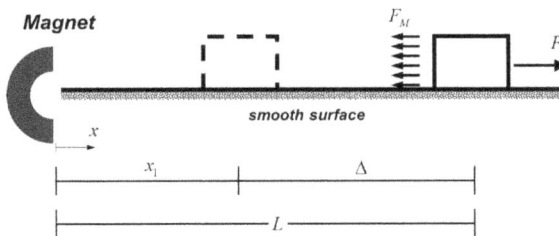

**FIGURE 5.18**    Rigid Block Subjected to a Magnetic Force.

One notices that:

$$\Delta = L - \sqrt[3]{\frac{k}{F_M}} = L - \sqrt[3]{\frac{k}{F}} \tag{5.107}$$

By substituting Eqs. (5.107) into (5.106) and imposing Eq. (5.105), the virtual work expression is rewritten as:

$$\delta W^* = \left( -\Delta + L - \sqrt[3]{\frac{k}{F}} \right) \delta F = 0 \tag{5.108}$$

Due to arbitrariness of $\delta F$, one has:

$$-\Delta + L - \sqrt[3]{\frac{k}{F}} = 0 \tag{5.109}$$

The above relation is simply the kinematical relation of this problem. Therefore, it can be stated that the virtual force equation is an alternative way to express the compatibility equation of the system.

The derivation of the virtual force principle for a rigid body under a system of virtual forces $\delta \vec{\mathbf{P}}$ starts with the assumption that the rigid body displaces through the real rigid-body movement. One has:

$$\delta W^* = \delta W^*_{ext} = \vec{\mathbf{u}} \cdot \left( \sum_{i=1}^{n} \delta \vec{\mathbf{P}}_i \right) + \vec{\omega} \cdot \left( \sum_{i=1}^{n} \delta \vec{\mathbf{P}}_i \times \vec{\mathbf{d}}_i \right) \tag{5.110}$$

After carrying out the scalar products in Eq. (5.110), one has:

$$\delta W^* = \delta W^*_{ext} = \left( \sum_{i=1}^{n} \delta P_i \right) u \cos \theta_i + \left( \sum_{i=1}^{n} \delta P_i \times d_i \right) \omega \cos \psi_i \tag{5.111}$$

If the virtual force system is selected such that they are in equilibrium:

$$\sum_{i=1}^{n} \delta P_i = 0 \text{ and } \sum_{i=1}^{n} \delta P_i \times d_i = 0 \tag{5.112}$$

Consequently, the virtual work of these equilibrated virtual forces under the real displacements vanishes:

$$\delta W^*_{ext} = 0 \tag{5.113}$$

Without additional comments, the virtual force principle for a rigid particle and a rigid body can be expressed in a verbal manner as:

*The real displacements of a rigid particle and a rigid body in reaching their compatible configuration are such that the total virtual work vanishes for any system of statically admissible virtual forces.*

The principle of virtual forces is not very useful for rigid body problems but becomes very useful for deformable body problems since it is an alternative way to express the compatibility equations of the system.

## 5.8  COMPLEMENTARY VIRTUAL WORK PRINCIPLE: DEFORMABLE BODIES

In the previous section, the complementary virtual work principle or, more specifically, the virtual force principle were derived and applied to rigid particles and rigid bodies. As mentioned previously, the virtual force principle is just an alternative way to describe the compatibility conditions. The application of the virtual force principle to rigid particles and rigid bodies is relatively useless. This is due to the fact that the crucial roles of material properties and stress-strain natures in determining the results are meaningless in problems of undeformable systems. However, the virtual force principle becomes a very powerful tool in structural analysis when deformable natures of systems are accounted for.

The verbal statement of the virtual force principle for deformable systems can simply be modified from that of the virtual displacement principle. This is done by replacing the terms *"stress (resultant stress)"* and *"force"* with *"strain (resultant strain)"* and *"displacement"* and interchanging the terms *"equilibrium"* with *"compatibility"*. Therefore, the virtual force principle can be stated as:

*The strains (resultant strains) and displacements in a deformable body are compatible and consistent with the geometric constraints if the summation of the total external complementary virtual work and the total internal complementary virtual work vanishes for any system of statically admissible virtual forces.*

The algebraic statement of the virtual force principle is:

$$\delta W^* = \delta W_{ext}^* + \delta W_{int}^* = 0 \tag{5.114}$$

or

$$\delta W^* = \delta W_{ext}^* - \delta U_{int}^* = 0 \tag{5.115}$$

where $\delta W_{int}^*$ is denoted for the complementary internal virtual work and $\delta U_{int}^*$ is denoted for the complementary virtual strain energy.

To account for deformable natures of systems (solids and structures), there is a need to distinguish between the complementary external and internal virtual work. The next two sections will be devoted to these two quantities.

## 5.9  COMPLEMENTARY INTERNAL VIRTUAL WORK AND COMPLEMENTARY VIRTUAL STRAIN ENERGY EXPRESSIONS: A FRAME MEMBER

Since frame structures are of great interest in this textbook, the primary modes of internal actions considered herein are axial, shear, and flexural responses. In the following subsections, the complementary virtual strain energy expression for each type of structural actions will be derived.

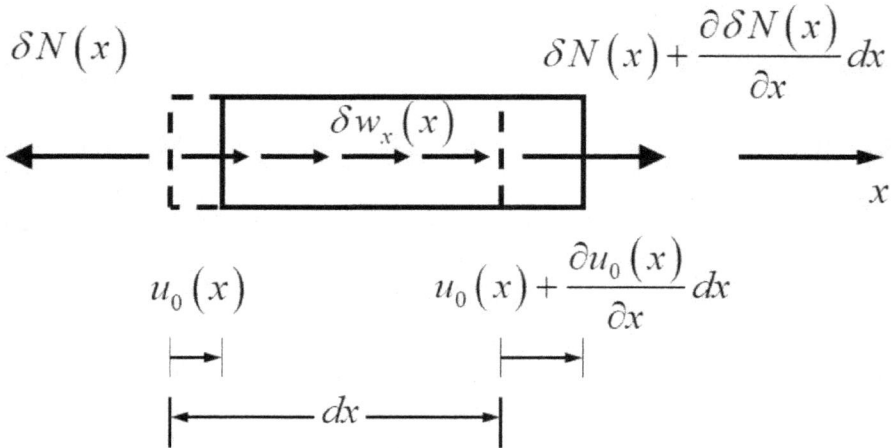

**FIGURE 5.19**  Complementary Virtual Strain Energy $\delta \bar{U}_{int}^{*}$ of an Infinitesimal Segment: Axial Response.

### 5.9.1 Axial Response: Virtual Force Principle

Consider the infinitesimal segment $dx$ of an axially loaded member of length $L$ as shown in Figure 5.19. This segment is in the compatible state with displacements $u_0(x)$ at the left end and $u_0(x) + \dfrac{\partial u_0(x)}{\partial x} dx$ at the right end. In Chapter 2, the axial strain compatible with these end displacements is defined as the rate of change in the axial displacement with respect to the axial coordinate.

$$\varepsilon_0(x) = \frac{\partial u_0(x)}{\partial x} \tag{5.116}$$

The whole infinitesimal segment is considered as a free body and is disturbed by the virtual force $\delta N(x)$ at the left end, by $\delta N(x) + \dfrac{\partial \delta N(x)}{\partial x} dx$ at the right end, and by the virtual external distributed load $\delta w_x(x)$ along the length. Consequently, the resulting virtual work is:

$$\delta W_{ext}^{*} = \left(\delta w_x(x) dx\right) u_0(x) - \delta N(x) u_0(x) + \left(\delta N(x) + \frac{\partial \delta N(x)}{\partial x} dx\right)\left(u_0(x) + \frac{\partial u_0(x)}{\partial x} dx\right) \tag{5.117}$$

$$\delta W_{ext}^{*} = \left(\delta w_x(x) dx\right) u_0(x) - \delta N(x) u_0(x) + \delta N(x) u_0(x) \quad +$$

$$\delta N(x)\frac{\partial u_0(x)}{\partial x} dx + \underbrace{\left(\frac{\partial \delta N(x)}{\partial x} dx\right)}_{-\delta w_x(x)} u_0(x) + \underbrace{\left(\frac{\partial \delta N(x)}{\partial x} dx\right)\left(\frac{\partial u_0(x)}{\partial x} dx\right)}_{2^{nd}\ order\ term} \tag{5.118}$$

By imposing the differential equilibrium equation (2/2.1) of a bar segment and neglecting the 2nd order term, Eq. (5.118) is written as:

$$\delta W_{ext}^* = \delta N(x) \frac{\partial u_0(x)}{\partial x} dx \tag{5.119}$$

Based on Eq. (5.115), one can conclude that:

$$\delta \bar{U}_{int}^* = \delta N(x) \frac{\partial u_0(x)}{\partial x} dx \tag{5.120}$$

where $\delta \bar{U}_{int}^*$ is denoted for the complementary virtual strain energy of an infinitesimal segment $dx$. Eq. (5.120) can also be written in term of strain as:

$$\delta \bar{U}_{int}^* = \delta N(x) \varepsilon_0(x) dx \tag{5.121}$$

For the whole length of an axially loaded member, the complementary virtual strain energy $\delta U_{int}^*$ is:

$$\delta U_{int}^* = \int_L \delta \bar{U}_{int}^* = \int_L \delta N(x) \varepsilon_0(x) dx \tag{5.122}$$

If only linearly elastic material is considered, the following section flexibility relation

$$\varepsilon_0(x) = \frac{N(x)}{EA(x)} \tag{5.123}$$

is substituted into Eq. (5.122) and the complementary virtual strain energy is entirely expressed in terms of internal axial force (resultant stress) as:

$$\delta U_{int}^* = \int_L \delta \bar{U}_{int}^* = \int_L \delta N(x) \left( \frac{1}{EA(x)} \right) N(x) dx \tag{5.124}$$

Based on Eq. (5.114), the complementary internal virtual work is written as:

$$\delta W_{int}^* = -\int_L \delta N(x) \varepsilon_0(x) dx \tag{5.125}$$

It is important to note that in the virtual force principle, the real displacement field $u_0(x)$ must be associated with the basic system (forces and deformations) in which there are no rigid body motions. Otherwise, it is not possible to establish the statically admissible virtual force field $\delta N(x)$.

### 5.9.2   SHEAR RESPONSE: VIRTUAL FORCE PRINCIPLE

Consider the infinitesimal segment $dx$ of a vertically loaded member of length $L$ as shown in Figure 5.20. This segment is in the compatible state with displacements $v_s(x)$ at the left end and $v_s(x) + \frac{\partial v_s(x)}{\partial x} dx$ at the right end. In Chapter 2, the shear strain is defined as the rate of change in the vertical displacement with respect to the axial coordinate.

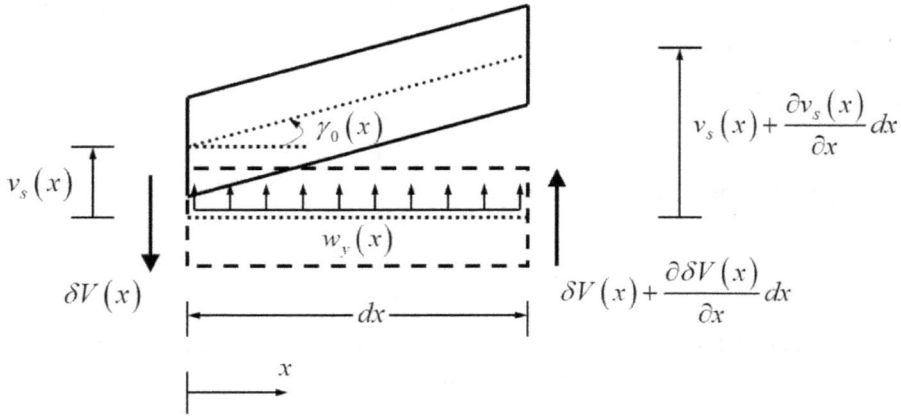

**FIGURE 5.20**   Complementary Virtual Strain Energy $\delta \bar{U}_{int}^{*}$ of an Infinitesimal Segment: Shear Response.

$$\gamma_0(x) = \frac{\partial v_s(x)}{\partial x} \tag{5.126}$$

The whole infinitesimal segment is considered as a free body and is disturbed by the virtual force $\delta V(x)$ at the left end, by $\delta V(x) + \dfrac{\partial \delta V(x)}{\partial x} dx$ at the right end, and by the virtual external distributed load $\delta w_y(x)$ along the length. Consequently, the resulting virtual work is:

$$\delta W_{ext}^{*} = \left(\delta w_y(x)dx\right)v_s(x) - \delta V(x)v_s(x) + \left(\delta V(x) + \frac{\partial \delta V(x)}{\partial x}dx\right)\left(v_s(x) + \frac{\partial v_s(x)}{\partial x}dx\right) \tag{5.127}$$

$$\delta W_{ext}^{*} = \left(\delta w_y(x)dx\right)v_s(x) - \delta V(x)v_s(x) + \delta V(x)v_s(x) \quad +$$

$$\delta V(x)\frac{\partial v_s(x)}{\partial x}dx + \underbrace{\frac{\partial \delta V(x)}{\partial x}dx}_{-\delta w_y(x)}v_s(x) + \underbrace{\left(\frac{\partial \delta V(x)}{\partial x}dx\right)\left(\frac{\partial v_s(x)}{\partial x}dx\right)}_{2^{nd}\ order\ term} \tag{5.128}$$

By imposing the vertical equilibrium equation (2/2.13) of a beam segment and neglecting the 2nd order term, Eq. (5.128) is written as:

$$\delta W_{ext}^{*} = \delta V(x)\frac{\partial v_s(x)}{\partial x}dx \tag{5.129}$$

Based on Eq. (5.115), it can be concluded that:

$$\delta \bar{U}_{int}^{*} = \delta V(x)\frac{\partial v_s(x)}{\partial x}dx \tag{5.130}$$

where $\delta \bar{U}_{int}^{*}$ is denoted for the complementary virtual strain energy of an infinitesimal segment $dx$. Eq. (5.130) can also be written in terms of strain as:

$$\delta \bar{U}_{int}^{*} = \delta V(x)\gamma_0(x)dx \tag{5.131}$$

For the whole length of a vertically loaded member, the complementary virtual strain energy $\delta U_{int}^{*}$ is:

$$\delta U_{int}^{*} = \int_{L} \delta \bar{U}_{int}^{*} = \int_{L} \delta V(x) \gamma_0(x) dx \qquad (5.132)$$

If only linearly elastic material is considered, the following section flexibility relation

$$\gamma_0(x) = \frac{V(x)}{GA_s(x)} \qquad (5.133)$$

is substituted into Eq. (5.132) and the complementary virtual strain energy is entirely expressed in terms of internal shear force (resultant stress) as:

$$\delta U_{int}^{*} = \int_{L} \delta \bar{U}_{int}^{*} = \int_{L} \delta V(x) \left( \frac{1}{GA_s(x)} \right) V(x) dx \qquad (5.134)$$

Based on Eq. (5.114), the complementary internal virtual work is written as:

$$\delta W_{int}^{*} = -\int_{L} \delta V(x) \gamma_0(x) dx \qquad (5.135)$$

It is important to note that in the virtual force principle, the real displacement field $v_s(x)$ must be associated with the basic system (forces and deformations) in which there are no rigid body motions. Otherwise, it is not possible to establish the statically admissible virtual force field $\delta V(x)$.

### 5.9.3 Flexural Response: Virtual Force Principle

Consider the infinitesimal segment $dx$ of a vertically loaded member of length $L$ as shown in Figure 5.21. This segment is in the compatible state with displacement $v_0(x)$ and rotation $\theta_z(x)$ at the left end and $v_0(x) + \dfrac{\partial v_0(x)}{\partial x} dx$ and $\theta_z(x) + \dfrac{\partial \theta_z(x)}{\partial x} dx$ at the right end. The whole infinitesimal segment is considered as a free body and is disturbed by the virtual force $\delta V(x)$ and virtual moment $\delta M(x)$ at the left end, by $\delta V(x) + \dfrac{\partial \delta V(x)}{\partial x} dx$ and $\delta M(x) + \dfrac{\partial \delta M(x)}{\partial x} dx$ at the right end, and by the virtual external distributed load $\delta w_y(x)$ along the length.

Consequently, the resulting virtual work is:

$$\begin{aligned}
\delta W_{ext}^{*} = {} & \left( \delta w_y(x) dx \right) v_0(x) - \delta V(x) v_0(x) + \\
& \left( \delta V(x) + \frac{\partial \delta V(x)}{\partial x} dx \right) \left( v_0(x) + \frac{\partial v_0(x)}{\partial x} dx \right) + \\
& \left( \delta M(x) + \frac{\partial \delta M(x)}{\partial x} dx \right) \left( \theta_z(x) + \frac{\partial \theta_z(x)}{\partial x} dx \right)
\end{aligned} \qquad (5.136)$$

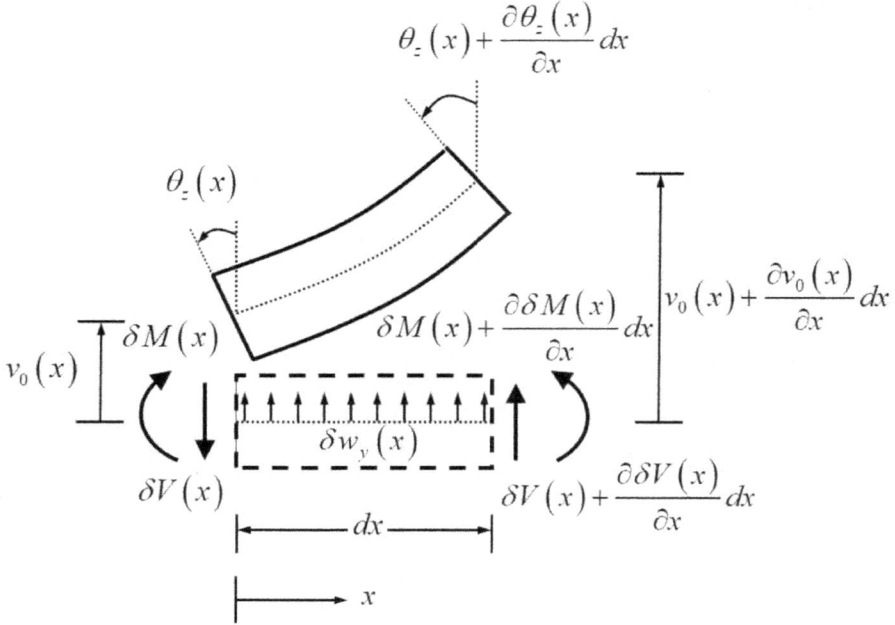

**FIGURE 5.21** Complementary Virtual Strain Energy $\delta\bar{U}^*_{int}$ of an Infinitesimal Segment: Flexural Response.

$$\delta W^*_{ext} = \begin{pmatrix} \left(\delta w_y(x)dx\right)v_0(x) - \delta V(x)v_0(x) - \delta M(x)\theta_z(x) + \delta V(x)v_0(x) \quad + \\[2mm] \delta V(x)\underbrace{\frac{\partial v_0(x)}{\partial x}}_{\theta_z(x)}dx + \left(\underbrace{\frac{\partial \delta V(x)}{\partial x}}_{-\delta w_y(x)}dx\right)v_0(x) + \underbrace{\left(\frac{\partial \delta V(x)}{\partial x}dx\right)\left(\frac{\partial v_0(x)}{\partial x}dx\right)}_{2^{nd}\ order\ term} + \\[2mm] \delta M(x)\theta_z(x) + \delta M(x)\frac{\partial \theta_z(x)}{\partial x}dx + \left(\underbrace{\frac{\partial \delta M(x)}{\partial x}}_{-\delta V(x)}dx\right)\theta_z(x) \quad + \\[2mm] \underbrace{\left(\frac{\partial \delta M(x)}{\partial x}dx\right)\left(\frac{\partial \theta_z(x)}{\partial x}dx\right)}_{2^{nd}\ order\ term} \end{pmatrix} \tag{5.137}$$

By imposing the differential equilibrium equations (2/2.13-14) of a beam segment, defining $\theta_z(x) = \dfrac{\partial v_0(x)}{\partial x}$, and neglecting the 2nd order terms, Eq. (5.137) is written as:

$$\delta W^*_{ext} = \delta M(x)\frac{\partial \theta_z(x)}{\partial x}dx \tag{5.138}$$

Based on Eq. (5.115), one can conclude that:

$$\delta \bar{U}_{int}^* = \delta M(x) \frac{\partial \theta_z(x)}{\partial x} dx \tag{5.139}$$

In Chapter 2, the flexural strain (curvature) is defined as the rate of change in the rotation with respect to the axial coordinate.

$$\kappa_0(x) = \frac{\partial \theta_z(x)}{\partial x} \tag{5.140}$$

Therefore, Eq. (5.139) is rewritten as:

$$\delta \bar{U}_{int}^* = \delta M(x) \kappa_0(x) dx \tag{5.141}$$

For the whole length of a vertically loaded member, the complementary virtual strain energy $\delta U_{int}^*$ is:

$$\delta U_{int}^* = \int_L \delta \bar{U}_{int}^* = \int_L \delta M(x) \kappa_0(x) dx \tag{5.142}$$

If only linearly elastic material is considered, the following section flexibility relation

$$\kappa_0(x) = \frac{M(x)}{IE(x)} \tag{5.143}$$

is substituted into Eq. (5.142) and the complementary virtual strain energy is expressed entirely in terms of internal moment (resultant stress) as:

$$\delta U_{int}^* = \int_L \delta \bar{U}_{int}^* = \int_L \delta M(x) \left( \frac{1}{IE(x)} \right) M(x) dx \tag{5.144}$$

Based on Eq. (5.114), the complementary internal virtual work is written as:

$$\delta W_{int}^* = -\int_L \delta M(x) \kappa_0(x) dx \tag{5.145}$$

It is important to note that in the virtual force principle, the real displacement $v_0(x)$ and rotation $\theta_z(x)$ fields must be associated with the basic system (forces and deformations) in which there are no rigid body motions. Otherwise, it is not possible to establish the statically admissible virtual force $\delta V(x)$ and virtual moment $\delta M(x)$ fields.

Eqs. (5.122), (5.142), and (5.132) form the complementary virtual strain energy expression for a planar Timoshenko frame member and can be written in the matrix form as:

$$\delta U_{int}^* = \int_L \delta \sigma_{TB}(x)^T \varepsilon_{TB}(x) dx \tag{5.146}$$

where the vector $\delta \sigma_{TB}(x)$ contains the virtual resultant stresses and is defined as:

$$\delta \sigma_{TB}(x) = \lfloor \delta N(x) \quad \delta M(x) \quad \delta V(x) \rfloor^T \tag{5.147}$$

and the vector $\varepsilon_{TB}(x)$ contains the real resultant strains and is defined as:

$$\varepsilon_{TB}(x) = \lfloor \varepsilon_0(x) \quad \kappa_0(x) \quad \gamma_0(x) \rfloor^T \tag{5.148}$$

If only linearly elastic material is considered, the following section flexibility matrix

$$\begin{Bmatrix} \varepsilon_0(x) \\ \kappa_0(x) \\ \gamma_0(x) \end{Bmatrix} = \begin{bmatrix} \dfrac{1}{EA(x)} & 0 & 0 \\ 0 & \dfrac{1}{EI(x)} & 0 \\ 0 & 0 & \dfrac{1}{GA_s(x)} \end{bmatrix} \begin{Bmatrix} N(x) \\ M(x) \\ V(x) \end{Bmatrix} \text{ or } \varepsilon_{TB}(x) = \mathbf{C}_{TB}(x)\sigma_{TB}(x) \tag{5.149}$$

is substituted into Eq. (5.146) and the complementary virtual strain energy is entirely expressed in terms of stresses (resultant stresses) as:

$$\delta U^*_{int} = \int_L \begin{Bmatrix} \delta N(x) \\ \delta M(x) \\ \delta V(x) \end{Bmatrix}^T \begin{bmatrix} \dfrac{1}{EA(x)} & 0 & 0 \\ 0 & \dfrac{1}{EI(x)} & 0 \\ 0 & 0 & \dfrac{1}{GA_s(x)} \end{bmatrix} \begin{Bmatrix} N(x) \\ M(x) \\ V(x) \end{Bmatrix} dx \tag{5.150}$$

or in the compact form as:

$$\delta U^*_{int} = \int_L \delta\sigma_{TB}(x)^T \mathbf{C}_{TB}(x)\sigma_{TB}(x)\,dx \tag{5.151}$$

For an Euler-Bernoulli frame member, the shear effects are ignored. Therefore, its complementary virtual strain energy is written as:

$$\delta U^*_{int} = \int_L \delta\sigma_{EB}(x)^T \varepsilon_{EB}(x)\,dx = \int_L \begin{Bmatrix} \delta N(x) \\ \delta M(x) \end{Bmatrix}^T \begin{Bmatrix} \varepsilon_0(x) \\ \kappa_0(x) \end{Bmatrix} dx \tag{5.152}$$

where the vector $\delta\sigma_{EB}(x)$ contains the virtual resultant stresses and is defined as:

$$\delta\sigma_{EB}(x) = \lfloor \delta N(x) \quad \delta M(x) \rfloor^T \tag{5.153}$$

and the vector $\varepsilon_{EB}(x)$ contains the real resultant strains and is defined as:

$$\varepsilon_{EB}(x) = \lfloor \varepsilon_0(x) \quad \kappa_0(x) \rfloor^T \tag{5.154}$$

If only linearly elastic material is considered, the following section flexibility matrix

$$\begin{Bmatrix} \varepsilon_0(x) \\ \kappa_0(x) \end{Bmatrix} = \begin{bmatrix} \dfrac{1}{EA(x)} & 0 \\ 0 & \dfrac{1}{EI(x)} \end{bmatrix} \begin{Bmatrix} N(x) \\ M(x) \end{Bmatrix} \text{ or } \varepsilon_{EB}(x) = \mathbf{C}_{EB}(x)\sigma_{EB}(x) \tag{5.155}$$

is substituted into Eq. (5.152) and the complementary virtual strain energy is entirely expressed in terms of stresses (resultant stresses) as:

$$\delta U_{int}^* = \int_L \begin{Bmatrix} \delta N(x) \\ \delta M(x) \end{Bmatrix}^T \begin{bmatrix} \dfrac{1}{EA(x)} & 0 \\ 0 & \dfrac{1}{EI(x)} \end{bmatrix} \begin{Bmatrix} N(x) \\ M(x) \end{Bmatrix} dx \tag{5.156}$$

or in the compact form as:

$$\delta U_{int}^* = \int_L \delta\sigma_{EB}(x)^T \mathbf{C}_{EB}(x)\sigma_{EB}(x)dx \tag{5.157}$$

## 5.10 COMPLEMENTARY EXTERNAL VIRTUAL WORK: A FRAME MEMBER

In the previous section, the expressions of the complementary virtual strain energy for several structural actions were derived. In this section, the main objective is to derive the expressions of the complementary external virtual work due to several external loadings.

Consider the bar member in its basic system shown in Figure 5.22 $a$. This member is in compatible state between the real deformation $v_1$ and the real displacement field $u_0(x)$. The complementary external virtual work done by the system of statically admissible virtual forces ($\delta q_1$ and $\delta w_x(x)$) on the real deformation $v_1$ and the real displacement field $u_0(x)$ is:

$$\delta W_{ext}^* = \delta q_1 v_1 + \int_L \delta w_x(x)u_0(x)dx \tag{5.158}$$

Consider the beam member in its basic system shown in Figure 5.22 $b$. This member is in compatible state between the real deformations $v_2$ and $v_3$ and the real displacement field $v_0(x)$. The

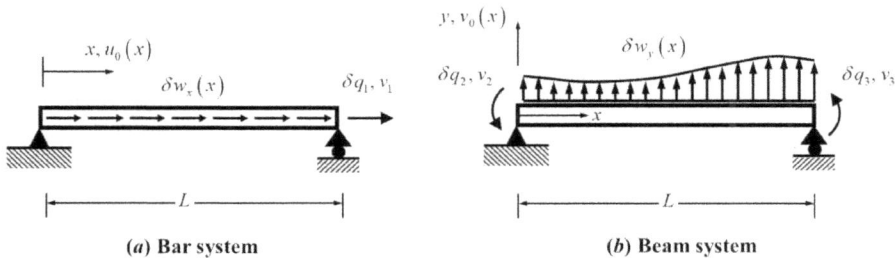

(a) Bar system          (b) Beam system

FIGURE 5.22 Complementary External Virtual Work: Bar and Beam Systems.

complementary external virtual work done by the system of statically admissible virtual forces $(\delta q_2, \delta q_3,$ and $\delta w_y(x))$ on the real deformations $(v_2$ and $v_3)$ and the real displacement field $v_0(x)$ is:

$$\delta W^*_{ext} = \delta q_2 v_2 + \delta q_3 v_3 + \int_L \delta w_y(x) v_0(x) dx \qquad (5.159)$$

Eqs. (5.158) and (5.159) form the complementary external virtual work expression for a plane frame member and can be written together as:

$$\delta W^*_{ext} = \delta q_1 v_1 + \delta q_2 v_2 + \delta q_3 v_3 + \int_L \delta w_x(x) u_0(x) dx + \int_L \delta w_y(x) v_0(x) dx \qquad (5.160)$$

or in the matrix form as:

$$\delta W^*_{ext} = \delta \mathbf{q}^T \mathbf{v} + \int_L \delta \mathbf{w}(x)^T \mathbf{u}(x) dx \qquad (5.161)$$

where the basic member deformation vector $\mathbf{v}$ is defined as $\lfloor v_1 \quad v_2 \quad v_3 \rfloor^T$; the basic member force vector $\mathbf{q}$ is defined as $\lfloor q_1 \quad q_2 \quad q_3 \rfloor^T$; the member displacement field vector $\mathbf{u}(x)$ associated with the basic system is defined as $\lfloor u_0(x) \quad v_0(x) \rfloor^T$; and the member load vector $\mathbf{w}(x)$ is defined as $\lfloor w_x(x) \quad w_y(x) \rfloor^T$.

## 5.11  VIRTUAL FORCE PRINCIPLE: A FRAME MEMBER

In the previous two sections, the expressions for the complementary virtual strain energy and the complementary external virtual work of a frame member under several loading actions were derived. In this section, the virtual force equations for several types of frame members are to be set up. It should be noted that the virtual force equation is an alternative way to express the compatibility equation.

### BAR MEMBER:

From Eqs. (5.158) and (5.122), the total complementary virtual work expression (5.115) for a bar member can be written as:

$$\delta W^* = \underbrace{\delta q_1 v_1 + \int_L \delta w_x(x) u_0(x) dx}_{\delta W^*_{ext}} - \underbrace{\int_L \delta N(x) \varepsilon_0(x) dx}_{\delta U^*_{int}} = 0 \qquad (5.162)$$

It is worth re-emphasizing that the displacement field $u_0(x)$ as well as the end deformation $v_1$ are associated with the basic system in which it is possible to establish a set of statically admissible virtual forces $(\delta q_1, \delta w_x(x),$ and $\delta N(x))$. Due to the arbitrary nature of virtual forces, the virtual force system can be selected such that $\delta w_x(x) = 0$ without any loss of its generality.

Therefore, Eq. (5.162) can be simplified as:

$$\delta W^* = \delta q_1 v_1 - \int_L \delta N(x) \varepsilon_0(x) dx = 0 \qquad (5.163)$$

## TIMOSHENKO FRAME MEMBER:

From Eqs. (5.160), (5.122), (5.142), and (5.132), the total complementary virtual work expression (5.115) for a Timoshenko frame member can be written as:

$$\delta W^* = \left[\begin{array}{c} \underbrace{\delta q_1 v_1 + \delta q_2 v_2 + \delta q_3 v_3 + \int_L \delta w_x (x) u_0 (x) dx + \int_L \delta w_y (x) v_0 (x) dx}_{\delta W^*_{ext}} \\ \left[ -\left| \underbrace{\int_L \delta N (x) \varepsilon_0 (x) dx + \int_L \delta M (x) \kappa_0 (x) dx + \int_L \delta V (x) \gamma_0 (x) dx}_{\delta U^*_{int}} \right| \right] \end{array}\right] = 0 \quad (5.164)$$

in the matrix form as:

$$\delta W^* = \underbrace{\delta \mathbf{q}^T \mathbf{v} + \int_L \delta \mathbf{w}(x)^T \mathbf{u}(x) dx}_{\delta W^*_{ext}} - \underbrace{\int_L \delta \sigma_{TB} (x)^T \varepsilon_{TB} (x) dx}_{\delta U^*_{int}} = 0 \quad (5.165)$$

It is worth re-emphasizing that the displacement fields $\mathbf{u}(x)$ as well as the end deformations $\mathbf{v}$ are associated with the basic system in which it is possible to establish a set of statically admissible virtual forces ($\delta \mathbf{q}$, $\delta \mathbf{w}(x)$, and $\delta \sigma_{TB} (x)$). Due to the arbitrary nature of virtual forces, the virtual force system can be selected such that $\delta \mathbf{w}(x) = 0$ without any loss of its generality.

Therefore, Eq. (5.165) can be simplified as:

$$\delta W^* = \delta \mathbf{q}^T \mathbf{v} - \int_L \delta \sigma_{TB} (x)^T \varepsilon_{TB} (x) dx = 0 \quad (5.166)$$

## EULER-BERNOULLI FRAME MEMBER:

From Eqs. (5.160), (5.122), and, (5.142), the total complementary virtual work expression (5.115) for an Euler-Bernoulli frame member can be written as:

$$\delta W^* = \left[\begin{array}{c} \underbrace{\delta q_1 v_1 + \delta q_2 v_2 + \delta q_3 v_3 + \int_L \delta w_x (x) u_0 (x) dx + \int_L \delta w_y (x) v_0 (x) dx}_{\delta W^*_{ext}} \\ \left[ -\left[ \underbrace{\int_L \delta N (x) \varepsilon_0 (x) dx + \int_L \delta M (x) \kappa_0 (x) dx}_{\delta U^*_{int}} \right] \right] \end{array}\right] = 0 \quad (5.167)$$

in the matrix form as:

$$\delta W^* = \underbrace{\delta \mathbf{q}^T \mathbf{v} + \int_L \delta \mathbf{w}(x)^T \mathbf{u}(x) dx}_{\delta W^*_{ext}} - \underbrace{\int_L \delta \sigma_{EB} (x)^T \varepsilon_{EB} (x) dx}_{\delta U^*_{int}} = 0 \quad (5.168)$$

It is worth re-emphasizing that the displacement fields $\mathbf{u}(x)$ as well as the end deformations $\mathbf{v}$ are associated with the basic system in which it is possible to establish a set of statically admissible virtual forces ($\delta\mathbf{q}$, $\delta\mathbf{w}(x)$, and $\delta\sigma_{EB}(x)$). Due to the arbitrary nature of virtual forces, the virtual force system can be selected such that $\delta\mathbf{w}(x) = 0$ without any loss of its generality.

Therefore, Eq. (5.168) can be simplified as:

$$\delta W^* = \delta\mathbf{q}^T\mathbf{v} - \int_L \delta\sigma_{EB}(x)^T \varepsilon_{EB}(x)\,dx = 0 \tag{5.169}$$

## 5.12  FEATURES OF STATICALLY ADMISSIBLE VIRTUAL FORCE SYSTEMS

Like the virtual displacement field, the virtual force field is not entirely arbitrary but has to satisfy a certain condition. This condition is known as a *"statically admissible"* condition. In other words, the fundamental requirement of any virtual force system is that it must meet both internal and external equilibrium conditions.

For example, the simply-supported beam (Figure 5.23) subjected to the virtual force $\delta P$ at mid-span is considered.

From the basic principle, it is noticed that the corresponding external virtual forces at ends $a$ and $b$ are:

$$\delta R_a = \delta P / 2 \text{ and } \delta R_b = \delta P / 2 \tag{5.170}$$

and the corresponding virtual shear expressions are:

$$\delta V(x) = -\delta P / 2 \text{ for } 0 \leq x \leq L / 2 \tag{5.171}$$

$$\delta V(x) = \delta P / 2 \text{ for } L / 2 \leq x \leq L \tag{5.172}$$

and the corresponding virtual moment expressions are:

$$\delta M(x) = \frac{\delta P x}{2} \text{ for } 0 \leq x \leq L / 2 \tag{5.173}$$

$$\delta M(x) = \left(1 - \frac{x}{L}\right)\frac{\delta P L}{2} \text{ for } L / 2 \leq x \leq L \tag{5.174}$$

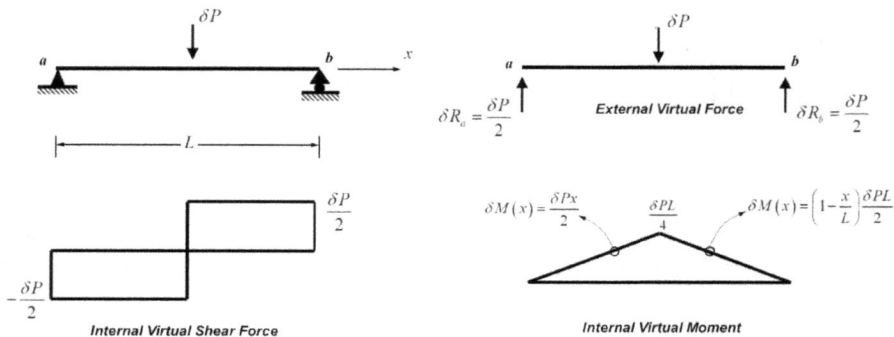

**FIGURE 5.23**  Statically Admissible Virtual Force System.

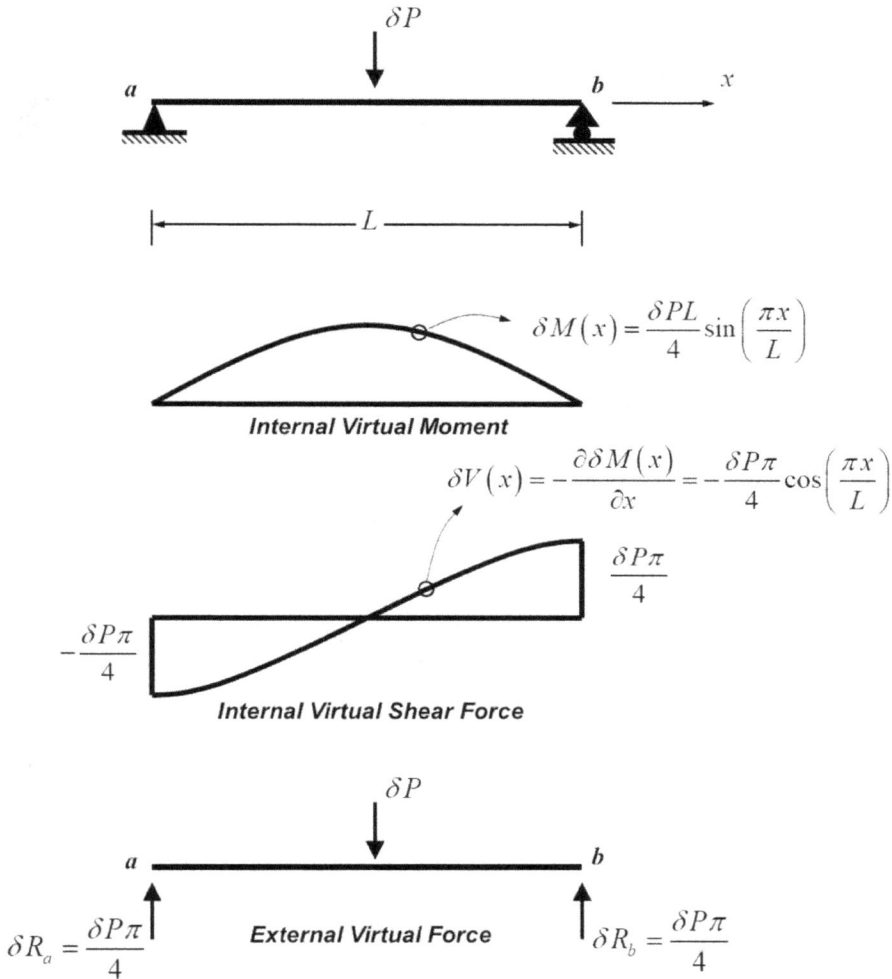

$$\delta M(x) = \frac{\delta PL}{4}\sin\left(\frac{\pi x}{L}\right)$$

**Internal Virtual Moment**

$$\delta V(x) = -\frac{\partial \delta M(x)}{\partial x} = -\frac{\delta P\pi}{4}\cos\left(\frac{\pi x}{L}\right)$$

$$\frac{\delta P\pi}{4}$$

$$-\frac{\delta P\pi}{4}$$

**Internal Virtual Shear Force**

$$\delta R_a = \frac{\delta P\pi}{4}\qquad \textbf{External Virtual Force}\qquad \delta R_b = \frac{\delta P\pi}{4}$$

**FIGURE 5.24** Inadmissible Virtual Force System.

Therefore, the external virtual forces ($\delta P$, $\delta R_a$, and $\delta R_b$) and internal virtual forces ($\delta V(x)$ and $\delta M(x)$) form the statically admissible set of virtual forces since they satisfy both external (global) and internal (local) equilibrium conditions. It is left to the reader to verify these equilibrium conditions.

It is now assumed the same beam (Figure 5.24) is subjected to the virtual force $\delta P$ at midspan. Assuming that the virtual moment expression is:

$$\delta M(x) = \frac{\delta PL}{4}\sin\left(\frac{\pi x}{L}\right) \tag{5.175}$$

and the corresponding virtual shear expression is:

$$\delta V(x) = -\frac{\partial \delta M(x)}{\partial x} = -\frac{\delta P\pi}{4}\cos\left(\frac{\pi x}{L}\right) \tag{5.176}$$

50 *kN*

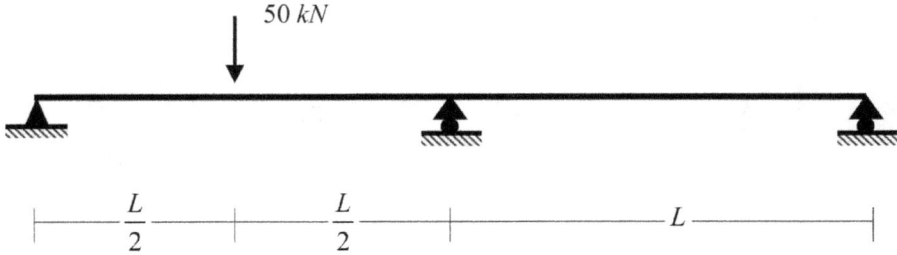

$$\frac{L}{2} \qquad \frac{L}{2} \qquad L$$

**FIGURE 5.25**   Statically Indeterminate Beams.

and the corresponding external virtual forces at ends $a$ and $b$ are:

$$\delta R_a = \delta P \pi / 4 \, \text{and} \, \delta R_b = \delta P \pi / 4 \tag{5.177}$$

Clearly, this system of external virtual forces ($\delta P$, $\delta R_a$, and $\delta R_b$) and internal virtual forces ($\delta V(x)$ and $\delta M(x)$) is not statically admissible since it violates both external (global) and internal (local) equilibrium conditions.

$$\delta R_a + \delta R_b - \delta P \neq 0 \, \text{and} \, \frac{\partial^2 M(x)}{\partial x^2} - \underbrace{w_y(x)}_{0} \neq 0 \tag{5.178}$$

Generally and especially for statically indeterminate structures, there is more than one choice of a statically admissible virtual force system. For example, the continuous beam shown in Figure 5.25. This beam is externally statically indeterminate. Therefore, there are infinite sets of statically admissible virtual force systems with given external forces acting on the beam. Figure 5.26 shows three examples of statically admissible virtual force systems for the beam in Figure 5.25. All of them satisfy external (global) as well as internal (local) equilibrium conditions.

In spite of infinite choices for statically admissible virtual force systems available for a statically indeterminate structure, it is wiser to use the statically determinate forms of the same structure such as the ones shown in Figure 5.26 $b$ and $c$. The statically indeterminate form of the structure shown in Figure 5.26 $a$ is not a wise choice for the virtual force system since it represents the structure that is being analyzed.

## EXAMPLE 5-3

For the cantilever Timoshenko beam shown in Figure 5.27, use the virtual force principle to determine the tip displacement.

**Solution:** Since the tip displacement is of interest, the external virtual force system of Figure 5.28 is applied:

$$\delta R_{ay} = -1.0; \, \delta M_{az} = -L; \, \text{and} \, \delta P = 1.0$$

The corresponding internal real deformations and internal virtual forces are shown in Figure 5.29.

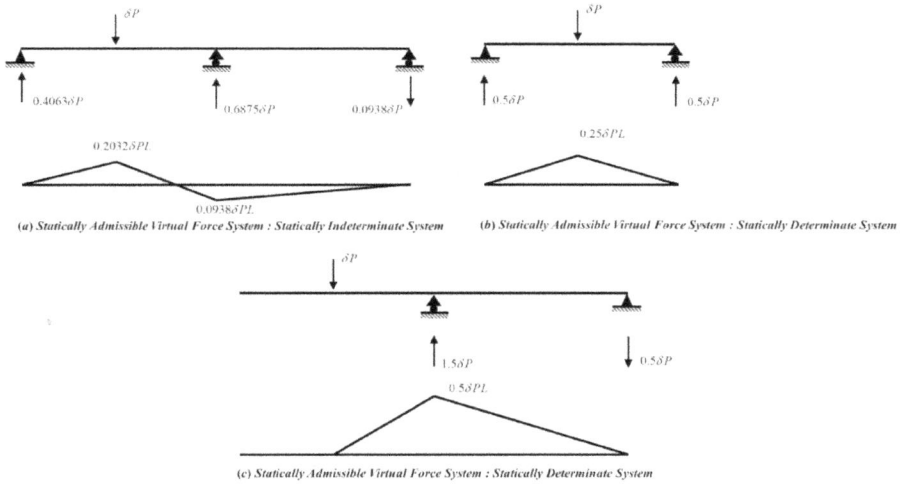

(a) Statically Admissible Virtual Force System : Statically Indeterminate System   (b) Statically Admissible Virtual Force System : Statically Determinate System

(c) Statically Admissible Virtual Force System : Statically Determinate System

**FIGURE 5.26** Statically Admissible Sets of Virtual Forces: Statically Indeterminate and Determinate Systems.

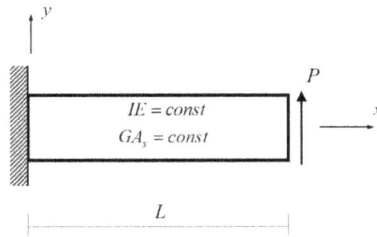

**FIGURE 5.27** Example 5.3.

The complementary external virtual work is

$$\delta W_{ext}^* = \underbrace{\delta P}_{1} \Delta_{tip} = \Delta_{tip}$$

The complementary internal virtual work is

$$\delta W_{int}^* = -\int_L \delta M(x)\kappa_0(x)dx - \int_L \delta V(x)\gamma_0(x)dx$$

$$= -\int_L \left(1 - \frac{x}{L}\right)\frac{PL}{IE}\left(1 - \frac{x}{L}\right)Ldx - \int_L \frac{P}{GA_s}(1.0)dx = -\frac{PL^3}{3IE} - \frac{PL}{GA_s}$$

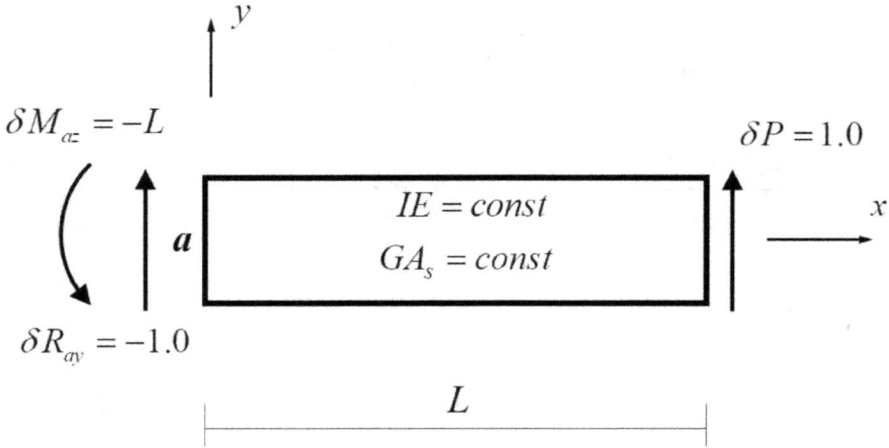

**FIGURE 5.28**   Example 5.3 (Continued).

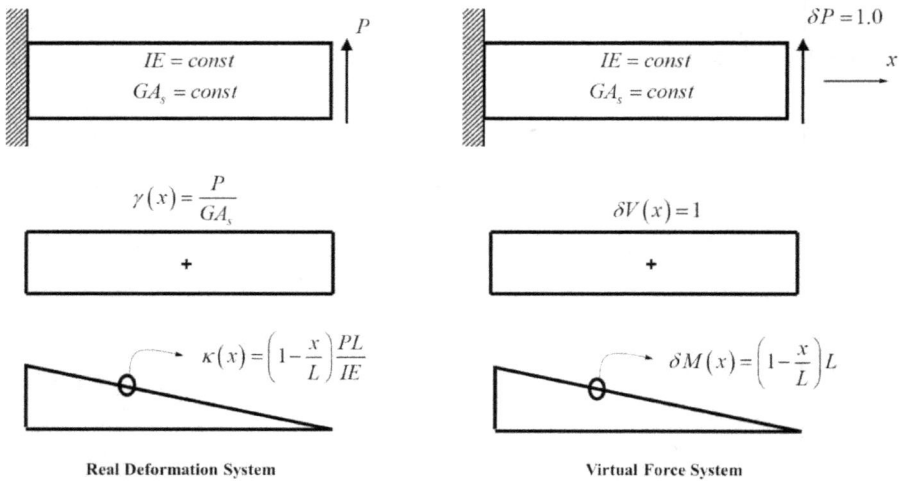

**FIGURE 5.29**   Example 5.3 (Continued).

The total complementary virtual work is:

$$\delta W^* = \delta W^*_{ext} + \delta W^*_{int} = \Delta_{tip} - \frac{PL^3}{3IE} - \frac{PL}{GA_s} = 0 \quad \Rightarrow \quad \Delta_{tip} = \frac{PL^3}{3IE} + \frac{PL}{GA_s} \;\uparrow$$

When this solution is compared to the "*exact*" one obtained from the strong form of the problem, it is found out that they are the same. This should not be surprised since there is no approximation has been made any so far. This is one of the attractive features of the virtual force principle. For the last twenty years, the author has noticed this desirable feature of the virtual force principle and has formulated several flexibility-based frame models within the framework of virtual force principle (e.g. Limkatanyu and Spacone 2006; Limkatanyu et al. 2012; Limkatanyu et al. 2023).

## 5.13 DERIVATION OF CONTRAGRADIENT RELATIONS BETWEEN FORCE AND DISPLACEMENT SYSTEMS

Before proceeding to the next sections on establishing the virtual work and complementary virtual work principles for a structure as a whole, it is desirable to provide the rational derivations of contragradient relations between force and displacement systems of the frame member. These relations have already been noticed in Chapters 3 and 4. They can be proven by the virtual work and complementary virtual work principles.

### 5.13.1 1ST ASPECT OF CONTRAGRADIENT RELATIONS: VIRTUAL FORCE PRINCIPLE

The first aspect that is about to investigate is on the selection of the basic member deformations $\mathbf{v}$ as conjugate work pairs of the basic member forces $\mathbf{q}$. As found in Chapters 3 and 4, it is much simpler to select the basic member forces by considering equilibrium equations than to select the basic member deformations by considering geometric compatibility equations. So far, there has been no proof provided on conjugate work natures between basic member forces (Chapter 3) and basic member deformations (Chapter 4) for the simply-supported beam system. This could be done with the help of the complementary virtual work principle.

The complete sets of end force and end displacements of a frame member shown in Figure 5.30 are considered. It is noted that there is no member load present in the system (Figure 5.30 $a$) since its effect can be included in the equivalent member end forces as discussed in Chapter 3.

Let member end displacements $\overline{\mathbf{U}}$ shown (Figure 5.30 $b$) be in the compatible state. This compatible system is then disturbed by a set of virtual member end forces $\delta\overline{\mathbf{P}}$. The external complementary virtual work conducted by the virtual forces $\delta\overline{\mathbf{P}}$ on the real displacements $\overline{\mathbf{U}}$ is defined as:

$$\delta W_{ext}^{*} = \delta\overline{\mathbf{P}}^{T}\overline{\mathbf{U}} = \delta\overline{P}_{1}\overline{U}_{1} + \delta\overline{P}_{2}\overline{U}_{2} + \delta\overline{P}_{3}\overline{U}_{3} + \delta\overline{P}_{4}\overline{U}_{4} + \delta\overline{P}_{5}\overline{U}_{5} + \delta\overline{P}_{6}\overline{U}_{6} \qquad (5.179)$$

It is recalled that the virtual member end forces $\delta\overline{\mathbf{P}}$ are not completely arbitrary but they have to form the statically admissible system. In other words, there are only three independent virtual forces $\delta\mathbf{q}$ and the remaining three end forces can simply be expressed in terms of these

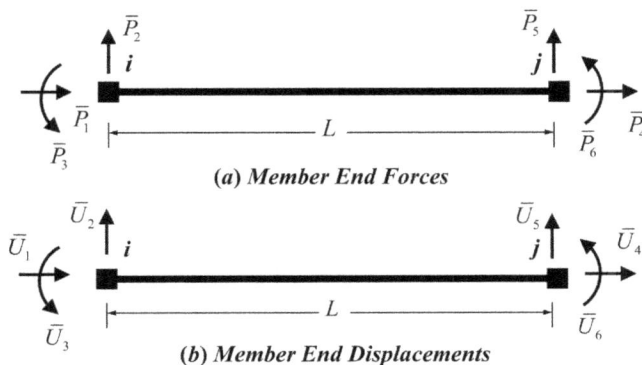

(a) Member End Forces

(b) Member End Displacements

FIGURE 5.30   Member End Forces and End Displacements: Local Reference System.

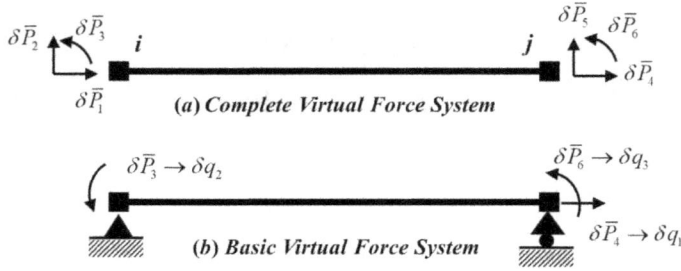

**FIGURE 5.31** Simply Supported Beam System: (a) Complete Virtual Force System; (b) Basic Virtual Force System.

independent virtual forces through whole member equilibrium conditions using the simply-supported beam as shown in Figure 5.31.

From Chapter 3, the statical relations (3/3.10) between the virtual forces in the complete and basic systems are:

$$\delta \bar{P}_1 = -\delta q_1; \quad \delta \bar{P}_2 = \delta q_2 / L + \delta q_3 / L; \quad \delta \bar{P}_3 = \delta q_2$$
$$\delta \bar{P}_4 = \delta q_1; \quad \delta \bar{P}_5 = -\delta q_2 / L - \delta q_3 / L; \quad \delta \bar{P}_6 = \delta q_3$$

(5.180)

in the matrix form as:

$$
\begin{Bmatrix} \delta \bar{P}_1 \\ \delta \bar{P}_2 \\ \delta \bar{P}_3 \\ \delta \bar{P}_4 \\ \delta \bar{P}_5 \\ \delta \bar{P}_6 \end{Bmatrix} =
\begin{bmatrix}
-1 & 0 & 0 \\
0 & 1/L & 1/L \\
0 & 1 & 0 \\
1 & 0 & 0 \\
0 & -1/L & -1/L \\
0 & 0 & 1
\end{bmatrix}
\begin{Bmatrix} \delta q_1 \\ \delta q_2 \\ \delta q_3 \end{Bmatrix} \text{ or } \delta \bar{\mathbf{P}} = \bar{\mathbf{b}}_B \delta \mathbf{q}
$$

(5.181)

Substituting Eqs. (5.180) into (5.179), one has:

$$
\begin{aligned}
\delta W^*_{ext} &= \delta \bar{\mathbf{P}}^T \bar{\mathbf{U}} = \delta \bar{P}_1 \bar{U}_1 + \delta \bar{P}_2 \bar{U}_2 + \delta \bar{P}_3 \bar{U}_3 + \delta \bar{P}_4 \bar{U}_4 + \delta \bar{P}_5 \bar{U}_5 + \delta \bar{P}_6 \bar{U}_6 \\
&= -\delta q_1 \bar{U}_1 + (\delta q_2 / L + \delta q_3 / L) \bar{U}_2 + \delta q_2 \bar{U}_3 + \delta q_1 \bar{U}_4 + (-\delta q_2 / L - \delta q_3 / L) \bar{U}_5 + \delta q_3 \bar{U}_6 \\
&= \delta q_1 (\bar{U}_4 - \bar{U}_1) + \delta q_2 (\bar{U}_3 + \bar{U}_2 / L - \bar{U}_5 / L) + (\bar{U}_6 + \bar{U}_2 / L - \bar{U}_5 / L) \delta q_3 \\
&= \delta q_1 v_1 + \delta q_2 v_2 + \delta q_3 v_3 = \delta \mathbf{q}^T \mathbf{v}
\end{aligned}
$$

(5.182)

From the last two relations of Eq. (5.182), one can conclude that the basic member deformations, which are the conjugate work pairs of the basic member forces, are defined in terms of the member end displacements as:

$$v_1 = (\bar{U}_4 - \bar{U}_1); \quad v_2 = \bar{U}_3 - (\bar{U}_5 - \bar{U}_2) / L; \text{ and } v_3 = \bar{U}_6 - (\bar{U}_5 - \bar{U}_2) / L$$

(5.183)

in the matrix form as:

$$\begin{Bmatrix} v_1 \\ v_2 \\ v_3 \end{Bmatrix} = \begin{bmatrix} -1 & 0 & 0 & 1 & 0 & 0 \\ 0 & 1/L & 1 & 0 & -1/L & 0 \\ 0 & 1/L & 0 & 0 & -1/L & 1 \end{bmatrix} \begin{Bmatrix} \bar{U}_1 \\ \bar{U}_2 \\ \bar{U}_3 \\ \bar{U}_4 \\ \bar{U}_5 \\ \bar{U}_6 \end{Bmatrix} \text{ or } \mathbf{v} = \bar{\mathbf{b}}_B^T \bar{\mathbf{U}} \qquad (5.184)$$

These relations are exactly the same as those obtained in Chapter 4 (4/4.22) from linearized geometrical considerations of the frame kinematics.

Another important conclusion drawn from the first and last relations of Eq. (5.182) is:

$$\delta \bar{\mathbf{P}}^T \bar{\mathbf{U}} = \delta \mathbf{q}^T \mathbf{v} \qquad (5.185)$$

This relation simply implies the invariance of the complementary virtual work whether it is defined with respect to the complete or basic member system. This conclusion is justified due to the fact that the virtual work is a scalar quantity and should be independent of any reference frame.

Based on observations made for the simply-supported beam system, one can now infer the preceding conclusions to other basic systems and to any set of forces and displacements for a frame member and later for a frame structure. Assume that the two force systems $\bar{\mathbf{P}}$ and $\mathbf{q}$ are related through the equilibrium matrix $\bar{\mathbf{b}}$:

$$\bar{\mathbf{P}} = \bar{\mathbf{b}} \mathbf{q} \qquad (5.186)$$

Let the two corresponding displacement systems be $\bar{\mathbf{U}}$ and $\mathbf{v}$, respectively. Since the complementary external virtual work is invariant to any particular reference system, the work equality relation can be written as:

$$\underbrace{\delta \bar{\mathbf{P}}^T \bar{\mathbf{U}}}_{\text{System I}} = \underbrace{\delta \mathbf{q}^T \mathbf{v}}_{\text{System II}} \qquad (5.187)$$

It is noted that the two virtual force systems are related through:

$$\delta \bar{\mathbf{P}} = \bar{\mathbf{b}} \delta \mathbf{q} \implies \delta \bar{\mathbf{P}}^T = \delta \mathbf{q}^T \bar{\mathbf{b}}^T \qquad (5.188)$$

Substituting Eqs. (5.188) into (5.187), one has:

$$\underbrace{\delta \mathbf{q}^T \bar{\mathbf{b}}^T \bar{\mathbf{U}}}_{\text{System I}} = \underbrace{\delta \mathbf{q}^T \mathbf{v}}_{\text{System II}} \implies \delta \mathbf{q}^T \left( \bar{\mathbf{b}}^T \bar{\mathbf{U}} - \mathbf{v} \right) = 0 \qquad (5.189)$$

Due to arbitrariness and independence of $\delta \mathbf{q}$, the following relation is obtained:

$$\mathbf{v} = \bar{\mathbf{b}}^T \bar{\mathbf{U}} \qquad (5.190)$$

Therefore, it can generally be stated that, if two force systems are related through equilibrium relation as $\bar{\mathbf{P}} = \bar{\mathbf{b}}\mathbf{q}$, then their associated displacement systems must meet the compatibility relation of $\mathbf{v} = \bar{\mathbf{b}}^T\bar{\mathbf{U}}$. The corresponding contragradient transformation is:

$$\bar{\mathbf{P}} = \bar{\mathbf{b}}\mathbf{q} \quad \Rightarrow \quad \mathbf{v} = \bar{\mathbf{b}}^T\bar{\mathbf{U}} \tag{5.191}$$

It should be said that the complementary virtual work equation is another way to represent the compatibility relations.

### 5.13.2   2ND ASPECT OF CONTRAGRADIENT RELATIONS: VIRTUAL DISPLACEMENT PRINCIPLE

The second aspect under current consideration is to prove that if two displacement systems are related through a compatibility relation as $\mathbf{v} = \bar{\mathbf{a}}\,\bar{\mathbf{U}}$, then their associated force systems must meet the equilibrium relation of $\bar{\mathbf{P}} = \bar{\mathbf{a}}^T\mathbf{q}$. This could be done with the help of the virtual work principle.

It is assumed that the basic set of member end forces $\mathbf{q}$ shown in Figure 5.32 $a$ is in equilibrium state. It is noted that there is no member load present in the system since its effect can be included in the equivalent member end forces as discussed in Chapter 3. This equilibrated system is then disturbed by a set of basic virtual member deformations $\delta\mathbf{v}$ (Figure 5.32 ($b$)). The external virtual work conducted by real forces $\mathbf{q}$ on virtual deformations $\delta\mathbf{v}$ is defined as:

$$\delta W_{ext} = \delta\mathbf{v}^T\mathbf{q} = \delta v_1 q_1 + \delta v_2 q_2 + \delta v_3 q_3 \tag{5.192}$$

Using the simply-supported beam as basic member system, one has the kinematical relations (4/4.21) between the virtual displacements in the basic and complete systems (Figure 5.32 $c$) as:

$$\delta v_1 = \delta\bar{U}_4 - \delta\bar{U}_1; \quad \delta v_2 = \delta\bar{U}_3 - \left(\delta\bar{U}_5 - \delta\bar{U}_2\right)/L; \text{and } \delta v_3 = \delta\bar{U}_6 - \left(\delta\bar{U}_5 - \delta\bar{U}_2\right)/L \tag{5.193}$$

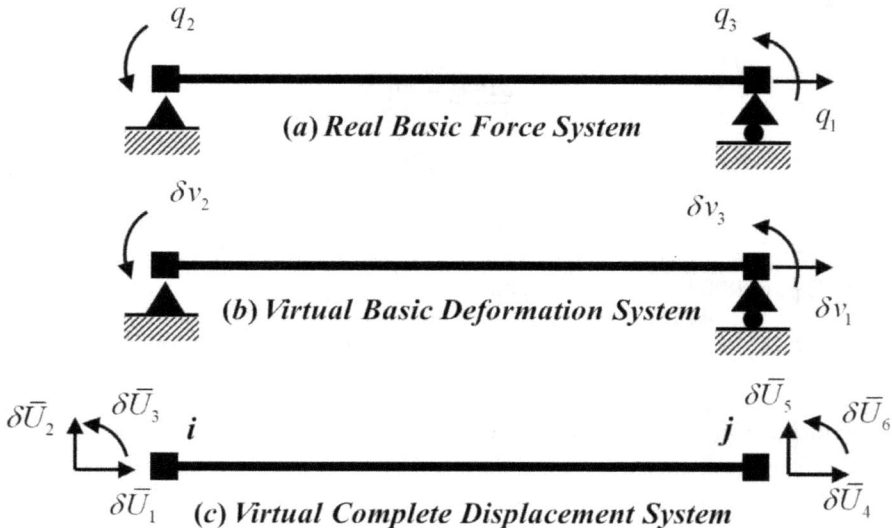

**FIGURE 5.32** Simply-Supported Beam System: ($a$) Real Basic Force System; ($b$) Virtual Basic Deformation System; ($c$) Virtual Complete Displacement System.

in the matrix form as:

$$
\begin{Bmatrix} \delta v_1 \\ \delta v_2 \\ \delta v_3 \end{Bmatrix} = \begin{bmatrix} -1 & 0 & 0 & 1 & 0 & 0 \\ 0 & 1/L & 1 & 0 & -1/L & 0 \\ 0 & 1/L & 0 & 0 & -1/L & 1 \end{bmatrix} \begin{Bmatrix} \delta\bar{U}_1 \\ \delta\bar{U}_2 \\ \delta\bar{U}_3 \\ \delta\bar{U}_4 \\ \delta\bar{U}_5 \\ \delta\bar{U}_6 \end{Bmatrix} \text{ or } \delta\mathbf{v} = \bar{\mathbf{a}}_B \delta\bar{\mathbf{U}} \tag{5.194}
$$

Substituting Eqs. (5.193) into (5.192), one has:

$$
\begin{aligned}
\delta W_{ext} &= \delta\mathbf{v}^T\mathbf{q} = \delta v_1 q_1 + \delta v_2 q_2 + \delta v_3 q_3 \\
&= \left(\delta\bar{U}_4 - \delta\bar{U}_1\right)q_1 + \left(\delta\bar{U}_3 - \left(\delta\bar{U}_5 - \delta\bar{U}_2\right)/L\right)q_2 + \left(\delta\bar{U}_6 - \left(\delta\bar{U}_5 - \delta\bar{U}_2\right)/L\right)q_3 \\
&= -q_1\delta\bar{U}_1 + \left(\left(q_2 + q_3\right)/L\right)\delta\bar{U}_2 + q_2\delta\bar{U}_3 + q_1\delta\bar{U}_4 + \left(\left(-q_2 - q_3\right)/L\right)\delta\bar{U}_5 + q_3\delta\bar{U}_6 \\
&= \bar{P}_1\delta\bar{U}_1 + \bar{P}_2\delta\bar{U}_2 + \bar{P}_3\delta\bar{U}_3 + \bar{P}_4\delta\bar{U}_4 + \bar{P}_5\delta\bar{U}_5 + \bar{P}_6\delta\bar{U}_6 = \delta\bar{\mathbf{U}}^T\bar{\mathbf{P}}
\end{aligned} \tag{5.195}
$$

From the last two relations of Eq. (5.195), it can be concluded that the member end forces, which are the conjugate work pairs of the member end displacements, are defined in terms of the basic member forces as:

$$
\begin{aligned}
\bar{P}_1 &= -q_1; \quad \bar{P}_2 = \left(q_2 + q_3\right)/L; \quad \bar{P}_3 = q_2 \\
\bar{P}_4 &= q_1; \quad \bar{P}_5 = -\left(q_2 + q_3\right)/L; \quad \bar{P}_6 = q_3
\end{aligned} \tag{5.196}
$$

in the matrix form as:

$$
\begin{Bmatrix} \bar{P}_1 \\ \bar{P}_2 \\ \bar{P}_3 \\ \bar{P}_4 \\ \bar{P}_5 \\ \bar{P}_6 \end{Bmatrix} = \begin{bmatrix} -1 & 0 & 0 \\ 0 & 1/L & 1/L \\ 0 & 1 & 0 \\ 1 & 0 & 0 \\ 0 & -1/L & -1/L \\ 0 & 0 & 1 \end{bmatrix} \begin{Bmatrix} q_1 \\ q_2 \\ q_3 \end{Bmatrix} \text{ or } \bar{\mathbf{P}} = \bar{\mathbf{a}}_B^T\mathbf{q} \tag{5.197}
$$

These relations are exactly the same as those obtained in Chapter 3 (3/3.10) from linearized equilibrium consideration of the frame statics.

Another important conclusion drawn from the first and last relations of Eq. (5.195) is:

$$
\delta\mathbf{v}^T\mathbf{q} = \delta\bar{\mathbf{U}}^T\bar{\mathbf{P}} \tag{5.198}
$$

This relation simply implies the invariance of the virtual work whether it is defined with respect to the basic or complete member system. This conclusion is justified due to the fact that the virtual work is a scalar quantity and should be independent of any reference frame.

Based on observations made for the simply-supported beam system, one can now infer the preceding conclusions to other basic systems and to any set of forces and displacements for a frame member and later for a frame structure. Assume that the two displacement systems $\mathbf{v}$ and $\bar{\mathbf{U}}$ are related through the compatibility matrix $\bar{\mathbf{a}}$:

$$
\mathbf{v} = \bar{\mathbf{a}}\,\bar{\mathbf{U}} \tag{5.199}
$$

Let the two corresponding force systems be $\mathbf{q}$ and $\overline{\mathbf{P}}$, respectively. Since the external virtual work is invariant to any particular reference system, the work equality relation can be written as:

$$\underbrace{\delta\mathbf{v}^T\mathbf{q}}_{System\ I} = \underbrace{\delta\overline{\mathbf{U}}^T\overline{\mathbf{P}}}_{System\ II} \tag{5.200}$$

It is noted that the two virtual displacement systems are related through:

$$\delta\mathbf{v} = \overline{\mathbf{a}}\,\delta\overline{\mathbf{U}} \quad\Rightarrow\quad \delta\mathbf{v}^T = \delta\overline{\mathbf{U}}^T\overline{\mathbf{a}}^T \tag{5.201}$$

Substituting Eqs. (5.201) into (5.200), one has:

$$\underbrace{\delta\overline{\mathbf{U}}^T\overline{\mathbf{a}}^T\mathbf{q}}_{System\ I} = \underbrace{\delta\overline{\mathbf{U}}^T\overline{\mathbf{P}}}_{System\ II} \quad\Rightarrow\quad \delta\overline{\mathbf{U}}^T\left(\overline{\mathbf{a}}^T\mathbf{q} - \overline{\mathbf{P}}\right) = 0 \tag{5.202}$$

Due to arbitrariness and independence of $\delta\overline{\mathbf{U}}$, the following relation is obtained:

$$\overline{\mathbf{P}} = \overline{\mathbf{a}}^T\mathbf{q} \tag{5.203}$$

Therefore, it can generally be stated that, if two displacement systems are related through a compatibility relation as $\mathbf{v} = \overline{\mathbf{a}}\,\overline{\mathbf{U}}$, then their associated force systems must meet the equilibrium relation of $\overline{\mathbf{P}} = \overline{\mathbf{a}}^T\mathbf{q}$. The corresponding contragradient transformation is:

$$\mathbf{v} = \overline{\mathbf{a}}\,\overline{\mathbf{U}} \quad\Rightarrow\quad \overline{\mathbf{P}} = \overline{\mathbf{a}}^T\mathbf{q} \tag{5.204}$$

It should be recited that the virtual work equation is another way to represent the equilibrium relations.

From Eqs. (5.191) and (5.204), it can be concluded that:

$$\overline{\mathbf{b}} = \overline{\mathbf{a}}^T \text{ and } \overline{\mathbf{b}}^T = \overline{\mathbf{a}} \tag{5.205}$$

## 5.14   VIRTUAL WORK PRINCIPLE: A FRAME STRUCTURE

Thus far, the virtual work principle for a frame member has been established. All of the essential parts have been armed, namely:

- The virtual displacement principle for a rigid particle and a rigid body.
- The virtual displacement principle for a frame member (deformable body) including both internal and external virtual work expressions.
- The contragradient transformation between two systems of forces.

In this section, the virtual work principle for a whole frame structure is to be discussed. As already mentioned, the frame structure is composed of nodes (rigid particles) and frame members (deformable links). The whole frame structure is subjected externally to prescribed nodal forces $\mathbf{F}_{free}$ (known) and prescribed distributed loads $\mathbf{w}(x)$ (known) along each member as well as to the restrained nodal forces $\mathbf{F}_{constr}$ (unknown) at supports. The whole frame structure can be regarded as a single deformable body if it is removed from its supports. To ensure that both global and local equilibrium conditions of the frame structure are satisfied, the nodal forces ($\mathbf{F}_{free}$ and $\mathbf{F}_{constr}$), member loads $\mathbf{w}(x)$, and internal force fields $\sigma(x)$ must be in equilibrium. Therefore, the virtual work principle developed for a deformable body can be applied to the frame structure as a whole, that is, the summation of external and internal virtual work vanishes. In this case, the external virtual work is defined as the work done by nodal forces and member loads on virtual displacements at global degrees of freedom, and the internal virtual work is defined as the work done by internal force fields on virtual member deformation fields. It should be kept in mind that the virtual member deformation fields must be compatible with the virtual displacements at global degrees of freedom. In other words, they must form a set of the kinematically admissible virtual displacement system.

Therefore, the external virtual work can be expressed as:

$$\delta W_{ext} = \delta \mathbf{U}_{free}^T \mathbf{F}_{free} + \delta \mathbf{U}_{constr}^T \mathbf{F}_{constr} + \sum_{i=1}^{N} \left( \int_{L_i} \delta \mathbf{u}(x)^T \mathbf{w}(x) dx \right) \tag{5.206}$$

where $L_i$ is the length of member $i$ and $N$ is the number of frame members. It is noted that the direct summation of external virtual work conducted by member loads $\mathbf{w}(x)$ on the virtual member displacement fields $\delta \mathbf{u}(x)$ over all members is possible due to the scalar nature of the work.

In a similar analogy, the total internal virtual work can be defined as the direct summation of internal virtual work over all members and is expressed as:

$$\delta W_{int} = -\delta U_{int} = -\sum_{i=1}^{N} \left( \int_{L_i} \delta \varepsilon(x)^T \boldsymbol{\sigma}(x) dx \right) \tag{5.207}$$

The virtual work equation (5.30) of the whole frame as a deformable body is:

$$\delta W = \underbrace{\delta \mathbf{U}_{free}^T \mathbf{F}_{free} + \delta \mathbf{U}_{constr}^T \mathbf{F}_{constr} + \sum_{i=1}^{N} \left( \int_{L_i} \delta \mathbf{u}(x)^T \mathbf{w}(x) dx \right)}_{\delta W_{ext}} \underbrace{- \sum_{i=1}^{N} \left( \int_{L_i} \delta \varepsilon(x)^T \boldsymbol{\sigma}(x) dx \right)}_{\delta W_{int}} = 0 \tag{5.208}$$

It is recalled that the virtual work principle (5.208) is an alternative way to express the equilibrium equations.

Moving all integral terms to the right hand side, Eq. (5.208) is rewritten as:

$$\delta \mathbf{U}_{free}^T \mathbf{F}_{free} + \delta \mathbf{U}_{constr}^T \mathbf{F}_{constr} = \sum_{i=1}^{N} \left( \int_{L_i} \delta \varepsilon(x)^T \boldsymbol{\sigma}(x) dx \right) - \sum_{i=1}^{N} \left( \int_{L_i} \delta \mathbf{u}(x)^T \mathbf{w}(x) dx \right) \tag{5.209}$$

Comparing the right hand side of Eq. (5.209) to Eqs. (5.96) and (5.98), it can be concluded that:

$$\sum_{i=1}^{N} \left( \int_{L_i} \delta \varepsilon(x)^T \boldsymbol{\sigma}(x) dx - \int_{L_i} \delta \mathbf{u}(x)^T \mathbf{w}(x) dx \right) = \sum_{i=1}^{N} \left( \delta \bar{\mathbf{U}}^{(i)T} \bar{\mathbf{P}}^{(i)} \right) \tag{5.210}$$

where the superscript $i$ is denoted for the member number. In the absence of member loads, the invariant transformation (5.200) of the work is recalled:

$$\delta \bar{\mathbf{U}}^{(i)T} \bar{\mathbf{P}}^{(i)} = \delta \mathbf{v}^{(i)T} \mathbf{q}^{(i)} \tag{5.211}$$

Substituting Eqs. (5.211) into (5.210) and (5.209), one has

$$\delta \mathbf{U}_{free}^T \mathbf{F}_{free} + \delta \mathbf{U}_{constr}^T \mathbf{F}_{constr} = \sum_{i=1}^{N} \delta \mathbf{v}^{(i)T} \mathbf{q}^{(i)} \tag{5.212}$$

It is recalled that the structural basic force vector $\mathbf{Q}$ (Chapter 3) and structural basic deformation vector $\mathbf{V}$ (Chapter 4) are defined as:

$$\mathbf{Q} = \lfloor \mathbf{q}^{(1)} \quad \mathbf{q}^{(2)} \quad \cdots \quad \mathbf{q}^{(k)} \quad \cdots \quad \mathbf{q}^{(N)} \rfloor^T_{Element}$$

$$= \lfloor \mathbf{Q}_1 \quad \mathbf{Q}_2 \quad \cdots \quad \mathbf{Q}_k \quad \cdots \quad \mathbf{Q}_N \rfloor^T_{Structure} \tag{5.213}$$

$$\mathbf{V} = \lfloor \mathbf{v}^{(1)} \quad \mathbf{v}^{(2)} \quad \cdots \quad \mathbf{v}^{(k)} \quad \cdots \quad \mathbf{v}^{(N)} \rfloor^T_{Element}$$

$$= \lfloor \mathbf{V}_1 \quad \mathbf{V}_2 \quad \cdots \quad \mathbf{V}_k \quad \cdots \quad \mathbf{V}_N \rfloor^T_{Structure} \tag{5.214}$$

Consequently, the right hand side of Eq. (5.212) can be written as the scalar product of $\delta \mathbf{V}$ and $\mathbf{Q}$.

$$\sum_{i=1}^{N} \delta \mathbf{v}^{(i)^T} \mathbf{q}^{(i)} = \delta \mathbf{V}^T \mathbf{Q} \tag{5.215}$$

Substituting Eqs. (5.215) into (5.212), one has

$$\delta \mathbf{U}^T_{free} \mathbf{F}_{free} + \delta \mathbf{U}^T_{constr} \mathbf{F}_{constr} = \delta \mathbf{V}^T \mathbf{Q} \tag{5.216}$$

Eq. (5.216) represents the structural equilibrium conditions between the external nodal forces ($\mathbf{F}_{free}$ and $\mathbf{F}_{constr}$) and internal forces ($\mathbf{Q}$) as long as the virtual deformations $\delta \mathbf{V}$ are compatible with the virtual displacements ($\delta \mathbf{U}_{free}$ and $\delta \mathbf{U}_{constr}$). It is noted that this relation is valid only when the member load is absent. However, the member-load effects can simply be treated as the equivalent nodal forces the virtual member displacement fields $\delta \mathbf{u}(x)$ has to be defined. This issue will be discussed later.

### EXAMPLE 5.4

The inextensible frame structure shown in Figure 5.33 $a$ is applied by a horizontal force at node 1.

Given the bending moment diagram shown in Figure 5.33 $b$, determine the magnitude of the horizontal force at node 1 by the virtual displacement principle.

**Solution:** Member, node, and DOFs numbering systems are shown in Figure 5.34 $a$ and the basic member deformation numbering system is shown in Figure 5.34 $b$.

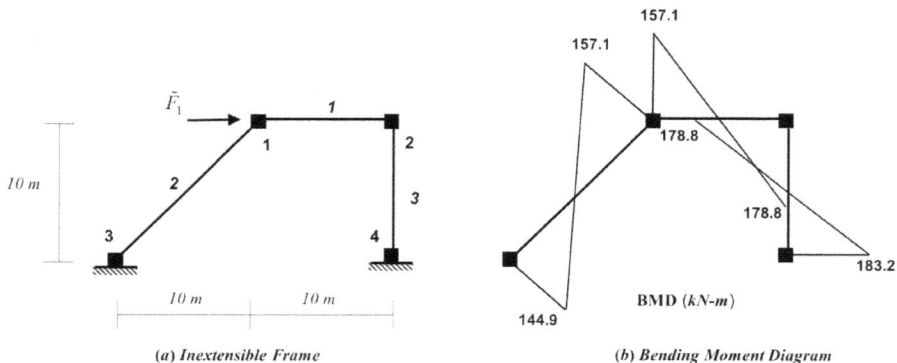

(a) Inextensible Frame                    (b) Bending Moment Diagram

**FIGURE 5.33**   Example 5.4.

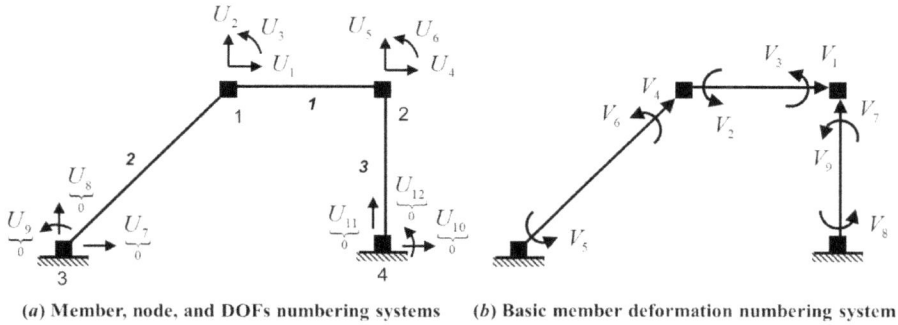

(a) Member, node, and DOFs numbering systems   (b) Basic member deformation numbering system

**FIGURE 5.34**   Example 5.4 (Continued).

Following the approach described in Chapter 4, the matrix compatibility equation of the whole structure at free degrees of freedom can be expressed in the compact format as:

$$\mathbf{V} = \mathbf{V}_{free} = \mathbf{A}_{free}\mathbf{U}_{free} \text{ or } \begin{Bmatrix} V_1 \\ V_2 \\ V_3 \\ V_4 \\ V_5 \\ V_6 \\ V_7 \\ V_8 \\ V_9 \end{Bmatrix} = \begin{bmatrix} -1 & 0 & 0 & 1 & 0 & 0 \\ 0 & 0.1 & 1 & 0 & -0.1 & 0 \\ 0 & 0.1 & 0 & 0 & -0.1 & 1 \\ 0.7071 & 0.7071 & 0 & 0 & 0 & 0 \\ 0.05 & -0.05 & 0 & 0 & 0 & 0 \\ 0.05 & -0.05 & 1 & 0 & 0 & 0 \\ 0 & 0 & 0 & 0 & 1 & 0 \\ 0 & 0 & 0 & 0.1 & 0 & 0 \\ 0 & 0 & 0 & 0.1 & 0 & 1 \end{bmatrix} \begin{Bmatrix} U_1 \\ U_2 \\ U_3 \\ U_4 \\ U_5 \\ U_6 \end{Bmatrix}$$

The kinematical relations between $\mathbf{U}_{free}$ and $\tilde{\mathbf{U}}$ corresponding to the inextensibility constraints (Figure 5.35) are written in matrix form as:

$$\mathbf{U}_{free} = \mathbf{A}_c\tilde{\mathbf{U}} \text{ or } \begin{Bmatrix} U_1 \\ U_2 \\ U_3 \\ U_4 \\ U_5 \\ U_6 \end{Bmatrix} = \begin{bmatrix} 1 & 0 & 0 \\ -1 & 0 & 0 \\ 0 & 1 & 0 \\ 1 & 0 & 0 \\ 0 & 0 & 0 \\ 0 & 0 & 1 \end{bmatrix} \begin{Bmatrix} \tilde{U}_1 \\ \tilde{U}_2 \\ \tilde{U}_3 \end{Bmatrix}$$

The matrix compatibility equation of the constrained structure is:

$$\mathbf{V}_{free} = \mathbf{A}_{free}\mathbf{U}_{free} = \mathbf{A}_{free}\mathbf{A}_c\tilde{\mathbf{U}} = \tilde{\mathbf{A}}_{free}\tilde{\mathbf{U}} \text{ or } \begin{Bmatrix} V_1 \\ V_2 \\ V_3 \\ V_4 \\ V_5 \\ V_6 \\ V_7 \\ V_8 \\ V_9 \end{Bmatrix} = \begin{bmatrix} 0 & 0 & 0 \\ -0.1 & 1 & 0 \\ -0.1 & 0 & 1 \\ 0 & 0 & 0 \\ 0.1 & 0 & 0 \\ 0.1 & 1 & 0 \\ 0 & 0 & 0 \\ 0.1 & 0 & 0 \\ 0.1 & 0 & 1 \end{bmatrix} \begin{Bmatrix} \tilde{U}_1 \\ \tilde{U}_2 \\ \tilde{U}_3 \end{Bmatrix}$$

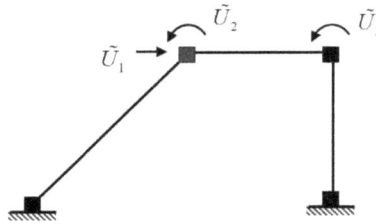

**FIGURE 5.35**   Example 5.4 (Continued).

It is clear that the basic deformations associated with the axial response are zero. Based on the bending moment diagram of Figure 5.33 $b$, the basic member force vector $\mathbf{Q}$ is defined as:

$$\mathbf{Q} = \begin{bmatrix} Q_1 & Q_2 & Q_3 & Q_4 & Q_5 & Q_6 & Q_7 & Q_8 & Q_9 \end{bmatrix}^T$$
$$= \begin{bmatrix} N.A. & -157.1 & -178.8 & N.A. & 144.9 & 157.1 & N.A. & 183.2 & 178.8 \end{bmatrix}^T$$

The following virtual nodal displacements are imposed on the constrained structure:

$$\delta \tilde{U}_1 = 1 \text{ and } \delta \tilde{U}_2 = \delta \tilde{U}_3 = 0$$

The corresponding nodal displacements of the unconstrained structure is shown in Figure 5.36 $a$ and the virtual basic deformations is shown in Figure 5.36 $b$.

The internal virtual work is: $\delta W_{int} = -\delta \mathbf{V}^T \mathbf{Q} = -100$

The external virtual work is: $\delta W_{ext} = \delta \tilde{U}_1 \tilde{F}_1 = 1.0 \times \tilde{F}_1$

The total virtual work is: $\delta W = \delta W_{ext} + \delta W_{int} = 1.0 \times \tilde{F}_1 - 100 = 0 \quad \Rightarrow \quad \tilde{F}_1 = 100 \, kN \quad \rightarrow$

It is left to the reader to verify this value of force by systematic considerations of nodal and member equilibriums. This process can be very tedious.

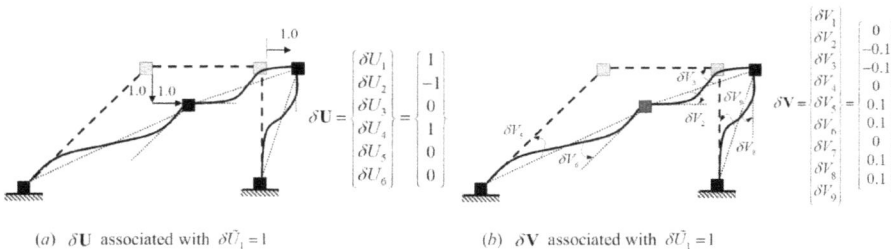

(a)  $\delta \mathbf{U}$ associated with $\delta \tilde{U}_1 = 1$                                    (b)  $\delta \mathbf{V}$ associated with $\delta \tilde{U}_1 = 1$

**FIGURE 5.36**   Example 5.4 (Continued).

## 5.15  THEORETICAL LOOKS AT THE VIRTUAL DISPLACEMENT FIELDS $\delta\mathbf{u}(x)$

To a certain extent, the continuity requirements of the virtual displacement fields $\delta\mathbf{u}(x)$ for several structural actions were discussed in the previous section. However, no theoretical view of the virtual displacement fields has been provided. These virtual fields play an essential role in evaluating the external virtual work done by the member loads $\mathbf{w}(x)$ as indicated in Eq. (5.208). Generally, the virtual displacement fields for relevant structural actions must satisfy their associated differential compatibility equations defined in Chapter 2 and specific displacement boundaries. The real displacement fields $\mathbf{u}(x)$ are also subjected to the same requirements. These issues for each type of frame members will be investigated in the following sub-sections.

### 5.15.1  Bar Member

It is recalled from Chapter 2 that the differential compatibility equation of a bar segment is

$$\varepsilon_0(x) = \frac{\partial u_0(x)}{\partial x} \tag{2/2.3}$$

The virtual displacement field $\delta u_0(x)$ also has to hold this relation. Therefore, it must be continuous and at least possess the continuous first derivative at every point inside the member domain. The most common choice of (virtual) displacement fields is polynomial since it is easy to evaluate, differentiate, and integrate both analytically and numerically. In this chapter, the simplest configuration of a bar member is of interest. The simplest bar model (Figure 5.37) has two end nodes and its relevant displacement field is a linear polynomial. Therefore, the Lagrange-type interpolation is used to represent the displacement field in terms of end displacements as:

$$u_0(x) = \left(1 - \frac{x}{L}\right)\bar{U}_1 + \frac{x}{L}\bar{U}_4 \tag{5.217}$$

### 5.15.2  Euler-Bernoulli Beam Member

It is recalled from Chapter 2 that the differential compatibility equation of an Euler-Bernoulli beam segment is

$$\kappa_0(x) = \frac{\partial^2 v_0(x)}{\partial x^2} \tag{2/2.17}$$

The virtual displacement field $\delta v_0(x)$ also has to hold this relation. Therefore, it must be continuous and at least possess the continuous second derivative at every point inside the member domain. The most common choice of (virtual) displacement fields is polynomial since it is easy to evaluate, differentiate, and integrate both analytically and numerically. In this chapter, the simplest configuration of an Euler-Bernoulli beam model is of interest. The simplest Euler-Bernoulli beam model (Figure 5.38) has two end nodes, and each end node has two boundary conditions namely, vertical displacement and rotation. Therefore, a cubic polynomial is the relevant choice for the displacement field. This polynomial function has to satisfy prescribed values for the function and its slope at end boundaries. Such a polynomial is known as *"Hermitian"* interpolation function and is expressed as:

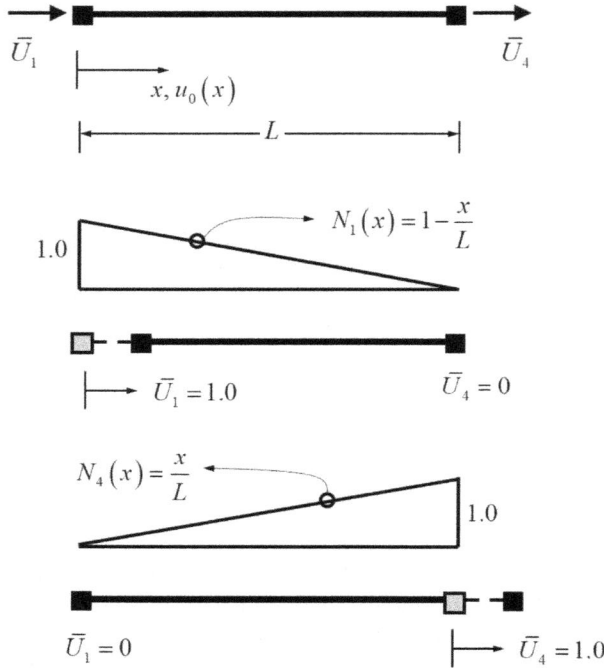

**FIGURE 5.37**  Displacement Field for 2-Node Bar Model.

$$v_0(x) = \left[1 - 3\left(\frac{x}{L}\right)^2 + 2\left(\frac{x}{L}\right)^3\right]\bar{U}_2 + \left[L\left(\frac{x}{L} - 2\left(\frac{x}{L}\right)^2 + \left(\frac{x}{L}\right)^3\right)\right]\bar{U}_3 +$$
$$\left[3\left(\frac{x}{L}\right)^2 - 2\left(\frac{x}{L}\right)^3\right]\bar{U}_5 + \left[L\left(-\left(\frac{x}{L}\right)^2 + \left(\frac{x}{L}\right)^3\right)\right]\bar{U}_6 \tag{5.218}$$

### 5.15.3 Timoshenko Beam Member

It is recalled from Chapter 2 that the differential compatibility equations of a Timoshenko beam segment are

$$\kappa_0(x) = \frac{\partial \alpha_0(x)}{\partial x} \text{ and } \gamma_0(x) = \frac{\partial v_0(x)}{\partial x} - \alpha_0(x) \tag{2/2.41}$$

Unlike the Euler-Bernoulli beam with single variable field $v_0(x)$, there are two independent variable fields, namely, the displacement field $v_0(x)$ and the flexural rotation field $\alpha_0(x)$. The most common choice of (virtual) variable fields is polynomial since it is easy to evaluate, differentiate, and integrate both analytically and numerically. In this chapter, the simplest configuration of a

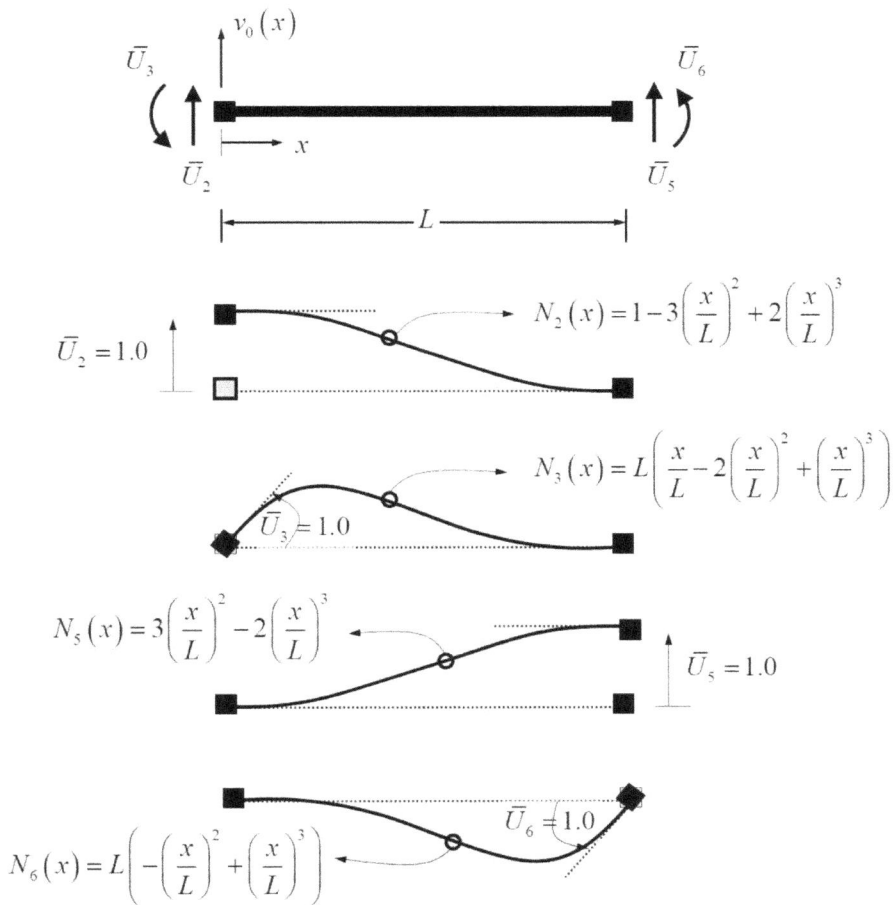

**FIGURE 5.38** Displacement Field for 2-Node Euler-Bernoulli Beam Model.

Timoshenko beam model is of interest. The simplest Timoshenko beam model (Figure 5.39) has two end nodes. Each variable field has to satisfy two specific function values at member ends and its relevant field is a linear polynomial. The Lagrange-type interpolation is used to represent the variable field in terms of end values as:

$$v_0(x) = \left(1 - \frac{x}{L}\right)\bar{U}_2 + \frac{x}{L}\bar{U}_5 \qquad (5.219)$$

$$\alpha_0(x) = \left(1 - \frac{x}{L}\right)\bar{U}_3 + \frac{x}{L}\bar{U}_6 \qquad (5.220)$$

The virtual variable fields ($\delta v_0(x)$ and $\delta\alpha_0(x)$) also have to hold these relations.

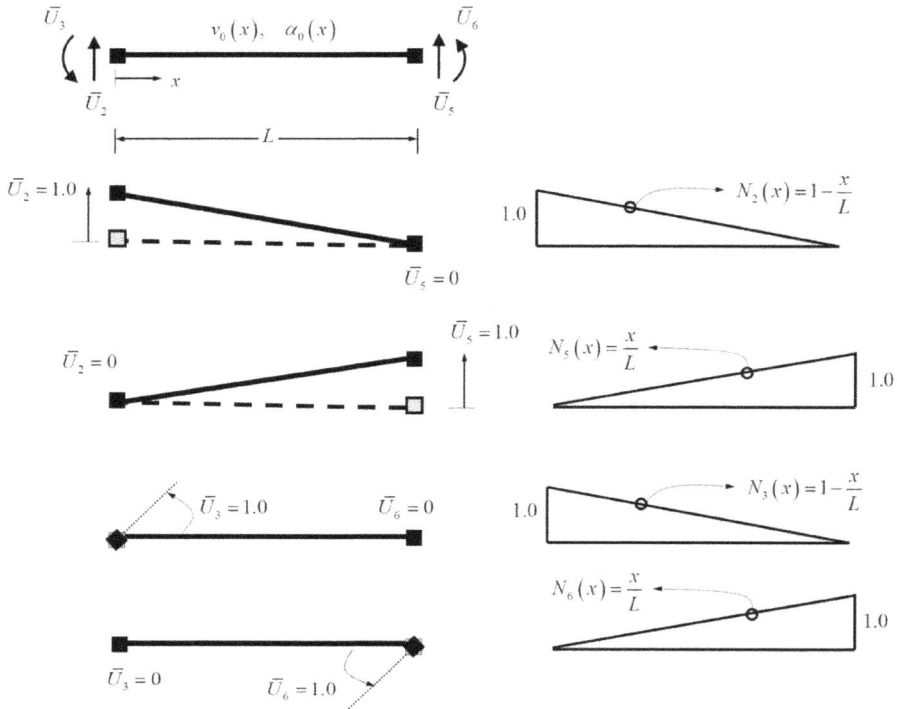

**FIGURE 5.39**   Displacement Field for 2-Node Timoshenko Beam Model.

## EXAMPLE 5.5

The inextensible frame structure shown in Figure 5.40 *a* is subjected only to a uniformly distributed load *w* along the inclined member.

Given the end moments shown in Figure 5.40 *b*, determine the magnitude of the uniformly distributed load *w* along the inclined member by the virtual displacement principle.

**Solution:** Member, node, and DOFs numbering systems are shown in Figure 5.41 *a* and the basic member deformation numbering system is shown in Figure 5.41 *b*.

Following the approach described in Chapter 4, the matrix compatibility equation of the whole structure at free degrees of freedom can be expressed in the compact format as:

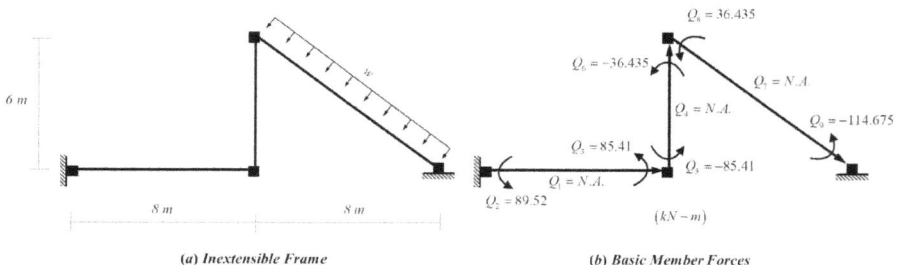

(a) Inextensible Frame                                    (b) Basic Member Forces

**FIGURE 5.40**   Example 5.5.

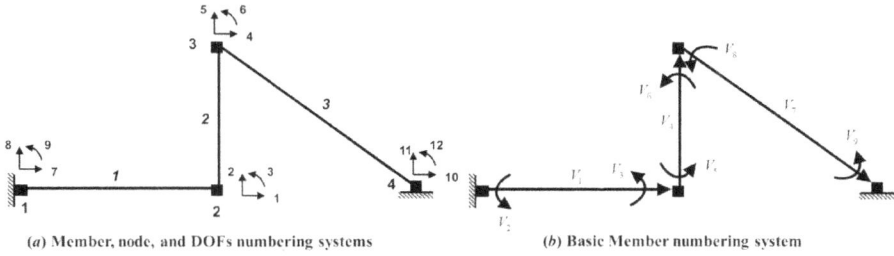

(a) Member, node, and DOFs numbering systems          (b) Basic Member numbering system

**FIGURE 5.41**   Example 5.5 (Continued).

$$
\begin{Bmatrix} V_1 \\ V_2 \\ V_3 \\ V_4 \\ V_5 \\ V_6 \\ V_7 \\ V_8 \\ V_9 \end{Bmatrix} = \begin{bmatrix} 1 & 0 & 0 & 0 & 0 & 0 \\ 0 & -0.125 & 0 & 0 & 0 & 0 \\ 0 & -0.125 & 1 & 0 & 0 & 0 \\ 0 & -1 & 0 & 0 & 1 & 0 \\ -0.1667 & 0 & 1 & 0.1667 & 0 & 0 \\ -0.1667 & 0 & 0 & 0.1667 & 0 & 1 \\ 0 & 0 & 0 & -0.8 & 0.6 & 0 \\ 0 & 0 & 0 & 0.06 & 0.08 & 1 \\ 0 & 0 & 0 & 0.06 & 0.08 & 0 \end{bmatrix} \begin{Bmatrix} U_1 \\ U_2 \\ U_3 \\ U_4 \\ U_5 \\ U_6 \end{Bmatrix} \text{ or } \mathbf{V}_{free} = \mathbf{A}_{free}\mathbf{U}_{free}
$$

The kinematical relations between $\mathbf{U}_{free}$ and $\tilde{\mathbf{U}}$ corresponding to the inextensibility constraints (Figure 5.42) are written in matrix form as:

$$
\begin{Bmatrix} U_1 \\ U_2 \\ U_3 \\ U_4 \\ U_5 \\ U_6 \end{Bmatrix} = \begin{bmatrix} 0 & 0 & 0 \\ 1 & 0 & 0 \\ 0 & 1 & 0 \\ 0.75 & 0 & 0 \\ 1 & 0 & 0 \\ 0 & 0 & 1 \end{bmatrix} \begin{Bmatrix} \tilde{U}_1 \\ \tilde{U}_2 \\ \tilde{U}_3 \end{Bmatrix} \text{ or } \mathbf{U}_{free} = \mathbf{A}_c\tilde{\mathbf{U}}
$$

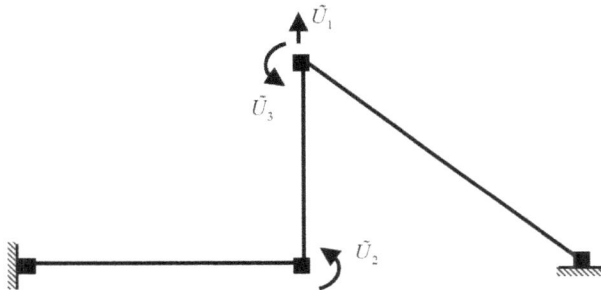

**FIGURE 5.42**   Example 5.5 (Continued).

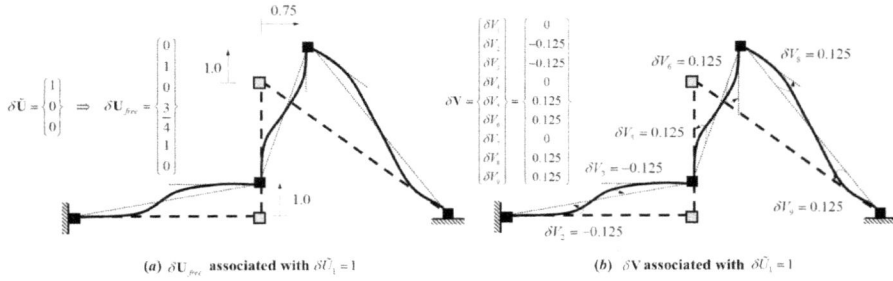

(a) $\delta\mathbf{U}_{free}$ associated with $\delta\tilde{U}_1 = 1$        (b) $\delta\mathbf{V}$ associated with $\delta\tilde{U}_1 = 1$

**FIGURE 5.43**    Example 5.5 (Continued).

The matrix compatibility equation of the constrained structure is:

$$\mathbf{V}_{free} = \mathbf{A}_{free}\mathbf{U}_{free} = \mathbf{A}_{free}\mathbf{A}_c\tilde{\mathbf{U}} = \tilde{\mathbf{A}}_{free}\tilde{\mathbf{U}} \text{ or } \begin{Bmatrix} V_1 \\ V_2 \\ V_3 \\ V_4 \\ V_5 \\ V_6 \\ V_7 \\ V_8 \\ V_9 \end{Bmatrix} = \begin{bmatrix} 0 & 0 & 0 \\ -0.125 & 0 & 0 \\ -0.125 & 1 & 0 \\ 0 & 0 & 0 \\ 0.125 & 1 & 0 \\ 0.125 & 0 & 1 \\ 0 & 0 & 0 \\ 0.125 & 0 & 1 \\ 0.125 & 0 & 0 \end{bmatrix} \begin{Bmatrix} U_1 \\ U_2 \\ U_3 \\ U_4 \\ U_5 \\ U_6 \end{Bmatrix}$$

It is clear that the basic deformations associated with the axial response are zero.

The following virtual nodal displacements are imposed on the constrained structure.

$$\delta\tilde{U}_1 = 1 \text{ and } \delta\tilde{U}_2 = \delta\tilde{U}_3 = 0$$

The corresponding nodal displacements of the unconstrained structure is shown in Figure 5.43 $a$ and the virtual basic deformations is shown in Figure 5.43 $b$.

Based on the end moments shown in Figure 5.40 $b$, the basic member force vector $\mathbf{Q}$ is defined as:

$$\mathbf{Q} = \begin{bmatrix} Q_1 & Q_2 & Q_3 & Q_4 & Q_5 & Q_6 & Q_7 & Q_8 & Q_9 \end{bmatrix}^T$$
$$= \begin{bmatrix} N.A. & 89.52 & 85.41 & N.A. & -85.41 & -36.44 & N.A. & 36.44 & -114.68 \end{bmatrix}^T$$

The internal virtual work is: $\delta W_{int} = -\delta\mathbf{V}^T\mathbf{Q} = 46.88$

The external virtual work is: $\delta W_{ext} = \delta\mathbf{U}_{free}^T \underbrace{\mathbf{F}_{free}}_{0} + \sum_{i=1}^{3}\int_{L_i}\delta\mathbf{u}(x)\mathbf{w}_i(x)dx = -\int_{L_3}\delta v_0(x)w\,dx$

Special care is needed in evaluating the external virtual work done by the distributed load $w$ on the virtual displacement field $\delta v_0(x)$ along member 3.

From Eq. (5.218), the virtual displacement field $\delta v_0(x)$ of member 3 can be expressed as:

$$\delta v_0^{(3)}(x) = \left[1 - 3\left(\frac{x}{L}\right)^2 + 2\left(\frac{x}{L}\right)^3\right]\delta \bar{U}_2^{(3)} + \left[L\left(\frac{x}{L} - 2\left(\frac{x}{L}\right)^2 + \left(\frac{x}{L}\right)^3\right)\right]\underbrace{\delta \bar{U}_3^{(3)}}_{0} +$$

$$\left[3\left(\frac{x}{L}\right)^2 - 2\left(\frac{x}{L}\right)^3\right]\underbrace{\delta \bar{U}_5^{(3)}}_{0} + \left[L\left(-\left(\frac{x}{L}\right)^2 + \left(\frac{x}{L}\right)^3\right)\right]\underbrace{\delta \bar{U}_6^{(3)}}_{0}$$

From Eq. (4/4.43), one has

$$\delta \bar{U}_2^{(3)} = -\sin\phi\, \delta U_1^{(3)} + \cos\phi\, \delta U_2^{(3)} = -(-0.6)\times 0.75 + 0.8\times 1 = 1.25$$

Therefore,

$$\delta v_0^{(3)}(x) = \left[1 - 3\left(\frac{x}{L}\right)^2 + 2\left(\frac{x}{L}\right)^3\right](1.25)$$

$$\delta W_{ext} = -\int_{L_3} \delta v_0^{(3)}(x)\, w\, dx = -1.25w\int_{L_3}\left(1 - 3\left(\frac{x}{L}\right)^2 + 2\left(\frac{x}{L}\right)^3\right)dx = -6.25w$$

The total virtual work is: $\delta W = \delta W_{ext} + \delta W_{int} = -6.25w + 46.88 = 0 \implies w = 7.5\ kN/m$

## 5.16 COMPLEMENTARY VIRTUAL WORK PRINCIPLE: A FRAME STRUCTURE

Up to now, the complementary virtual work principle for a frame member has been established. All of the necessary ingredients have been armed namely,

- The virtual force principle for a rigid particle and a rigid body.
- The virtual force principle for a frame member (deformable body) including both complementary internal and external virtual work expressions.
- The contragradient transformation between two systems of displacements.

In this section, the complementary virtual work principle for a whole frame structure is to be set up. As already mentioned, the frame structure consists of nodes (rigid particles) and frame members (deformable links). The whole frame structure is subjected to prescribed nodal displacements $\mathbf{U}_{constr}$ (known) while the unconstrained degrees of freedom move through the free nodal displacements $\mathbf{U}_{free}$ (unknown). The internal deformation fields $\varepsilon(x)$ must be compatible with these external displacements ($\mathbf{U}_{free}$ and $\mathbf{U}_{constr}$). Since the whole frame structure can be simply regarded as a single deformable body, the complementary virtual work principle developed for a deformable body can be applied to the frame structure as a whole, that is, the summation of complementary external and internal virtual work vanishes. In this case, the complementary external virtual work is defined as the work done by virtual nodal forces and virtual member loads on actual displacements at global degrees of freedom, and the complementary internal virtual work is defined as the work done by virtual internal force fields on actual member deformation fields. It should be kept in mind that the virtual member force fields must be in equilibrium with the virtual external forces at global degrees of freedom. In other words, they must form a set of the statically admissible virtual force system.

Therefore, the complementary external virtual work can be expressed as:

$$\delta W_{ext}^* = \delta \mathbf{F}_{free}^T \mathbf{U}_{free} + \delta \mathbf{F}_{constr}^T \mathbf{U}_{constr} + \sum_{i=1}^{N} \left( \int_{L_i} \delta \mathbf{w}^T(x) \mathbf{u}(x) dx \right) \tag{5.221}$$

It is noted that the direct summation of complementary external virtual work conducted by virtual member loads $\delta \mathbf{w}(x)$ on the real member displacement fields $\mathbf{u}(x)$ over all members is possible due to the scalar nature of the work.

In the similar analogy, the total complementary internal virtual work can be defined as the direct summation of complementary internal virtual work over all members and is expressed as:

$$\delta W_{int}^* = -\delta U_{int}^* = -\sum_{i=1}^{N} \left( \int_{L_i} \delta \boldsymbol{\sigma}(x)^T \boldsymbol{\varepsilon}(x) dx \right) \tag{5.222}$$

The complementary virtual work equation (5.114) of the whole frame as a deformable body is:

$$\delta W^* = \underbrace{\delta \mathbf{F}_{free}^T \mathbf{U}_{free} + \delta \mathbf{F}_{constr}^T \mathbf{U}_{constr} + \sum_{i=1}^{N} \left( \int_{L_i} \delta \mathbf{w}(x)^T \mathbf{u}(x) dx \right)}_{\delta W_{ext}^*} \underbrace{- \sum_{i=1}^{N} \left( \int_{L_i} \delta \boldsymbol{\sigma}(x)^T \boldsymbol{\varepsilon}(x) dx \right)}_{\delta W_{int}^*} = 0 \tag{5.223}$$

Due to the arbitrary nature of virtual forces, the virtual force system can be selected such that $\delta \mathbf{w}(x) = 0$ without any loss of its generality. Therefore, Eq. (5.223) is modified as:

$$\delta W^* = \delta \mathbf{F}_{free}^T \mathbf{U}_{free} + \delta \mathbf{F}_{constr}^T \mathbf{U}_{constr} - \sum_{i=1}^{N} \left( \int_{L_i} \delta \boldsymbol{\sigma}(x)^T \boldsymbol{\varepsilon}(x) dx \right) = 0 \tag{5.224}$$

It is recalled that the virtual work principle (5.224) is an alternative way to express the compatibility equations. Moving the integral term to the right hand side, Eq. (5.224) is rewritten as:

$$\delta \mathbf{F}_{free}^T \mathbf{U}_{free} + \delta \mathbf{F}_{constr}^T \mathbf{U}_{constr} = \sum_{i=1}^{N} \left( \int_{L_i} \delta \boldsymbol{\sigma}(x)^T \boldsymbol{\varepsilon}(x) dx \right) \tag{5.225}$$

Comparing the right hand side of Eq. (5.225) to Eqs. (5.166) and (5.169), it can be concluded that:

$$\sum_{i=1}^{N} \left( \int_{L_i} \delta \boldsymbol{\sigma}(x)^T \boldsymbol{\varepsilon}(x) dx \right) = \sum_{i=1}^{N} \left( \delta \mathbf{q}^{(i)^T} \mathbf{v}^{(i)} \right) \tag{5.226}$$

It is recalled that the structural basic force vector $\mathbf{Q}$ (Chapter 3) and structural basic deformation vector $\mathbf{V}$ (Chapter 4) are defined in Eqs. (5.213) and (5.214), respectively. Consequently, the right hand side of Eq. (5.226) can be written as the scalar product of $\delta \mathbf{Q}$ and $\mathbf{V}$.

$$\sum_{i=1}^{N} \delta \mathbf{q}^{(i)^T} \mathbf{v}^{(i)} = \delta \mathbf{Q}^T \mathbf{V} \tag{5.227}$$

Substituting Eqs. (5.227) into (5.225) leads to:

$$\delta \mathbf{F}_{free}^T \mathbf{U}_{free} + \delta \mathbf{F}_{constr}^T \mathbf{U}_{constr} = \delta \mathbf{Q}^T \mathbf{V} \tag{5.228}$$

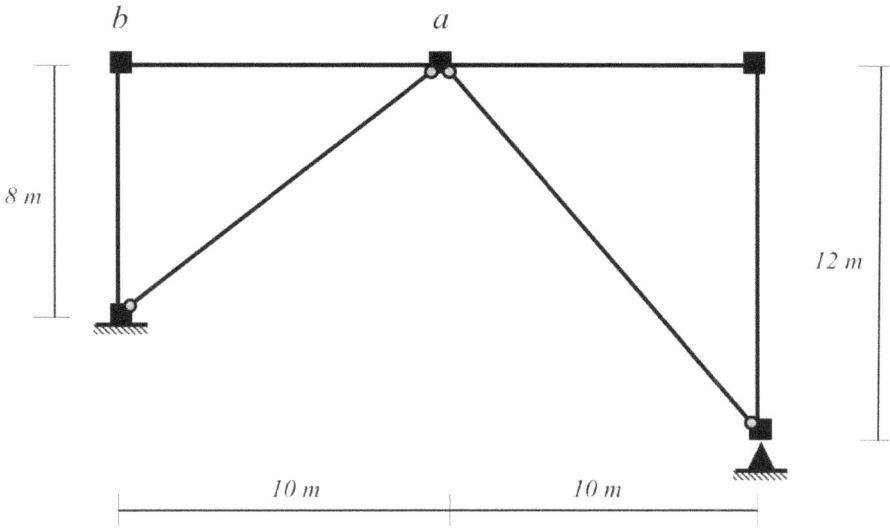

**FIGURE 5.44**   Example 5.6.

Eq. (5.228) is an alternative way to express the compatibility relations between the structural basic deformations $\mathbf{V}$ and structural nodal displacements $\mathbf{U}_{free}$ and $\mathbf{U}_{constr}$ as long as the internal virtual forces $\delta\mathbf{Q}$ and external virtual forces $\delta\mathbf{F}_{free}$ and $\delta\mathbf{F}_{constr}$ form the set of the statically admissible virtual force system.

## EXAMPLE 5.6

Given that the structural basic deformations of the braced frame structure shown in Figure 5.44 are presented in Figure 5.45:

Use the virtual force principle to determine the vertical displacement at node $a$ and horizontal displacement at node $b$.

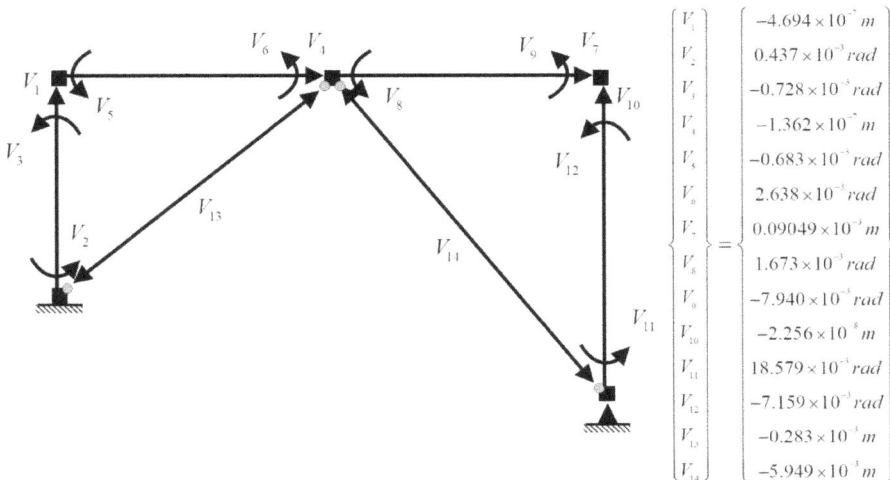

$$\begin{Bmatrix} V_1 \\ V_2 \\ V_3 \\ V_4 \\ V_5 \\ V_6 \\ V_7 \\ V_8 \\ V_9 \\ V_{10} \\ V_{11} \\ V_{12} \\ V_{13} \\ V_{14} \end{Bmatrix} = \begin{Bmatrix} -4.694 \times 10^{-2}\,m \\ 0.437 \times 10^{-3}\,rad \\ -0.728 \times 10^{-3}\,rad \\ -1.362 \times 10^{-2}\,m \\ -0.683 \times 10^{-3}\,rad \\ 2.638 \times 10^{-3}\,rad \\ 0.09049 \times 10^{-3}\,m \\ 1.673 \times 10^{-3}\,rad \\ -7.940 \times 10^{-3}\,rad \\ -2.256 \times 10^{-5}\,m \\ 18.579 \times 10^{-3}\,rad \\ -7.159 \times 10^{-3}\,rad \\ -0.283 \times 10^{-4}\,m \\ -5.949 \times 10^{-3}\,m \end{Bmatrix}$$

**FIGURE 5.45**   Example 5.6 (Continued).

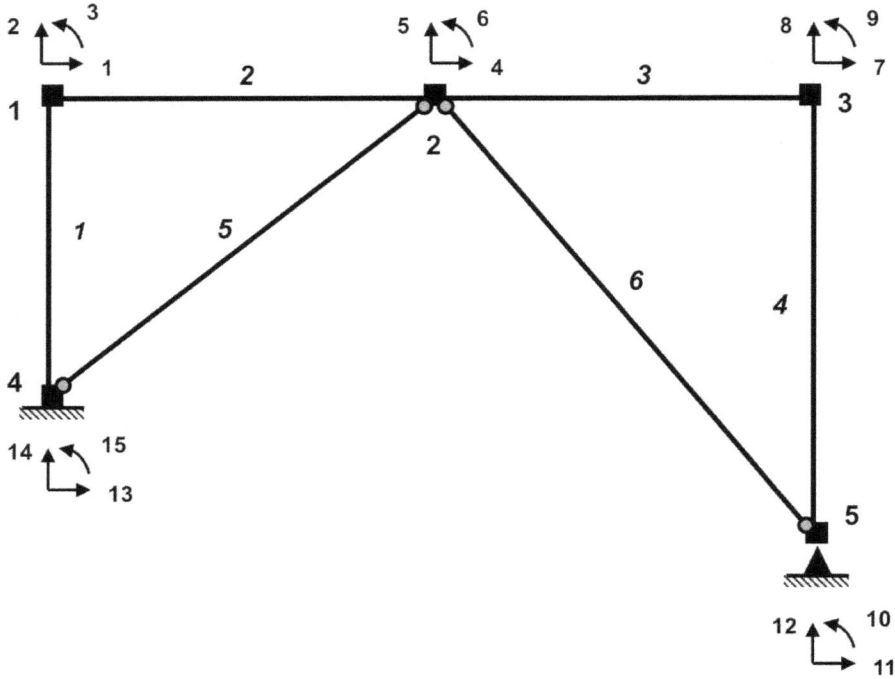

**FIGURE 5.46**  Example 5.6 (Continued).

**Solution:** Member, node, and DOFs numbering systems are shown in Figure 5.46.

To determine the vertical displacement at node $a$, the virtual force system of Figure 5.47 is used.

The external virtual force ($\delta F_5$, $\delta F_{10}$, $\delta F_{11}$, $\delta F_{12}$, $\delta F_{13}$, $\delta F_{14}$, and $\delta F_{15}$) and internal virtual force ($\delta Q_{13}$ and $\delta Q_{14}$) systems form the set of statically admissible system since they satisfy both internal and external equilibrium conditions.

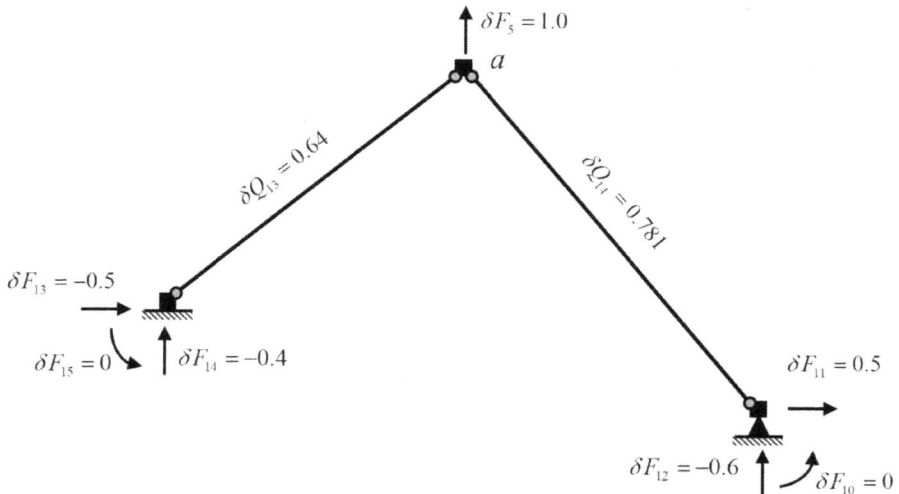

**FIGURE 5.47**  Example 5.6 (Continued).

The external virtual force vector $\delta\mathbf{F}$ is defined as:

$$\delta\mathbf{F} = \begin{bmatrix} 0 & 0 & 0 & 0 & 1 & 0 & 0 & 0 & 0 & 0 & 0.5 & -0.6 & -0.5 & -0.4 & 0 \end{bmatrix}^T$$

The internal virtual force vector $\delta\mathbf{Q}$ is defined as:

$$\delta\mathbf{Q} = \begin{bmatrix} 0 & 0 & 0 & 0 & 0 & 0 & 0 & 0 & 0 & 0 & 0 & 0 & 0.64 & 0.781 \end{bmatrix}^T$$

The complementary external virtual work is:

$$\delta W_{ext}^* = \delta\mathbf{F}^T\mathbf{U} = U_5$$

where

$$\mathbf{U} = \begin{bmatrix} U_1 & U_2 & U_3 & U_4 & U_5 & U_6 & U_7 & U_8 & U_9 & U_{10} & \underset{0}{U_{11}} & \underset{0}{U_{12}} & \underset{0}{U_{13}} & \underset{0}{U_{14}} & \underset{0}{U_{15}} \end{bmatrix}^T$$

The complementary internal virtual work is:

$$\delta W_{int}^* = -\delta\mathbf{Q}^T\mathbf{V} = 4.827 \times 10^{-3}$$

The total complementary virtual work is:

$$\delta W^* = \delta W_{ext}^* + \delta W_{int}^* = U_5 + 4.827 \times 10^{-3} = 0 \quad \Rightarrow \quad U_5 = -4.827 \times 10^{-3} = 4.827 \times 10^{-3} \; m \downarrow$$

It is noted that this virtual force system is the statically determinate part of the whole frame. Therefore, it is relatively easy to analyze this virtual force system.

To determine the horizontal displacement at node $b$, the virtual force system of Figure 5.48 is used:

The external virtual force ($\delta F_1$, $\delta F_{13}$, $\delta F_{14}$, and $\delta F_{15}$) and internal virtual force ($\delta Q_1$, $\delta Q_2$, and $\delta Q_3$) systems form the set of a statically admissible system, since they satisfy both internal and external equilibrium conditions.

The external virtual force vector $\delta\mathbf{F}$ is defined as:

$$\delta\mathbf{F} = \begin{bmatrix} 1 & 0 & 0 & 0 & 0 & 0 & 0 & 0 & 0 & 0 & 0 & 0 & -1 & 0 & 8 \end{bmatrix}^T$$

The internal virtual force vector $\delta\mathbf{Q}$ is defined as:

$$\delta\mathbf{Q} = \begin{bmatrix} 0 & 8 & 0 & 0 & 0 & 0 & 0 & 0 & 0 & 0 & 0 & 0 & 0 & 0 \end{bmatrix}^T$$

The complementary external virtual work is:

$$\delta W_{ext}^* = \delta\mathbf{F}^T\mathbf{U} = U_1$$

The complementary internal virtual work is:

$$\delta W_{int}^* = -\delta\mathbf{Q}^T\mathbf{V} = -3.496 \times 10^{-3}$$

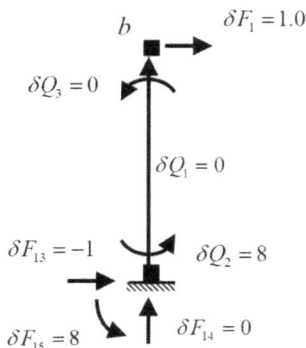

**FIGURE 5.48**  Example 5.6 (Continued).

The total complementary virtual work is:

$$\delta W^* = \delta W_{ext}^* + \delta W_{int}^* = U_1 - 3.496 \times 10^{-3} = 0 \quad \Rightarrow \quad U_1 = 3.496 \times 10^{-3} = 3.496 \times 10^{-3} \ m \ \rightarrow$$

## 5.17  EXERCISES

**Problem 5.1:** The complex truss of Figure 5.49 is subjected to the concentrated force $F_{4Y}$ at node 4.

Given the corresponding member forces shown in Figure 5.50, verify that the magnitudes of $F_{4Y}$ is 100 $kN$ by the virtual displacement principle.

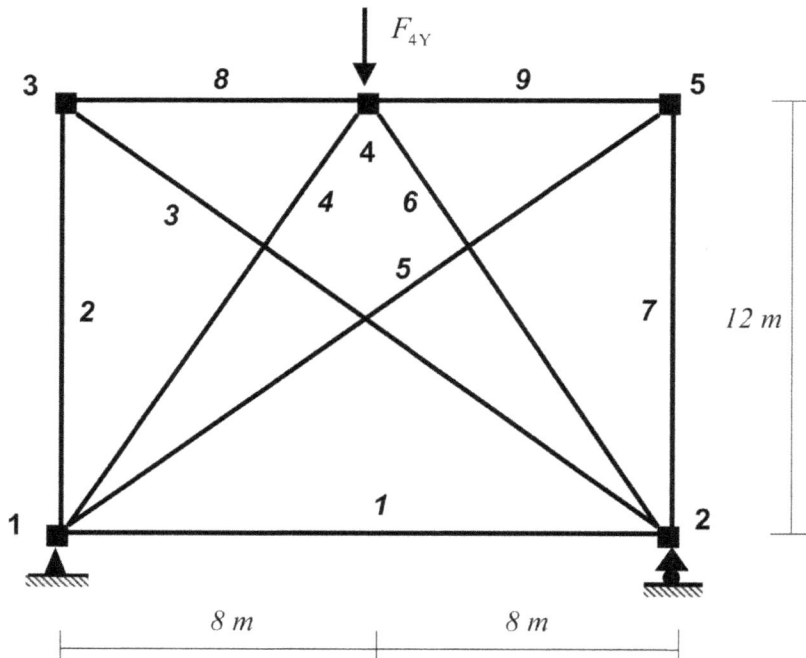

**FIGURE 5.49**  Problem 5.1.

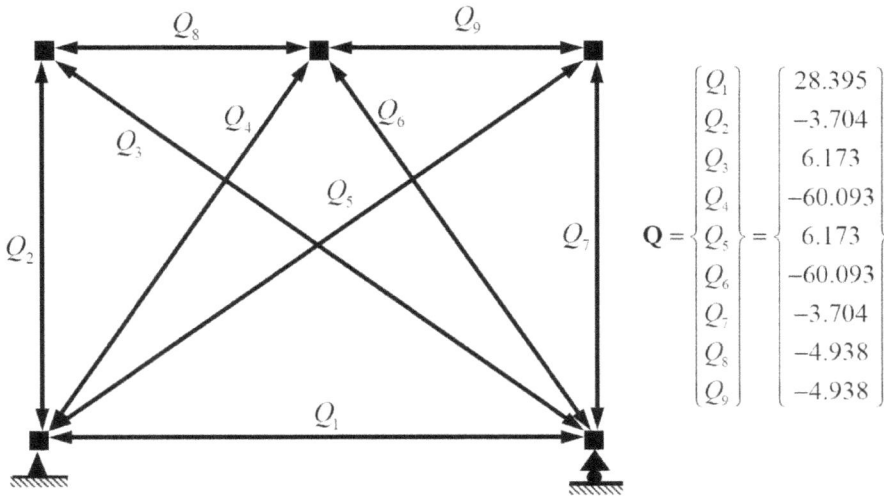

$$\mathbf{Q} = \begin{Bmatrix} Q_1 \\ Q_2 \\ Q_3 \\ Q_4 \\ Q_5 \\ Q_6 \\ Q_7 \\ Q_8 \\ Q_9 \end{Bmatrix} = \begin{Bmatrix} 28.395 \\ -3.704 \\ 6.173 \\ -60.093 \\ 6.173 \\ -60.093 \\ -3.704 \\ -4.938 \\ -4.938 \end{Bmatrix}$$

**FIGURE 5.50**  Problem 5.1 (Continued).

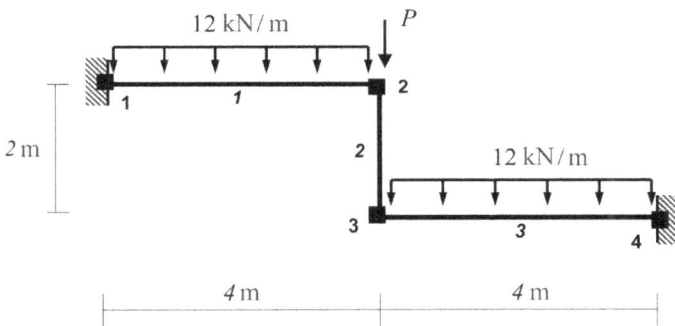

**FIGURE 5.51**  Problem 5.2.

**Problem 5.2:** The inextensible frame shown in Figure 5.51 is subjected to the uniformly distributed load along members 1 and 3 and nodal force at node 2.

Given the corresponding end moments shown in Figure 5.52, verify that the magnitude of the vertical force $P$ at node 2 is 10 $kN$ by the virtual displacement principle.

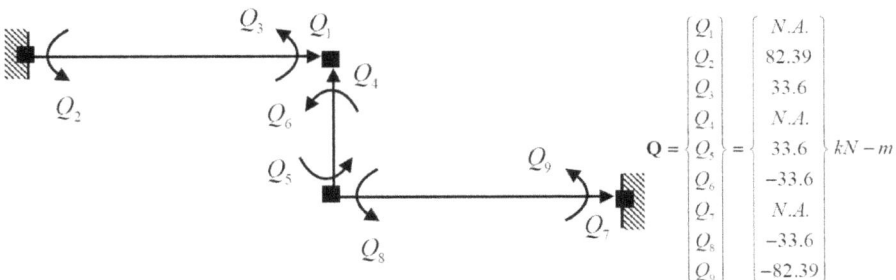

$$\mathbf{Q} = \begin{Bmatrix} Q_1 \\ Q_2 \\ Q_3 \\ Q_4 \\ Q_5 \\ Q_6 \\ Q_7 \\ Q_8 \\ Q_9 \end{Bmatrix} = \begin{Bmatrix} N.A. \\ 82.39 \\ 33.6 \\ N.A. \\ 33.6 \\ -33.6 \\ N.A. \\ -33.6 \\ -82.39 \end{Bmatrix} kN-m$$

**FIGURE 5.52**  Problem 5.2 (Continued).

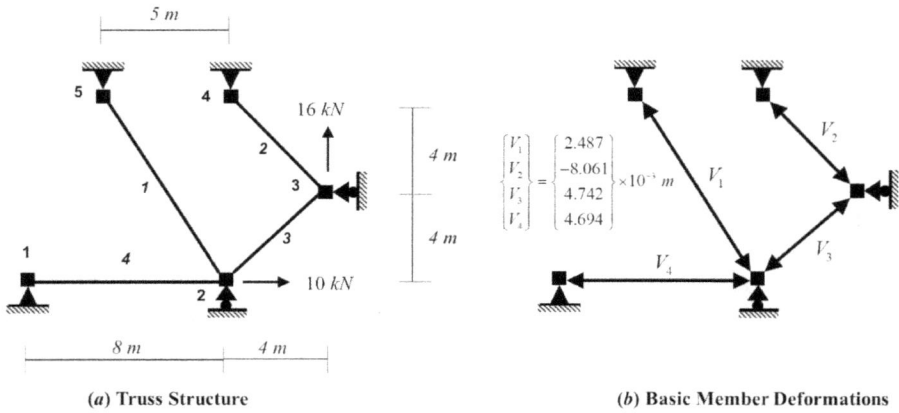

(a) Truss Structure

(b) Basic Member Deformations

**FIGURE 5.53**   Problem 5.3.

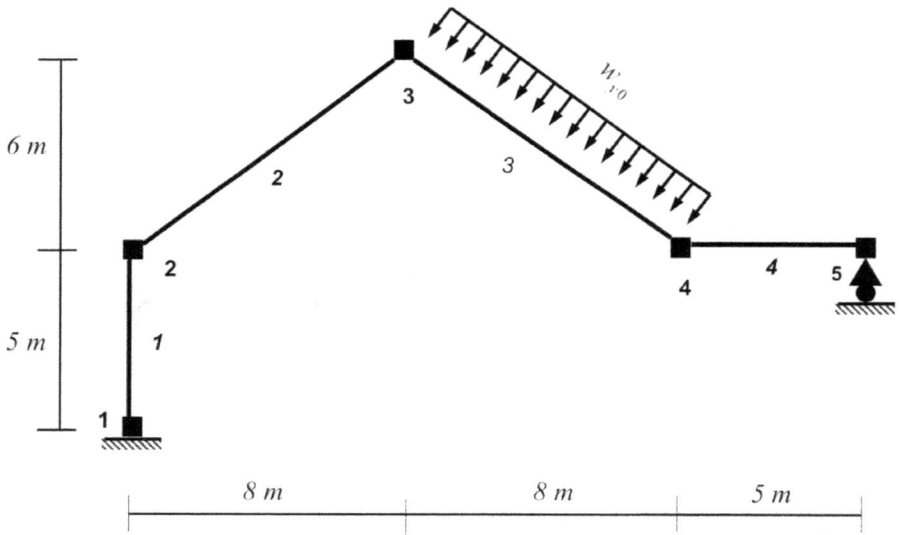

**FIGURE 5.54**   Problem 5.4.

**Problem 5.3:** The truss structure is subjected to the forces at nodes 2 and 3 as shown in Figure 5.53 a.

Given the corresponding member deformations shown in Figure 5.53 b, determine the horizontal displacement at node 2 and the vertical displacement at node 3 by the virtual force principle.

**Problem 5.4:** The gable frame shown in Figure 5.54 is subjected to the uniformly distributed load $w_{y0}$ along member 3.

Given the corresponding basic member deformations shown Figure 5.55, verify by the virtual force principle that

- The horizontal displacement at node 4 is $0.004795\ m$   ←
- The vertical displacement at node 3 is $0.006665\ m$   ↓

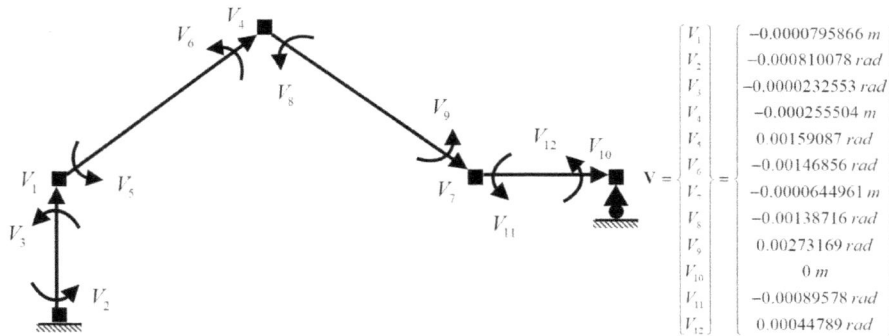

**FIGURE 5.55**   Problem 5.4 (Continued).

## REFERENCES

Dawe, D.J.1984. *Matrix and finite element displacement analysis of structures.* Clarendon Press.

Elias, Z.M. 1986. *Theory and methods of structural analysis.* John Wiley and Son Inc.

Filippou, F.C. 2002. CE 220 Course Notes: Structural Analysis: Theory and Applications. Department of Civil and Environmental Engineering, University of California, Berkeley, USA.

Ghali, A. and A.M. Neville. 2017. *Structural analysis: A unified classical and matrix approach (7th Edition).* CRC Press.

Kanchi. M.B. 1993. *Matrix methods of structural analysis (2nd Edition).* Wiley Eastern Limited.

Limkatanyu, S. and E. Spacone. 2006. Frame element with lateral deformable supports: formulation and numerical validation. *Computers and Structures* 84(13-14): 942-954.

Limkatanyu, S., K. Kuntiyawichai, E. Spacone, and M. Kwon. 2012. Natural stiffness matrix for beams on Winkler foundation: exact force-based derivation. *Structural Engineering and Mechanics* 42(1): 39-53.

Limkatanyu, S., W. Prachasaree, G. Kaewkulchai, and E. Spacone. 2014. Unification of mixed Euler-Bernoulli-von Karman planar frame model and corotational approach. *Mechanics Based Design of Structures and Machines: An International Journal* 42(4): 419-441.

Limkatanyu, S., W. Sae-Long, H.M. Sedighi, et al. 2023. Flexibility-based stress-driven nonlocal frame element: formulation and applications. *Engineering with Computers* 39(1): 399-417.

Livesley, R.K. 1974. *Matrix methods of structural analysis.* Pergamon Press.

Lukkunaprasit, P. 1984. *Structural analysis.* The Engineering Institute of Thailand (EIT).

McGuire, W., R.H. Gallagher, and R.D. Ziemian. 2000. *Matrix structural analysis (2nd Edition).* John Wiley & Sons Inc.

Neal, B.G. 1964. *Structural theorems and their applications.* Pergomon Press.

Przemieniecki, J.S. 1985. *Theory of matrix structural analysis.* Dover Publications Inc.

Spacone, E. 2003. CVEN 5525 Course Notes: Matrix Structural Analysis. Department of Civil, Environmental, and Architectural Engineering, University of Colorado, Boulder, USA.

Washizu, K. 1982. *Variational methods in elasticity and plasticity (3rd Edition).* Pergomon Press.

Whittaker, A. 2001. CIE 423 Course Notes: Structures III. Department of Civil, Structural, and Environmental Engineering, State University of New York, Bufflao, USA.

# 6 The Stiffness Method I
## *Conceptual Approach*

## 6.1 INTRODUCTION

Thus far, any matrix approach to analyze a frame structure has not been developed, though it is the main topic in this textbook. However, almost all of the fundamental ingredients on hand to develop such methods of structural analysis are ready, and they are:

- Equilibrium requirements from Chapter 3.
- Geometric compatibility requirements from Chapter 4.
- Strong-form solutions of frame members from Chapter 2 to obtain basic member properties.
- Virtual displacement and virtual force principles from Chapter 5.

The first three are the governing requirements that need to be satisfied, and the last one is the tool allowing those three governing requirements to be written together.

Generally, there are two matrix approaches available in structural analysis: "The Stiffness Method" and "The Flexibility Method". A brief introduction to these methods was given in Chapter 1.

The main objective of this chapter is to formulate the equilibrium equations with respect to structural degrees of freedom in matrix form. These matrix relations are known as "basic stiffness equations" or "stiffness equations" in short. The term "basic" infers that they are formulated based on their basic member systems (e.g. a simply supported beam) in which rigid body modes are eliminated. The computer-oriented version of the stiffness method will be discussed in the next chapter. The fundamental solution obtained from solving stiffness equations is the set of nodal displacements caused by a given loading (both mechanical and non-mechanical). This explains a lot why the stiffness method is also known as "the displacement method".

## 6.2 "EXACT" BASIC STIFFNESS MATRICES: STRONG-FORM APPROACH

In Chapter 2, the basic stiffness matrix for a prismatic Timoshenko beam member was derived and discussed in Example 2.4. This basic stiffness matrix is claimed to be "exact" in the sense that it is obtained by analytically solving its relevant governing differential equations. Following the same approach, the basic stiffness matrices for other types of frame members discussed in Chapter 2 can be derived, and it is worth giving them here.

### 6.2.1 PRISMATIC BAR MEMBER

The basic stiffness equation of a prismatic bar member (Figure 6.1) is written as:

$$q_1 = \frac{EA}{L} v_1 \text{ or } \mathbf{q} = \mathbf{k}_{Bar} \mathbf{v} \tag{6.1}$$

DOI: 10.1201/9781003595458-6

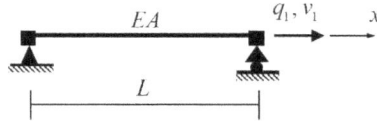

FIGURE 6.1    Prismatic Bar Member: Basic System.

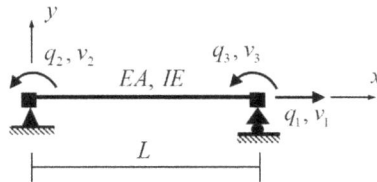

FIGURE 6.2    Prismatic Euler-Bernoulli Frame Member: Simply Supported System.

### 6.2.2    PRISMATIC EULER-BERNOULLI FRAME MEMBER: SIMPLY SUPPORTED SYSTEM

The basic stiffness equation of an Euler-Bernoulli frame member in the simply supported system (Figure 6.2) is written as:

$$\begin{Bmatrix} q_1 \\ q_2 \\ q_3 \end{Bmatrix} = \begin{bmatrix} \dfrac{EA}{L} & 0 & 0 \\ 0 & \dfrac{4IE}{L} & \dfrac{2IE}{L} \\ 0 & \dfrac{2IE}{L} & \dfrac{4IE}{L} \end{bmatrix} \begin{Bmatrix} v_1 \\ v_2 \\ v_3 \end{Bmatrix} \text{ or } \mathbf{q} = \overline{\mathbf{k}}_{EB}^{B}\mathbf{v} \tag{6.2}$$

### 6.2.3    PRISMATIC TIMOSHENKO FRAME MEMBER: SIMPLY SUPPORTED SYSTEM

The basic stiffness equation of a Timoshenko frame member in the simply supported system (Figure 6.3) is written as:

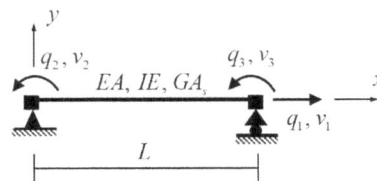

FIGURE 6.3    Prismatic Timoshenko Frame Member: Simply Supported System.

$$\begin{Bmatrix} q_1 \\ q_2 \\ q_3 \end{Bmatrix} = \begin{bmatrix} \dfrac{EA}{L} & 0 & 0 \\[2ex] 0 & \dfrac{(4+\Phi)}{(1+\Phi)}\dfrac{IE}{L} & \dfrac{(2-\Phi)}{(1+\Phi)}\dfrac{IE}{L} \\[2ex] 0 & \dfrac{(2-\Phi)}{(1+\Phi)}\dfrac{IE}{L} & \dfrac{(4+\Phi)}{(1+\Phi)}\dfrac{IE}{L} \end{bmatrix} \begin{Bmatrix} v_1 \\ v_2 \\ v_3 \end{Bmatrix} \text{ or } \mathbf{q} = \bar{\mathbf{k}}_{TB}^{B}\mathbf{v} \qquad (6.3)$$

where $\Phi = (12IE)/(GA_s L^2)$ is a dimensionless measure of the flexural-to-shear rigidity relation. If the frame is slender, the shear stiffness is very small compared to the flexural stiffness, then $\Phi \to 0$ and the stiffness matrix (6.3) of the Timoshenko frame eventually approaches that of the Euler-Bernoulli frame (6.2).

## 6.3   COMPLETE STIFFNESS MATRICES BY INJECTING RIGID-BODY-MODES INTO BASIC STIFFNESS MATRICES

In the previous section, several basic stiffness matrices for several types of common frame members were given. These stiffness matrices are exact in the sense that they are based on the strong-form solutions to their relevant governing differential equations. In this section, the stiffness matrices for those types of frame members with respect to their complete systems are given. The stiffness matrix with respect to the complete system is essential in the computer-oriented version of the stiffness method (direct stiffness method) as will be discussed in the next chapter. The complete stiffness matrices can be obtained readily from their basic stiffness matrices through equilibrium and compatibility transformation matrices as given in Chapters 3 and 4, respectively. Even though more direct methods are available, this approach is much simpler and plays a crucial role in the flexibility-based formulation of the frame model (Limkatanyu 2002). The physics behind this approach is similar to injecting the rigid-body modes into the basic system.

The complete stiffness matrix for a frame member generally relates the nodal forces $\bar{\mathbf{P}} = \lfloor \bar{\mathbf{P}}_1 \quad \bar{\mathbf{P}}_2 \rfloor^T$ in terms of its nodal displacements $\bar{\mathbf{U}} = \lfloor \bar{\mathbf{U}}_1 \quad \bar{\mathbf{U}}_2 \rfloor^T$. Usually, they are written in the following form:

$$\bar{\mathbf{P}} = \begin{Bmatrix} \bar{\mathbf{P}}_1 \\ \bar{\mathbf{P}}_2 \end{Bmatrix} = \begin{bmatrix} \bar{\mathbf{K}}_{11} & \bar{\mathbf{K}}_{12} \\ \bar{\mathbf{K}}_{21} & \bar{\mathbf{K}}_{22} \end{bmatrix} \begin{Bmatrix} \bar{\mathbf{U}}_1 \\ \bar{\mathbf{U}}_2 \end{Bmatrix} = \bar{\mathbf{K}}\bar{\mathbf{U}} \qquad (6.4)$$

Equation (6.4) is the stiffness-matrix expression for a frame member having two nodes, one at each end. The sizes of the vectors and submatrices depend on the frame member in question. Nonetheless, the general nature of the stiffness-matrix expression of Eq. (6.4) can be observed from the generic configuration of a frame member in Figure 6.4.

It is clearly noticed that submatrices $\bar{\mathbf{K}}_{11}$ and $\bar{\mathbf{K}}_{21}$ represent the end forces solely due to the nodal displacements $\bar{\mathbf{U}}_1$. Similarly, $\bar{\mathbf{K}}_{12}$ and $\bar{\mathbf{K}}_{22}$ are the end forces due to the nodal displacements $\bar{\mathbf{U}}_2$ only. Based on Betti's Reciprocal Theorem or Maxwell's Reciprocal Theorem, it can be proved that the stiffness matrix is always symmetric for the conservative system.

$$\bar{\mathbf{K}}_{12} = \bar{\mathbf{K}}_{21}^T \qquad (6.5)$$

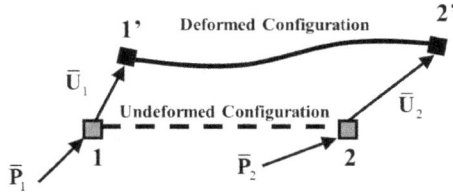

**FIGURE 6.4**   Generic Frame Member: Nodal Forces and Displacements.

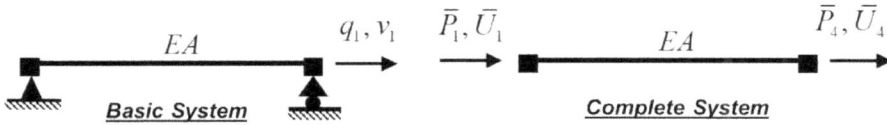

**FIGURE 6.5**   Prismatic Bar Member: Basic System vs. Complete System.

It is crucial to point out that the complete stiffness matrix $\overline{\mathbf{K}}$ is singular. In other words, there is no inverse for $\overline{\mathbf{K}}$. The physical meaning of this singularity is that the nodal displacements $\overline{\mathbf{U}}$ cannot be expressed in terms of the nodal forces $\overline{\mathbf{P}}$. This is justified since the complete stiffness matrix is obtained by injecting the rigid-body modes into the basic stiffness matrix.

### 6.3.1   Prismatic Bar Member

Consider the basic and complete systems of a prismatic bar member shown in Figure 6.5. End forces between these two systems are related through the matrix equilibrium equations (3/3.50) discussed in Chapter 3.

$$\left\{\begin{matrix} \overline{P}_1 \\ \overline{P}_4 \end{matrix}\right\} = \begin{bmatrix} -1 \\ 1 \end{bmatrix} q_1 \text{ or } \overline{\mathbf{P}} = \begin{bmatrix} \overline{\mathbf{b}}_1^{truss} \\ \overline{\mathbf{b}}_2^{truss} \end{bmatrix} \mathbf{q} = \overline{\mathbf{b}}^{truss} \mathbf{q} \tag{3/3.50}$$

The basic member deformation $v_1$ and member end displacements $\overline{U}_1$ and $\overline{U}_4$ are related through the matrix compatibility equations (4/4.76) discussed in Chapter 4.

$$v_1 = \begin{bmatrix} -1 & 1 \end{bmatrix} \left\{\begin{matrix} \overline{U}_1 \\ \overline{U}_4 \end{matrix}\right\} \text{ or } \mathbf{v} = \begin{bmatrix} \overline{\mathbf{a}}_1^{truss} & \overline{\mathbf{a}}_2^{truss} \end{bmatrix} \left\{\begin{matrix} \overline{U}_1 \\ \overline{U}_4 \end{matrix}\right\} = \overline{\mathbf{a}}^{truss} \overline{\mathbf{U}} \tag{4/4.76}$$

From invariant property of the external virtual work (5/5.200), it can be concluded that:

$$\underbrace{\delta \mathbf{v}^T \mathbf{q}}_{\text{Basic System}} = \underbrace{\delta \overline{\mathbf{U}}^T \overline{\mathbf{P}}}_{\text{Complete System}} \tag{6.6}$$

Substituting Eqs. (4/4.76) into (6.6), one has

$$\underbrace{\delta \overline{\mathbf{U}}^T}_{\text{Arbitrary}} \left[ \overline{\mathbf{P}} - \overline{\mathbf{a}}^{truss^T} \mathbf{q} \right] = 0 \implies \overline{\mathbf{P}} = \overline{\mathbf{a}}^{truss^T} \mathbf{q} \tag{6.7}$$

Substituting Eqs. (6.1) into (6.7), one has

$$\overline{\mathbf{P}} = \overline{\mathbf{a}}^{truss^T} \mathbf{k}_{Bar} \mathbf{v} \tag{6.8}$$

Substituting Eqs. (4/4.76) into (6.8), one has the stiffness relation for a complete system as:

$$\overline{\mathbf{P}} = \overline{\mathbf{a}}^{truss^T} \mathbf{k}_{Bar} \overline{\mathbf{a}}^{truss} \overline{\mathbf{U}} = \overline{\mathbf{K}}_{Bar} \overline{\mathbf{U}} \tag{6.9}$$

where the complete member stiffness $\overline{\mathbf{K}}_{Bar}$ is defined as

$$\overline{\mathbf{K}}_{Bar} = \overline{\mathbf{a}}^{truss^T} \mathbf{k}_{Bar} \overline{\mathbf{a}}^{truss} \tag{6.10}$$

This form of transformation is called "congruent" transformation.

Explicitly, the complete member stiffness $\overline{\mathbf{K}}_{Bar}$ is expressed as:

$$\overline{\mathbf{K}}_{Bar} = \begin{bmatrix} \overline{\mathbf{a}}_1^{truss^T} \mathbf{k}_{Bar} \overline{\mathbf{a}}_1^{truss} & \overline{\mathbf{a}}_1^{truss^T} \mathbf{k}_{Bar} \overline{\mathbf{a}}_2^{truss} \\ \overline{\mathbf{a}}_2^{truss^T} \mathbf{k}_{Bar} \overline{\mathbf{a}}_1^{truss} & \overline{\mathbf{a}}_2^{truss^T} \mathbf{k}_{Bar} \overline{\mathbf{a}}_2^{truss} \end{bmatrix} = \frac{EA}{L} \begin{bmatrix} 1 & -1 \\ -1 & 1 \end{bmatrix} \tag{6.11}$$

The matrix $\overline{\mathbf{K}}_{Bar}$ is singular since it contains the rigid body mode. Its eigenvalues $\lambda_i$ and corresponding eigenvectors $\phi_i$ are:

$$\lambda_1 = \frac{2EA}{L} \text{ and } \lambda_2 = 0 \tag{6.12}$$

$$\phi_1 = \lfloor -1 \quad 1 \rfloor^T \text{ and } \phi_2 = \lfloor 1 \quad 1 \rfloor^T \tag{6.13}$$

It is noticed that the matrix $\overline{\mathbf{K}}_{Bar}$ is semi-positive definite and has one zero eigenvalue which corresponds to the number of rigid body mode.

As shown in Figure 6.6, the first mode is associated with the axial deformation and the work needed to deform the bar with this mode is:

$$W_I = \frac{1}{2} \phi_1^T \overline{\mathbf{K}}_{Bar} \phi_1 = \frac{2EA}{L} \tag{6.14}$$

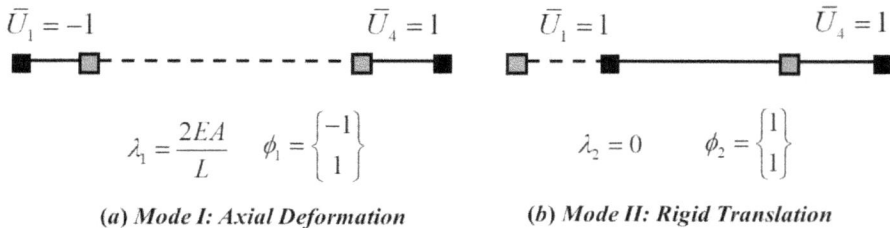

$\overline{U}_1 = -1$        $\overline{U}_4 = 1$        $\overline{U}_1 = 1$        $\overline{U}_4 = 1$

$\lambda_1 = \dfrac{2EA}{L} \quad \phi_1 = \begin{Bmatrix} -1 \\ 1 \end{Bmatrix}$        $\lambda_2 = 0 \quad \phi_2 = \begin{Bmatrix} 1 \\ 1 \end{Bmatrix}$

*(a) Mode I: Axial Deformation*        *(b) Mode II: Rigid Translation*

**FIGURE 6.6**   Eigenvalues and Eigenvectors of the Stiffness Matrix $\overline{\mathbf{K}}_{Bar}$.

The second mode is associated with the rigid-body translation, and the work needed to displace the bar with this mode is:

$$W_{II} = \frac{1}{2} \phi_2^T \overline{\mathbf{K}}_{Bar} \phi_2 = 0 \tag{6.15}$$

## 6.3.2 PRISMATIC EULER-BERNOULLI FRAME MEMBER: SIMPLY SUPPORTED SYSTEM

Consider the basic and complete systems of a prismatic Euler-Bernoulli member shown in Figure 6.7. End forces between these two systems are related through the matrix equilibrium equations (3/3.11) discussed in Chapter 3.

$$\left\{ \begin{matrix} \overline{\mathbf{P}}_1 \\ \overline{\mathbf{P}}_2 \end{matrix} \right\} = \left[ \begin{matrix} \overline{\mathbf{b}}_{1B} \\ \overline{\mathbf{b}}_{2B} \end{matrix} \right] \mathbf{q} \text{ or } \overline{\mathbf{P}} = \overline{\mathbf{b}}_B \mathbf{q} \tag{3/3.11}$$

The basic member deformations $\mathbf{v}$ and member end displacements $\overline{\mathbf{U}}$ are related through the matrix compatibility equations (4/4.23) discussed in Chapter 4.

$$\mathbf{v} = \left[ \overline{\mathbf{a}}_{1B} \quad \overline{\mathbf{a}}_{2B} \right] \left\{ \begin{matrix} \overline{\mathbf{U}}_1 \\ \overline{\mathbf{U}}_2 \end{matrix} \right\} \text{ or } \mathbf{v} = \overline{\mathbf{a}}_B \overline{\mathbf{U}} \tag{4/4.23}$$

From the invariant property of the external virtual work (5/5.200), it can be concluded that:

$$\underbrace{\delta \mathbf{v}^T \mathbf{q}}_{Basic\ System} = \underbrace{\delta \overline{\mathbf{U}}^T \overline{\mathbf{P}}}_{Complete\ System} \tag{6.16}$$

Substituting Eqs. (4/4.23) into (6.16), one has

$$\underbrace{\delta \overline{\mathbf{U}}^T}_{Arbitrary} \left[ \overline{\mathbf{P}} - \overline{\mathbf{a}}_B^T \mathbf{q} \right] = 0 \quad \Rightarrow \quad \overline{\mathbf{P}} = \overline{\mathbf{a}}_B^T \mathbf{q} \tag{6.17}$$

Substituting Eqs. (6.2) into (6.17), one has

$$\overline{\mathbf{P}} = \overline{\mathbf{a}}_B^T \overline{\mathbf{k}}_{EB}^B \mathbf{v} \tag{6.18}$$

Substituting Eqs. (4/4.23) into (6.18), one gains

$$\overline{\mathbf{P}} = \overline{\mathbf{a}}_B^T \overline{\mathbf{k}}_{EB}^B \overline{\mathbf{a}}_B \overline{\mathbf{U}} = \overline{\mathbf{K}}_{EB} \overline{\mathbf{U}} \tag{6.19}$$

where the complete member stiffness $\overline{\mathbf{K}}_{EB}$ is defined as

$$\overline{\mathbf{K}}_{EB} = \overline{\mathbf{a}}_B^T \overline{\mathbf{k}}_{EB}^B \overline{\mathbf{a}}_B \tag{6.20}$$

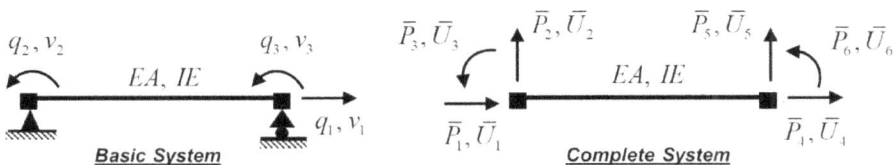

**FIGURE 6.7** Prismatic Euler-Bernoulli Frame Member: Simply Supported System vs. Complete System.

This form of transformation is called the "congruent" transformation.

Explicitly, the complete member stiffness $\bar{\mathbf{K}}_{EB}$ is expressed as:

$$\bar{\mathbf{K}}_{EB} = \begin{bmatrix} \bar{\mathbf{a}}_{1B}^T \bar{\mathbf{k}}_{EB}^B \bar{\mathbf{a}}_{1B} & \bar{\mathbf{a}}_{1B}^T \bar{\mathbf{k}}_{EB}^B \bar{\mathbf{a}}_{2B} \\ \bar{\mathbf{a}}_{2B}^T \bar{\mathbf{k}}_{EB}^B \bar{\mathbf{a}}_{1B} & \bar{\mathbf{a}}_{2B}^T \bar{\mathbf{k}}_{EB}^B \bar{\mathbf{a}}_{2B} \end{bmatrix} \tag{6.21}$$

$$\bar{\mathbf{K}}_{EB} = \begin{bmatrix} \dfrac{EA}{L} & 0 & 0 & -\dfrac{EA}{L} & 0 & 0 \\ 0 & 12\dfrac{EI}{L^3} & 6\dfrac{EI}{L^2} & 0 & -12\dfrac{EI}{L^3} & 6\dfrac{EI}{L^2} \\ 0 & 6\dfrac{EI}{L^2} & 4\dfrac{EI}{L} & 0 & -6\dfrac{EI}{L^2} & 2\dfrac{EI}{L} \\ -\dfrac{EA}{L} & 0 & 0 & \dfrac{EA}{L} & 0 & 0 \\ 0 & -12\dfrac{EI}{L^3} & -6\dfrac{EI}{L^2} & 0 & 12\dfrac{EI}{L^3} & -6\dfrac{EI}{L^2} \\ 0 & 6\dfrac{EI}{L^2} & 2\dfrac{EI}{L} & 0 & -6\dfrac{EI}{L^2} & 4\dfrac{EI}{L} \end{bmatrix} \tag{6.22}$$

The matrix $\bar{\mathbf{K}}_{EB}$ is singular since it contains the rigid body mode. Its eigenvalues $\lambda_i$ and corresponding eigenvectors $\phi_i$ are:

$$\lambda_1 = \frac{2EA}{L}; \lambda_2 = \frac{2IE}{L}; \lambda_3 = \frac{6\left(4IE + IEL^2\right)}{L^3}; \text{and } \lambda_4 = \lambda_5 = \lambda_6 = 0 \tag{6.23}$$

$$\phi_1 = \lfloor -1 \ \ 0 \ \ 0 \ \ 1 \ \ 0 \ \ 0 \rfloor^T; \phi_2 = \lfloor 0 \ \ 0 \ \ -1 \ \ 0 \ \ 0 \ \ 1 \rfloor^T; \phi_3 = \left\lfloor 0 \ \ \frac{2}{L} \ \ 1 \ \ 0 \ \ -\frac{2}{L} \ \ 1 \right\rfloor^T;$$

$$\phi_4 = \lfloor 1 \ \ 0 \ \ 0 \ \ 1 \ \ 0 \ \ 0 \rfloor^T; \ \ \phi_5 = \lfloor 0 \ \ 1 \ \ 0 \ \ 0 \ \ 1 \ \ 0 \rfloor^T; \ \ \phi_6 = \lfloor 0 \ \ -L \ \ 1 \ \ 0 \ \ 0 \ \ 1 \rfloor^T \tag{6.24}$$

It is noticed that the matrix $\bar{\mathbf{K}}_{EB}$ is semi-positive definite and has three zero eigenvalues, which correspond to the number of rigid body modes.

As shown in Figure 6.8, the first deformable mode is associated with the axial deformation and the work needed to deform the beam with this mode is:

$$W_I = \frac{1}{2}\phi_1^T \bar{\mathbf{K}}_{EB}\phi_1 = \frac{2EA}{L} \tag{6.25}$$

The second deformable mode is associated with pure bending, and the work needed to deform the beam with this mode is:

$$W_{II} = \frac{1}{2}\phi_2^T \bar{\mathbf{K}}_{EB}\phi_2 = \frac{2IE}{L} \tag{6.26}$$

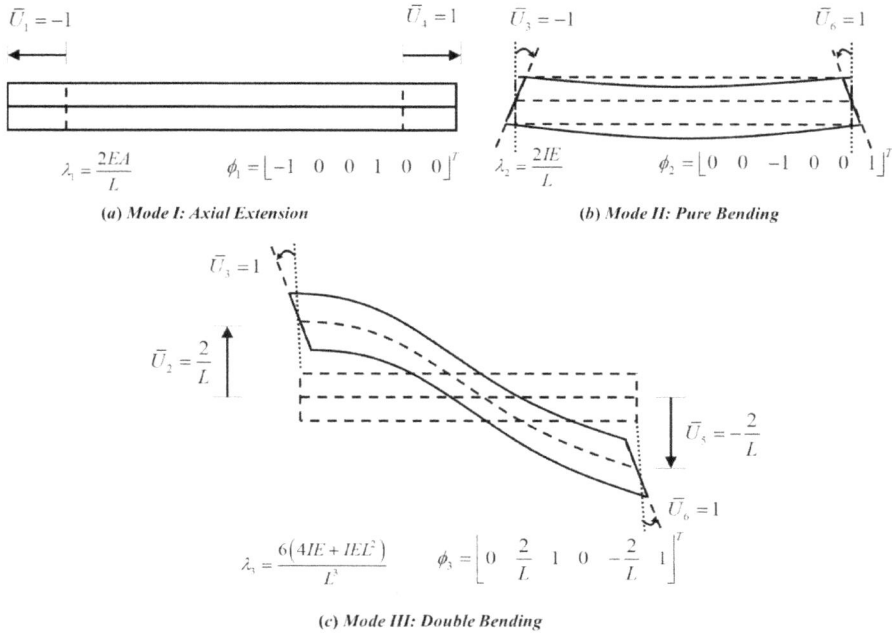

$\bar{U}_1 = -1$
$\bar{U}_4 = 1$
$\bar{U}_3 = -1$
$\bar{U}_6 = 1$

$\lambda_1 = \dfrac{2EA}{L}$      $\phi_1 = \lfloor -1 \quad 0 \quad 0 \quad 1 \quad 0 \quad 0 \rfloor^T$

$\lambda_2 = \dfrac{2IE}{L}$      $\phi_2 = \lfloor 0 \quad 0 \quad -1 \quad 0 \quad 0 \quad 1 \rfloor^T$

*(a) Mode I: Axial Extension*                    *(b) Mode II: Pure Bending*

$\bar{U}_3 = 1$

$\bar{U}_2 = \dfrac{2}{L}$

$\bar{U}_4 = -\dfrac{2}{L}$

$\bar{U}_6 = 1$

$\lambda_1 = \dfrac{6\left(4IE + IEL^2\right)}{L^5}$      $\phi_3 = \left\lfloor 0 \quad \dfrac{2}{L} \quad 1 \quad 0 \quad -\dfrac{2}{L} \quad 1 \right\rfloor^T$

*(c) Mode III: Double Bending*

**FIGURE 6.8**   Deformation Modes: Euler-Bernoulli Frame Member.

The third deformable mode is associated with double bending, and the work needed to deform the beam with this mode is:

$$W_{III} = \frac{1}{2}\phi_3^T \bar{\mathbf{K}}_{EB}\phi_3 = \frac{6IE\left(4+L^2\right)^2}{L^5} \tag{6.27}$$

As shown in Figure 6.9, the fourth mode is associated with rigid horizontal translation, and the work needed to displace the beam with this mode is:

$$W_{IV} = \frac{1}{2}\phi_4^T \bar{\mathbf{K}}_{EB}\phi_4 = 0 \tag{6.28}$$

The fifth mode is associated with rigid vertical translation and the work needed to displace the beam with this mode is:

$$W_V = \frac{1}{2}\phi_5^T \bar{\mathbf{K}}_{EB}\phi_5 = 0 \tag{6.29}$$

The sixth mode is associated with rigid rotation and the work needed to displace the beam with this mode is:

$$W_{VI} = \frac{1}{2}\phi_6^T \bar{\mathbf{K}}_{EB}\phi_6 = 0 \tag{6.30}$$

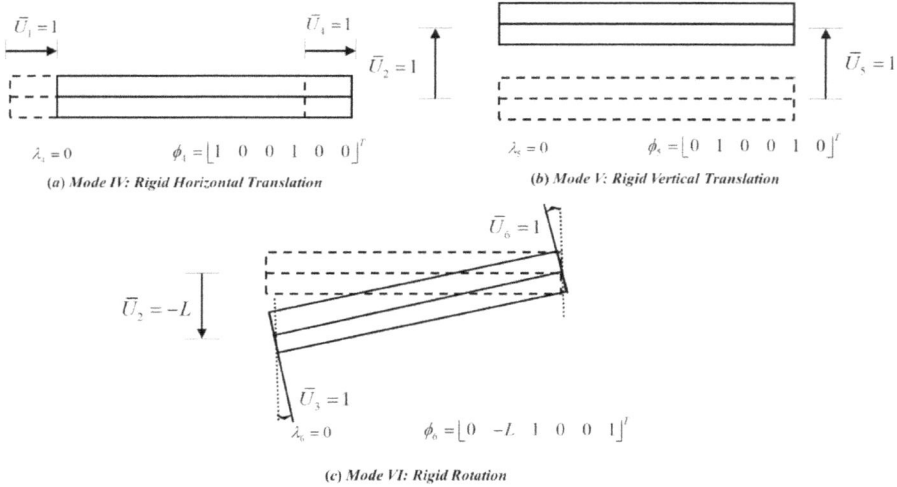

(a) Mode IV: Rigid Horizontal Translation

$\lambda_4 = 0$    $\phi_4 = \lfloor 1 \ 0 \ 0 \ 1 \ 0 \ 0 \rfloor^T$

(b) Mode V: Rigid Vertical Translation

$\lambda_5 = 0$    $\phi_5 = \lfloor 0 \ 1 \ 0 \ 0 \ 1 \ 0 \rfloor^T$

(c) Mode VI: Rigid Rotation

$\lambda_6 = 0$    $\phi_6 = \lfloor 0 \ -L \ 1 \ 0 \ 0 \ 1 \rfloor^T$

**FIGURE 6.9**   Rigid Body Modes: Euler-Bernoulli Frame Member.

### 6.3.3   PRISMATIC TIMOSHENKO FRAME MEMBER: SIMPLY SUPPORTED SYSTEM

Consider the basic and complete systems of a prismatic Timoshenko frame member shown in Figure 6.10. End forces between these two systems are related through the matrix equilibrium equations (3/3.11) discussed in Chapter 3. The basic member deformations $\mathbf{v}$ and member end displacements $\overline{\mathbf{U}}$ are related through the matrix compatibility equations (4/4.23) discussed in Chapter 4. It is worth repeating them here.

$$\begin{Bmatrix} \overline{\mathbf{P}}_1 \\ \overline{\mathbf{P}}_2 \end{Bmatrix} = \begin{bmatrix} \overline{\mathbf{b}}_{1B} \\ \overline{\mathbf{b}}_{2B} \end{bmatrix} \mathbf{q} \text{ or } \overline{\mathbf{P}} = \overline{\mathbf{b}}_B \mathbf{q} \tag{3/3.11}$$

$$\mathbf{v} = \begin{bmatrix} \overline{\mathbf{a}}_{1B} & \overline{\mathbf{a}}_{2B} \end{bmatrix} \begin{Bmatrix} \overline{\mathbf{U}}_1 \\ \overline{\mathbf{U}}_2 \end{Bmatrix} \text{ or } \mathbf{v} = \overline{\mathbf{a}}_B \overline{\mathbf{U}} \tag{4/4.23}$$

From the invariant property of the external virtual work (5/5.200), it can be concluded that:

$$\underbrace{\delta \mathbf{v}^T \mathbf{q}}_{Basic\ System} = \underbrace{\delta \overline{\mathbf{U}}^T \overline{\mathbf{P}}}_{Complete\ System} \tag{6.31}$$

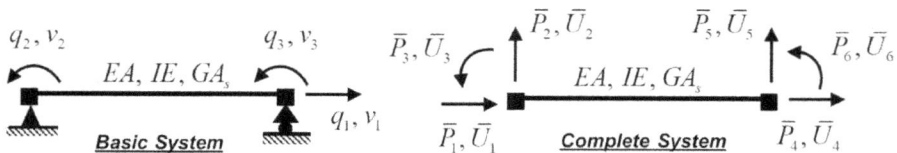

**FIGURE 6.10**   Prismatic Timoshenko Frame Member: Simply Supported System vs. Complete System.

Substituting Eqs. (4/4.23) into (6.31), one has

$$\underbrace{\delta\bar{\mathbf{U}}^T}_{Arbitrary}\left[\bar{\mathbf{P}}-\bar{\mathbf{a}}_B^T\mathbf{q}\right]=0 \quad\Rightarrow\quad \bar{\mathbf{P}}=\bar{\mathbf{a}}_B^T\mathbf{q} \tag{6.32}$$

Substituting Eqs. (6.3) into (6.17), one gains

$$\bar{\mathbf{P}}=\bar{\mathbf{a}}_B^T\bar{\mathbf{k}}_{TB}^B\mathbf{v} \tag{6.33}$$

Substituting Eqs. (4/4.23) into (6.33), one has

$$\bar{\mathbf{P}}=\bar{\mathbf{a}}_B^T\bar{\mathbf{k}}_{TB}^B\bar{\mathbf{a}}_B\bar{\mathbf{U}}=\bar{\mathbf{K}}_{TB}\bar{\mathbf{U}} \tag{6.34}$$

where the complete member stiffness $\bar{\mathbf{K}}_{TB}$ is defined as

$$\bar{\mathbf{K}}_{TB}=\bar{\mathbf{a}}_B^T\bar{\mathbf{k}}_{TB}^B\bar{\mathbf{a}}_B \tag{6.35}$$

Explicitly, the complete member stiffness $\bar{\mathbf{K}}_{TB}$ is

$$\bar{\mathbf{K}}_{TB}=\begin{bmatrix}\bar{\mathbf{a}}_{1B}^T\bar{\mathbf{k}}_{TB}^B\bar{\mathbf{a}}_{1B} & \bar{\mathbf{a}}_{1B}^T\bar{\mathbf{k}}_{TB}^B\bar{\mathbf{a}}_{2B} \\ \bar{\mathbf{a}}_{2B}^T\bar{\mathbf{k}}_{TB}^B\bar{\mathbf{a}}_{1B} & \bar{\mathbf{a}}_{2B}^T\bar{\mathbf{k}}_{TB}^B\bar{\mathbf{a}}_{2B}\end{bmatrix} \tag{6.36}$$

$$\bar{\mathbf{K}}_{TB}=\begin{bmatrix} \dfrac{EA}{L} & 0 & 0 & -\dfrac{EA}{L} & 0 & 0 \\[2mm] 0 & 12\dfrac{EI}{L^3(1+\Phi)} & 6\dfrac{EI}{L^2(1+\Phi)} & 0 & -12\dfrac{EI}{L^3(1+\Phi)} & 6\dfrac{EI}{L^2(1+\Phi)} \\[2mm] 0 & 6\dfrac{EI}{L^2(1+\Phi)} & \dfrac{(4+\Phi)EI}{(1+\Phi)L} & 0 & -6\dfrac{EI}{L^2(1+\Phi)} & \dfrac{(2-\Phi)EI}{(1+\Phi)L} \\[2mm] -\dfrac{EA}{L} & 0 & 0 & \dfrac{EA}{L} & 0 & 0 \\[2mm] 0 & -12\dfrac{EI}{L^3(1+\Phi)} & -6\dfrac{EI}{L^2(1+\Phi)} & 0 & 12\dfrac{EI}{L^3(1+\Phi)} & -6\dfrac{EI}{L^2(1+\Phi)} \\[2mm] 0 & 6\dfrac{EI}{L^2(1+\Phi)} & \dfrac{(2-\Phi)EI}{(1+\Phi)L} & 0 & -6\dfrac{EI}{L^2(1+\Phi)} & \dfrac{(4+\Phi)EI}{(1+\Phi)L} \end{bmatrix} \tag{6.37}$$

It is noticed that the matrix $\bar{\mathbf{K}}_{TB}$ is singular since it contains the rigid body mode. Its eigenvalues $\lambda_i$ and corresponding eigenvectors $\phi_i$ are:

$$\lambda_1=\frac{2EA}{L};\lambda_2=\frac{2IE}{L};\lambda_3=\frac{6(4IE+IEL^2)}{L^3(1+\Phi)};\text{and }\lambda_4=\lambda_5=\lambda_6=0 \tag{6.38}$$

$$\phi_1 = \lfloor -1 \quad 0 \quad 0 \quad 1 \quad 0 \quad 0 \rfloor^T ; \phi_2 = \lfloor 0 \quad 0 \quad -1 \quad 0 \quad 0 \quad 1 \rfloor^T ; \phi_3 = \lfloor 0 \quad \frac{2}{L} \quad 1 \quad 0 \quad -\frac{2}{L} \quad 1 \rfloor^T ;$$

$$\phi_4 = \lfloor 1 \quad 0 \quad 0 \quad 1 \quad 0 \quad 0 \rfloor^T ; \quad \phi_5 = \lfloor 0 \quad 1 \quad 0 \quad 0 \quad 1 \quad 0 \rfloor^T ; \quad \phi_6 = \lfloor 0 \quad -L \quad 1 \quad 0 \quad 0 \quad 1 \rfloor^T$$

$$(6.39)$$

As expected, there are three zero eigenvalues corresponding to three rigid body modes. All eigen-vectors are the same as those of the prismatic Euler-Bernoulli frame stiffness matrix (Figure 6.8 and Figure 6.9). This is justified since both types of the beam member should present the same deformation and rigid-body modes. Furthermore, it is interesting to observe that only the first two eigenvalues ($\lambda_1$ and $\lambda_2$) are the same as those of the prismatic Euler-Bernoulli frame stiff-ness matrix. This is reasonable since those two eigenvalues and their corresponding eigenvectors ($\phi_1$ and $\phi_2$) are associated with the axial deformation and pure bending modes (Figure 6.8a and b) in which shear forces are zero. However, the third eigenvalue $\lambda_3$ is different from that of the prismatic Euler-Bernoulli frame stiffness matrix. This is expected since the deformation mode (Figure 6.8c) corresponding to this eigenvalue is in a double bending state in which shear force exists.

## 6.4  STIFFNESS TRANSFORMATIONS

In the previous section, the stiffness transformations were used to derive the complete stiffness matrices from the basic stiffness matrices of several member types. This transformation type may be called "Rigid-Body-Mode" transformation. In this section, other types of stiffness transformations are to be discussed. It will be shown that the stiffness transformation relations can be derived from the virtual work principle.

### 6.4.1  COORDINATE TRANSFORMATION

In the previous section, complete stiffness matrices of several frame members with respect to their local reference system were obtained. Now, it is desirable to transform those stiffness matrices from local ($x - y - z$) to global (X - Y - Z) reference systems. This can be achieved by coordinate transformation with the help of the virtual work principle.

Consider the 2-D frame member shown in Figure 6.11. The member-end forces $\bar{\mathbf{P}}$ and displacements $\bar{\mathbf{U}}$ with respect to the local reference axes are defined as:

$$\bar{\mathbf{P}} = \lfloor \bar{P}_1 \quad \bar{P}_2 \quad \bar{P}_3 \quad \bar{P}_4 \quad \bar{P}_5 \quad \bar{P}_6 \rfloor^T \text{ and } \bar{\mathbf{U}} = \lfloor \bar{U}_1 \quad \bar{U}_2 \quad \bar{U}_3 \quad \bar{U}_4 \quad \bar{U}_5 \quad \bar{U}_6 \rfloor^T \quad (6.40)$$

The member-end forces $\mathbf{P}$ and displacements $\mathbf{U}$ with respect to the global reference axes are defined as:

$$\mathbf{P} = \lfloor P_1 \quad P_2 \quad P_3 \quad P_4 \quad P_5 \quad P_6 \rfloor^T \text{ and } \mathbf{U} = \lfloor U_1 \quad U_2 \quad U_3 \quad U_4 \quad U_5 \quad U_6 \rfloor^T \quad (6.41)$$

The local end displacements $\bar{\mathbf{U}}$ and global end displacements $\mathbf{U}$ are related through the matrix compatibility equations (4/4.47) discussed in Chapter 4. It is worth repeating them here.

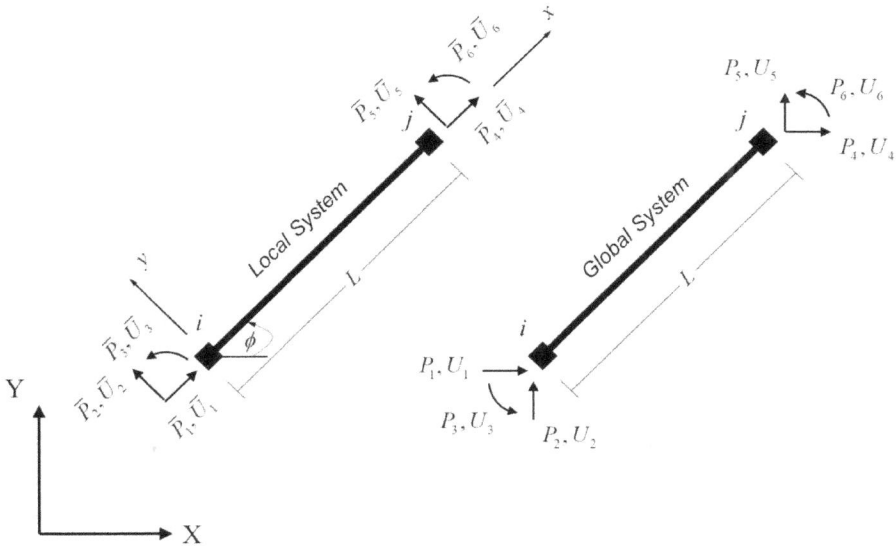

**FIGURE 6.11**    2-D Frame Coordinate Transformation: Local System vs. Global System.

$$\begin{Bmatrix} \bar{U}_1 \\ \bar{U}_2 \\ \bar{U}_3 \\ \bar{U}_4 \\ \bar{U}_5 \\ \bar{U}_6 \end{Bmatrix} = \begin{bmatrix} \cos\phi & \sin\phi & 0 & 0 & 0 & 0 \\ -\sin\phi & \cos\phi & 0 & 0 & 0 & 0 \\ 0 & 0 & 1 & 0 & 0 & 0 \\ 0 & 0 & 0 & \cos\phi & \sin\phi & 0 \\ 0 & 0 & 0 & -\sin\phi & \cos\phi & 0 \\ 0 & 0 & 0 & 0 & 0 & 1 \end{bmatrix} \begin{Bmatrix} U_1 \\ U_2 \\ U_3 \\ U_4 \\ U_5 \\ U_6 \end{Bmatrix} \text{ or } \bar{\mathbf{U}} = \mathbf{\Gamma}_{ROT}\mathbf{U} \qquad (4/4.47)$$

where

$$\mathbf{\Gamma}_{ROT} = \begin{bmatrix} \mathbf{T}_{ROT} & \mathbf{0} \\ \mathbf{0} & \mathbf{T}_{ROT} \end{bmatrix} \qquad (4/4.48)$$

From invariant property of the external virtual work (5/5.200), it can be concluded that:

$$\underbrace{\delta\bar{\mathbf{U}}^T\bar{\mathbf{P}}}_{\text{Local System}} = \underbrace{\delta\mathbf{U}^T\mathbf{P}}_{\text{Global System}} \qquad (6.42)$$

Substituting Eqs. (4/4.47) into (6.42), one has

$$\underbrace{\delta\mathbf{U}^T}_{\text{Arbitrary}}\left[\mathbf{P} - \mathbf{\Gamma}_{ROT}^T\bar{\mathbf{P}}\right] = 0 \quad \Rightarrow \quad \mathbf{P} = \mathbf{\Gamma}_{ROT}^T\bar{\mathbf{P}} \qquad (6.43)$$

It is recalled that the local member stiffness equation is:

$$\bar{\mathbf{P}} = \bar{\mathbf{K}}\bar{\mathbf{U}} \qquad (6.44)$$

Substituting Eqs. (6.44) into (6.43), one has

$$\mathbf{P} = \mathbf{\Gamma}_{ROT}^{T} \overline{\mathbf{K}} \overline{\mathbf{U}} \tag{6.45}$$

Substituting Eqs. (4/4.47) into (6.45), one obtains

$$\mathbf{P} = \mathbf{\Gamma}_{ROT}^{T} \overline{\mathbf{K}} \mathbf{\Gamma}_{ROT} \mathbf{U} = \mathbf{K}\mathbf{U} \tag{6.46}$$

Therefore, the relation between local and global stiffness matrices is:

$$\mathbf{K} = \mathbf{\Gamma}_{ROT}^{T} \overline{\mathbf{K}} \mathbf{\Gamma}_{ROT} \tag{6.47}$$

For a prismatic Euler-Bernoulli frame member, the global stiffness matrix $\mathbf{K}_{EB}$ can be determined as:

$$\mathbf{K}_{EB} = \mathbf{\Gamma}_{ROT}^{T} \overline{\mathbf{K}}_{EB} \mathbf{\Gamma}_{ROT} \tag{6.48}$$

Explicitly,

$$\mathbf{K}_{EB} = \begin{bmatrix} c^2\dfrac{EA}{L}+s^2\dfrac{12EI}{L^3} & cs\dfrac{EA}{L}-cs\dfrac{12EI}{L^3} & -s\dfrac{6EI}{L^2} & -c^2\dfrac{EA}{L}-s^2\dfrac{12EI}{L^3} & -cs\dfrac{EA}{L}+cs\dfrac{12EI}{L^3} & -s\dfrac{6EI}{L^2} \\[2mm] cs\dfrac{EA}{L}-cs\dfrac{12EI}{L^3} & s^2\dfrac{EA}{L}+c^2\dfrac{12EI}{L^3} & c\dfrac{6EI}{L^2} & -cs\dfrac{EA}{L}+cs\dfrac{12EI}{L^3} & -s^2\dfrac{EA}{L}-c^2\dfrac{12EI}{L^3} & c\dfrac{6EI}{L^2} \\[2mm] -s\dfrac{6EI}{L^2} & c\dfrac{6EI}{L^2} & \dfrac{4EI}{L} & s\dfrac{6EI}{L^2} & -c\dfrac{6EI}{L^2} & \dfrac{2EI}{L} \\[2mm] -c^2\dfrac{EA}{L}-s^2\dfrac{12EI}{L^3} & -cs\dfrac{EA}{L}+cs\dfrac{12EI}{L^3} & s\dfrac{6EI}{L^2} & c^2\dfrac{EA}{L}+s^2\dfrac{12EI}{L^3} & cs\dfrac{EA}{L}-cs\dfrac{12EI}{L^3} & s\dfrac{6EI}{L^2} \\[2mm] -cs\dfrac{EA}{L}+cs\dfrac{12EI}{L^3} & -s^2\dfrac{EA}{L}-c^2\dfrac{12EI}{L^3} & -c\dfrac{6EI}{L^2} & cs\dfrac{EA}{L}-cs\dfrac{12EI}{L^3} & s^2\dfrac{EA}{L}+c^2\dfrac{12EI}{L^3} & -c\dfrac{6EI}{L^2} \\[2mm] -s\dfrac{6EI}{L^2} & c\dfrac{6EI}{L^2} & \dfrac{2EI}{L} & s\dfrac{6EI}{L^2} & -c\dfrac{6EI}{L^2} & \dfrac{4EI}{L} \end{bmatrix} \tag{6.49}$$

where

$$c = \cos\phi \text{ and } s = \sin\phi \tag{6.50}$$

The eigenvalues of the matrix $\mathbf{K}_{EB}$ are:

$$\lambda_1 = \frac{2EA}{L}; \lambda_2 = \frac{2IE}{L}; \lambda_3 = \frac{6\left(4IE + IEL^2\right)}{L^3}; \text{and } \lambda_4 = \lambda_5 = \lambda_6 = 0 \tag{6.51}$$

It is noticed that the matrix $\mathbf{K}_{EB}$ is semi-positive definite and gives the proper number of zero eigenvalues associated with rigid body modes. Furthermore, their non-zero eigenvalues are the same as those of the local stiffness matrix $\overline{\mathbf{K}}_{EB}$. This is reasonable since the magnitude of the eigenvalues is related to the amount of work which is invariant to any reference frame.

For a prismatic bar member, the global stiffness matrix $\mathbf{K}_{Bar}$ can be determined as:

$$\mathbf{K}_{Bar} = \mathbf{\Gamma}_{ROT}^{Bar\ T} \overline{\mathbf{K}}_{Bar} \mathbf{\Gamma}_{ROT}^{Bar} = \frac{EA}{L} \begin{bmatrix} c^2 & cs & -c^2 & -cs \\ cs & s^2 & -cs & -s^2 \\ -c^2 & -cs & c^2 & cs \\ -cs & -s^2 & cs & s^2 \end{bmatrix} \tag{6.52}$$

where

$$\mathbf{\Gamma}_{ROT}^{Bar} = \begin{bmatrix} c & s & 0 & 0 \\ 0 & 0 & c & s \end{bmatrix} \tag{6.53}$$

The eigenvalues of the matrix $\mathbf{K}_{Bar}$ are:

$$\lambda_1 = \frac{2EA}{L} \text{ and } \lambda_2 = \lambda_3 = \lambda_4 = 0 \tag{6.54}$$

It is noticed that the matrix $\mathbf{K}_{Bar}$ is semi-positive definite and gives the proper number of zero eigenvalues associated with rigid body modes. Furthermore, their non-zero eigenvalue is the same as that of the local stiffness matrix $\overline{\mathbf{K}}_{Bar}$. This is justified since the magnitude of the eigenvalues is related to the amount of work which is invariant to any reference frame.

For other frame members, the same steps can be followed to obtain the global stiffness matrices from local stiffness matrices.

### 6.4.2 Rigid-End-Zone Transformation

There are several situations where the Rigid-End-Zone transformation plays a crucial role in analyzing structures.

For example,

1. *Finite Joint Size* (Figure 6.12): In reality, a structural joint (node) is not a mathematical point as normally presumed in a structural model. When a joint is adequately represented by a point, this implies that the additional length of columns and beams (typically half the column or beam depth) sufficiently makes up for the effects of a finite joint. For most applications, this hypothesis is valid. However, as the depth of a structural member becomes large or the joint stiffness becomes large, this hypothesis breaks down.
2. *Rigid Floor Diaphragms* (Figure 6.13): Typically, deformations in the floor and roof diaphragm of a building are small compared with the drift between neighboring stories. If this is the case, the diaphragm can be taken as effectively rigid, thus resulting in a considerable reduction in the number of degrees of freedom. Many commercial programs (e.g. SAP 2000) make use of this assumption in reducing the computational effort.
3. *Rigid Link between Slab and Beam* (Figure 6.14): When the whole system of slab and supporting beams is analyzed, it is essential to take into account the offsets between the nodes of the frame (beam) and of the plate (slab).
4. *Nodal Offsets* (Figure 6.15): Generally, it is the top of a beam and not its center lines that share the same elevation in the floor system. Therefore, adjacent beams with different depths will result in nodal offsets.

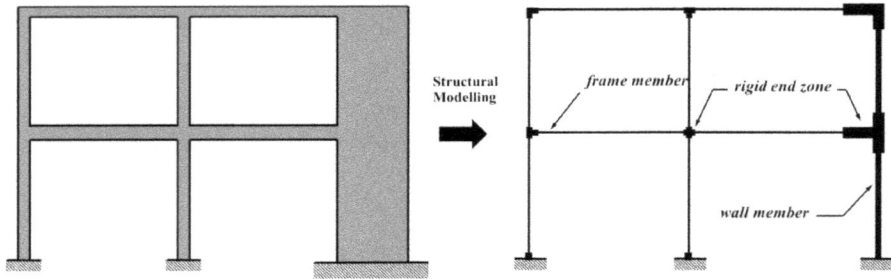

**FIGURE 6.12**   Frame-Wall Structure with Finite Joints.

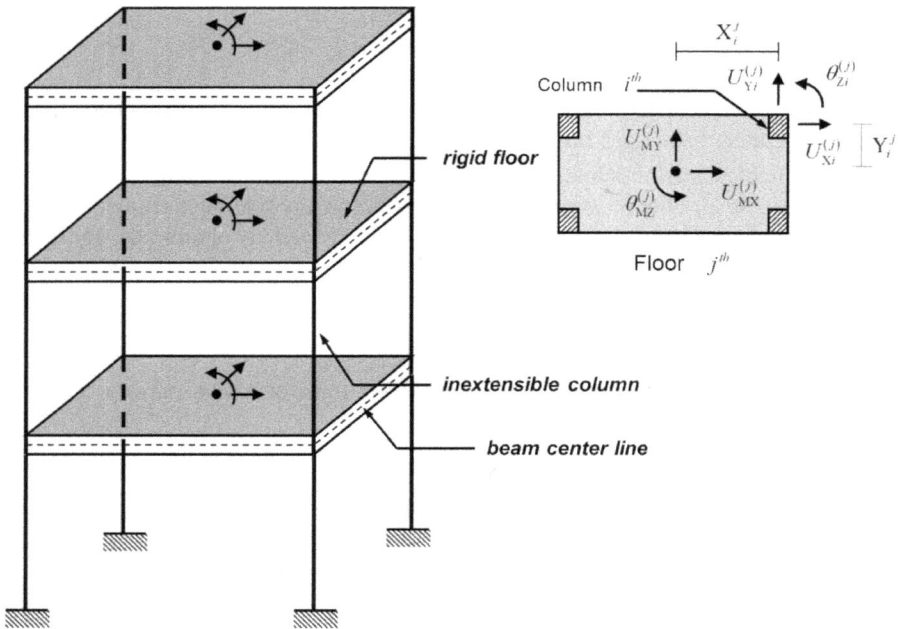

**FIGURE 6.13**   Multi-Storey Frame with Rigid Floor Diaphragms and Inextensible Columns.

The Rigid-End-Zone transformation can be used to account for the aforementioned effects due to the presence of rigid portions. The corresponding stiffness matrix transformation can be obtained by considering the geometrical relations and employing the virtual displacement principle.

A general case of a 2-D frame member with rigid end zones shown in Figure 6.16 is now considered.

The end displacements $\bar{\mathbf{U}}$ and end forces $\bar{\mathbf{P}}$ of the middle flexible frame are written in the vector form as:

$$\bar{\mathbf{U}} = \lfloor \bar{U}_1 \quad \bar{U}_2 \quad \bar{U}_3 \quad \bar{U}_4 \quad \bar{U}_5 \quad \bar{U}_6 \rfloor^T \tag{6.55}$$

$$\bar{\mathbf{P}} = \lfloor \bar{P}_1 \quad \bar{P}_2 \quad \bar{P}_3 \quad \bar{P}_4 \quad \bar{P}_5 \quad \bar{P}_6 \rfloor^T \tag{6.56}$$

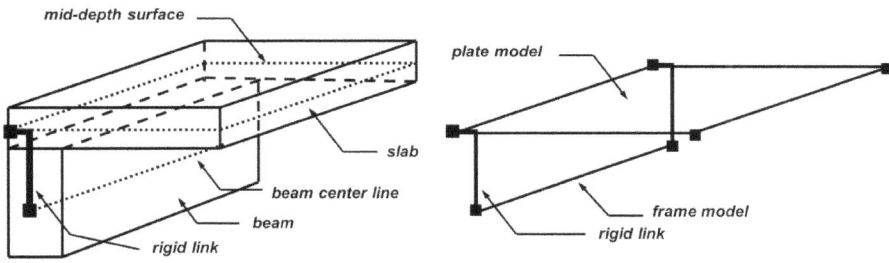

**FIGURE 6.14** Rigid Links between Plate Model (Slab) and Frame Model (Beam).

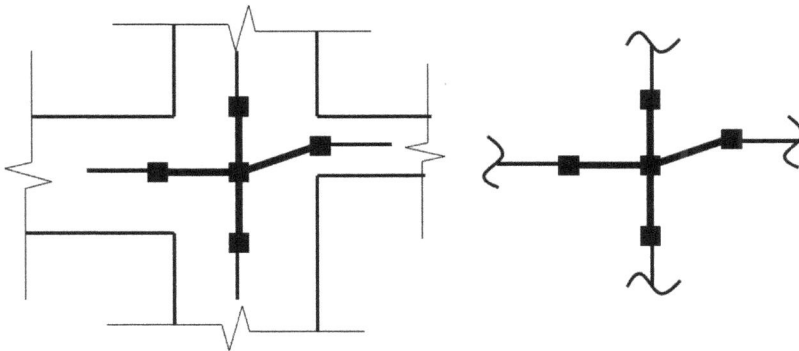

**FIGURE 6.15** Finite Joint with Beam Offsets.

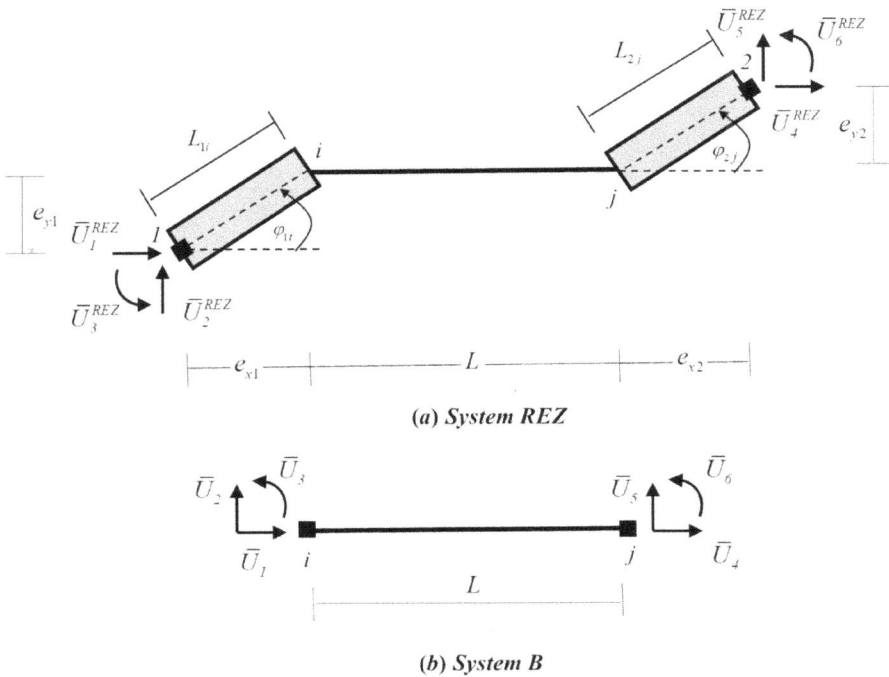

(a) *System REZ*

(b) *System B*

**FIGURE 6.16** 2-D Frame Member: (a) with Rigid End Zones; (b) without Rigid End Zones.

The stiffness relation between these two end quantities is:

$$\overline{\mathbf{P}} = \left\{ \frac{\overline{\mathbf{P}}_1}{\overline{\mathbf{P}}_2} \right\} = \begin{bmatrix} \overline{\mathbf{K}}_{11} & \overline{\mathbf{K}}_{12} \\ \overline{\mathbf{K}}_{21} & \overline{\mathbf{K}}_{22} \end{bmatrix} \left\{ \frac{\overline{\mathbf{U}}_1}{\overline{\mathbf{U}}_2} \right\} = \overline{\mathbf{K}}\overline{\mathbf{U}} \tag{6.57}$$

It is noted that the middle frame member can be one of the frame members developed previously.

The end displacements $\overline{\mathbf{U}}_{REZ}$ and end forces $\overline{\mathbf{P}}_{REZ}$ at rigid ends are written in the vector form as:

$$\overline{\mathbf{U}}_{REZ} = \left\lfloor \overline{U}_1^{REZ} \quad \overline{U}_2^{REZ} \quad \overline{U}_3^{REZ} \quad \overline{U}_4^{REZ} \quad \overline{U}_5^{REZ} \quad \overline{U}_6^{REZ} \right\rfloor^T \tag{6.58}$$

$$\overline{\mathbf{P}}_{REZ} = \left\lfloor \overline{P}_1^{REZ} \quad \overline{P}_2^{REZ} \quad \overline{P}_3^{REZ} \quad \overline{P}_4^{REZ} \quad \overline{P}_5^{REZ} \quad \overline{P}_6^{REZ} \right\rfloor^T \tag{6.59}$$

The compatibility relations between end displacements $\overline{\mathbf{U}}$ and $\overline{\mathbf{U}}_{REZ}$ can be obtained by considering the kinematics of the left and right rigid ends.

From a kinematical consideration (Figure 6.17*a*) of the left rigid end, the following relations are obtained:

$$\begin{aligned}
\overline{U}_1 &= \overline{U}_1^{REZ} - \left( L_{1i} \cos\left(\varphi_{1i}\right) - L_{1i} \cos\left(\varphi_{1i} + \overline{U}_3\right) \right) \\
\overline{U}_2 &= \overline{U}_2^{REZ} + L_{1i} \sin\left(\varphi_{1i} + \overline{U}_3\right) - L_{1i} \sin\left(\varphi_{1i}\right) \\
\overline{U}_3 &= \overline{U}_3^{REZ}
\end{aligned} \tag{6.60}$$

These nonlinear kinematical relations are the same as the rigid-body constraints (4/4.125) obtained in Chapter 4. Since main interests are only limited to linear geometric structures, Eq. (6.60) could be linearized by expanding all trigonometric functions and using the following linearization:

$$\cos\overline{U}_3 \approx 1 \text{ and } \sin\overline{U}_3 \approx \overline{U}_3 \tag{6.61}$$

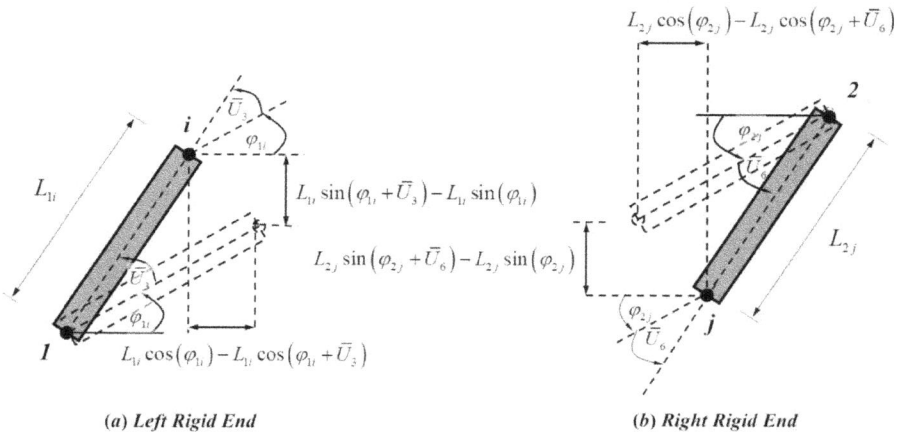

(a) *Left Rigid End*                    (b) *Right Rigid End*

**FIGURE 6.17**    Kinematics of Rigid Ends:

Therefore, Eq. (6.60) can be rewritten as:

$$\bar{U}_1 = \bar{U}_1^{REZ} - \bar{U}_3 L_{1i} \sin\left(\varphi_{1i}\right); \quad \bar{U}_2 = \bar{U}_2^{REZ} + \bar{U}_3 L_{1i} \cos\left(\varphi_{1i}\right); \text{and } \bar{U}_3 = \bar{U}_3^{REZ} \qquad (6.62)$$

and expressed in the matrix form as:

$$\begin{Bmatrix} \bar{U}_1 \\ \bar{U}_2 \\ \bar{U}_3 \end{Bmatrix} = \begin{bmatrix} 1 & 0 & -L_{1i}\sin\left(\varphi_{1i}\right) \\ 0 & 1 & L_{1i}\cos\left(\varphi_{1i}\right) \\ 0 & 0 & 1 \end{bmatrix} \begin{Bmatrix} \bar{U}_1^{REZ} \\ \bar{U}_2^{REZ} \\ \bar{U}_3^{REZ} \end{Bmatrix} \text{ or } \bar{\mathbf{U}}_1 = \mathbf{\Gamma}_{REZ1} \bar{\mathbf{U}}_1^{REZ} \qquad (6.63)$$

Similarly, from a kinematical consideration (Figure 6.17*b*) of the right rigid end, the following relations are obtained:

$$\begin{aligned} \bar{U}_4 &= \bar{U}_4^{REZ} + \left(L_{2j}\cos\left(\varphi_{2j}\right) - L_{2j}\cos\left(\varphi_{2j} + \bar{U}_6\right)\right) \\ \bar{U}_5 &= \bar{U}_5^{REZ} - L_{2j}\sin\left(\varphi_{2j} + \bar{U}_6\right) + L_{2j}\sin\left(\varphi_{2j}\right) \\ \bar{U}_6 &= \bar{U}_6^{REZ} \end{aligned} \qquad (6.64)$$

Eq. (6.64) could be linearized as:

$$\bar{U}_4 = \bar{U}_4^{REZ} + \bar{U}_6 L_{2j} \sin\left(\varphi_{2j}\right); \quad \bar{U}_5 = \bar{U}_5^{REZ} - \bar{U}_6 L_{2j} \cos\left(\varphi_{2j}\right); \text{and } \bar{U}_6 = \bar{U}_6^{REZ} \qquad (6.65)$$

and expressed in the matrix form as:

$$\begin{Bmatrix} \bar{U}_4 \\ \bar{U}_5 \\ \bar{U}_6 \end{Bmatrix} = \begin{bmatrix} 1 & 0 & L_{2j}\sin\left(\varphi_{2j}\right) \\ 0 & 1 & -L_{2j}\cos\left(\varphi_{2j}\right) \\ 0 & 0 & 1 \end{bmatrix} \begin{Bmatrix} \bar{U}_4^{REZ} \\ \bar{U}_5^{REZ} \\ \bar{U}_6^{REZ} \end{Bmatrix} \text{ or } \bar{\mathbf{U}}_2 = \mathbf{\Gamma}_{REZ2} \bar{\mathbf{U}}_2^{REZ} \qquad (6.66)$$

Combining Eqs. (6.63) and (6.66), the matrix compatibility relation between $\bar{\mathbf{U}}$ and $\bar{\mathbf{U}}_{REZ}$ is:

$$\begin{Bmatrix} \bar{U}_1 \\ \bar{U}_2 \\ \bar{U}_3 \\ \bar{U}_4 \\ \bar{U}_5 \\ \bar{U}_6 \end{Bmatrix} = \begin{bmatrix} 1 & 0 & -L_{1i}\sin\left(\varphi_{1i}\right) & 0 & 0 & 0 \\ 0 & 1 & L_{1i}\cos\left(\varphi_{1i}\right) & 0 & 0 & 0 \\ 0 & 0 & 1 & 0 & 0 & 0 \\ 0 & 0 & 0 & 1 & 0 & L_{2j}\sin\left(\varphi_{2j}\right) \\ 0 & 0 & 0 & 0 & 1 & -L_{2j}\cos\left(\varphi_{2j}\right) \\ 0 & 0 & 0 & 0 & 0 & 1 \end{bmatrix} \begin{Bmatrix} \bar{U}_1^{REZ} \\ \bar{U}_2^{REZ} \\ \bar{U}_3^{REZ} \\ \bar{U}_4^{REZ} \\ \bar{U}_5^{REZ} \\ \bar{U}_6^{REZ} \end{Bmatrix} \text{ or } \bar{\mathbf{U}} = \mathbf{\Gamma}_{REZ}\bar{\mathbf{U}}_{REZ} \qquad (6.67)$$

With reference to Figure 6.16, the Rigid-End-Zone transformation matrix can be expressed in terms of offset dimensions as:

$$\Gamma_{REZ} = \begin{bmatrix} 1 & 0 & -e_{y1} & 0 & 0 & 0 \\ 0 & 1 & e_{x1} & 0 & 0 & 0 \\ 0 & 0 & 1 & 0 & 0 & 0 \\ 0 & 0 & 0 & 1 & 0 & e_{y2} \\ 0 & 0 & 0 & 0 & 1 & -e_{x2} \\ 0 & 0 & 0 & 0 & 0 & 1 \end{bmatrix} \tag{6.68}$$

In terms of nodal coordinates, the offset dimensions can be defined as:

$$e_{x1} = x_i - x_1; \quad e_{y1} = y_i - y_1; \quad e_{x2} = x_2 - x_j; \text{and } e_{y2} = y_2 - y_j; \tag{6.69}$$

From the invariant property of the external virtual work (5/5.200), it can be concluded that:

$$\underbrace{\delta \bar{\mathbf{U}}_{REZ}^T \bar{\mathbf{P}}_{REZ}}_{System\ REZ} = \underbrace{\delta \bar{\mathbf{U}}^T \bar{\mathbf{P}}}_{System\ B} \tag{6.70}$$

Substituting Eqs. (6.67) into (6.70), one has

$$\underbrace{\delta \bar{\mathbf{U}}_{REZ}^T}_{Arbitrary} \left[ \bar{\mathbf{P}}_{REZ} - \Gamma_{REZ}^T \bar{\mathbf{P}} \right] = 0 \quad \Rightarrow \quad \bar{\mathbf{P}}_{REZ} = \Gamma_{REZ}^T \bar{\mathbf{P}} \tag{6.71}$$

Equation (6.71) simply represents the equilibrium relations between two end-force systems $\bar{\mathbf{P}}$ and $\bar{\mathbf{P}}_{REZ}$.

Substituting Eqs. (6.57) into (6.71), one has

$$\bar{\mathbf{P}}_{REZ} = \Gamma_{REZ}^T \bar{\mathbf{K}} \bar{\mathbf{U}} \tag{6.72}$$

Substituting Eqs. (6.67) into (6.72), one obtains

$$\bar{\mathbf{P}}_{REZ} = \Gamma_{REZ}^T \bar{\mathbf{K}} \Gamma_{REZ} \bar{\mathbf{U}}_{REZ} = \bar{\mathbf{K}}_{REZ} \bar{\mathbf{U}}_{REZ} \tag{6.73}$$

Therefore, the transformation relation accounting for the presence of rigid end zones is:

$$\bar{\mathbf{K}}_{REZ} = \Gamma_{REZ}^T \bar{\mathbf{K}} \Gamma_{REZ} = \begin{bmatrix} \Gamma_{REZ1}^T \bar{\mathbf{K}}_{11} \Gamma_{REZ1} & \Gamma_{REZ1}^T \bar{\mathbf{K}}_{12} \Gamma_{REZ2} \\ \Gamma_{REZ2}^T \bar{\mathbf{K}}_{21} \Gamma_{REZ1} & \Gamma_{REZ2}^T \bar{\mathbf{K}}_{22} \Gamma_{REZ2} \end{bmatrix} \tag{6.74}$$

In the case of a prismatic Euler-Bernoulli frame, its member stiffness matrix including rigid end zones is:

$$\bar{\mathbf{K}}_{REZ}^{EB} = \Gamma_{REZ}^T \bar{\mathbf{K}}_{EB} \Gamma_{REZ}$$

For other types of middle frame members, their stiffness matrices can be transformed in a similar fashion.

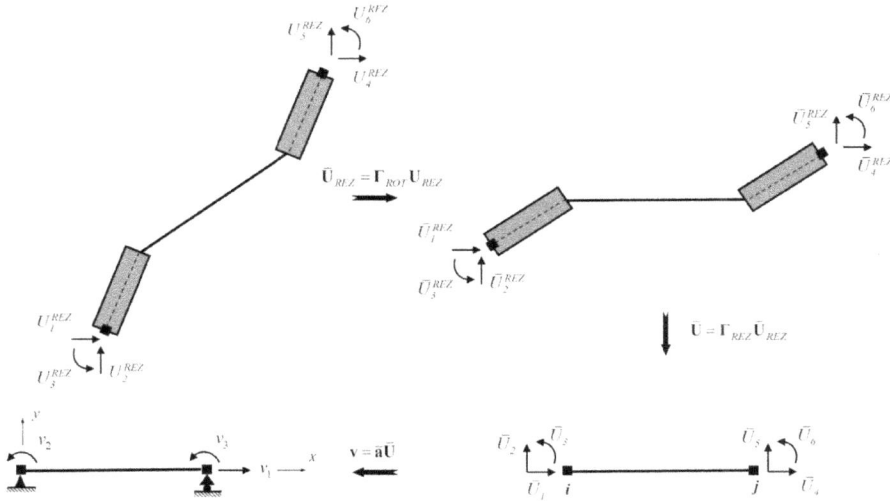

**FIGURE 6.18** Schematic Representation of Displacement Transformations.

## 6.5 SUMMARY OF STIFFNESS TRANSFORMATIONS

So far, this chapter has dealt with three member displacement transformations, namely:

- Rigid-Body-Mode transformation
- Coordinate transformation
- Rigid-End-Zone transformation

All of these transformations can be written together as:

$$\mathbf{v} = \overline{\mathbf{a}}\,\Gamma_{REZ}\Gamma_{ROT}\mathbf{U}_{REZ} = \Gamma\mathbf{U}_{REZ} \qquad (6.75)$$

and are schematically represented in Figure 6.18.

From the virtual work principle, the corresponding force transformations are expressed as:

$$\mathbf{P}_{REZ} = \Gamma_{ROT}^{T}\Gamma_{REZ}^{T}\overline{\mathbf{a}}^{T}\mathbf{q} = \Gamma^{T}\mathbf{q} \qquad (6.76)$$

and are schematically represented in Figure 6.19.

Consequently, the stiffness matrix transformation can be written together as:

$$\mathbf{P}_{REZ} = \Gamma_{ROT}^{T}\Gamma_{REZ}^{T}\overline{\mathbf{a}}^{T}\overline{\mathbf{k}}\overline{\mathbf{a}}\,\Gamma_{REZ}\Gamma_{ROT}\mathbf{U}_{REZ} = \Gamma^{T}\overline{\mathbf{k}}\Gamma\mathbf{U}_{REZ} = \mathbf{K}_{REZ}\mathbf{U}_{REZ} \qquad (6.77)$$

where

$$\mathbf{K}_{REZ} = \Gamma_{ROT}^{T}\Gamma_{REZ}^{T}\overline{\mathbf{a}}^{T}\overline{\mathbf{k}}\overline{\mathbf{a}}\,\Gamma_{REZ}\Gamma_{ROT} = \Gamma^{T}\overline{\mathbf{k}}\Gamma \qquad (6.78)$$

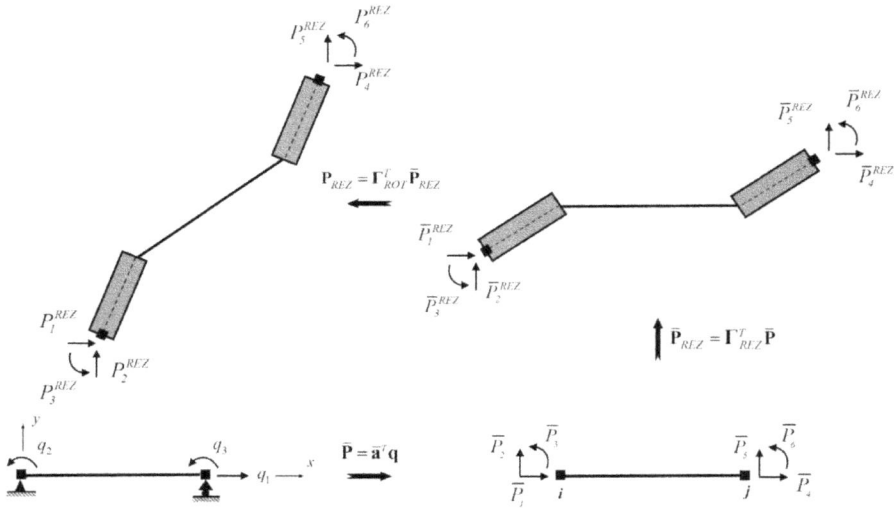

**FIGURE 6.19**  Schematic Representation of Force Transformations.

## 6.6   STIFFNESS METHOD: CONCEPTUAL APPROACH

It is the main objective of this chapter to develop the stiffness equation of a frame structure. The basic solution of this equation is the set of nodal displacements caused by given loadings (both mechanical and non-mechanical ones). The member forces and support reactions are then determined based on these nodal displacements using subsidiary procedures. The main task in the stiffness method is to establish the structural stiffness matrix $\mathbf{K}$. In this section, the so-called "conceptual" approach will be used to construct the structural stiffness matrix. The general steps in this approach are as follows:

1. The structure is rendered kinematically determinate by locking all nodes against displacements.
2. Each locked node is imposed a unit displacement corresponding to each displacement degree of freedom. By definition, the induced nodal forces needed to hold this displaced structural shape are *stiffness coefficients*. It is noted that only a single displacement at a time is imposed in computing the stiffness coefficients while others remain locked.
3. All equilibrium equations associated with free degrees of freedom are written together and then solved to yield the nodal displacements.

It is important to note that pursuing the aforementioned procedures, the governing equations of the structure are satisfied in the following hierarchy:

I. Compatibility Conditions: Member deformations must be geometrically compatible with imposed nodal displacements.
II. Constitutive Conditions: Stress-strain relation or force-deformation relation must be held.
III. Equilibrium Conditions: Nodal and member forces must be in an equilibrated state.

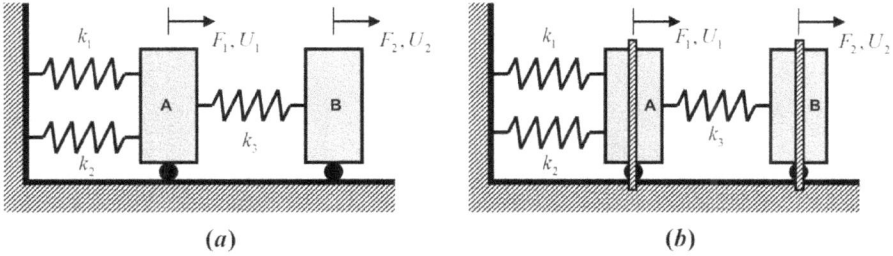

FIGURE 6.20   Example 6.1.

## EXAMPLE 6.1

Consider a simple arrangement of three springs as shown in Figure 6.20a. Use the conceptual approach to construct the stiffness equation relating the forces $F_1$ and $F_2$ to the displacements $U_1$ and $U_2$.

**Solution:** The system is first rendered kinematically determinate by locking all nodes against displacements (Figure 6.20b) and then is imposed with a unit displacement of $U_1$ (Figure 6.21a).

From a geometrical consideration, the deformations of springs 1, 2, and 3 are:

$$\Delta_1 = \Delta_2 = 1.0 \text{ and } \Delta_3 = -1.0$$

From the constitutive relation of each spring, the corresponding internal force in each spring can be determined as:

$$N_1 = k_1\Delta_1 = k_1; \quad N_2 = k_2\Delta_2 = k_2; \text{ and } N_3 = k_3\Delta_3 = -k_3$$

The forces needed to hold this displaced shape are determined by considering equilibrium conditions of blocks A and B (Figure 6.21 b). Therefore, $K_{11}$ and $K_{21}$ are:

$$K_{11} = F_1 = k_1 + k_2 + k_3 \text{ and } K_{21} = F_2 = -k_3$$

Similarly, the system is then imposed with a unit displacement of $U_2$ (Figure 6.22a).

From a geometrical consideration, the deformations of springs 1, 2, and 3 are:

$$\Delta_1 = \Delta_2 = 0 \text{ and } \Delta_3 = 1.0$$

From the constitutive relation of each spring, the corresponding internal force in each spring can be determined as:

$$N_1 = k_1\Delta_1 = 0; \quad N_2 = k_2\Delta_2 = 0; \text{ and } N_3 = k_3\Delta_3 = k_3$$

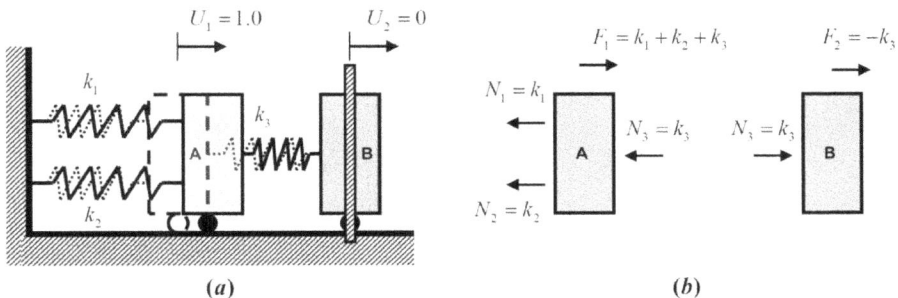

FIGURE 6.21   Example 6.1 (Continued).

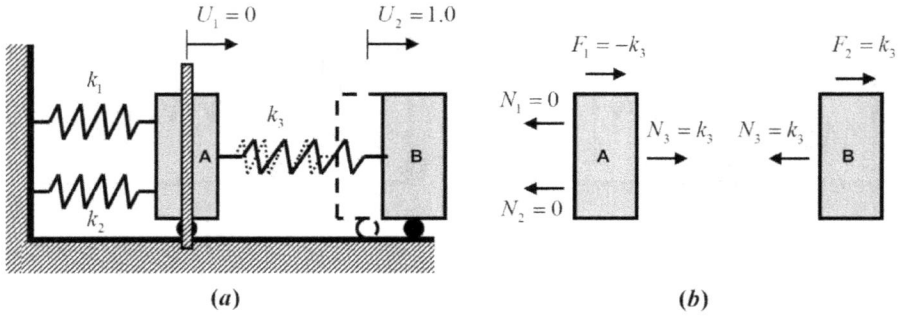

FIGURE 6.22   Example 6.1 (Continued).

The forces needed to hold this displaced shape are determined by considering equilibrium conditions of blocks A and B (Figure 6.22 b). Therefore, $K_{12}$ and $K_{22}$ are:

$$K_{12} = F_1 = -k_3 \text{ and } K_{22} = F_2 = k_3$$

From the principle of superposition, two equilibrium equations associated with $U_1$ and $U_2$ can be written as:

$$F_1 = \left(k_1 + k_2 + k_3\right)U_1 - k_3 U_2 \text{ and } F_2 = -k_3 U_1 + k_3 U_2$$

in the matrix form as:

$$\left\{ \begin{matrix} F_1 \\ F_2 \end{matrix} \right\} = \left[ \begin{matrix} k_1 + k_2 + k_3 & -k_3 \\ -k_3 & k_3 \end{matrix} \right] \left\{ \begin{matrix} U_1 \\ U_2 \end{matrix} \right\} \text{ or } \mathbf{F} = \mathbf{KU}$$

## EXAMPLE 6.2

For the frame of Example 4.1, use the conceptual approach to construct the structural stiffness equation associated with free degrees of freedom and determine nodal displacements, member end forces, and support reactions. For all members, $EA = 10 \times 10^3 \ kN$ and $IE = 10^3 \ kN-m^2$.

**Solution:** From Example 4.1, the matrix compatibility equation of the structure associated with free degrees of freedom is:

$$\left\{ \begin{matrix} V_1 \\ V_2 \\ V_3 \\ V_4 \\ V_5 \\ V_6 \\ V_7 \\ V_8 \\ V_9 \end{matrix} \right\} = \left[ \begin{matrix} 0.6 & 0.8 & 0 & 0 & 0 & 0 \\ 0.16 & -0.12 & 0 & 0 & 0 & 0 \\ 0.16 & -0.12 & 1 & 0 & 0 & 0 \\ -1 & 0 & 0 & 1 & 0 & 0 \\ 0 & 0.2 & 1 & 0 & -0.2 & 0 \\ 0 & 0.2 & 0 & 0 & -0.2 & 1 \\ 0 & 0 & 0 & 0 & 1 & 0 \\ 0 & 0 & 0 & 0.25 & 0 & 0 \\ 0 & 0 & 0 & 0.25 & 0 & 1 \end{matrix} \right] \left\{ \begin{matrix} U_1 \\ U_2 \\ U_3 \\ U_4 \\ U_5 \\ U_6 \end{matrix} \right\} \text{ or } \mathbf{V} = \mathbf{V}_{free} = \mathbf{A}_{free}\mathbf{U}_{free}$$

The structure is first rendered kinematically determinate by locking all nodes against displacements (Figure 6.24a). Then each locked node is imposed by unit amounts corresponding to the unknown displacements.

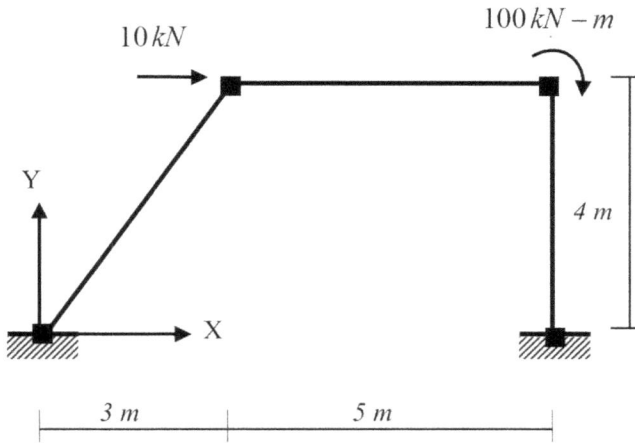

**FIGURE 6.23** Example 6.2.

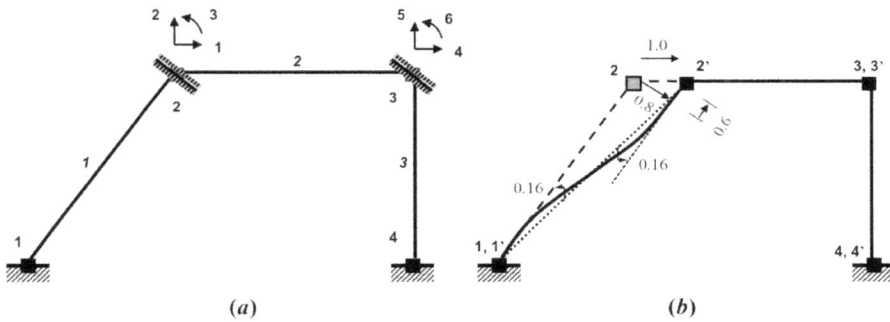

(a)                                                    (b)

**FIGURE 6.24** Example 6.2 (Continued).

For example, the first degree of freedom is released while others remain locked ($U_1 = 1.0$) as shown in Figure 6.24 b.

From the structural compatibility matrix $\mathbf{A}_{free}$, the associated deformations of member 1 are shown in Figure 6.25 and the corresponding basic member forces are:

$$
\begin{Bmatrix} Q_1 \\ Q_2 \\ Q_3 \end{Bmatrix} = \begin{bmatrix} 2000 & 0 & 0 \\ 0 & 800 & 400 \\ 0 & 400 & 800 \end{bmatrix} \begin{Bmatrix} 0.6 \\ 0.16 \\ 0.16 \end{Bmatrix} = \begin{Bmatrix} 1200 \\ 192 \\ 192 \end{Bmatrix}
$$

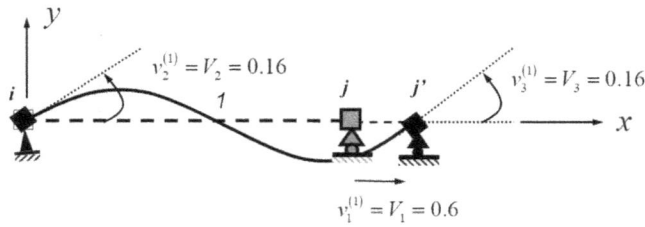

**FIGURE 6.25**  Example 6.2 (Continued).

The corresponding member end forces with respect to the global reference system are:

$$
\begin{Bmatrix} P_1^{(1)} \\ P_2^{(1)} \\ P_3^{(1)} \\ P_4^{(1)} \\ P_5^{(1)} \\ P_6^{(1)} \end{Bmatrix} =
\begin{bmatrix}
-0.6 & -0.16 & -0.16 \\
-0.8 & 0.12 & 0.12 \\
0 & 1 & 0 \\
0.6 & 0.16 & 0.16 \\
0.8 & -0.12 & -0.12 \\
0 & 0 & 1
\end{bmatrix}
\begin{Bmatrix} Q_1 \\ Q_2 \\ Q_3 \end{Bmatrix} =
\begin{Bmatrix} -781.44 \\ -913.92 \\ 192 \\ 781.44 \\ 913.92 \\ 192 \end{Bmatrix}
$$

The associated deformations of member 2 are shown in Figure 6.26 and the corresponding basic member forces are:

$$
\begin{Bmatrix} Q_4 \\ Q_5 \\ Q_6 \end{Bmatrix} =
\begin{bmatrix}
2000 & 0 & 0 \\
0 & 800 & 400 \\
0 & 400 & 800
\end{bmatrix}
\begin{Bmatrix} -1 \\ 0 \\ 0 \end{Bmatrix} =
\begin{Bmatrix} -2000 \\ 0 \\ 0 \end{Bmatrix}
$$

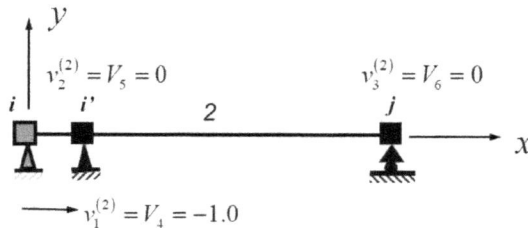

**FIGURE 6.26**  Example 6.2 (Continued).

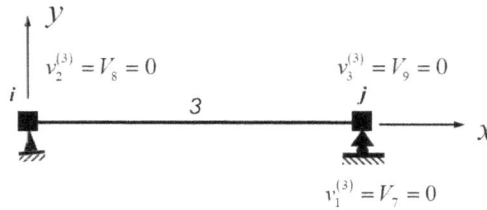

**FIGURE 6.27**  Example 6.2 (Continued).

The corresponding member end forces with respect to the global reference system are:

$$\begin{Bmatrix} P_1^{(2)} \\ P_2^{(2)} \\ P_3^{(2)} \\ P_4^{(2)} \\ P_5^{(2)} \\ P_6^{(2)} \end{Bmatrix} = \begin{bmatrix} -1 & 0 & 0 \\ 0 & 0.2 & 0.2 \\ 0 & 1 & 0 \\ 1 & 0 & 0 \\ 0 & -0.2 & -0.2 \\ 0 & 0 & 1 \end{bmatrix} \begin{Bmatrix} -2000 \\ 0 \\ 0 \end{Bmatrix} = \begin{Bmatrix} 2000 \\ 0 \\ 0 \\ -2000 \\ 0 \\ 0 \end{Bmatrix}$$

The associated deformations of member 3 are are shown in Figure 6.27 and the corresponding basic member forces are:

$$\lfloor Q_7 \quad Q_8 \quad Q_9 \rfloor^T = \lfloor 0 \quad 0 \quad 0 \rfloor^T$$

The corresponding member end forces with respect to the global reference system are:

$$\lfloor P_1^{(3)} \quad P_2^{(3)} \quad P_3^{(3)} \quad P_4^{(3)} \quad P_5^{(3)} \quad P_6^{(3)} \rfloor^T = \lfloor 0 \quad 0 \quad 0 \quad 0 \quad 0 \quad 0 \rfloor^T$$

Consequently, the constrained forces needed to hold this displaced shape ($U_1 = 1.0$) are shown in Figure 6.28.

Applying the same procedure to other free DOFs, the equilibrium equation associated with each degree of freedom can be written as:

$$10 \quad = 2781.44U_1 + 913.92U_2 + 192U_3 - 2000U_4$$
$$0 \quad = 913.92U_1 + 1410.56U_2 + 96U_3 - 96U_5 + 240U_6$$
$$0 \quad = 192U_1 + 96U_2 + 1600U_3 - 240U_5 + 400U_6$$
$$0 \quad = -2000U_1 + 2187.5U_4 + 375U_6$$
$$0 \quad = -96U_2 - 240U_3 + 2596U_5 - 240U_6$$
$$-100 = 240U_2 + 400U_3 + 375U_4 - 240U_5 + 1800U_6$$

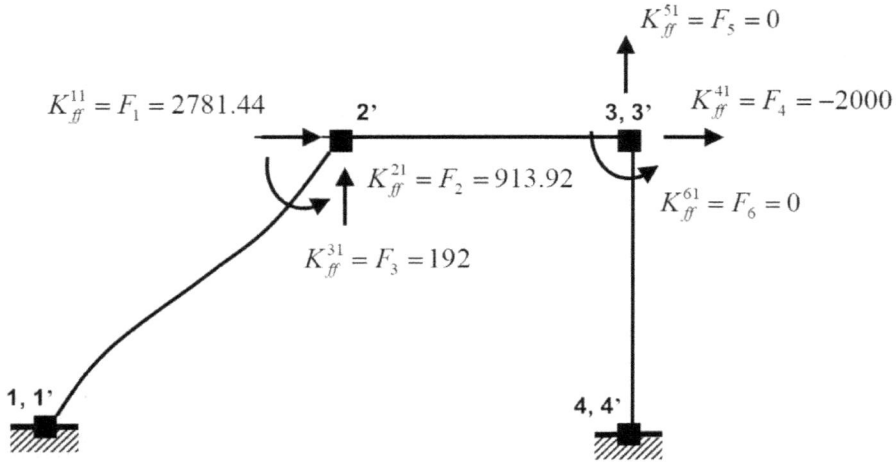

**FIGURE 6.28** Example 6.2 (Continued).

These equations can be written in the matrix form as:

$$
\mathbf{F}_{free} = \mathbf{K}_{ff}\mathbf{U}_{free} \text{ or } 
\begin{Bmatrix} 10 \\ 0 \\ 0 \\ 0 \\ 0 \\ -100 \end{Bmatrix}
=
\begin{bmatrix}
2781.44 & 913.92 & 192 & -2000 & 0 & 0 \\
913.92 & 1410.56 & 96 & 0 & -96 & 240 \\
192 & 96 & 1600 & 0 & -240 & 400 \\
-2000 & 0 & 0 & 2187.5 & 0 & 375 \\
0 & -96 & -240 & 0 & 2596 & -240 \\
0 & 240 & 400 & 375 & -240 & 1800
\end{bmatrix}
\begin{Bmatrix} U_1 \\ U_2 \\ U_3 \\ U_4 \\ U_5 \\ U_6 \end{Bmatrix}
$$

This is the structural stiffness relation associated with free degrees of freedom, and the set of nodal displacements can be solved from this relation as:

$$\mathbf{U}_{free} = \left[\mathbf{K}_{ff}\right]^{-1}\mathbf{F}_{free} = \lfloor 0.06082 \quad -0.02883 \quad 0.01071 \quad 0.06743 \quad -0.006455 \quad -0.06900 \rfloor^T$$

The basic member deformations are recovered by substituting this set of nodal displacements into the structural compatibility relation.

$$
\mathbf{V} = \mathbf{A}_{free}\mathbf{U}_{free} = 
\begin{Bmatrix} V_1 \\ V_2 \\ V_3 \\ V_4 \\ V_5 \\ V_6 \\ V_7 \\ V_8 \\ V_9 \end{Bmatrix}
=
\begin{bmatrix}
0.6 & 0.8 & 0 & 0 & 0 & 0 \\
0.16 & -0.12 & 0 & 0 & 0 & 0 \\
0.16 & -0.12 & 1 & 0 & 0 & 0 \\
-1 & 0 & 0 & 1 & 0 & 0 \\
0 & 0.2 & 1 & 0 & -0.2 & 0 \\
0 & 0.2 & 0 & 0 & -0.2 & 1 \\
0 & 0 & 0 & 0 & 1 & 0 \\
0 & 0 & 0 & 0.25 & 0 & 0 \\
0 & 0 & 0 & 0.25 & 0 & 1
\end{bmatrix}
\begin{Bmatrix} 0.06082 \\ -0.02883 \\ 0.01071 \\ 0.06743 \\ -0.006455 \\ -0.06900 \end{Bmatrix}
=
\begin{Bmatrix} 0.01342 \\ 0.01319 \\ 0.02390 \\ 0.006616 \\ 0.006238 \\ -0.07348 \\ -0.006455 \\ 0.01686 \\ -0.05214 \end{Bmatrix}
$$

The basic forces of each member are determined from the basic stiffness equation of each member.

For example, the basic forces of member 1 are:

$$\mathbf{Q}_1 = \mathbf{k}_1 \mathbf{V}_1 \text{ or } \begin{Bmatrix} Q_1 \\ Q_2 \\ Q_3 \end{Bmatrix} = \begin{bmatrix} 2000 & 0 & 0 \\ 0 & 800 & 400 \\ 0 & 400 & 800 \end{bmatrix} \begin{Bmatrix} 0.01342 \\ 0.01319 \\ 0.02390 \end{Bmatrix} = \begin{Bmatrix} 26.85 \, kN \\ 20.11 \, kN - m \\ 24.4 \, kN - m \end{Bmatrix}$$

The complete forces of each member are determined from the equilibrium equations of each member.

For example, the complete forces of member 1 are:

$$\mathbf{P}^{(1)} = \mathbf{b}_B^{(1)} \mathbf{Q}_1 \text{ or } \begin{Bmatrix} P_1^{(1)} \\ P_2^{(1)} \\ P_3^{(1)} \\ P_4^{(1)} \\ P_5^{(1)} \\ P_6^{(1)} \end{Bmatrix} = \begin{bmatrix} -0.6 & -0.16 & -0.16 \\ -0.8 & 0.12 & 0.12 \\ 0 & 1 & 0 \\ 0.6 & 0.16 & 0.16 \\ 0.8 & -0.12 & -0.12 \\ 0 & 0 & 1 \end{bmatrix} \begin{Bmatrix} 26.85 \\ 20.11 \\ 24.4 \end{Bmatrix} = \begin{Bmatrix} -23.23 \, kN \\ -16.14 \, kN \\ 20.11 \, kN - m \\ 23.23 \, kN \\ 16.14 \, kN \\ 24.4 \, kN - m \end{Bmatrix}$$

The reaction forces are shown in Figure 6.29.

Till now, frame structures subjected to only nodal loads have been considered. This type of loading action can be treated with ease since they are associated directly with structural degrees of freedom. Therefore, it is relatively simple to include them in the structural stiffness equations.

Now, it is desirable to extend the "conceptual" stiffness method to analyze frame structures under the actions of both nodal and member loads. The general steps discussed previously remain the same. The member-load effects can be simply treated as equivalent nodal loads. Therefore, it is only necessary to modify the structural load vector.

Consider the frame structure shown in Figure 6.30a. This structure is subjected to both nodal and member loads. The external loads acting on the actual structure (Figure 6.30a) can be decomposed into two nodal loading systems as shown in Figure 6.30b and c, respectively. The first system (Figure 6.30b) is comprised of nodal loads required to restrain all nodes against movements due to member loads. This set of restrained nodal loads can be determined by first locking all

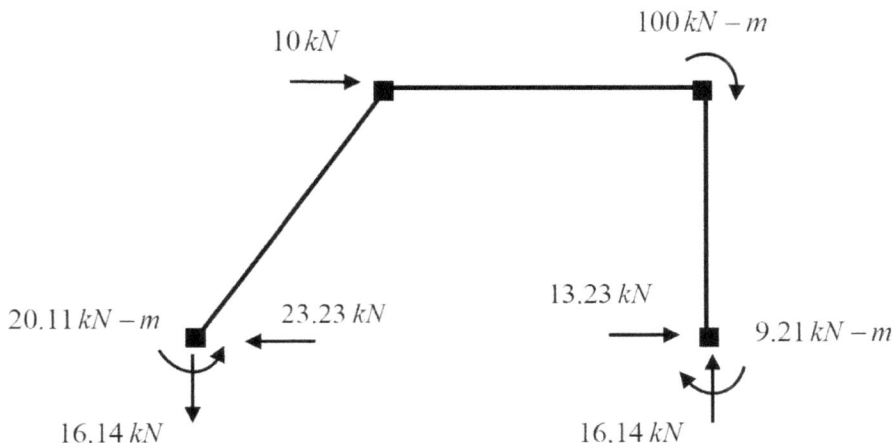

**FIGURE 6.29** Example 6.2 (Continued).

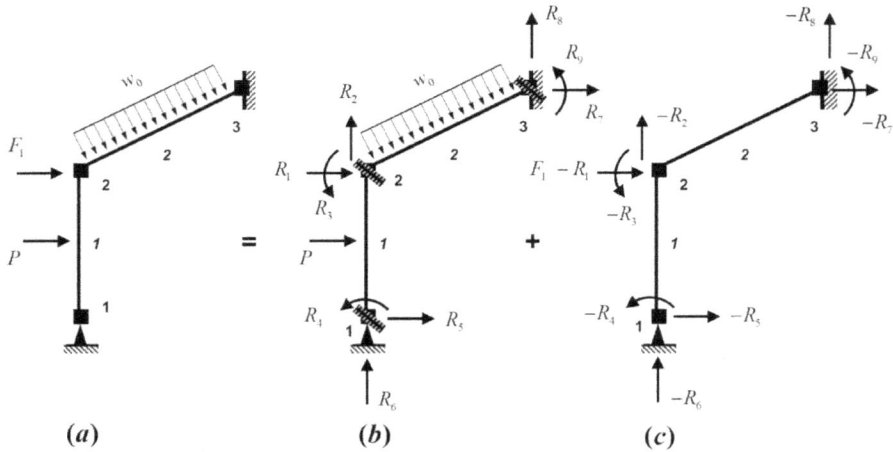

**FIGURE 6.30**  Superposition of the Restrained Structure and the Structure Subjected to Equivalent Nodal Loads.

nodes against movements and then considering equilibrium conditions at all locked nodes. The second system (Figure 6.30c) consists of all nodal loads acting on the actual structure and the opposite nodal loads from the first loading system. These nodal loads are called "equivalent" nodal loads. Based on the principle of superposition, the internal forces and deformations at any cross section of a member and displacement components of any point of the structure shown in Figure 6.30a are the sum of the associated quantities of the frames shown in Figure 6.30b and c.

It should be noted that the restrained nodal loads of the first system are not necessarily induced by mechanical loads acting along members. They can be caused by the following disturbances:

A. The change of temperature in the corresponding member of the actual structure.
B. The corresponding members of the actual structure not fitting together.
C. Support movements.

The general procedures to compute the equivalent nodal loads imposed on the structures can be summarized as follows:

1. Compute the fixed-end forces $\overline{\mathbf{P}}_F^{(i)}$ of each member subjected to member loads with respect to its local reference axes (see Appendix I).
2. Transform the fixed-end forces from local to global reference axes using the rotational matrix (3/3.24).

$$\mathbf{P}_F^{(i)} = \mathbf{\Gamma}_{ROT}^{(i)}{}^T \overline{\mathbf{P}}_F^{(i)} \tag{6.79}$$

3. Establish the restrained nodal loads needed to prevent all nodes against movement. These loads are computed by considering nodal equilibriums under the actions of fixed-end forces. It is noted that the restrained nodal loads are stored in the vector $\mathbf{R}$ and are partitioned according to the free and constrained degrees of freedom as:

$$\mathbf{R} = \lfloor \mathbf{R}_{free} \quad \mathbf{R}_{constr} \rfloor^T \tag{6.80}$$

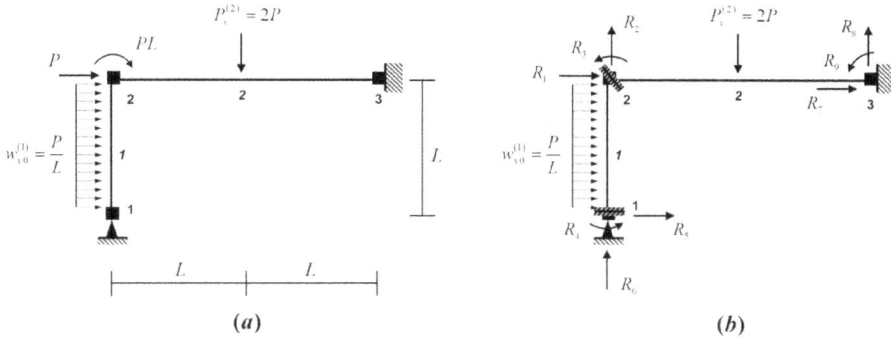

**FIGURE 6.31**    Example 6.3.

4. Define the equivalent nodal load vector as:

$$\mathbf{F}^E = \mathbf{F} - \mathbf{R} = \left\{ \begin{array}{c} \mathbf{F}^E_{free} \\ \mathbf{F}^E_{constr} \end{array} \right\} = \left\{ \begin{array}{c} \mathbf{F}_{free} \\ \mathbf{F}_{constr} \end{array} \right\} - \left\{ \begin{array}{c} \mathbf{R}_{free} \\ \mathbf{R}_{constr} \end{array} \right\} \tag{6.81}$$

## EXAMPLE 6.3

Construct the equivalent nodal load vector $\mathbf{F}^E$ for the frame structure shown in Figure 6.31a.

**Solution:** The structure is first rendered kinematically determinate by locking all nodes against displacements (Figure 6.31b). The fixed end force vector for each member shown in Figure 6.32 with respect to its local reference axes is:

$$\overline{\mathbf{P}}_F^{(1)} = \left\lfloor 0 \quad \frac{P}{2} \quad \frac{PL}{12} \quad 0 \quad \frac{P}{2} \quad -\frac{PL}{12} \right\rfloor^T; \quad \overline{\mathbf{P}}_F^{(2)} = \left\lfloor 0 \quad P \quad \frac{PL}{2} \quad 0 \quad P \quad -\frac{PL}{2} \right\rfloor^T;$$

The rotational transformation matrix of each member is used to transform the fixed-end forces from local to global reference axes.

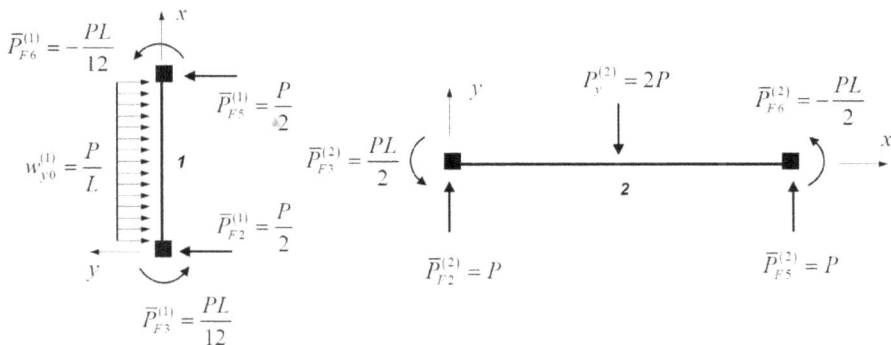

**FIGURE 6.32**    Example 6.3 (Continued).

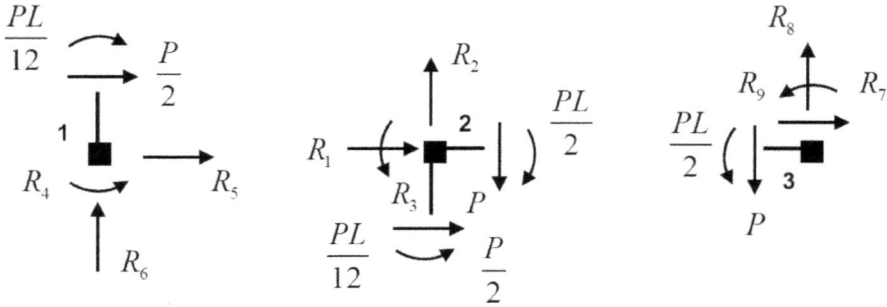

**FIGURE 6.33** Example 6.3 (Continued).

$$\mathbf{P}_F^{(1)} = {\mathbf{\Gamma}_{ROT}^{(1)}}^T \bar{\mathbf{P}}_F^{(1)} = \left\lfloor -\frac{P}{2} \quad 0 \quad \frac{PL}{12} \quad -\frac{P}{2} \quad 0 \quad -\frac{PL}{12} \right\rfloor^T \text{ and}$$

$$\mathbf{P}_F^{(2)} = {\mathbf{\Gamma}_{ROT}^{(2)}}^T \bar{\mathbf{P}}_F^{(2)} = \left\lfloor 0 \quad P \quad \frac{PL}{2} \quad 0 \quad P \quad -\frac{PL}{2} \right\rfloor^T$$

The restrained load vector **R** is constructed by considering nodal equilibriums under the actions of fixed-end forces (Figure 6.33).

$$\mathbf{R} = \left\lfloor \mathbf{R}_{free} \quad \mathbf{R}_{constr} \right\rfloor^T$$

where

$$\mathbf{R}_{free} = \left\lfloor -\frac{P}{2} \quad P \quad \frac{5PL}{12} \quad \frac{PL}{12} \right\rfloor^T \text{ and } \mathbf{R}_{constr} = \left\lfloor -\frac{P}{2} \quad 0 \quad 0 \quad P \quad -\frac{PL}{2} \right\rfloor^T$$

It is noted that this is the set of nodal forces needed to prevent all nodes from moving. They can be shown in Figure 6.34a.

The nodal load vector is defined as:

$$\mathbf{F} = \left\lfloor \mathbf{F}_{free} \quad \mathbf{F}_{constr} \right\rfloor^T$$

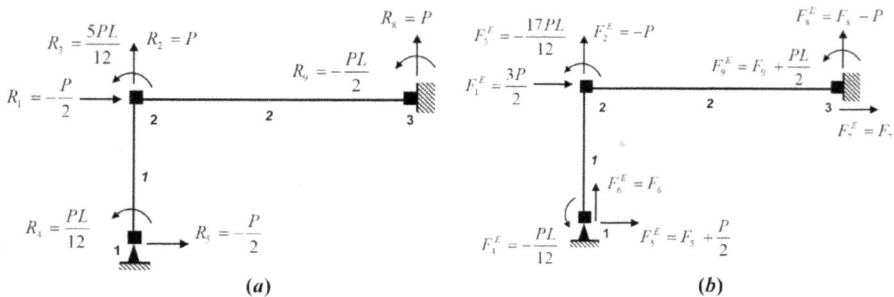

**FIGURE 6.34** Example 6.3 (Continued).

where

$$\mathbf{F}_{free} = \lfloor P \quad 0 \quad -PL \quad 0 \rfloor^T \text{ and } \mathbf{F}_{constr} = \lfloor F_5 \quad F_6 \quad F_7 \quad F_8 \quad F_9 \rfloor^T$$

The equivalent nodal load vector (Figure 6.34 *b*) is defined as:

$$\mathbf{F}^E = \mathbf{F} - \mathbf{R} = \left\{ \begin{array}{c} \mathbf{F}^E_{free} \\ \mathbf{F}^E_{constr} \end{array} \right\} = \left\{ \begin{array}{c} \mathbf{F}_{free} \\ \mathbf{F}_{constr} \end{array} \right\} - \left\{ \begin{array}{c} \mathbf{R}_{free} \\ \mathbf{R}_{constr} \end{array} \right\}$$

where

$$\mathbf{F}^E_{free} = \left| \frac{3P}{2} \quad -P \quad \frac{-17PL}{12} \quad -\frac{PL}{12} \right|^T \text{ and } \mathbf{F}^E_{constr} = \left| F_5 + \frac{P}{2} \quad F_6 \quad F_7 \quad F_8 - P \quad F_9 + \frac{P}{2} \right|^T$$

## EXAMPLE 6.4

The frame structure of Example 6.2 is subjected to member and nodal loads as shown in Figure 6.35. Use the stiffness approach to determine nodal displacements and member end forces.

**Solution:** The structure is rendered kinematically determinate by locking all nodes against displacements. As a result, the nodal loads needed to prevent each node from movement are determined by considering nodal equilibrium conditions.

The fixed-end forces for each member are first determined as shown in Figure 6.36.

Member 1: $\mathbf{P}_F^{(1)} = \overline{\mathbf{P}}_F^{(1)} = \mathbf{0}$

Member 2: $\overline{\mathbf{P}}_F^{(2)} = \lfloor 0 \quad 250 \quad 208.33 \quad 0 \quad 250 \quad -208.33 \rfloor^T$

$$\mathbf{P}_F^{(2)} = \Gamma_{ROT}^{(2)\,T} \overline{\mathbf{P}}_F^{(2)} = \lfloor 0 \quad 250 \quad 208.33 \quad 0 \quad 250 \quad -208.33 \rfloor^T$$

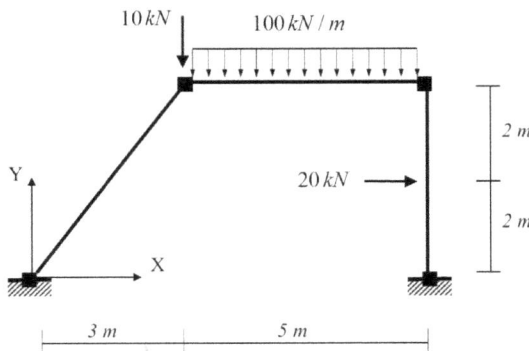

**FIGURE 6.35** Example 6.4.

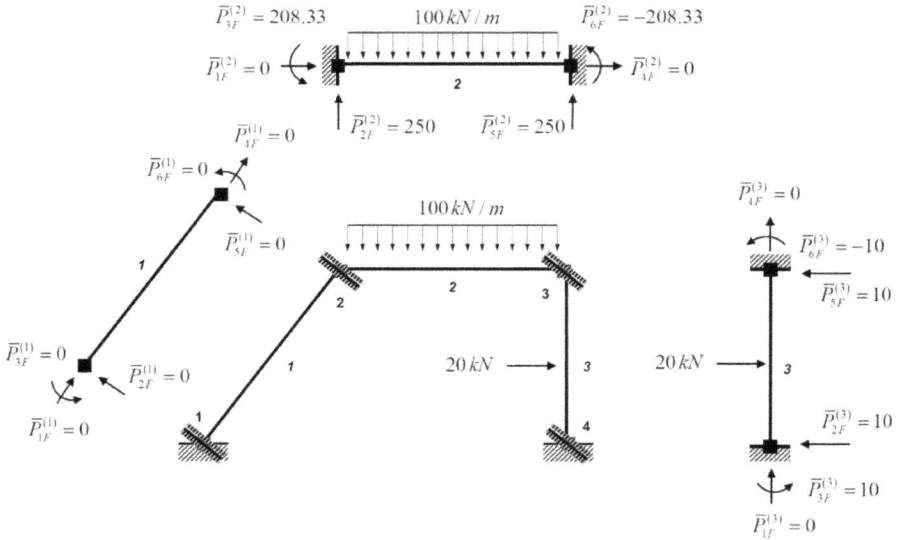

**FIGURE 6.36** Example 6.4 (Continued).

Member 3: $\overline{\mathbf{P}}_F^{(3)} = \lfloor 0 \quad 10 \quad 10 \quad 0 \quad 10 \quad -10 \rfloor^T$

$$\mathbf{P}_F^{(3)} = \mathbf{\Gamma}_{ROT}^{(3)}{}^T \overline{\mathbf{P}}_F^{(3)} = \lfloor -10 \quad 0 \quad 10 \quad -10 \quad 0 \quad -10 \rfloor^T$$

The restrained nodal forces needed to prevent all nodes from movement are determined by considering nodal equilibrium conditions as shown in Figure 6.37.

The restrained nodal forces (Figure 6.38) are collected in the vector form as:

$$\mathbf{R} = \lfloor \mathbf{R}_{free} \quad \mathbf{R}_{constr} \rfloor^T$$

where

$$\mathbf{R}_{free} = \lfloor 0 \quad 250 \quad 208.33 \quad -10 \quad 250 \quad -218.33 \rfloor^T \text{ and } \mathbf{R}_{constr} = \lfloor 0 \quad 0 \quad 0 \quad -10 \quad 0 \quad 10 \rfloor^T$$

The stiffness equations of the structure associated with free degrees of freedom $(\mathbf{F}_{free}^E = \mathbf{F}_{free} - \mathbf{R}_{free} = \mathbf{K}_{ff}\mathbf{U}_{free})$ are:

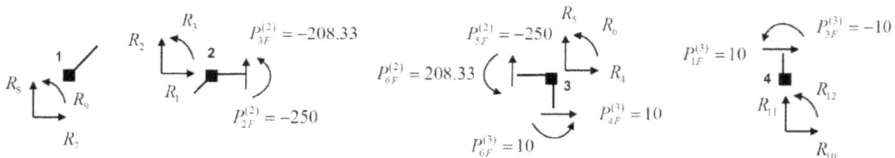

**FIGURE 6.37** Example 6.4 (Continued).

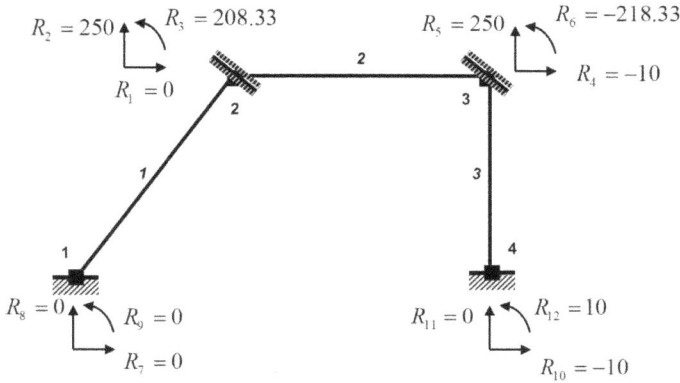

**FIGURE 6.38** Example 6.4 (Continued).

$$\begin{Bmatrix} 0 \\ -260 \\ -208.33 \\ 10 \\ -250 \\ 218.33 \end{Bmatrix} = \begin{bmatrix} 2781.44 & 913.92 & 192 & -2000 & 0 & 0 \\ 913.92 & 1410.56 & 96 & 0 & -96 & 240 \\ 192 & 96 & 1600 & 0 & -240 & 400 \\ -2000 & 0 & 0 & 2187.5 & 0 & 375 \\ 0 & -96 & -240 & 0 & 2596 & -240 \\ 0 & 240 & 400 & 375 & -240 & 1800 \end{bmatrix} \begin{Bmatrix} U_1 \\ U_2 \\ U_3 \\ U_4 \\ U_5 \\ U_6 \end{Bmatrix}$$

$$\mathbf{U}_{free} = \left[\mathbf{K}_{ff}\right]^{-1}\left(\mathbf{F}_{free} - \mathbf{R}_{free}\right) = \left[\mathbf{K}_{ff}\right]^{-1}\mathbf{F}_{free}^{E}$$

$$\mathbf{U}_{free} = \lfloor U_1 \quad U_2 \quad U_3 \quad U_4 \quad U_5 \quad U_6 \rfloor^{T}$$

$$= \lfloor 0.523 \quad -0.539 \quad -0.211 \quad 0.461 \quad -0.124 \quad 0.127 \rfloor^{T}$$

The basic member deformations are recovered by substituting this set of nodal displacements into the structural compatibility relation.

$$\begin{Bmatrix} V_1 \\ V_2 \\ V_3 \\ V_4 \\ V_5 \\ V_6 \\ V_7 \\ V_8 \\ V_9 \end{Bmatrix} = \begin{bmatrix} 0.6 & 0.8 & 0 & 0 & 0 & 0 \\ 0.16 & -0.12 & 0 & 0 & 0 & 0 \\ 0.16 & -0.12 & 1 & 0 & 0 & 0 \\ -1 & 0 & 0 & 1 & 0 & 0 \\ 0 & 0.2 & 1 & 0 & -0.2 & 0 \\ 0 & 0.2 & 0 & 0 & -0.2 & 1 \\ 0 & 0 & 0 & 0 & 1 & 0 \\ 0 & 0 & 0 & 0.25 & 0 & 0 \\ 0 & 0 & 0 & 0.25 & 0 & 1 \end{bmatrix} \begin{Bmatrix} 0.523 \\ -0.539 \\ -0.211 \\ 0.461 \\ -0.124 \\ 0.127 \end{Bmatrix} = \begin{Bmatrix} -0.117\,m \\ 0.148\,rad \\ -0.06267\,rad \\ -0.06215\,m \\ -0.2942\,rad \\ 0.04442\,rad \\ -0.124\,m \\ 0.1153\,rad \\ 0.2428\,rad \end{Bmatrix}$$

The complete forces of each member are determined from the equilibrium equations of each member.

For example, the complete forces of member 1 are:

$$\mathbf{P}^{(1)} = \mathbf{b}_B^{(1)} \mathbf{Q}_1 + \mathbf{P}_F^{(1)} = \mathbf{b}_B^{(1)} \overline{\mathbf{k}}_1 \mathbf{V}_1 + \mathbf{P}_F^{(1)}$$

$$
\begin{Bmatrix} P_1^{(1)} \\ P_2^{(1)} \\ P_3^{(1)} \\ P_4^{(1)} \\ P_5^{(1)} \\ P_6^{(1)} \end{Bmatrix} =
\begin{bmatrix} -0.6 & -0.16 & -0.16 \\ -0.8 & 0.12 & 0.12 \\ 0 & 1 & 0 \\ 0.6 & 0.16 & 0.16 \\ 0.8 & -0.12 & -0.12 \\ 0 & 0 & 1 \end{bmatrix}
\begin{bmatrix} 2000 & 0 & 0 \\ 0 & 800 & 400 \\ 0 & 400 & 800 \end{bmatrix}
\begin{Bmatrix} -0.117 \\ 0.148 \\ -0.06267 \end{Bmatrix} +
\begin{Bmatrix} 0 \\ 0 \\ 0 \\ 0 \\ 0 \\ 0 \end{Bmatrix} =
\begin{Bmatrix} 124.3 \; kN \\ 200.1 \; kN \\ 93.7 \; kN-m \\ -124.3 \; kN \\ -200.1 \; kN \\ 9.3 \; kN-m \end{Bmatrix}
$$

## 6.7   STIFFNESS METHOD: SEMI-AUTOMATIC APPROACH

In the previous section, the structural stiffness matrix was constructed using the conceptual approach. This approach to constructing the structural stiffness matrix is relatively straightforward but it could be very tedious even for a small frame structure (e.g. the frame structure in Example 6.2). Consequently, the more computerized approach is preferable. Such an approach is known as the "semi-automatic" approach and can be derived from the virtual displacement principle. It is recalled that the basic equations of the stiffness method are derived from three sets of governing equations, namely:

1. *Equilibrium Equations*: Structural equilibrium equations were explored in Chapter 3 and are written as:

$$\mathbf{F} = \mathbf{BQ} + \mathbf{F}^0 \text{ or } \begin{Bmatrix} \mathbf{F}_{free} \\ \hline \mathbf{F}_{constr} \end{Bmatrix} = \begin{bmatrix} \mathbf{B}_{free} \\ \hline \mathbf{B}_{constr} \end{bmatrix} \mathbf{Q} + \begin{Bmatrix} \mathbf{F}_{free}^0 \\ \hline \mathbf{F}_{constr}^0 \end{Bmatrix} \tag{3/3.43}$$

2. *Compatibility Equations*: Structural compatibility equations were explored in Chapter 4 and are written as:

$$\mathbf{V} = \mathbf{AU} \text{ or } \mathbf{V} = \mathbf{V}_{free} + \mathbf{V}_{constr} = \mathbf{A}_{free}\mathbf{U}_{free} + \mathbf{A}_{constr}\mathbf{U}_{constr} \tag{4/4.56}$$

3. *Basic Member Stiffness Equations*: Basic member stiffness equations were discussed earlier in this chapter and partly in Chapter 2.

For an individual member, they are written as:

$$\mathbf{q}^{(i)} = \mathbf{k}^{(i)}\mathbf{v}^{(i)} + \mathbf{q}_0^{(i)} + \mathbf{q}_w^{(i)} \tag{6.82}$$

For all members, they are written together as:

$$\mathbf{Q} = \mathbf{kV} + \mathbf{Q}_0 + \mathbf{Q}_w \tag{6.83}$$

$$
\begin{Bmatrix} \mathbf{Q}_1 \\ \mathbf{Q}_2 \\ \vdots \\ \mathbf{Q}_m \end{Bmatrix} = \begin{bmatrix} \mathbf{k}^{(1)} & \mathbf{0} & \cdots & \mathbf{0} \\ \mathbf{0} & \mathbf{k}^{(2)} & \cdots & \mathbf{0} \\ \vdots & \vdots & \ddots & \vdots \\ \mathbf{0} & \mathbf{0} & \cdots & \mathbf{k}^{(m)} \end{bmatrix} \begin{Bmatrix} \mathbf{V}_1 \\ \mathbf{V}_2 \\ \vdots \\ \mathbf{V}_m \end{Bmatrix} + \begin{Bmatrix} \mathbf{Q}_0^1 \\ \mathbf{Q}_0^2 \\ \vdots \\ \mathbf{Q}_0^m \end{Bmatrix} + \begin{Bmatrix} \mathbf{Q}_w^1 \\ \mathbf{Q}_w^2 \\ \vdots \\ \mathbf{Q}_w^m \end{Bmatrix}
\tag{6.84}
$$

where $m$ is the number of members; the vector $\mathbf{q}_0^{(i)}$ contains the basic member forces due to initial deformations; and the vector $\mathbf{q}_w^{(i)}$ contains the basic member forces due to member loads.

All of these governing equations can be written together through the virtual displacement principle. From the invariant property of the external virtual work (5/5.200), it can be concluded that:

$$
\delta \mathbf{V}^T \mathbf{Q} = \delta \mathbf{U}^T \left( \mathbf{F} - \mathbf{F}^0 \right)
\tag{6.85}
$$

Substituting Eqs. (4/4.56) into (6.85), one has

$$
\underbrace{\delta \mathbf{U}^T}_{Arbitrary} \left[ \left( \mathbf{F} - \mathbf{F}^0 \right) - \mathbf{A}^T \mathbf{Q} \right] = 0 \quad \Rightarrow \quad \mathbf{F} - \mathbf{F}^0 = \mathbf{A}^T \mathbf{Q}
\tag{6.86}
$$

Substituting Eqs. (6.83) into (6.86), one obtains

$$
\mathbf{F} - \mathbf{F}^0 = \mathbf{A}^T \mathbf{k} \mathbf{V} + \mathbf{A}^T \mathbf{Q}_0 + \mathbf{A}^T \mathbf{Q}_w
\tag{6.87}
$$

Substituting Eqs. (4/4.56) into (6.87), one gains

$$
\mathbf{F} - \mathbf{F}^0 = \mathbf{A}^T \mathbf{k} \mathbf{A} \mathbf{U} + \mathbf{A}^T \mathbf{Q}_0 + \mathbf{A}^T \mathbf{Q}_w
\tag{6.88}
$$

Eq. (6.88) represents the structural stiffness equations and can be partitioned following free and constrained degrees of freedom as:

$$
\mathbf{F}_{free}^E = \mathbf{F}_{free} - \left( \mathbf{F}_{free}^0 + \mathbf{K}_{fc} \mathbf{U}_{constr} + \mathbf{A}_{free}^T \mathbf{Q}_0 + \mathbf{A}_{free}^T \mathbf{Q}_w \right) = \mathbf{K}_{ff} \mathbf{U}_{free}
\tag{6.89}
$$

$$
\mathbf{F}_{constr}^E = \mathbf{F}_{constr} - \left( \mathbf{F}_{constr}^0 + \mathbf{K}_{cc} \mathbf{U}_{constr} + \mathbf{A}_{constr}^T \mathbf{Q}_0 + \mathbf{A}_{constr}^T \mathbf{Q}_w \right) = \mathbf{K}_{cf} \mathbf{U}_{free}
\tag{6.90}
$$

where

$$
\mathbf{K}_{ff} = \mathbf{A}_{free}^T \mathbf{k} \mathbf{A}_{free} \quad \mathbf{K}_{fc} = \mathbf{A}_{free}^T \mathbf{k} \mathbf{A}_{constr} ; \mathbf{K}_{cf} = \mathbf{A}_{constr}^T \mathbf{k} \mathbf{A}_{free} ; \text{and} \; \mathbf{K}_{cc} = \mathbf{A}_{constr}^T \mathbf{k} \mathbf{A}_{constr}
\tag{6.91}
$$

The terms in parentheses of Eqs. (6.89) and (6.90) represent the restrained nodal forces associated with free and constrained degrees of freedom, respectively.

$$
\mathbf{R}_{free} = \mathbf{F}_{free}^0 + \mathbf{K}_{fc} \mathbf{U}_{constr} + \underbrace{\mathbf{A}_{free}^T}_{\mathbf{B}_{free}} \mathbf{Q}_0 + \underbrace{\mathbf{A}_{free}^T}_{\mathbf{B}_{free}} \mathbf{Q}_w
$$

$$
\mathbf{R}_{constr} = \mathbf{F}_{constr}^0 + \mathbf{K}_{cc} \mathbf{U}_{constr} + \underbrace{\mathbf{A}_{constr}^T}_{\mathbf{B}_{constr}} \mathbf{Q}_0 + \underbrace{\mathbf{A}_{constr}^T}_{\mathbf{B}_{constr}} \mathbf{Q}_w
\tag{6.92}
$$

$$F = BQ + F^0 = A^T Q + F^0$$

Equilibrium

Q: Basic Forces

$$Q = kV + Q_0 + Q_w$$    Members                    Structure    $F^E = KU$

V: Basic Deformations

Compatibility

$$V = AU$$

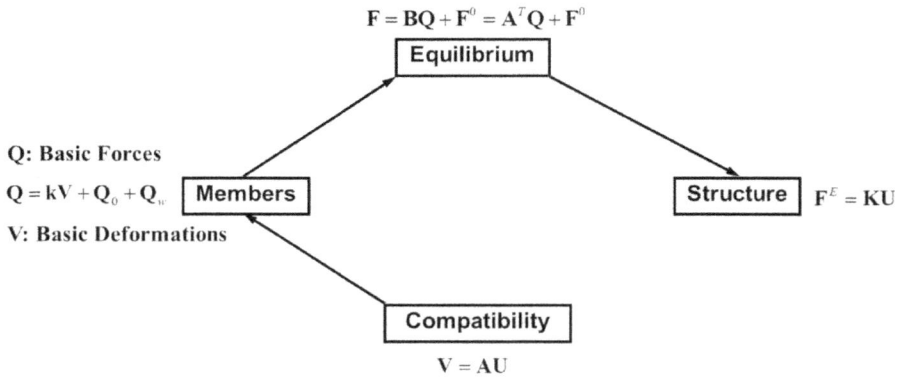

**FIGURE 6.39**  Schematic Representation of the Stiffness Method.

A set of unknown nodal displacements $U_{free}$ is determined by solving Eq. (6.89). Then, the support reaction forces $F_{constr}$ are determined from Eq. (6.90). The schematic representation of the stiffness method is presented in Figure 6.39.

The cooking-recipe steps in the semi-automated stiffness approach are summarized as follows:

Step 1: Construct compatibility matrix:

$$A = \begin{bmatrix} A_{free} & A_{constr} \end{bmatrix}$$

Step 2: Establish equilibrium equations through the virtual displacement principle:

$$F_{free} = A^T_{free} Q + F^0_{free}$$

Step 3: Construct the structural stiffness matrix and the equivalent load vector:

$$K_{ff} = A^T_{free} k A_{free} \text{ and } F^E_{free} = F_{free} - R_{free}$$

Step 4: Solve the structural stiffness equations for the unknown nodal displacements at free degrees of freedom:

$$F^E_{free} = K_{ff} U_{free} \quad \rightarrow \quad U_{free} = \begin{bmatrix} K_{ff} \end{bmatrix}^{-1} F^E_{free}$$

Step 5: Compute member deformations:

$$V = A_{free} U_{free} + A_{constr} U_{constr}$$

Step 6: Compute basic member forces:

$$\textit{Structural-wise: } Q = kV + Q_0 + Q_w$$

$$\textit{Member-wise: } q^{(i)} = k^{(i)} v^{(i)} + q_0^{(i)} + q_w^{(i)}$$

Step 7: Determine other forces by considering member and nodal equilibriums (e.g. shear or axial forces if there are inextensible and/or inflexible constraints).

Step 8: Determine support reactions and verify the global equilibrium

$$F_{constr} = A^T_{constr} Q + F^0_{constr}$$

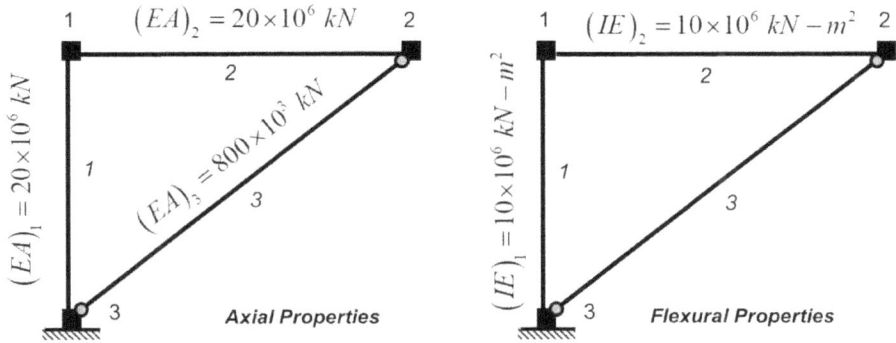

**FIGURE 6.40**   Example 6.5.

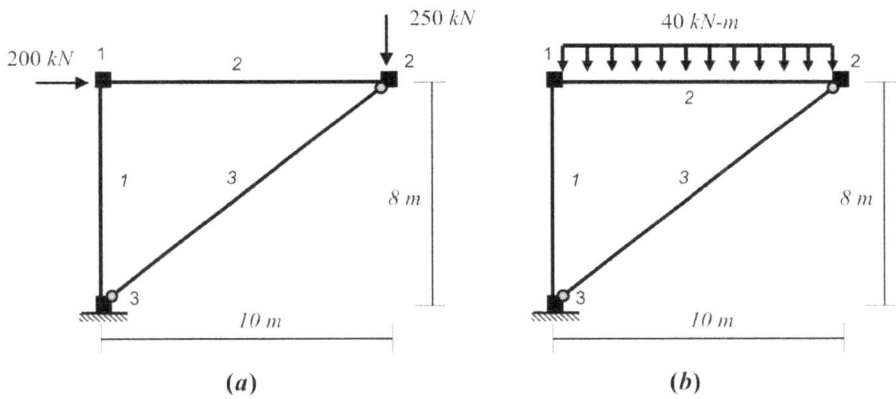

**FIGURE 6.41**   Example 6.5 (Continued).

## EXAMPLE 6.5

Analyze the braced frame shown in Figure 6.40 under the following loading conditions:

(i) The applied nodal forces shown in Figure 6.41a.
(ii) Member 2 is subjected to the uniformly distributed load $w_y = -40\ kN/m$ shown in Figure 6.41 b.

## SOLUTION

**Condition (i):** The structural equilibrium equations are first established. DOFs numbering system, basic force system, and basic deformation system are shown in Figure 6.42.

Following the approach described in Chapter 3, the structural equilibrium equations are:

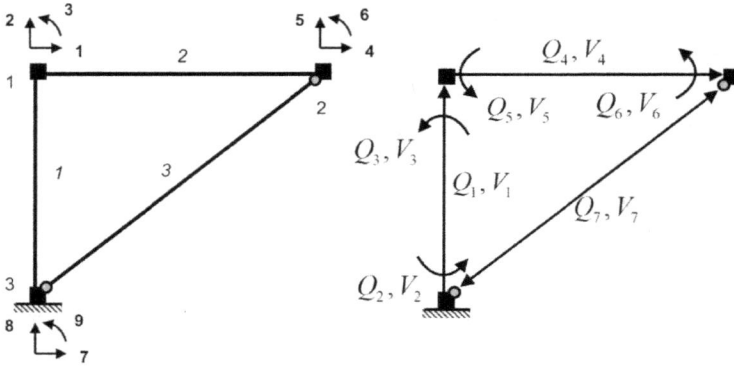

Member, Node, and DOFs Numbering Systems   Basic-Force and Basic-Deformation Systems

**FIGURE 6.42**   Example 6.5 (Continued).

$$\begin{Bmatrix} 200 \\ 0 \\ 0 \\ 0 \\ -250 \\ 0 \\ F_7 \\ F_8 \\ F_9 \end{Bmatrix} = \begin{bmatrix} 0 & 0.125 & 0.125 & -1 & 0 & 0 & 0 \\ 1 & 0 & 0 & 0 & 0.1 & 0.1 & 0 \\ 0 & 0 & 1 & 0 & 1 & 0 & 0 \\ 0 & 0 & 0 & 1 & 0 & 0 & 0.781 \\ 0 & 0 & 0 & 0 & -0.1 & -0.1 & 0.625 \\ 0 & 0 & 0 & 0 & 0 & 1 & 0 \\ 0 & -0.125 & 0.125 & 0 & 0 & 0 & -0.781 \\ -1 & 0 & 0 & 0 & 0 & 0 & -0.625 \\ 0 & 1 & 0 & 0 & 0 & 0 & 0 \end{bmatrix} \begin{Bmatrix} Q_1 \\ Q_2 \\ Q_3 \\ Q_4 \\ Q_5 \\ Q_6 \\ Q_7 \end{Bmatrix} \text{ or }$$

$$\mathbf{F} = \begin{Bmatrix} \mathbf{F}_{free} \\ \mathbf{F}_{constr} \end{Bmatrix} = \begin{bmatrix} \mathbf{B}_{free} \\ \mathbf{B}_{constr} \end{bmatrix} \mathbf{Q}$$

Following the approach described in Chapter 4, the structural compatibility equations are:

$$\begin{Bmatrix} V_1 \\ V_2 \\ V_3 \\ V_4 \\ V_5 \\ V_6 \\ V_7 \end{Bmatrix} = \begin{bmatrix} 0 & 1 & 0 & 0 & 0 & 0 \\ 0.125 & 0 & 0 & 0 & 0 & 0 \\ 0.125 & 0 & 1 & 0 & 0 & 0 \\ -1 & 0 & 0 & 1 & 0 & 0 \\ 0 & 0.1 & 1 & 0 & -0.1 & 0 \\ 0 & 0.1 & 0 & 0 & -0.1 & 1 \\ 0 & 0 & 0 & 0.781 & 0.625 & 0 \end{bmatrix} \begin{Bmatrix} U_1 \\ U_2 \\ U_3 \\ U_4 \\ U_5 \\ U_6 \end{Bmatrix} \text{ or } \mathbf{V} = \begin{bmatrix} \mathbf{A}_{free} | \mathbf{A}_{constr} \end{bmatrix} \begin{Bmatrix} \mathbf{U}_{free} \\ \mathbf{U}_{constr} \end{Bmatrix}$$

Finally, the stiffness equation for the basic system of each frame member is considered and grouped in the matrix form as:

$$\begin{Bmatrix} Q_1 \\ Q_2 \\ Q_3 \\ Q_4 \\ Q_5 \\ Q_6 \\ Q_7 \end{Bmatrix} = 10^6 \times \begin{bmatrix} 2.5 & 0 & 0 & 0 & 0 & 0 & 0 \\ 0 & 5 & 2.5 & 0 & 0 & 0 & 0 \\ 0 & 2.5 & 5 & 0 & 0 & 0 & 0 \\ 0 & 0 & 0 & 2 & 0 & 0 & 0 \\ 0 & 0 & 0 & 0 & 4 & 2 & 0 \\ 0 & 0 & 0 & 0 & 2 & 4 & 0 \\ 0 & 0 & 0 & 0 & 0 & 0 & 0.06247 \end{bmatrix} \begin{Bmatrix} V_1 \\ V_2 \\ V_3 \\ V_4 \\ V_5 \\ V_6 \\ V_7 \end{Bmatrix} \text{ or } \mathbf{Q} = \mathbf{kV}_{free}$$

It is clear that these stiffness equations are uncoupled since all of the off-diagonal terms in matrix $\mathbf{k}$ are zero. Physically, it can be interpreted that all of the members are disconnected.

From the contragradient relations between the statical and kinematical systems, the structural stiffness equations corresponding to free degrees of freedom are written in the matrix form as:

$$\mathbf{F}_{free} = \mathbf{A}_{free}^T \mathbf{kA}_{free} = \mathbf{K}_{ff} \mathbf{U}_{free}$$

This type of transformation is known as the "congruent" transformation.

The structural stiffness equations at free degrees of freedom are:

$$\begin{Bmatrix} 200 \\ 0 \\ 0 \\ 0 \\ -250 \\ 0 \end{Bmatrix} = 10^6 \times \begin{bmatrix} 2.234 & 0 & 0.9375 & -2 & 0 & 0 \\ 0 & 2.62 & 0.6 & 0 & -0.12 & 0.6 \\ 0.9375 & 0.6 & 9 & 0 & -0.6 & 2 \\ -2 & 0 & 0 & 2.0381 & 0.03049 & 0 \\ 0 & -0.12 & -0.6 & 0.03049 & 0.1444 & -0.6 \\ 0 & 0.6 & 2 & 0 & -0.6 & 4 \end{bmatrix} \begin{Bmatrix} U_1 \\ U_2 \\ U_3 \\ U_4 \\ U_5 \\ U_6 \end{Bmatrix}$$

The unknown displacements are determined from the structural stiffness equation as:

$$\mathbf{U}_{free} = \left[ \mathbf{K}_{ff} \right]^{-1} \mathbf{F}_{free}$$
$$= 10^{-3} \times \lfloor 9.264 \; m \quad -0.01946 \; m \quad -1.834 \; rad \quad 9.390 \; m \quad -19.985 \; m \quad -2.078 \; rad \rfloor^T$$

The member deformations are recovered from the compatibility relations as:

$$\mathbf{V}_{free} = \mathbf{A}_{free} \mathbf{U}_{free}$$
$$= 10^{-4} \times \lfloor -0.1946 \, m \quad 11.58 \, rad \quad -6.763 \, rad \quad 1.258 \, m \quad 1.622 \, rad \quad -0.8109 \, rad \quad -515.7 \, m \rfloor^T$$

The basic member forces are determined from the member stiffness equations:

$$\mathbf{Q}_{free} = \mathbf{kV}_{free} = \lfloor -48.7 \, kN \quad 4100 \, kN - m \quad -486.6 \, kN - m \quad 251.6 \, kN \quad 486.6 \, kN - m \quad 0 \quad -322.2 \, kN \rfloor^T$$

The reaction forces are computed from the equilibrium equations associated with constrained degrees of freedom:

$$\mathbf{F}_{constr} = \mathbf{B}_{constr} \mathbf{Q} = \lfloor F_7 \quad F_8 \quad F_9 \rfloor^T = \lfloor -200 \, kN \quad 250 \, kN \quad 4100 \, kN - m \rfloor^T$$

**Condition (*ii*):** In this load case, there are no nodal loads but member 2 is subjected to the uniformly distributed load.

From Chapter 3, the member load vector due to the presence of member loads is:

$$\mathbf{P}_0^{(2)} = \lfloor 0 \quad 200 \quad 0 \quad 0 \quad 200 \quad 0 \rfloor^T$$

Consequently, the equivalent nodal load vector due to the presence of member loads is:

$$\mathbf{F}_{free}^0 = \lfloor 0 \quad 200 \quad 0 \quad 0 \quad 200 \quad 0 \rfloor^T$$

The basic forces of member 2 due to the presence of member loads are:

$$\mathbf{q}_w^{(2)} = \left\lfloor 0 \quad -\frac{w_y L_2^{\,2}}{12} \quad \frac{w_y L_2^{\,2}}{12} \right\rfloor^T = \lfloor 0 \quad 333.33 \quad -333.33 \rfloor^T$$

Consequently, the basic structural forces due to member load are:

$$\mathbf{Q}_w = \lfloor 0 \quad 0 \quad 0 \quad 0 \quad 333.33 \quad -333.33 \quad 0 \rfloor^T$$

and the equivalent nodal load vector $\mathbf{F}_w$ is

$$\mathbf{F}_w = \mathbf{B}_{free}\mathbf{Q}_w = \lfloor 0 \quad 0 \quad 333.33 \quad 0 \quad 0 \quad -333.33 \rfloor^T$$

The equivalent nodal load vector is

$$\mathbf{F}_{free}^0 + \mathbf{F}_w = \lfloor 0 \quad 200 \quad 333.33 \quad 0 \quad 200 \quad -333.33 \rfloor^T$$

The unknown displacements are determined from the structural stiffness equation as:

$$\mathbf{U}_{free} = -\left[\mathbf{K}_{ff}\right]^{-1}\left(\mathbf{F}_{free}^0 + \mathbf{F}_w\right)$$
$$= 10^{-3} \times \lfloor 5.025\ m \quad -0.1085\ m \quad -1.085\ rad \quad 5.105\ m \quad -11.66\ m \quad -1.107\ rad \rfloor^T$$

The member deformations are recovered from the compatibility relations:

$$\mathbf{V}_{free} = \mathbf{A}_{free}\mathbf{U}_{free}$$
$$= 10^{-4} \times \lfloor -1.085\ m \quad 6.281\ rad \quad -4.564\ rad \quad 0.8049\ m \quad 0.7056\ rad \quad 0.4805\ rad \quad -33\ m \rfloor^T$$

The basic member forces are determined from the member stiffness equations:

$$\mathbf{Q} = \mathbf{k}\mathbf{V}_{free} + \mathbf{Q}_w$$
$$= \lfloor -271.2\ kN \quad 2000\ kN-m \quad -711.7\ kN-m \quad 160.99\ kN \quad 711.7\ kN-m \quad 0 \quad -206.13\ kN \rfloor^T$$

The reaction forces are computed from the equilibrium equations associated with constrained degrees of freedom:

$$\mathbf{F}_{constr} = \mathbf{B}_{constr}\mathbf{Q} = \lfloor F_7 \quad F_8 \quad F_9 \rfloor^T = \lfloor 0 \quad 400\ kN \quad 2000\ kN-m \rfloor^T$$

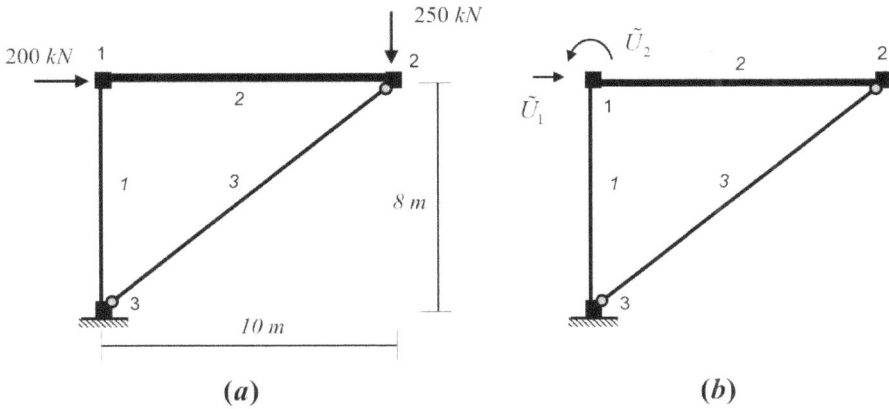

**FIGURE 6.43**  Example 6.6.

## EXAMPLE 6.6

For the braced frame of Example 6.5, analyze it under the following constraints (Figure 6.43 *a*):

- Member 1 is considered inextensible while member 2 is considered rigid.

**Solution:** Under the assumption that member 1 is considered inextensible and member 2 is considered rigid, the following multi-freedom constraints can be imposed:
Member 1: Inextensible

$$U_2 - \underbrace{U_8}_{0} = 0 \quad \Rightarrow \quad U_2 = 0 \qquad\qquad (a)$$

Member 2: Rigid

$$U_4 - U_1 = 0 \quad \Rightarrow \quad U_4 = U_1 \qquad\qquad (b)$$

$$U_5 - 10U_3 = 0 \quad \Rightarrow \quad U_5 = 10U_3 \qquad\qquad (c)$$

$$U_6 - U_3 = 0 \quad \Rightarrow \quad U_6 = U_3 \qquad\qquad (d)$$

From Eqs. (*a*)-(*d*), $U_1$ and $U_3$ are selected as masters while $U_4$, $U_5$, and $U_6$ are selected as slaves. Consequently, the new set of independent free nodal displacements $\tilde{\mathbf{U}}_{free}$ (Figure 6.43 *b*) is defined as:

$$\tilde{\mathbf{U}}_{free} = \lfloor U_1 \quad U_3 \rfloor^T = \lfloor \tilde{U}_1 \quad \tilde{U}_2 \rfloor^T$$

A matrix transformation equation relating $\mathbf{U}_{free}$ to $\tilde{\mathbf{U}}_{free}$ is written as:

$$
\begin{Bmatrix} U_1 \\ U_2 \\ U_3 \\ U_4 \\ U_5 \\ U_6 \end{Bmatrix} =
\begin{bmatrix} 1 & 0 \\ 0 & 0 \\ 0 & 1 \\ 1 & 0 \\ 0 & 10 \\ 0 & 1 \end{bmatrix}
\begin{Bmatrix} \tilde{U}_1 \\ \tilde{U}_2 \end{Bmatrix}
\quad \text{or} \quad \mathbf{U}_{free} = \mathbf{A}_c \tilde{\mathbf{U}}_{free}
$$

From the contragradient relations between statical and kinematical systems, the matrix transformation equation relating the nodal forces $\tilde{\mathbf{F}}_{free}$ to the nodal forces $\mathbf{F}_{free}$ is:

$$
\begin{Bmatrix} \tilde{F}_1 \\ \tilde{F}_2 \end{Bmatrix} =
\begin{bmatrix} 1 & 0 & 0 & 1 & 0 & 0 \\ 0 & 0 & 1 & 0 & 10 & 1 \end{bmatrix}
\begin{Bmatrix} F_1 \\ F_2 \\ F_3 \\ F_4 \\ F_5 \\ F_6 \end{Bmatrix}
\quad \text{or} \quad \tilde{\mathbf{F}}_{free} = \mathbf{A}_c^T \mathbf{F}_{free}
$$

This relation can easily be derived through the virtual displacement principle. Substituting the kinematical relation into the compatibility relation leads to:

$$
\mathbf{V} = \mathbf{V}_{free} = \mathbf{A}_{free}\mathbf{U}_{free} = \mathbf{A}_{free}\mathbf{A}_c\tilde{\mathbf{U}}_{free} = \tilde{\mathbf{A}}_{free}\tilde{\mathbf{U}}_{free}
\quad \Rightarrow \quad
\begin{Bmatrix} V_1 \\ V_2 \\ V_3 \\ V_4 \\ V_5 \\ V_6 \\ V_7 \end{Bmatrix} =
\begin{bmatrix} 0 & 0 \\ 0.125 & 0 \\ 0.125 & 1 \\ 0 & 0 \\ 0 & 0 \\ 0 & 0 \\ 0.781 & 6.25 \end{bmatrix}
\begin{Bmatrix} \tilde{U}_1 \\ \tilde{U}_2 \end{Bmatrix}
$$

Clearly, the basic deformation related to axial deformation of member 1 vanishes. Similarly, the basic deformations of rigid member 2 are zero.

From the congruent transformation law, the constrained structural stiffness equations corresponding to free degrees of freedom are written in the matrix form as:

$$
\tilde{\mathbf{F}}_{free} = \tilde{\mathbf{K}}_{ff} \tilde{\mathbf{U}}_{free}
$$

where

$$
\tilde{\mathbf{K}}_{ff} = \tilde{\mathbf{A}}_{free}^T \mathbf{k} \tilde{\mathbf{A}}_{free} =
\begin{bmatrix} 272479 & 1.242 \times 10^6 \\ 1.242 \times 10^6 & 7.4402 \times 10^6 \end{bmatrix}
$$

The load vector due to the multi-freedom constraints is:

$$
\tilde{\mathbf{F}}_{free} = \mathbf{A}_c^T \mathbf{F}_{free} = \mathbf{B}_c \mathbf{F}_{free} = \lfloor 200 \quad -2500 \rfloor^T
$$

Explicitly, the structural stiffness equations at free degrees of freedom are:

$$\begin{Bmatrix} 200 \\ -2500 \end{Bmatrix} = \begin{bmatrix} 272479 & 1.242\times10^6 \\ 1.242\times10^6 & 7.4402\times10^6 \end{bmatrix} \begin{Bmatrix} \tilde{U}_1 \\ \tilde{U}_2 \end{Bmatrix}$$

The unknown displacements are determined from the structural stiffness equation as:

$$\tilde{\mathbf{U}}_{free} = \lfloor \tilde{U}_1 \quad \tilde{U}_2 \rfloor^T = \lfloor 9.498\times10^{-3}\,m \quad -1.922\times10^{-3}\,rad \rfloor^T$$

The displacement degrees of freedom $\mathbf{U}_{free}$ are recovered from the kinematical constrained relation as:

$$\begin{Bmatrix} U_1 \\ U_2 \\ U_3 \\ U_4 \\ U_5 \\ U_6 \end{Bmatrix} = \begin{bmatrix} 1 & 0 \\ 0 & 0 \\ 0 & 1 \\ 1 & 0 \\ 0 & 10 \\ 0 & 1 \end{bmatrix} \begin{Bmatrix} 9.498\times10^{-3}\,m \\ -1.922\times10^{-3}\,rad \end{Bmatrix} = \begin{Bmatrix} 9.498\times10^{-3}\,m \\ 0 \\ -1.922\times10^{-3}\,rad \\ 9.498\times10^{-3}\,m \\ -1.922\times10^{-2}\,m \\ -1.922\times10^{-3}\,rad \end{Bmatrix} \text{ or } \mathbf{U}_{free} = \mathbf{A}_c\tilde{\mathbf{U}}_{free}$$

The member deformations are recovered from the compatibility relations as:

$$\begin{Bmatrix} V_1 \\ V_2 \\ V_3 \\ V_4 \\ V_5 \\ V_6 \\ V_7 \end{Bmatrix} = \begin{bmatrix} 0 & 1 & 0 & 0 & 0 & 0 \\ 0.125 & 0 & 0 & 0 & 0 & 0 \\ 0.125 & 0 & 1 & 0 & 0 & 0 \\ -1 & 0 & 0 & 1 & 0 & 0 \\ 0 & 0.1 & 1 & 0 & -0.1 & 0 \\ 0 & 0.1 & 0 & 0 & -0.1 & 1 \\ 0 & 0 & 0 & 0.781 & 0.625 & 0 \end{bmatrix} \begin{Bmatrix} 9.498\times10^{-3}\,m \\ 0 \\ -1.922\times10^{-3}\,rad \\ 9.498\times10^{-3}\,m \\ -1.922\times10^{-2}\,m \\ -1.922\times10^{-3}\,rad \end{Bmatrix} = \begin{Bmatrix} 0 \\ 1.1873\times10^{-3}\,rad \\ -7.348\times10^{-4}\,rad \\ 0 \\ 0 \\ 0 \\ -4.595\times10^{-3}\,m \end{Bmatrix}$$

The basic member forces are determined from the member stiffness equations:

$$\mathbf{Q} = \mathbf{k}\mathbf{V}_{free} = \lfloor N.A. \quad 4100\,kN-m \quad -705.91\,kN-m \quad N.A. \quad N.A. \quad N.A. \quad -287.05\,kN \rfloor^T$$

It is noted that reaction forces cannot be recovered from the equilibrium equations at constrained degrees of freedom ($\mathbf{F}_{constr} = \mathbf{B}_{constr}\mathbf{Q}$) due to multi-freedom constraints. However, they can be computed by considering nodal and member equilibriums in a successive manner as shown in Figure 6.44.

From the equilibrium considerations on nodes and members, the basic member forces related to the multi-freedom constraints are:

$$Q_1 = -70.68\,kN; \quad Q_4 = 224.26\,kN; \quad Q_5 = 705.91\,kN\text{-}m; \text{ and } Q_6 = 0\,kN\text{-}m$$

and support reactions are:

$$\mathbf{F}_{constr} = \lfloor F_7 \quad F_8 \quad F_9 \rfloor^T = \lfloor -200\,kN \quad 250\,kN \quad 4100\,kN-m \rfloor^T$$

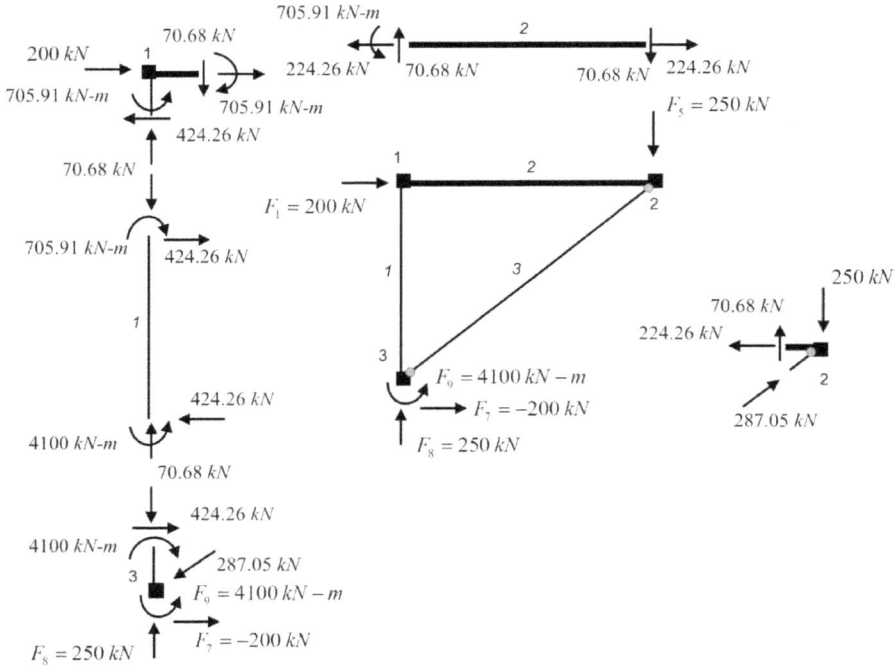

**FIGURE 6.44**   Example 6.6 (Continued).

## EXAMPLE 6.7

Consider the simple 1-story garage structure shown in Figure 6.45. Three shear walls carry the garage roof. The wall stiffness is shown in Figure 6.45. The shear walls have no out-of-plane stiffness.

How many degrees of freedom does the structure have?

What are the structural displacements under the applied load shown in Figure 6.45?

**Solution:** The displacements at point $M$ are selected as generalized displacements. The Rigid-End-Zone transformation can be used to obtain each wall displacement with respect to the master point $M$, as shown in Figure 6.46a.

$$U_{1x} = \mathbf{\Gamma}_{REZ}^{(1)} \mathbf{U}_M = \lfloor 0 \quad 1 \quad 2.5 \rfloor \begin{Bmatrix} U_X \\ U_Y \\ U_Z \end{Bmatrix}; \quad U_{2x} = \mathbf{\Gamma}_{REZ}^{(2)} \mathbf{U}_M = \lfloor 1 \quad 0 \quad -4 \rfloor \begin{Bmatrix} U_X \\ U_Y \\ U_Z \end{Bmatrix};$$

$$U_{3x} = \mathbf{\Gamma}_{REZ}^{(3)} \mathbf{U}_M = \lfloor 0 \quad 1 \quad -2.5 \rfloor \begin{Bmatrix} U_X \\ U_Y \\ U_Z \end{Bmatrix}$$

The stiffness matrix of each wall is

$$\text{Wall 1: } I_1 = \frac{1 \times 8^3}{12} = 42.67 \; m^4; \; G = \frac{E}{2(1+\upsilon)} = \frac{E}{2(1+0.2)} = 0.41667E$$

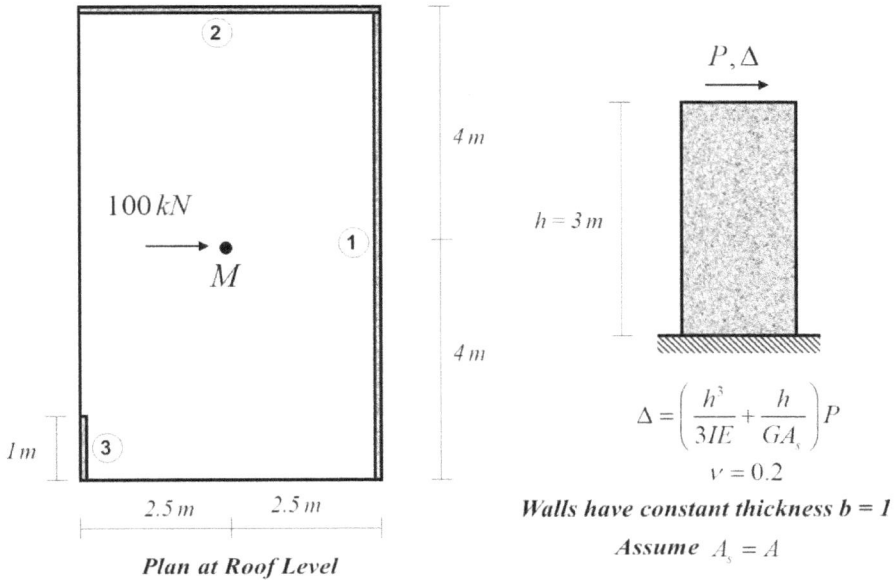

**FIGURE 6.45**   Example 6.7.

$$f_1 = \left(\frac{h^3}{3I_1E} + \frac{h}{GA_{s1}}\right) = \left(\frac{3^3}{3(42.67)E} + \frac{3}{(0.41667E)(8\times1)}\right) = \frac{1.11}{E} \Rightarrow k_1 = 0.9E$$

Wall 2:          $I_2 = \dfrac{1\times5^3}{12} = 10.42\ m^4;\ G = \dfrac{E}{2(1+\upsilon)} = \dfrac{E}{2(1+0.2)} = 0.41667E$

$$f_2 = \left(\frac{h^3}{3I_2E} + \frac{h}{GA_{s2}}\right) = \left(\frac{3^3}{3(10.42)E} + \frac{3}{(0.41667E)(5\times1)}\right) = \frac{2.304}{E} \Rightarrow k_2 = 0.4341E$$

Wall 3:          $I_3 = \dfrac{1\times1^3}{12} = 0.08333\ m^4;\ G = \dfrac{E}{2(1+\upsilon)} = \dfrac{E}{2(1+0.2)} = 0.41667E$

$$f_3 = \left(\frac{h^3}{3I_3E} + \frac{h}{GA_{s3}}\right) = \left(\frac{3^3}{3(0.08333)E} + \frac{3}{(0.41667E)(1\times1)}\right) = \frac{115.2}{E} \Rightarrow k_3 = 8.68\times10^{-3}E$$

The Rigid-End-Zone transformation can be used to obtain the stiffness matrix of each wall with respect to the master point $M$ in Figure 6.46a.

Wall 1:          $\mathbf{K}_{M1} = {\Gamma_{REZ}^{(1)}}^T k_1 \Gamma_{REZ}^{(1)} = E\begin{bmatrix} 0 & 0 & 0 \\ 0 & 0.9 & 2.25 \\ 0 & 2.25 & 5.625 \end{bmatrix}$

Wall 2:          $\mathbf{K}_{M2} = {\Gamma_{REZ}^{(2)}}^T k_2 \Gamma_{REZ}^{(2)} = E\begin{bmatrix} 0.4341 & 0 & -1.7364 \\ 0 & 0 & 0 \\ -1.7364 & 0 & 6.9456 \end{bmatrix}$

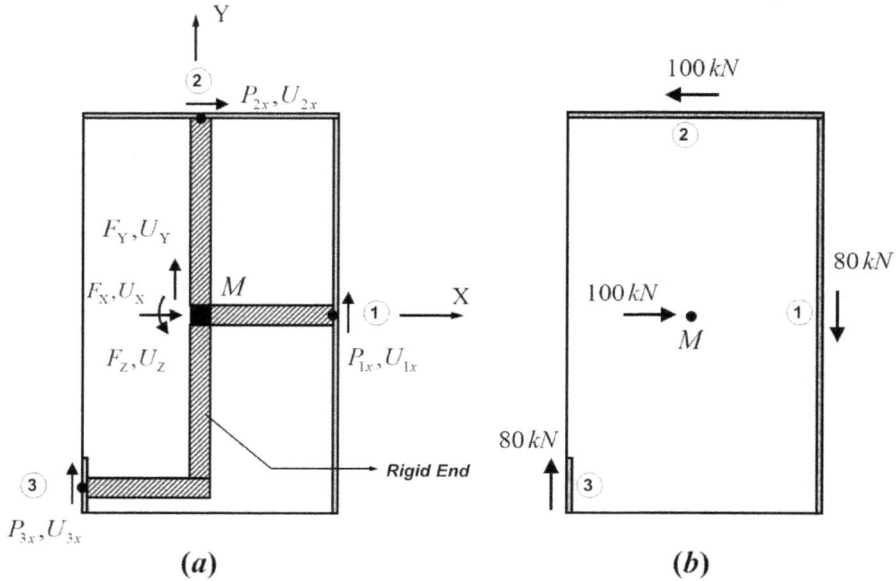

**FIGURE 6.46**   Example 6.7 (Continued).

Wall 3:
$$\mathbf{K}_{M3} = {\Gamma_{REZ}^{(3)}}^{T} k_3 \Gamma_{REZ}^{(3)} = E \begin{bmatrix} 0 & 0 & 0 \\ 0 & 0.000868 & -0.00217 \\ 0 & -0.00217 & 0.005425 \end{bmatrix}$$

Therefore, the stiffness matrix of this roof slab is:

$$\mathbf{K}_M = \mathbf{K}_{M1} + \mathbf{K}_{M2} + \mathbf{K}_{M3} = E \begin{bmatrix} 0.4341 & 0 & -1.7364 \\ 0 & 0.900868 & 2.24783 \\ -1.7364 & 2.24783 & 12.576 \end{bmatrix}$$

The load vector $\mathbf{F}_M$ is $\mathbf{F}_M = \lfloor 100 \quad 0 \quad 0 \rfloor^T$

The displacement vector $\mathbf{U}_M$ is $\mathbf{U}_M = \left[ \mathbf{K}_M \right]^{-1} \mathbf{F}_M = \left| \dfrac{74034.2}{E} \quad \dfrac{-46038.5}{E} \quad \dfrac{18457}{E} \right|^T$

The deformation of each wall is

$$U_{1x} = \Gamma_{REZ}^{(1)} \mathbf{U}_M = \frac{88.89}{E}; \quad U_{2x} = \Gamma_{REZ}^{(2)} \mathbf{U}_M = \frac{230.362}{E}; \quad U_{3x} = \Gamma_{REZ}^{(3)} \mathbf{U}_M = \frac{-92165.9}{E}$$

The force in each wall shown in Figure 6.46*b* is

$$F_{1x} = k_1 U_{1x} = \left( 0.9E \right) \left( \frac{88.89}{E} \right) = 80 \; kN$$

$$F_{2x} = k_2 U_{2x} = \left( 0.4341E \right) \left( \frac{230.362}{E} \right) = 100 \; kN$$

$$F_{3x} = k_3 U_{3x} = \left( 0.868 \times 10^{-3} E \right) \left( \frac{-92165.9}{E} \right) = -80 \; kN$$

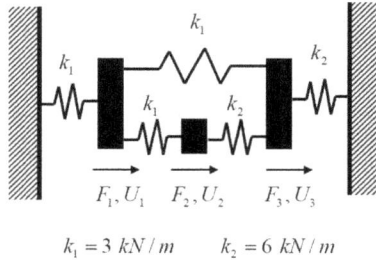

FIGURE 6.47 Problem 6.1.

## 6.8 EXERCISES

**Problem 6.1:** The idealized spring-block structure shown in Figure 6.47 has forces and displacements only in the horizontal direction. For $F_1 = 20 \, kN$, $F_2 = 40 \, kN$, and $F_3 = -30 \, kN$, use the conceptual stiffness to determine the displacements of the three blocks and the forces in five springs.

**Problem 6.2:** Compute and show a sketch of the equivalent nodal loads of the frame structure in Figure 6.48 for the combination of the following disturbances:

I. The external loads are shown in Figure 6.48.
II. A settlement of 25 $mm$ of support $1$.
III. A temperature of the external fibers $T_e = 30 \, C^0$ and of the internal fibers $T_i = -5 \, C^0$. The temperature during construction was $T_0 = 10 \, C^0$.

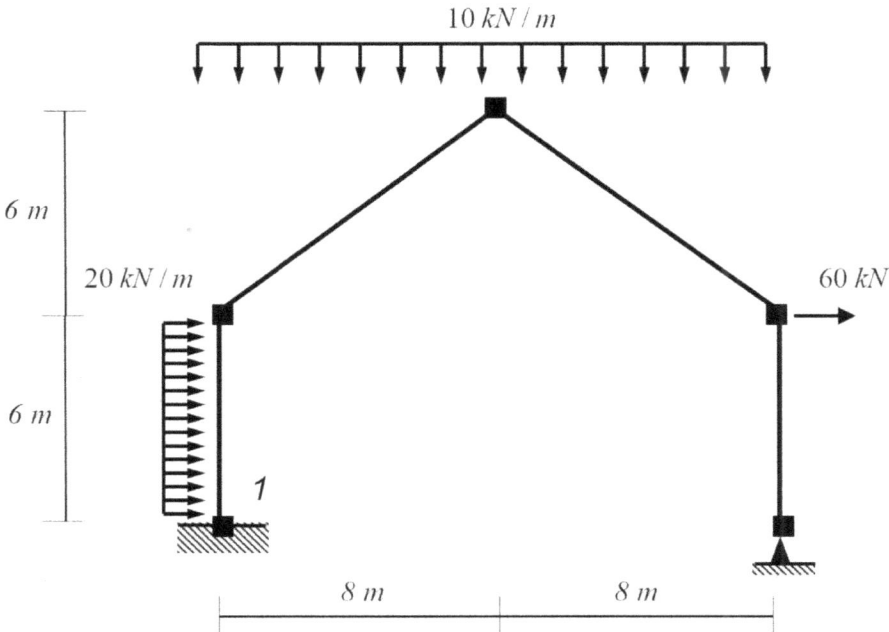

FIGURE 6.48 Problem 6.2.

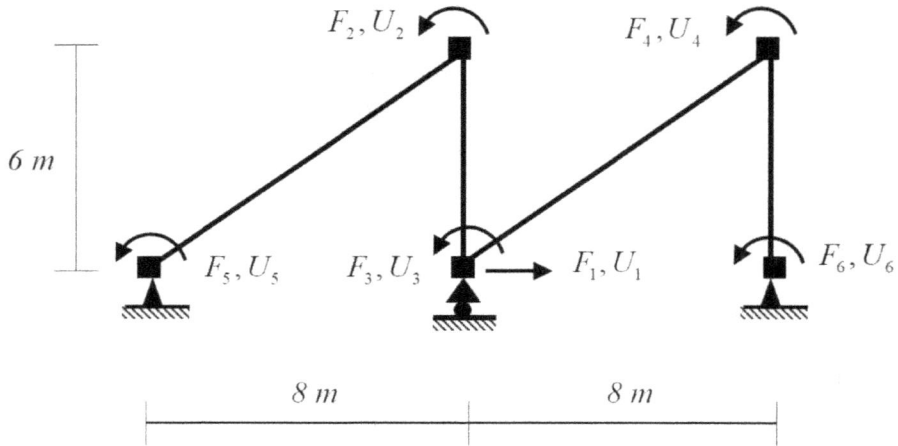

**FIGURE 6.49** Problem 6.3

The members of the structure are made of the same material ($E = 210$ *GPa* and$\alpha_T = 10^{-5} / C^0$) and have the same constant cross section ($I = 117.7 \times 10^6$ *mm*$^4$; $A = 6.26 \times 10^3$ *mm*$^2$; $h = 330$ *mm*;).

**Problem 6.3:** The planar inextensible frame shown in Figure 6.49 has six displacement degrees of freedom. Assume that all members have the same flexural rigidity of $IE = 45$ $kN - m^2$. Derive the structural stiffness matrix associated with degrees of freedom shown using the semi-automatic approach.

**Problem 6.4:** All members of the truss shown in Figure 6.50 have axial stiffness $EA = 50 \times 10^3$ $kN$. Member *a* is heated up by $\alpha \Delta T = 0.003$.

- Derive the structural stiffness matrix associated with degrees of freedom shown using the semi-automatic approach.
- Determine initial member forces as well as nodal displacements.

**Problem 6.5:** Use the semi-automatic stiffness approach to solve the internal forces and support reactions of the inextensible frame shown in Figure 6.51. Assume that member properties are: $IE_1 = 10^6$ $kN - m^2$; $IE_2 = 5 \times 10^6$ $kN - m^2$; and $IE_3 = 2 \times 10^6$ $kN - m^2$.

**Problem 6.6:** Use Eq. (6.92) to compute the restrained forces **R** of the frame in Example 6.4 and verify that they are the same as those computed in Example 6.4.

**Problem 6.7:** Analyze the frame shown in Figure 6.52 if the flexural stiffness *IE* of the girders is $200 \times 10^3$ $kN - m^2$. All columns are infinitely rigid.

**Problem 6.8:** For a single-story (cantilever) shear wall building shown in Figure 6.53, find the $3 \times 3$ stiffness matrix with respect to the three degrees of freedom ($U_X, U_Y$, and $\theta_Z$) by

(i) The conceptual approach.
(ii) The rigid-end-zone transformation matrix.

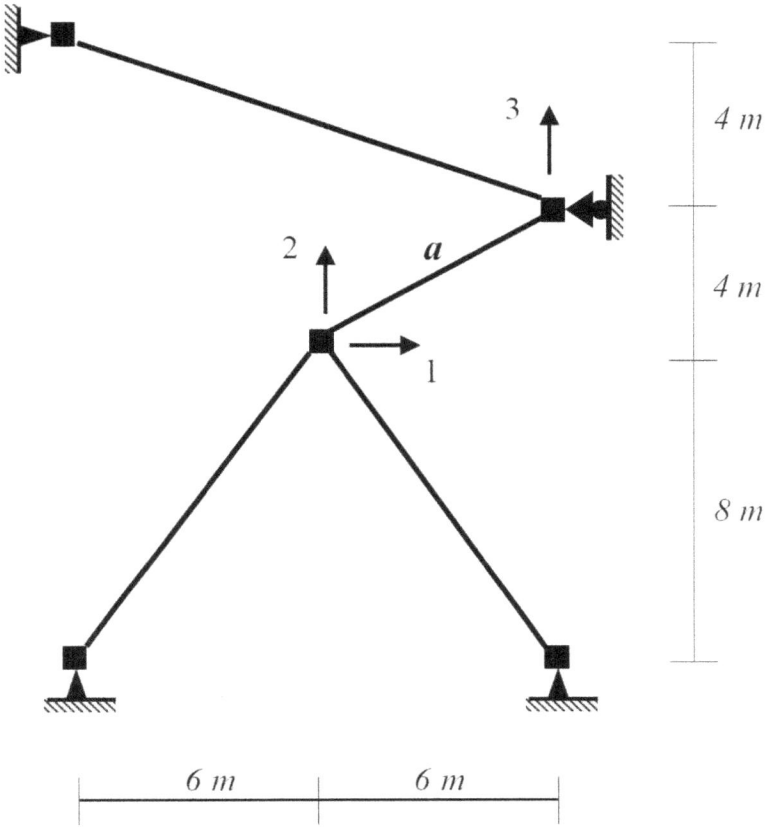

**FIGURE 6.50**    Problem 6.4.

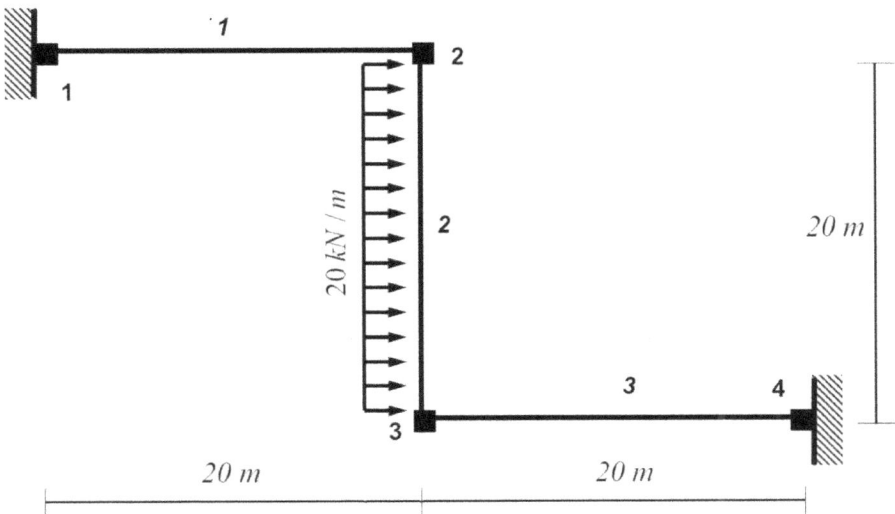

**FIGURE 6.51**    Problem 6.5.

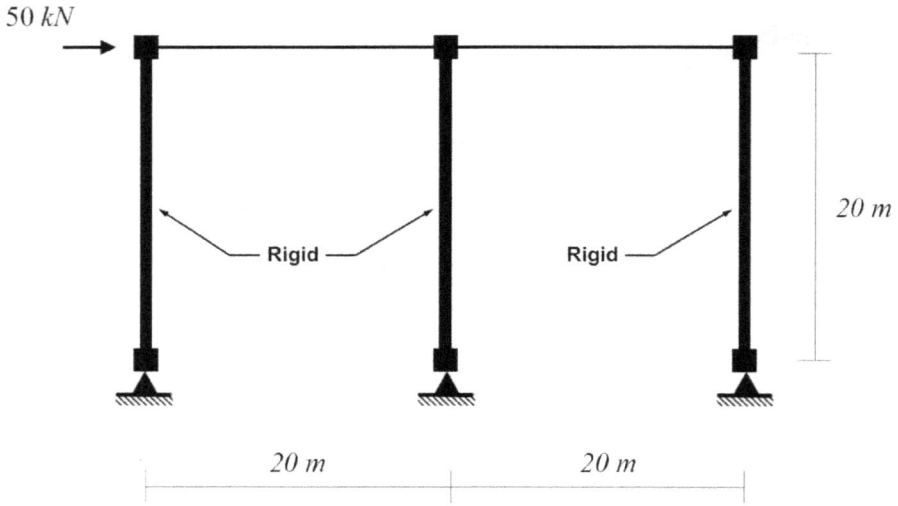

**FIGURE 6.52**   Problem 6.7.

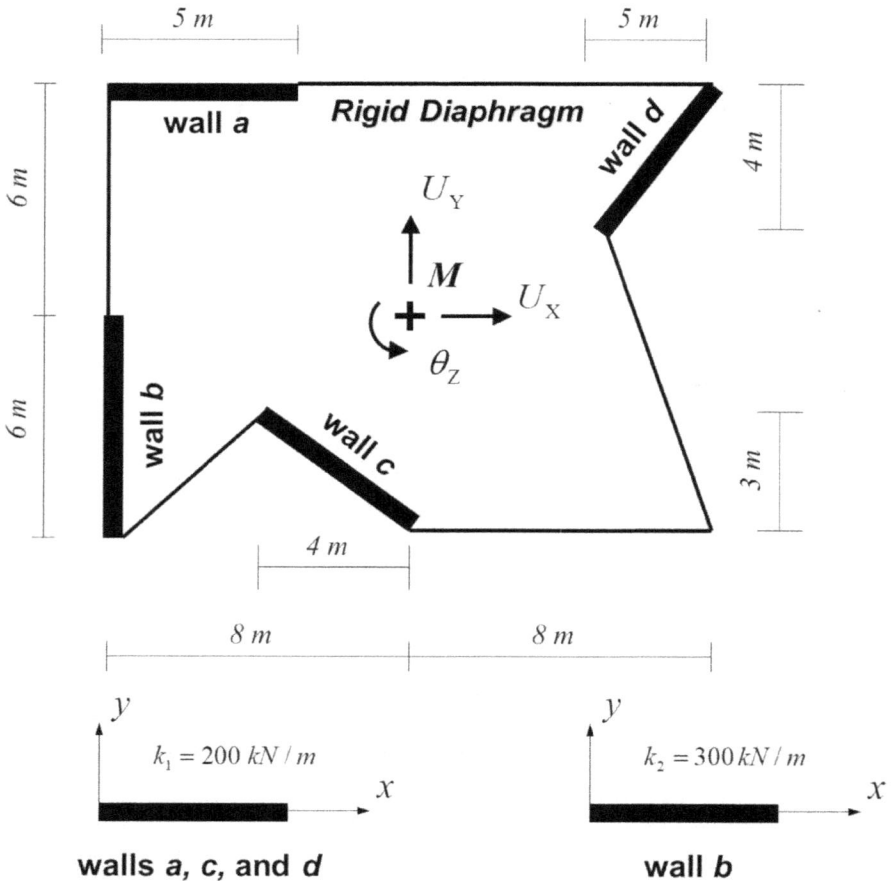

**FIGURE 6.53**   Problem 6.8.

## REFERENCES

Armenakas, A.E. 1991. *Modern structural analysis: The matrix method approach.* McGraw-Hill Inc.

Dawe, D.J. 1984. *Matrix and finite element displacement analysis of structures.* Clarendon Press.

Elias, Z.M. 1986. *Theory and methods of structural analysis.* John Wiley and Son Inc.

Felton, L.P. and R.B. Nelson. 1997. *Matrix structural analysis.* John Wiley & Sons Inc.

Filippou, F.C. 2002. *CE 220 course notes: Structural analysis: Theory and applications.* Department of Civil and Environmental Engineering, University of California.

Ghali, A. and A.M. Neville. 2017. *Structural analysis: A unified classical and matrix approach* (7th Edition). CRC Press.

Kanchi, M.B. 1993. *Matrix methods of structural analysis* (2nd Edition). Wiley Eastern Limited.

Limkatanyu, S. 2002. *Reinforced concrete models with bond-interfaces for the nonlinear static and dynamic analysis of reinforced concrete frame structures.* PhD diss. University of Colorado, Boulder.

Lukkunaprasit, P. 1984. *Structural analysis.* The Engineering Institute of Thailand (EIT).

McGuire, W., R.H. Gallagher, and R.D. Ziemian. 2000. *Matrix structural analysis* (2nd Edition). John Wiley & Sons Inc.

Przemieniecki, J.S. 1985. *Theory of matrix structural analysis.* Dover Publications Inc.

Rajasekaran, S. and G. Sankarasubramanian. 2006. *Computational structural mechanics.* Prentice-Hall of India Private Limited.

SAP 2000. User's Manual: Volumes I, II, and III. Swanson Analysis Systems User's Manual. Computers and Structures, Inc., Berkeley, California, USA.

Spacone, E. 2003. *CVEN 5525 course notes: Matrix structural analysis.* Department of Civil, Environmental, and Architectural Engineering, University of Colorado.

Strang, G. 1998. *Linear algebra and its applications* (3rd Edition). Saunders College Publishing.

Weaver, W. and J.M. Gere. 1990. *Matrix analysis of framed structures.* Chapman & Hall, Boca Raton.

Whittaker, A. 2001. *CIE 423 course notes: Structures III.* Department of Civil, Structural, and Environmental Engineering, State University of New York.

# 7 The Stiffness Method II
## *Direct Stiffness Approach*

## 7.1 INTRODUCTION

In Chapter 6, the stiffness method was first formulated through the principle of superposition. This method was referred to as the "conceptual" stiffness method. Then, the method was formalized and extended with help from the virtual displacement principle, hence leading to the so-called "semi-automatic" stiffness method. In this method, the structural stiffness matrix was assembled indirectly from stiffness matrices of individual members through congruent transformation. However, neither the conceptual nor semi-automatic approaches would be conducive to computer programming. A more suitable one evolved from the basic ideas of the two aforementioned stiffness approaches by adding a few computer-oriented thoughts. This method is known as the "direct" stiffness method. The term "direct" implies that the structural stiffness matrix can be directly assembled from member stiffness matrices. The assembling procedure of member stiffness matrices used in the direct stiffness method is very systematic and straightforward. This is the main reason why the direct stiffness method is well-suited to computer programming. It should be noted that almost all of the structural-analysis software available in markets (academic, research, and practical-engineering communities) is implemented based on the direct stiffness approach.

The main objectives of this chapter are to introduce the fundamental concepts of the direct stiffness method and to provide the computerized way of thinking necessary for implementing the direct stiffness method. Before pursuing further the direct stiffness method in this chapter, it is worthwhile discussing a brief history of stiffness-method developments.

The theoretical foundation of the stiffness method was established by George A. Maney (1888–1947), who invented the slope-deflection method in a 1915 University of Minnesota engineering publication. This method was an extension of earlier works on secondary stresses by Manderla and Mohr. Until the invention of the moment distribution method in 1930 by Hardy Cross, the slope-deflection method had been the most prevailing method for analyzing continuous beams and frames in the United States for almost 15 years. The slope-deflection method could be regarded as an ancestor of the stiffness method used nowadays.

In the precomputer era, the major drawback of this earlier matrix approach was that there was a need to obtain the direct solution of simultaneous algebraic equations. This became a formidable task with structures possessing many unknown displacements. Thanks to the invention of computers in the late 1940s, structural analysis moved from manual calculation to computerized approaches. In 1953, R.K. Livesley developed the matrix stiffness method. In 1956, M.J. Turner, R.W. Clough, H.C. Martin, and J.L. Topp derived stiffness matrices for the truss and frame elements through finite-element formulation and introduced the now-popular direct stiffness method for assembling the structural stiffness matrix. In 1960, J.H. Argyris and S. Kelsey presented a formulation of matrix methods based on the energy principle. Since the 1950s, the matrix stiffness method has been developed at a tremendous speed.

DOI: 10.1201/9781003595458-7

## 7.2  THE COMPLETE STIFFNESS EQUATIONS OF A FRAME STRUCTURE

It is recalled from Chapter 6 that the complete stiffness equations of a frame structure are written in the matrix form as:

$$\mathbf{F}^E = \mathbf{F} - \mathbf{R} = \mathbf{K}\mathbf{U} \tag{7.1}$$

where the vector $\mathbf{F}$ contains applied nodal loads; the vector $\mathbf{R}$ contains restrained nodal forces associated with member loads defined in Chapter 6 (6/6.80 or 6/6.92); the vector $\mathbf{F}^E$ contains equivalent nodal loads of the structure discussed in Chapter 6 (6/6.81 or 6/6.89–90); the vector $\mathbf{U}$ collects displacement components of all structural nodes including those inhibited against movements by support conditions; and the matrix $\mathbf{K}$ is the complete stiffness matrix of the structure. The unknown parameters in the structural stiffness Eq. (7.1) consist of nodal forces corresponding to constrained degrees of freedom (reactive forces) and nodal displacements corresponding to free degrees of freedom. For each specified (known) component of nodal forces, there is an associated unknown component of nodal displacements. Similarly, for each specified (known) component of nodal displacements, there is a corresponding unknown component of nodal forces. Therefore, the total number of unknown parameters in the complete stiffness equations of any given structure is invariant to the change in support conditions. In other words, the summation of the statical and kinematical degrees of indeterminacy is a constant quantity for any structure. Therefore, the total number of unknown parameters of any structure is equal to the number of its complete stiffness equations. However, the unknown parameters cannot be obtained directly by solving Eq. (7.1) since the complete stiffness matrix $\mathbf{K}$ is generally singular at the initial stage. This is justified since there is no support condition provided to the structure, hence implying the structure is infinitely flexible. In order to solve the unknown quantities, it is necessary to partition the complete structural stiffness equations (7.1) into unconstrained and constrained degrees of freedom as:

$$\begin{Bmatrix} \mathbf{F}^E_{free} \\ \mathbf{F}^E_{constr} \end{Bmatrix} = \begin{Bmatrix} \mathbf{F}_{free} \\ \mathbf{F}_{constr} \end{Bmatrix} - \begin{Bmatrix} \mathbf{R}_{free} \\ \mathbf{R}_{constr} \end{Bmatrix} = \begin{bmatrix} \mathbf{K}_{ff} & \mathbf{K}_{fc} \\ \mathbf{K}_{cf} & \mathbf{K}_{cc} \end{bmatrix} \begin{Bmatrix} \mathbf{U}_{free} \\ \mathbf{U}_{constr} \end{Bmatrix} \tag{7.2}$$

This partitioning step was accomplished in previous chapters by wisely numbering the degree-of-freedom system at the beginning stage (e.g. Example 6.2). Generally, in computer implementation, this step is not necessary and can be done once the assembling process of the structural stiffness matrix is completed. This issue will be visited in the next chapter. Eq. (7.2) can be split into two stiffness relations as:

$$\underbrace{\mathbf{F}^E_{free}}_{known} = \underbrace{\mathbf{F}_{free}}_{known} - \underbrace{\mathbf{R}_{free}}_{known} = \mathbf{K}_{ff}\underbrace{\mathbf{U}_{free}}_{unknown} + \mathbf{K}_{fc}\underbrace{\mathbf{U}_{constr}}_{known} \tag{7.3}$$

$$\underbrace{\mathbf{F}^E_{constr}}_{unknown} = \underbrace{\mathbf{F}_{constr}}_{unknown} - \underbrace{\mathbf{R}_{constr}}_{known} = \mathbf{K}_{cf}\underbrace{\mathbf{U}_{free}}_{unknown} + \mathbf{K}_{cc}\underbrace{\mathbf{U}_{constr}}_{known} \tag{7.4}$$

Typically, if there is no support movement, constrained displacement degrees of freedom $\mathbf{U}_{constr}$ are equal to zero. However, if a support movement exists and its actions have not been included in the restrained forces $\mathbf{R}$, constrained displacement degrees of freedom $\mathbf{U}_{constr}$ are not equal to zero. The unknown nodal displacements $\mathbf{U}_{free}$ can be solved from Eq. (7.3). Then, these now-known nodal displacements are substituted in Eq. (7.4) to yield the unknown nodal forces $\mathbf{F}_{constr}$.

## 7.3 DERIVATION OF THE COMPLETE STIFFNESS EQUATIONS OF A FRAME STRUCTURE: MEMBER-WISE ASSEMBLING TECHNIQUE

The complete stiffness equations of a frame structure can be derived based on the three basic ingredients:

1. *Equilibrium:* The nodal equilibrium equations of a structure can be written in the matrix form as:

$$\mathbf{F} = \mathbf{H}^{(1)}\mathbf{P}^{(1)} + \mathbf{H}^{(2)}\mathbf{P}^{(2)} + \ldots + \mathbf{H}^{(k)}\mathbf{P}^{(k)} + \ldots + \mathbf{H}^{(m)}\mathbf{P}^{(m)} = \mathbf{HP} \qquad (7.5)$$

where $\mathbf{H}$ is known as the structural equilibrium mapping matrix and contains equilibrium mapping submatrices $\mathbf{H}^{(k)}$ of $m$ members; $\mathbf{P}$ is known as the structural end force vector and contains end force subvectors $\mathbf{P}^{(k)}$ of $m$ members.

$$\mathbf{H} = \begin{bmatrix} \mathbf{H}^{(1)} & \mathbf{H}^{(2)} & \ldots & \mathbf{H}^{(k)} & \ldots & \mathbf{H}^{(m)} \end{bmatrix} \qquad (7.6)$$

$$\mathbf{P} = \begin{bmatrix} \mathbf{P}^{(1)} & \mathbf{P}^{(2)} & \ldots & \mathbf{P}^{(k)} & \ldots & \mathbf{P}^{(m)} \end{bmatrix}^{T} \qquad (7.7)$$

In Chapter 3, the approach to establish the matrix $\mathbf{H}^{(k)}$ for member $k$ based on its mapping vector $LM_k$ was presented. It is noted that the number of rows of Eq. (7.5) is equal to the number of all possible components of nodal forces including those directly resisted by supports.

2. *Compatibility:* The end displacements of a member must be compatible with the structural nodal displacements. This requirement leads to the member compatibility relations written in the matrix form as:

$$\mathbf{U}^{(k)} = \mathbf{G}^{(k)}\mathbf{U} \text{ for member } k \qquad (7.8)$$

where $\mathbf{U}^{(k)}$ contains end displacements of member $k$; $\mathbf{G}^{(k)}$ is the compatibility mapping matrix of member $k$; and $\mathbf{U}$ contains all nodal displacements including those restrained by support conditions.

For the whole structure, all member compatibility matrices are grouped together as:

$$\Delta = \mathbf{GU} \qquad (7.9)$$

where $\Delta$ is known as the structural end displacement vector and contains end displacement subvectors $\mathbf{U}^{(k)}$ of $m$ members; $\mathbf{G}$ is known as the structural compatibility mapping matrix and contains compatibility mapping submatrices $\mathbf{G}^{(k)}$ of $m$ members.

$$\Delta = \begin{bmatrix} \mathbf{U}^{(1)} & \mathbf{U}^{(2)} & \ldots & \mathbf{U}^{(k)} & \ldots & \mathbf{U}^{(m)} \end{bmatrix}^{T} \qquad (7.10)$$

$$\mathbf{G} = \begin{bmatrix} \mathbf{G}^{(1)} & \mathbf{G}^{(2)} & \cdots & \mathbf{G}^{(k)} & \cdots & \mathbf{G}^{(m)} \end{bmatrix}^T \qquad (7.11)$$

In Chapter 4, the approach to establish the matrix $\mathbf{G}^{(k)}$ for member $k$ based on its mapping vector $LM_k$ was presented. It is not surprising that the following contragradient holds:

$$\mathbf{G} = \mathbf{H}^T \qquad (7.12)$$

Eq. (7.12) can be easily proved through the virtual displacement principle.

3. *Member Stiffness Relations:* The response of each member is described by its stiffness matrix and fixed-end load vector. The stiffness equations of member $k$ are expressed as:

$$\mathbf{P}^{(k)} = \mathbf{K}^{(k)}\mathbf{U}^{(k)} + \mathbf{P}_F^{(k)} \qquad (7.13)$$

where $\mathbf{P}_F^{(k)}$ is the fixed-end load vector and $\mathbf{K}^{(k)}$ is the global member stiffness matrix. In Chapter 6, the approach to establish the fixed-end load vector $\mathbf{P}_F^{(k)}$ was presented.

For the whole structure, all member stiffness matrices are written together as:

$$\begin{Bmatrix} \mathbf{P}^{(1)} \\ \mathbf{P}^{(2)} \\ \vdots \\ \mathbf{P}^{(k)} \\ \vdots \\ \mathbf{P}^{(m)} \end{Bmatrix} = \begin{bmatrix} \mathbf{K}^{(1)} & \mathbf{0} & \cdots & \mathbf{0} & \cdots & \mathbf{0} \\ \mathbf{0} & \mathbf{K}^{(2)} & \cdots & \mathbf{0} & \cdots & \mathbf{0} \\ \vdots & \vdots & \ddots & & & \vdots \\ \mathbf{0} & \mathbf{0} & & \mathbf{K}^{(k)} & & \mathbf{0} \\ \vdots & \vdots & & & \ddots & \vdots \\ \mathbf{0} & \mathbf{0} & \cdots & \mathbf{0} & \cdots & \mathbf{K}^{(m)} \end{bmatrix} \begin{Bmatrix} \mathbf{U}^{(1)} \\ \mathbf{U}^{(2)} \\ \vdots \\ \mathbf{U}^{(k)} \\ \vdots \\ \mathbf{U}^{(m)} \end{Bmatrix} + \begin{Bmatrix} \mathbf{P}_F^{(1)} \\ \mathbf{P}_F^{(2)} \\ \vdots \\ \mathbf{P}_F^{(k)} \\ \vdots \\ \mathbf{P}_F^{(m)} \end{Bmatrix} \qquad (7.14)$$

or in the compact form as:

$$\mathbf{P} = \hat{\mathbf{K}}\Delta + \hat{\mathbf{P}}_F \qquad (7.15)$$

where $\hat{\mathbf{K}}$ is called the "unassembled" structural stiffness matrix.

As shown in Chapter 6, all of these governing equations can be written together through the virtual displacement principle.

From the invariant property of the external virtual work (5/5.200), it can be concluded that:

$$\delta\Delta^T\mathbf{P} = \delta\mathbf{U}^T\mathbf{F} \qquad (7.16)$$

Substituting Eqs. (7.9) into (7.16), one has

$$\underbrace{\delta\mathbf{U}^T}_{\text{Arbitrary}} \begin{bmatrix} \mathbf{F} - \mathbf{G}^T\mathbf{P} \end{bmatrix} = 0 \quad \Rightarrow \quad \mathbf{F} = \mathbf{G}^T\mathbf{P} \qquad (7.17)$$

Comparing Eqs. (7.17) with (7.5), it becomes clear why Eq. (7.12) holds.

Substituting Eqs. (7.15) into (7.17), one has

$$\mathbf{F} = \mathbf{G}^T \left( \hat{\mathbf{K}} \Delta + \hat{\mathbf{P}}_F \right) \tag{7.18}$$

Substituting Eqs. (7.9) into (7.18), one obtains

$$\mathbf{F} = \mathbf{G}^T \left( \hat{\mathbf{K}} \mathbf{G} \mathbf{U} + \hat{\mathbf{P}}_F \right) = \mathbf{G}^T \hat{\mathbf{K}} \mathbf{G} \mathbf{U} + \mathbf{G}^T \hat{\mathbf{P}}_F \tag{7.19}$$

Comparing Eqs. (7.19) with (7.1), the complete structural stiffness matrix $\mathbf{K}$ is defined as:

$$\mathbf{K} = \mathbf{G}^T \hat{\mathbf{K}} \mathbf{G} \tag{7.20}$$

and the restrained nodal force vector $\mathbf{R}$ is defined as:

$$\mathbf{R} = \mathbf{G}^T \hat{\mathbf{P}}_F \tag{7.21}$$

Due to the diagonalized property of matrix $\hat{\mathbf{K}}$, Eq. (7.20) can alternatively be expressed as:

$$\mathbf{K} = \sum_{k=1}^{m} \tilde{\tilde{\mathbf{K}}}^{(k)} = \sum_{k=1}^{m} \mathbf{G}^{(k)^T} \mathbf{K}^{(k)} \mathbf{G}^{(k)} \tag{7.22}$$

Similarly, the alternative form of Eq. (7.21) is written as:

$$\mathbf{R} = \sum_{k=1}^{m} \tilde{\tilde{\mathbf{R}}}^{(k)} = \sum_{k=1}^{m} \mathbf{G}^{(k)^T} \mathbf{P}_F^{(k)} \tag{7.23}$$

Eqs. (7.22) and (7.23) are the backbone relations of the direct stiffness method and nicely reveal the physical as well as assembling natures of this analysis method. Eq. (7.22) implies that the structural stiffness matrix is simply a summation of member stiffness matrices. This direct feature of the stiffness summation is very attractive for computer implementation since the assembling process for the structural stiffness matrix is just a summation process of member stiffness matrices.

The augmented member stiffness matrix $\tilde{\tilde{\mathbf{K}}}^{(k)}$ in Eq. (7.22) has the same size as the structural stiffness matrix $\mathbf{K}$. Physically, it can be interpreted as the structural stiffness matrix when only the contribution of member $k$ is accounted for. Therefore, the summation of all member contributions to the structural stiffness matrix is comparable to constantly updating the structural stiffness matrix member by member (member-wise assembly).

Likewise, Eq. (7.23) implies that the restrained nodal force vector is simply a summation of member fixed-end load vectors. The augmented nodal force vector $\tilde{\tilde{\mathbf{R}}}^{(k)}$ in Eq. (7.23) has the same size as the structural nodal load vector $\mathbf{F}$. Physically, it can be interpreted as the restrained nodal force vector when only the contribution of member $k$ is accounted for.

Algorithms for constructing the matrices $\mathbf{H}$, $\mathbf{G}$, and $\hat{\mathbf{K}}$ are quite straightforward and can be implemented with any computer language (e.g. MATLAB, C++, Fortran). The structural stiffness matrix and restrained nodal force vector are obtained by carrying out the matrix multiplications in Eqs. (7.22) and (7.23), respectively. However, this way of assembling stiffness matrices and force vectors has found very little practical uses. This is due to the fact that the matrices $\mathbf{H}$, $\mathbf{G}$, and $\hat{\mathbf{K}}$ are sparsely populated and are very large for complex structural systems.

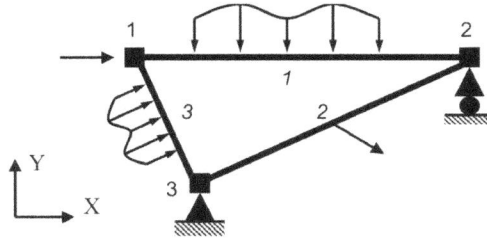

**FIGURE 7.1**    A Prototypical Frame: Nodal Loads, Member Loads, and Support Conditions.

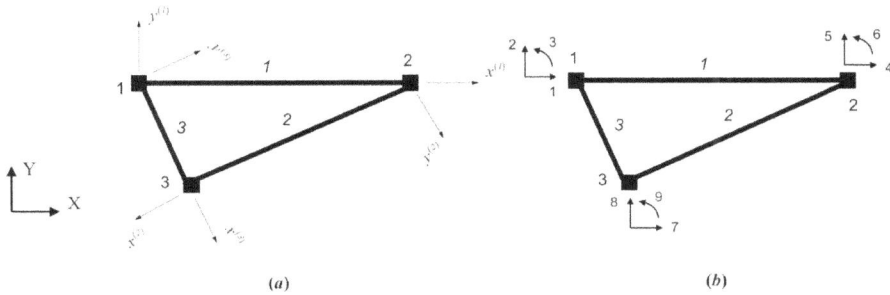

(a)                                                          (b)

**FIGURE 7.2**    A Prototypical Frame: (a) Local Reference Axes; (b) Nodal, Member, and DOF Numbering Systems.

The frame structure shown in Figure 7.1 will be used to illustrate that the process of assembling member stiffness matrices and member force vectors is much simpler than the one discussed above.

Referred to Figure 7.2, the mapping vector for each member is:

$$LM_1 = \begin{Bmatrix} 1 \\ 2 \\ 3 \\ 4 \\ 5 \\ 6 \end{Bmatrix}_1 \; ; \quad LM_2 = \begin{Bmatrix} 4 \\ 5 \\ 6 \\ 7 \\ 8 \\ 9 \end{Bmatrix}_2 \; ; \quad LM_3 = \begin{Bmatrix} 1 \\ 2 \\ 3 \\ 7 \\ 8 \\ 9 \end{Bmatrix}_3 \qquad (7.24)$$

Following the rules described in Chapter 3, the nodal equilibrium equations of the structure are expressed in the matrix form as:

$$\begin{Bmatrix} \mathbf{F}_1 \\ \mathbf{F}_2 \\ \mathbf{F}_3 \end{Bmatrix} = \begin{bmatrix} \mathbf{I} & \mathbf{0} & \mathbf{0} & \mathbf{0} & \mathbf{I} & \mathbf{0} \\ \mathbf{0} & \mathbf{I} & \mathbf{I} & \mathbf{0} & \mathbf{0} & \mathbf{0} \\ \mathbf{0} & \mathbf{0} & \mathbf{0} & \mathbf{I} & \mathbf{0} & \mathbf{I} \end{bmatrix} \begin{Bmatrix} \mathbf{P}_1^{(1)} \\ \mathbf{P}_2^{(1)} \\ \mathbf{P}_1^{(2)} \\ \mathbf{P}_2^{(2)} \\ \mathbf{P}_1^{(3)} \\ \mathbf{P}_2^{(3)} \end{Bmatrix} \text{ or } \mathbf{F} = \mathbf{HP} \tag{7.25}$$

The equilibrium mapping matrix for each member is:

$$\mathbf{H}^{(1)} = \begin{bmatrix} \mathbf{I} & \mathbf{0} \\ \mathbf{0} & \mathbf{I} \\ \mathbf{0} & \mathbf{0} \end{bmatrix}; \ \mathbf{H}^{(2)} = \begin{bmatrix} \mathbf{0} & \mathbf{0} \\ \mathbf{I} & \mathbf{0} \\ \mathbf{0} & \mathbf{I} \end{bmatrix}; \ \mathbf{H}^{(3)} = \begin{bmatrix} \mathbf{I} & \mathbf{0} \\ \mathbf{0} & \mathbf{0} \\ \mathbf{0} & \mathbf{I} \end{bmatrix}; \tag{7.26}$$

Following the rules described in Chapter 4, all member compatibility equations are expressed in the matrix form as:

$$\begin{Bmatrix} \mathbf{U}_1^{(1)} \\ \mathbf{U}_2^{(1)} \\ \mathbf{U}_1^{(2)} \\ \mathbf{U}_2^{(2)} \\ \mathbf{U}_1^{(3)} \\ \mathbf{U}_2^{(3)} \end{Bmatrix} = \begin{bmatrix} \mathbf{I} & \mathbf{0} & \mathbf{0} \\ \mathbf{0} & \mathbf{I} & \mathbf{0} \\ \mathbf{0} & \mathbf{I} & \mathbf{0} \\ \mathbf{0} & \mathbf{0} & \mathbf{I} \\ \mathbf{I} & \mathbf{0} & \mathbf{0} \\ \mathbf{0} & \mathbf{0} & \mathbf{I} \end{bmatrix} \begin{Bmatrix} \mathbf{U}_1 \\ \mathbf{U}_2 \\ \mathbf{U}_3 \end{Bmatrix} \text{ or } \Delta = \mathbf{GU} \tag{7.27}$$

The compatibility mapping matrix for each member is:

$$\mathbf{G}^{(1)} = \mathbf{H}^{(1)^T} = \begin{bmatrix} \mathbf{I} & \mathbf{0} & \mathbf{0} \\ \mathbf{0} & \mathbf{I} & \mathbf{0} \end{bmatrix}; \mathbf{G}^{(2)} = \mathbf{H}^{(2)^T} = \begin{bmatrix} \mathbf{0} & \mathbf{I} & \mathbf{0} \\ \mathbf{0} & \mathbf{0} & \mathbf{I} \end{bmatrix}; \mathbf{G}^{(3)} = \mathbf{H}^{(3)^T} = \begin{bmatrix} \mathbf{I} & \mathbf{0} & \mathbf{0} \\ \mathbf{0} & \mathbf{0} & \mathbf{I} \end{bmatrix}; \tag{7.28}$$

The unassembled stiffness equations of the structure are written as:

$$\begin{Bmatrix} \mathbf{P}_1^{(1)} \\ \mathbf{P}_2^{(1)} \\ \mathbf{P}_1^{(2)} \\ \mathbf{P}_2^{(2)} \\ \mathbf{P}_1^{(3)} \\ \mathbf{P}_2^{(3)} \end{Bmatrix} = \begin{bmatrix} \mathbf{K}_{11}^{(1)} & \mathbf{K}_{12}^{(1)} & \mathbf{0} & \mathbf{0} & \mathbf{0} & \mathbf{0} \\ \mathbf{K}_{21}^{(1)} & \mathbf{K}_{22}^{(1)} & \mathbf{0} & \mathbf{0} & \mathbf{0} & \mathbf{0} \\ \mathbf{0} & \mathbf{0} & \mathbf{K}_{11}^{(2)} & \mathbf{K}_{12}^{(2)} & \mathbf{0} & \mathbf{0} \\ \mathbf{0} & \mathbf{0} & \mathbf{K}_{21}^{(2)} & \mathbf{K}_{22}^{(2)} & \mathbf{0} & \mathbf{0} \\ \mathbf{0} & \mathbf{0} & \mathbf{0} & \mathbf{0} & \mathbf{K}_{11}^{(3)} & \mathbf{K}_{12}^{(3)} \\ \mathbf{0} & \mathbf{0} & \mathbf{0} & \mathbf{0} & \mathbf{K}_{21}^{(3)} & \mathbf{K}_{22}^{(3)} \end{bmatrix} \begin{Bmatrix} \mathbf{U}_1^{(1)} \\ \mathbf{U}_2^{(1)} \\ \mathbf{U}_1^{(2)} \\ \mathbf{U}_2^{(2)} \\ \mathbf{U}_1^{(3)} \\ \mathbf{U}_2^{(3)} \end{Bmatrix} + \begin{Bmatrix} \mathbf{P}_{F1}^{(1)} \\ \mathbf{P}_{F2}^{(1)} \\ \mathbf{P}_{F1}^{(2)} \\ \mathbf{P}_{F2}^{(2)} \\ \mathbf{P}_{F1}^{(3)} \\ \mathbf{P}_{F2}^{(3)} \end{Bmatrix} \text{ or } \mathbf{P} = \hat{\mathbf{K}}\Delta + \hat{\mathbf{P}}_F \tag{7.29}$$

The stiffness equations for each member are:

Member 1:

$$\begin{Bmatrix} \mathbf{P}_1^{(1)} \\ \mathbf{P}_2^{(1)} \end{Bmatrix} = \begin{bmatrix} \mathbf{K}_{11}^{(1)} & \mathbf{K}_{12}^{(1)} \\ \mathbf{K}_{21}^{(1)} & \mathbf{K}_{22}^{(1)} \end{bmatrix} \begin{Bmatrix} \mathbf{U}_1^{(1)} \\ \mathbf{U}_2^{(1)} \end{Bmatrix} + \begin{Bmatrix} \mathbf{P}_{F1}^{(1)} \\ \mathbf{P}_{F2}^{(1)} \end{Bmatrix} \text{ or } \mathbf{P}^{(1)} = \mathbf{K}^{(1)}\mathbf{U}^{(1)} + \mathbf{P}_F^{(1)} \tag{7.30}$$

Member 2:

$$\begin{Bmatrix} \mathbf{P}_1^{(2)} \\ \mathbf{P}_2^{(2)} \end{Bmatrix} = \begin{bmatrix} \mathbf{K}_{11}^{(2)} & \mathbf{K}_{12}^{(2)} \\ \mathbf{K}_{21}^{(2)} & \mathbf{K}_{22}^{(2)} \end{bmatrix} \begin{bmatrix} \mathbf{K}_1^{(2)} \\ \mathbf{U}_2^{(2)} \end{bmatrix} + \begin{Bmatrix} \mathbf{P}_{F1}^{(2)} \\ \mathbf{P}_{F2}^{(2)} \end{Bmatrix} \text{ or } \mathbf{P}^{(2)} = \mathbf{K}^{(2)} \mathbf{U}^{(2)} + \mathbf{P}_F^{(2)} \quad (7.31)$$

Member 3:

$$\begin{Bmatrix} \mathbf{P}_1^{(3)} \\ \mathbf{P}_2^{(3)} \end{Bmatrix} = \begin{bmatrix} \mathbf{K}_{11}^{(3)} & \mathbf{K}_{12}^{(3)} \\ \mathbf{K}_{21}^{(3)} & \mathbf{K}_{22}^{(3)} \end{bmatrix} \begin{bmatrix} \mathbf{U}_1^{(3)} \\ \mathbf{U}_2^{(3)} \end{bmatrix} + \begin{Bmatrix} \mathbf{P}_{F1}^{(3)} \\ \mathbf{P}_{F2}^{(3)} \end{Bmatrix} \text{ or } \mathbf{P}^{(3)} = \mathbf{K}^{(3)} \mathbf{U}^{(3)} + \mathbf{P}_F^{(3)} \quad (7.32)$$

From Eq. (7.22), the augmented stiffness matrix for each member is:
Member 1:

$$\breve{\mathbf{K}}^{(1)} = \mathbf{G}^{(1)^T} \mathbf{K}^{(1)} \mathbf{G}^{(1)} = \begin{bmatrix} \mathbf{I} & \mathbf{0} \\ \mathbf{0} & \mathbf{I} \\ \mathbf{0} & \mathbf{0} \end{bmatrix} \begin{bmatrix} \mathbf{K}_{11}^{(1)} & \mathbf{K}_{12}^{(1)} \\ \mathbf{K}_{21}^{(1)} & \mathbf{K}_{22}^{(1)} \end{bmatrix} \begin{bmatrix} \mathbf{I} & \mathbf{0} & \mathbf{0} \\ \mathbf{0} & \mathbf{I} & \mathbf{0} \end{bmatrix} = \begin{bmatrix} \mathbf{K}_{11}^{(1)} & \mathbf{K}_{12}^{(1)} & \mathbf{0} \\ \mathbf{K}_{21}^{(1)} & \mathbf{K}_{22}^{(1)} & \mathbf{0} \\ \mathbf{0} & \mathbf{0} & \mathbf{0} \end{bmatrix} \quad (7.33)$$

Member 2:

$$\breve{\mathbf{K}}^{(2)} = \mathbf{G}^{(2)^T} \mathbf{K}^{(2)} \mathbf{G}^{(2)} = \begin{bmatrix} \mathbf{0} & \mathbf{0} \\ \mathbf{I} & \mathbf{0} \\ \mathbf{0} & \mathbf{I} \end{bmatrix} \begin{bmatrix} \mathbf{K}_{11}^{(2)} & \mathbf{K}_{12}^{(2)} \\ \mathbf{K}_{21}^{(2)} & \mathbf{K}_{22}^{(2)} \end{bmatrix} \begin{bmatrix} \mathbf{0} & \mathbf{I} & \mathbf{0} \\ \mathbf{0} & \mathbf{0} & \mathbf{I} \end{bmatrix} = \begin{bmatrix} \mathbf{0} & \mathbf{0} & \mathbf{0} \\ \mathbf{0} & \mathbf{K}_{11}^{(2)} & \mathbf{K}_{12}^{(2)} \\ \mathbf{0} & \mathbf{K}_{21}^{(2)} & \mathbf{K}_{22}^{(2)} \end{bmatrix} \quad (7.34)$$

Member 3:

$$\breve{\mathbf{K}}^{(3)} = \mathbf{G}^{(3)^T} \mathbf{K}^{(3)} \mathbf{G}^{(3)} = \begin{bmatrix} \mathbf{I} & \mathbf{0} \\ \mathbf{0} & \mathbf{0} \\ \mathbf{0} & \mathbf{I} \end{bmatrix} \begin{bmatrix} \mathbf{K}_{11}^{(3)} & \mathbf{K}_{12}^{(3)} \\ \mathbf{K}_{21}^{(3)} & \mathbf{K}_{22}^{(3)} \end{bmatrix} \begin{bmatrix} \mathbf{I} & \mathbf{0} & \mathbf{0} \\ \mathbf{0} & \mathbf{0} & \mathbf{I} \end{bmatrix} = \begin{bmatrix} \mathbf{K}_{11}^{(3)} & \mathbf{0} & \mathbf{K}_{12}^{(3)} \\ \mathbf{0} & \mathbf{0} & \mathbf{0} \\ \mathbf{K}_{21}^{(3)} & \mathbf{0} & \mathbf{K}_{22}^{(3)} \end{bmatrix} \quad (7.35)$$

From Eq. (7.23), the augmented restrained nodal force vector for each member is:
Member 1:

$$\breve{\mathbf{R}}^{(1)} = \mathbf{G}^{(1)^T} \mathbf{P}_F^{(1)} = \begin{bmatrix} \mathbf{I} & \mathbf{0} \\ \mathbf{0} & \mathbf{I} \\ \mathbf{0} & \mathbf{0} \end{bmatrix} \begin{Bmatrix} \mathbf{P}_{F1}^{(1)} \\ \mathbf{P}_{F2}^{(1)} \end{Bmatrix} \rightarrow \breve{\mathbf{R}}^{(1)} = \begin{Bmatrix} \mathbf{P}_{F1}^{(1)} \\ \mathbf{P}_{F2}^{(1)} \\ \mathbf{0} \end{Bmatrix} \begin{matrix} 1 \\ 2 \\ 3 \end{matrix} \quad (7.36)$$

Member 2:

$$\breve{\mathbf{R}}^{(2)} = \mathbf{G}^{(2)^T} \mathbf{P}_F^{(2)} = \begin{bmatrix} \mathbf{0} & \mathbf{0} \\ \mathbf{I} & \mathbf{0} \\ \mathbf{0} & \mathbf{I} \end{bmatrix} \begin{Bmatrix} \mathbf{P}_{F1}^{(2)} \\ \mathbf{P}_{F2}^{(2)} \end{Bmatrix} \rightarrow \breve{\mathbf{R}}^{(2)} = \begin{Bmatrix} \mathbf{0} \\ \mathbf{P}_{F1}^{(2)} \\ \mathbf{P}_{F2}^{(2)} \end{Bmatrix} \begin{matrix} 1 \\ 2 \\ 3 \end{matrix} \quad (7.37)$$

Member 3:

$$\breve{\mathbf{R}}^{(3)} = {\mathbf{G}^{(3)}}^T \mathbf{P}_F^{(3)} = \begin{bmatrix} \mathbf{I} & \mathbf{0} \\ \mathbf{0} & \mathbf{0} \\ \mathbf{0} & \mathbf{I} \end{bmatrix} \begin{Bmatrix} \mathbf{P}_{F1}^{(3)} \\ \mathbf{P}_{F2}^{(3)} \end{Bmatrix} \rightarrow \breve{\mathbf{R}}^{(3)} = \begin{Bmatrix} \mathbf{P}_{F1}^{(3)} \\ \mathbf{0} \\ \mathbf{P}_{F2}^{(3)} \end{Bmatrix} \begin{matrix} 1 \\ 2 \\ 3 \end{matrix} \qquad (7.38)$$

Member 1 having its ends $i$ and $j$ connected to structural nodes 1 and 2, respectively, is considered. It is observed from Eq. (7.33) that submatrices $\mathbf{K}_{11}^{(1)}$, $\mathbf{K}_{12}^{(1)}$, $\mathbf{K}_{21}^{(1)}$, and $\mathbf{K}_{22}^{(1)}$ appear at locations associated with nodes 1 and 2 of the structural stiffness matrix. Similar conclusions can be drawn for members 2 and 3.

Generally speaking:

For a member $k$ having its ends $i$ and $j$ connected to structural nodes $M$ and $N$, respectively, the submatrices $\mathbf{K}_{11}^{(k)}$, $\mathbf{K}_{12}^{(k)}$, $\mathbf{K}_{21}^{(k)}$, and $\mathbf{K}_{22}^{(k)}$ appear at locations associated with nodes $M$ and $N$ of the structural stiffness matrix, as shown in Eq. (7.39).

$$\breve{\mathbf{K}}^{(k)} = {\mathbf{G}^{(k)}}^T \mathbf{K}^{(k)} \mathbf{G}^{(k)} = \begin{bmatrix} & \overset{M}{\vdots} & & \overset{N}{\vdots} & \\ \cdots & \mathbf{K}_{11}^{(k)} & \cdots & \mathbf{K}_{12}^{(k)} & \cdots \\ & \vdots & & \vdots & \\ & \vdots & & \vdots & \\ & \vdots & & \vdots & \\ \cdots & \mathbf{K}_{21}^{(k)} & \cdots & \mathbf{K}_{22}^{(k)} & \cdots \\ & \vdots & & \vdots & \end{bmatrix} \begin{matrix} \\ M \\ \\ \\ \\ N \\ \end{matrix} \qquad (7.39)$$

From Eq. (7.22), the complete stiffness matrix of the frame in Figure 7.1 is:

$$\mathbf{K} = \sum_{k=1}^{3} \breve{\mathbf{K}}^{(k)} = \sum_{k=1}^{3} {\mathbf{G}^{(k)}}^T \mathbf{K}^{(k)} \mathbf{G}^{(k)} = \begin{bmatrix} \mathbf{K}_{11}^{(1)} + \mathbf{K}_{11}^{(3)} & \mathbf{K}_{12}^{(1)} & \mathbf{K}_{12}^{(3)} \\ \mathbf{K}_{21}^{(1)} & \mathbf{K}_{22}^{(1)} + \mathbf{K}_{11}^{(2)} & \mathbf{K}_{12}^{(2)} \\ \mathbf{K}_{21}^{(3)} & \mathbf{K}_{21}^{(2)} & \mathbf{K}_{22}^{(2)} + \mathbf{K}_{22}^{(3)} \end{bmatrix} \qquad (7.40)$$

Similarly, it is observed from Eq. (7.36) that subvectors $\mathbf{P}_{F1}^{(1)}$ and $\mathbf{P}_{F2}^{(1)}$ appear at locations associated with nodes 1 and 2 of the restrained nodal force vector. A similar conclusion can be drawn for members 2 and 3.

Generally speaking:

For a member $k$ having its ends $i$ and $j$ connected to structural nodes $M$ and $N$, respectively, the subvectors $\mathbf{P}_{F1}^{(k)}$ and $\mathbf{P}_{F2}^{(k)}$ appear at locations associated with nodes $M$ and $N$ of the restrained nodal force vector as shown in Eq. (7.41).

$$\breve{\mathbf{R}}^{(k)} = {\mathbf{G}^{(k)}}^T \mathbf{P}_F^{(k)} = \begin{bmatrix} \cdots & \overset{M}{\mathbf{P}_{F1}^{(k)}} & \cdots & \cdots & \cdots & \overset{N}{\mathbf{P}_{F2}^{(k)}} & \cdots \end{bmatrix}^T \qquad (7.41)$$

From Eq. (7.23), the restrained nodal force vector of the frame in Figure 7.1 is:

$$\mathbf{R} = \sum_{k=1}^{3} \breve{\mathbf{R}}^{(k)} = \sum_{k=1}^{3} {\mathbf{G}^{(k)}}^T \mathbf{P}_F^{(k)} = \begin{bmatrix} \mathbf{P}_{F1}^{(1)} + \mathbf{P}_{F1}^{(3)} & \mathbf{P}_{F2}^{(1)} + \mathbf{P}_{F1}^{(2)} & \mathbf{P}_{F2}^{(2)} + \mathbf{P}_{F2}^{(3)} \end{bmatrix}^T \qquad (7.42)$$

## Example 7.1

For the frame structure shown in Figure 7.3, establish its complete stiffness matrix and equivalent load vector.

Given that the local stiffness matrix for members 1 and 2 is:

$$\overline{\mathbf{K}}^{(1)} = \overline{\mathbf{K}}^{(2)} = \frac{IE}{125} \begin{bmatrix} 1000 & 0 & 0 & -1000 & 0 & 0 \\ 0 & 12 & 30 & 0 & -12 & 30 \\ 0 & 30 & 100 & 0 & -30 & 50 \\ -1000 & 0 & 0 & 1000 & 0 & 0 \\ 0 & -12 & -30 & 0 & 12 & -30 \\ 0 & 30 & 50 & 0 & -30 & 100 \end{bmatrix}$$

**Solution:** The local reference axes for members 1 and 2 are shown in Figure 7.4a while the DOF, member, and nodal numbering systems and member mapping vectors are shown in Figure 7.4b:

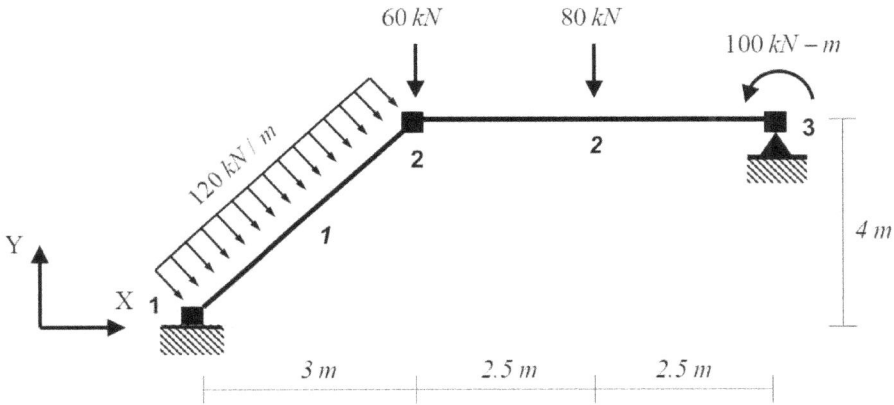

**FIGURE 7.3** Example 7.1.

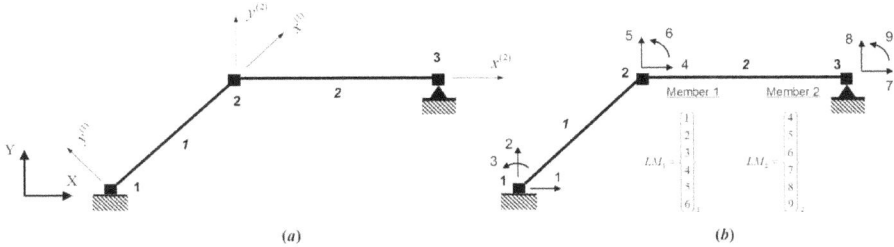

**FIGURE 7.4** Example 7.1 (Continued).

Based on mapping vectors, the structural equilibrium mapping matrix **H** is:

$$
\mathbf{H} =
\begin{bmatrix}
1 & 0 & 0 & 0 & 0 & 0 & 0 & 0 & 0 & 0 & 0 & 0 \\
0 & 1 & 0 & 0 & 0 & 0 & 0 & 0 & 0 & 0 & 0 & 0 \\
0 & 0 & 1 & 0 & 0 & 0 & 0 & 0 & 0 & 0 & 0 & 0 \\
0 & 0 & 0 & 1 & 0 & 0 & 1 & 0 & 0 & 0 & 0 & 0 \\
0 & 0 & 0 & 0 & 1 & 0 & 0 & 1 & 0 & 0 & 0 & 0 \\
0 & 0 & 0 & 0 & 0 & 1 & 0 & 0 & 1 & 0 & 0 & 0 \\
0 & 0 & 0 & 0 & 0 & 0 & 0 & 0 & 0 & 1 & 0 & 0 \\
0 & 0 & 0 & 0 & 0 & 0 & 0 & 0 & 0 & 0 & 1 & 0 \\
0 & 0 & 0 & 0 & 0 & 0 & 0 & 0 & 0 & 0 & 0 & 1
\end{bmatrix}
$$

Based on mapping vectors, the structural compatibility mapping matrix **G** is:

$$
\mathbf{G} =
\begin{bmatrix}
1 & 0 & 0 & 0 & 0 & 0 & 0 & 0 & 0 \\
0 & 1 & 0 & 0 & 0 & 0 & 0 & 0 & 0 \\
0 & 0 & 1 & 0 & 0 & 0 & 0 & 0 & 0 \\
0 & 0 & 0 & 1 & 0 & 0 & 0 & 0 & 0 \\
0 & 0 & 0 & 0 & 1 & 0 & 0 & 0 & 0 \\
0 & 0 & 0 & 0 & 0 & 1 & 0 & 0 & 0 \\
0 & 0 & 0 & 1 & 0 & 0 & 0 & 0 & 0 \\
0 & 0 & 0 & 0 & 1 & 0 & 0 & 0 & 0 \\
0 & 0 & 0 & 0 & 0 & 1 & 0 & 0 & 0 \\
0 & 0 & 0 & 0 & 0 & 0 & 1 & 0 & 0 \\
0 & 0 & 0 & 0 & 0 & 0 & 0 & 1 & 0 \\
0 & 0 & 0 & 0 & 0 & 0 & 0 & 0 & 1
\end{bmatrix}
$$

## Step I

The nodal equilibrium equations of the frame are written in the matrix form as:

$$
\begin{Bmatrix}
F_1 \\ F_2 \\ F_3 \\ 0 \\ -60 \\ 0 \\ F_7 \\ F_8 \\ 100
\end{Bmatrix}
=
\begin{bmatrix}
1 & 0 & 0 & 0 & 0 & 0 & 0 & 0 & 0 & 0 & 0 & 0 \\
0 & 1 & 0 & 0 & 0 & 0 & 0 & 0 & 0 & 0 & 0 & 0 \\
0 & 0 & 1 & 0 & 0 & 0 & 0 & 0 & 0 & 0 & 0 & 0 \\
0 & 0 & 0 & 1 & 0 & 0 & 1 & 0 & 0 & 0 & 0 & 0 \\
0 & 0 & 0 & 0 & 1 & 0 & 0 & 1 & 0 & 0 & 0 & 0 \\
0 & 0 & 0 & 0 & 0 & 1 & 0 & 0 & 1 & 0 & 0 & 0 \\
0 & 0 & 0 & 0 & 0 & 0 & 0 & 0 & 0 & 1 & 0 & 0 \\
0 & 0 & 0 & 0 & 0 & 0 & 0 & 0 & 0 & 0 & 1 & 0 \\
0 & 0 & 0 & 0 & 0 & 0 & 0 & 0 & 0 & 0 & 0 & 1
\end{bmatrix}
\begin{Bmatrix}
P_1^{(1)} \\ P_2^{(1)} \\ P_3^{(1)} \\ P_4^{(1)} \\ P_5^{(1)} \\ P_6^{(1)} \\ P_1^{(2)} \\ P_2^{(2)} \\ P_3^{(2)} \\ P_4^{(2)} \\ P_5^{(2)} \\ P_6^{(2)}
\end{Bmatrix}
\quad\text{or}\quad \mathbf{F} = \mathbf{HP}
$$

## Step II

The nodal compatibility equations of the frame are written in the matrix form as:

$$
\begin{Bmatrix} U_1^{(1)} \\ U_2^{(1)} \\ U_3^{(1)} \\ U_4^{(1)} \\ U_5^{(1)} \\ U_6^{(1)} \\ U_1^{(2)} \\ U_2^{(2)} \\ U_3^{(2)} \\ U_4^{(2)} \\ U_5^{(2)} \\ U_6^{(2)} \end{Bmatrix}
=
\begin{bmatrix}
1 & 0 & 0 & 0 & 0 & 0 & 0 & 0 & 0 \\
0 & 1 & 0 & 0 & 0 & 0 & 0 & 0 & 0 \\
0 & 0 & 1 & 0 & 0 & 0 & 0 & 0 & 0 \\
0 & 0 & 0 & 1 & 0 & 0 & 0 & 0 & 0 \\
0 & 0 & 0 & 0 & 1 & 0 & 0 & 0 & 0 \\
0 & 0 & 0 & 0 & 0 & 1 & 0 & 0 & 0 \\
0 & 0 & 0 & 1 & 0 & 0 & 0 & 0 & 0 \\
0 & 0 & 0 & 0 & 1 & 0 & 0 & 0 & 0 \\
0 & 0 & 0 & 0 & 0 & 1 & 0 & 0 & 0 \\
0 & 0 & 0 & 0 & 0 & 0 & 1 & 0 & 0 \\
0 & 0 & 0 & 0 & 0 & 0 & 0 & 1 & 0 \\
0 & 0 & 0 & 0 & 0 & 0 & 0 & 0 & 1
\end{bmatrix}
\begin{Bmatrix} U_1 \\ U_2 \\ U_3 \\ U_4 \\ U_5 \\ U_6 \\ U_7 \\ U_8 \\ U_9 \end{Bmatrix}
\qquad \text{or} \qquad \Delta = \mathbf{GU}
$$

## Step III

The fixed-end force vector and local stiffness matrix for each member are set up.

Member 1 (Figure 7.5a):

$$
\overline{\mathbf{P}}_F^{(1)} = \lfloor 0 \quad 300 \quad 250 \quad 0 \quad 300 \quad -250 \rfloor^T
$$

$$
\overline{\mathbf{K}}^{(1)} = \frac{IE}{125}
\begin{bmatrix}
1000 & 0 & 0 & -1000 & 0 & 0 \\
0 & 12 & 30 & 0 & -12 & 30 \\
0 & 30 & 100 & 0 & -30 & 50 \\
-1000 & 0 & 0 & 1000 & 0 & 0 \\
0 & -12 & -30 & 0 & 12 & -30 \\
0 & 30 & 50 & 0 & -30 & 100
\end{bmatrix}
$$

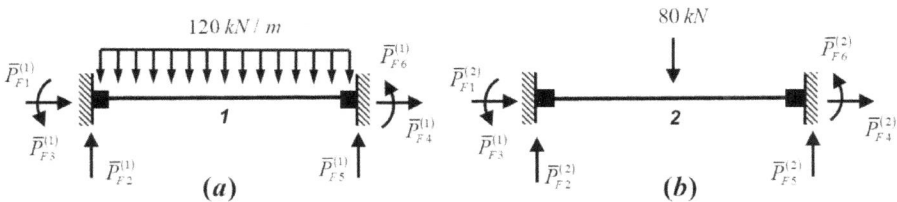

**FIGURE 7.5**  Example 7.1 (Continued).

Member 2 (Figure 7.5*b*):

$$\bar{\mathbf{P}}_F^{(2)} = \lfloor 0 \quad 40 \quad 50 \quad 0 \quad 40 \quad -50 \rfloor^T$$

$$\bar{\mathbf{K}}^{(2)} = \frac{IE}{125} \begin{bmatrix} 1000 & 0 & 0 & -1000 & 0 & 0 \\ 0 & 12 & 30 & 0 & -12 & 30 \\ 0 & 30 & 100 & 0 & -30 & 50 \\ -1000 & 0 & 0 & 1000 & 0 & 0 \\ 0 & -12 & -30 & 0 & 12 & -30 \\ 0 & 30 & 50 & 0 & -30 & 100 \end{bmatrix}$$

## STEP IV

The stiffness matrix and fixed-end force vector for each member are transformed from local to global reference systems.

## Member 1:

$$\Gamma_{ROT}^{(1)} = \begin{bmatrix} 0.6 & 0.8 & 0 & 0 & 0 & 0 \\ -0.8 & 0.6 & 0 & 0 & 0 & 0 \\ 0 & 0 & 1 & 0 & 0 & 0 \\ 0 & 0 & 0 & 0.6 & 0.8 & 0 \\ 0 & 0 & 0 & -0.8 & 0.6 & 0 \\ 0 & 0 & 0 & 0 & 0 & 1 \end{bmatrix}$$

$$\mathbf{K}^{(1)} = \Gamma_{ROT}^{(1)}{}^T \bar{\mathbf{K}}^{(1)} \Gamma_{ROT}^{(1)} = \frac{IE}{125} \begin{bmatrix} 367.68 & 474.24 & -24 & -367.68 & 474.24 & -24 \\ 474.24 & 644.32 & 18 & -474.24 & -644.32 & 18 \\ -24 & 18 & 100 & 24 & -18 & 50 \\ -367.68 & -474.24 & 24 & 367.68 & 474.24 & 24 \\ 474.24 & -644.32 & -18 & 474.24 & 644.32 & -18 \\ -24 & 18 & 50 & 24 & -18 & 100 \end{bmatrix}$$

$$\mathbf{P}_F^{(1)} = \Gamma_{ROT}^{(1)}{}^T \bar{\mathbf{P}}_F^{(1)} = \lfloor -240 \quad 180 \quad 250 \quad -240 \quad 180 \quad -250 \rfloor^T$$

## Member 2:

$$\Gamma_{ROT}^{(2)} = \begin{bmatrix} 1 & 0 & 0 & 0 & 0 & 0 \\ 0 & 1 & 0 & 0 & 0 & 0 \\ 0 & 0 & 1 & 0 & 0 & 0 \\ 0 & 0 & 0 & 1 & 0 & 0 \\ 0 & 0 & 0 & 0 & 1 & 0 \\ 0 & 0 & 0 & 0 & 0 & 1 \end{bmatrix}$$

$$\mathbf{K}^{(2)} = \Gamma_{ROT}^{(2)}{}^T \, \overline{\mathbf{K}}^{(2)} \Gamma_{ROT}^{(2)} = \frac{IE}{125} \begin{bmatrix} 1000 & 0 & 0 & -1000 & 0 & 0 \\ 0 & 12 & 30 & 0 & -12 & 30 \\ 0 & 30 & 100 & 0 & -30 & 50 \\ -1000 & 0 & 0 & 1000 & 0 & 0 \\ 0 & -12 & -30 & 0 & 12 & -30 \\ 0 & 30 & 50 & 0 & -30 & 100 \end{bmatrix}$$

$$\mathbf{P}_F^{(2)} = \Gamma_{ROT}^{(2)}{}^T \, \overline{\mathbf{P}}_F^{(2)} = \lfloor 0 \quad 40 \quad 50 \quad 0 \quad 40 \quad -50 \rfloor^T$$

Therefore, the unassembled fixed-end force vector is:

$$\hat{\mathbf{P}}_F = \lfloor -240 \quad 180 \quad 250 \quad -240 \quad 180 \quad -250 \quad 0 \quad 40 \quad 50 \quad 0 \quad 40 \quad -50 \rfloor^T$$

The unassembled structural stiffness matrix is:

$$\hat{\mathbf{K}} = \begin{bmatrix} \mathbf{K}^{(1)} & \mathbf{0} \\ \mathbf{0} & \mathbf{K}^{(2)} \end{bmatrix}$$

## STEP V

The applied nodal load vector is:

$$\mathbf{F} = \lfloor F_1 \quad F_2 \quad F_3 \quad 0 \quad -60 \quad 0 \quad F_7 \quad F_8 \quad 100 \rfloor^T$$

The equivalent nodal load vector (Figure 7.6) is:

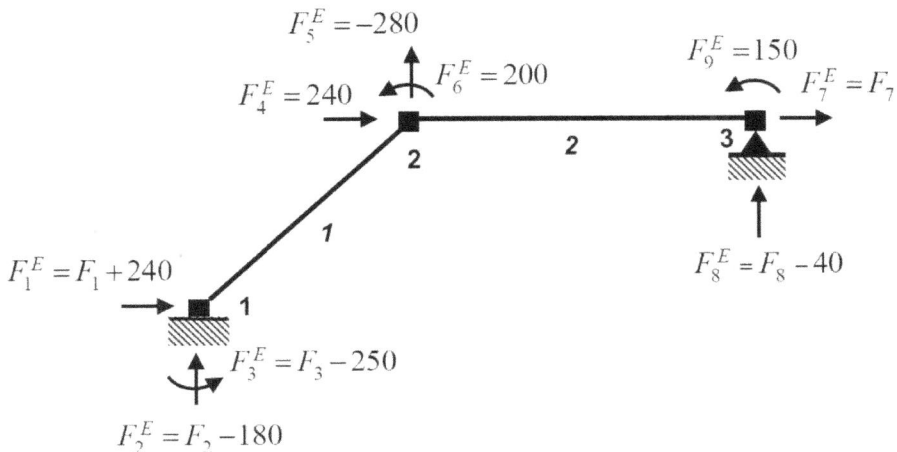

FIGURE 7.6   Example 7.1 (Continued).

$$\mathbf{F}^E = \mathbf{F} - \mathbf{H}\hat{\mathbf{P}}_F = \begin{Bmatrix} F_1 \\ F_2 \\ F_3 \\ 0 \\ -60 \\ 0 \\ F_7 \\ F_8 \\ 100 \end{Bmatrix} - \begin{bmatrix} 1 & 0 & 0 & 0 & 0 & 0 & 0 & 0 & 0 & 0 & 0 & 0 \\ 0 & 1 & 0 & 0 & 0 & 0 & 0 & 0 & 0 & 0 & 0 & 0 \\ 0 & 0 & 1 & 0 & 0 & 0 & 0 & 0 & 0 & 0 & 0 & 0 \\ 0 & 0 & 0 & 1 & 0 & 0 & 1 & 0 & 0 & 0 & 0 & 0 \\ 0 & 0 & 0 & 0 & 1 & 0 & 0 & 1 & 0 & 0 & 0 & 0 \\ 0 & 0 & 0 & 0 & 0 & 1 & 0 & 0 & 1 & 0 & 0 & 0 \\ 0 & 0 & 0 & 0 & 0 & 0 & 0 & 0 & 0 & 1 & 0 & 0 \\ 0 & 0 & 0 & 0 & 0 & 0 & 0 & 0 & 0 & 0 & 1 & 0 \\ 0 & 0 & 0 & 0 & 0 & 0 & 0 & 0 & 0 & 0 & 0 & 1 \end{bmatrix} \begin{Bmatrix} -240 \\ 180 \\ 250 \\ -240 \\ 180 \\ -250 \\ 0 \\ 40 \\ 50 \\ 0 \\ 40 \\ -50 \end{Bmatrix} = \begin{Bmatrix} F_1 + 240 \\ F_2 - 180 \\ F_3 - 250 \\ 240 \\ -280 \\ 200 \\ F_7 \\ F_8 - 40 \\ 150 \end{Bmatrix}$$

## STEP VI

The complete stiffness matrix of the frame is assembled as:

$$\mathbf{K} = \mathbf{G}^T \hat{\mathbf{K}} \mathbf{G} = \frac{IE}{125} \begin{bmatrix} 367.68 & 474.24 & -24 & -367.68 & -474.24 & -24 & 0 & 0 & 0 \\ 474.24 & 644.32 & 18 & -474.24 & -644.32 & 18 & 0 & 0 & 0 \\ -24 & 18 & 100 & 24 & -18 & 50 & 0 & 0 & 0 \\ -367.68 & -474.24 & 24 & 1367.68 & 474.24 & 24 & -1000 & 0 & 0 \\ -474.24 & -644.32 & -18 & 474.24 & 656.32 & 12 & 0 & -12 & 30 \\ -24 & 18 & 50 & 24 & 12 & 200 & 0 & -30 & 50 \\ 0 & 0 & 0 & -1000 & 0 & 0 & 1000 & 0 & 0 \\ 0 & 0 & 0 & 0 & -12 & -30 & 0 & 12 & -30 \\ 0 & 0 & 0 & 0 & 30 & 50 & 0 & -30 & 100 \end{bmatrix}$$

The complete stiffness equations of the frame are expressed in the matrix form as:

$$\begin{Bmatrix} F_1 + 240 \\ F_2 - 180 \\ F_3 - 250 \\ 240 \\ -280 \\ 200 \\ F_7 \\ F_8 - 40 \\ 150 \end{Bmatrix} = \frac{IE}{125} \begin{bmatrix} 367.68 & 474.24 & -24 & -367.68 & -474.24 & -24 & 0 & 0 & 0 \\ 474.24 & 644.32 & 18 & -474.24 & -644.32 & 18 & 0 & 0 & 0 \\ -24 & 18 & 100 & 24 & -18 & 50 & 0 & 0 & 0 \\ -367.68 & -474.24 & 24 & 1367.68 & 474.24 & 24 & -1000 & 0 & 0 \\ -474.24 & -644.32 & -18 & 474.24 & 656.32 & 12 & 0 & -12 & 30 \\ -24 & 18 & 50 & 24 & 12 & 200 & 0 & -30 & 50 \\ 0 & 0 & 0 & -1000 & 0 & 0 & 1000 & 0 & 0 \\ 0 & 0 & 0 & 0 & -12 & -30 & 0 & 12 & -30 \\ 0 & 0 & 0 & 0 & 30 & 50 & 0 & -30 & 100 \end{bmatrix} \begin{Bmatrix} U_1 = 0 \\ U_2 = 0 \\ U_3 = 0 \\ U_4 \\ U_5 \\ U_6 \\ U_7 = 0 \\ U_8 = 0 \\ U_9 \end{Bmatrix}$$

or    $\mathbf{F}^E = \mathbf{K}\mathbf{U}$

This system of equations contains 9 unknowns (4 nodal displacement plus 5 reactive forces) with 9 independent equations. However, the stiffness matrix $\mathbf{K}$ is singular since no support conditions are imposed. After rearranging rows and columns, this system of equations can be partitioned following free and constrained degrees of freedom to solve unknown quantities. It is interesting to observe that unknown forces on the left-hand side correspond to known displacements on the right-hand side and inversely, known forces on the left-hand side correspond to unknown displacements on the right-hand side. This simply implies that at any given degree of freedom, either force or displacement has to be prescribed.

## EXAMPLE 7.2

For the truss structure shown in Figure 7.7,

- Establish its complete stiffness matrix.
- Establish its equivalent nodal load vector accounting for an increase in temperature $\Delta T = 10\sqrt{2} \, C^0$ of member 2.

Assume that Young's modulus of the material is $200 \, GPa$ and the coefficient of thermal expansion is $6.5 \times 10^{-6} / C^0$.

**Solution:** The local reference, DOF, member, and nodal numbering systems and member mapping vectors are shown in Figure 7.8.

## STEP I

Based on each member mapping vector, the equilibrium mapping matrix for each member is:

$$\mathbf{H}^{(1)} = \begin{bmatrix} 0 & 0 & 1 & 0 \\ 0 & 0 & 0 & 1 \\ 0 & 0 & 0 & 0 \\ 0 & 0 & 0 & 0 \\ 1 & 0 & 0 & 0 \\ 0 & 1 & 0 & 0 \end{bmatrix}; \quad \mathbf{H}^{(2)} = \begin{bmatrix} 0 & 0 & 0 & 0 \\ 0 & 0 & 0 & 0 \\ 0 & 0 & 1 & 0 \\ 0 & 0 & 0 & 1 \\ 1 & 0 & 0 & 0 \\ 0 & 1 & 0 & 0 \end{bmatrix}; \quad \mathbf{H}^{(3)} = \begin{bmatrix} 1 & 0 & 0 & 0 \\ 0 & 1 & 0 & 0 \\ 0 & 0 & 1 & 0 \\ 0 & 0 & 0 & 1 \\ 0 & 0 & 0 & 0 \\ 0 & 0 & 0 & 0 \end{bmatrix};$$

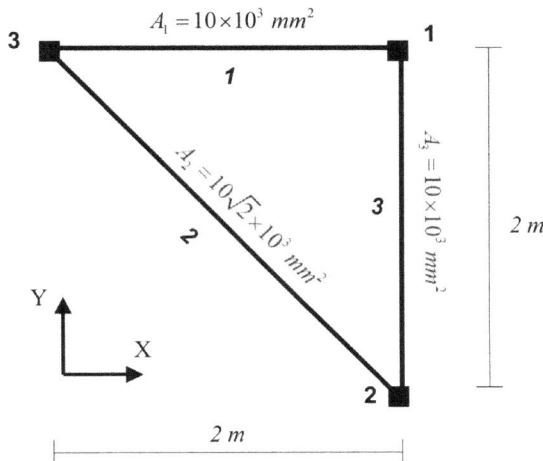

**FIGURE 7.7** Example 7.2.

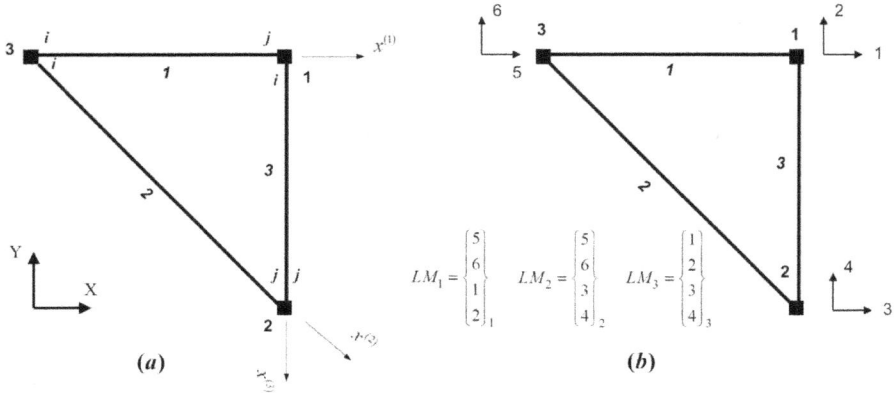

**FIGURE 7.8**  Example 7.2 (Continued).

The structural equilibrium mapping matrix **H** is:

$$\mathbf{H} = \begin{bmatrix} \mathbf{H}^{(1)} & \mathbf{H}^{(2)} & \mathbf{H}^{(3)} \end{bmatrix}$$

## STEP II

Based on each member mapping vector, the compatibility mapping matrix for each member is:

$$\mathbf{G}^{(1)} = \begin{bmatrix} 0 & 0 & 0 & 0 & 1 & 0 \\ 0 & 0 & 0 & 0 & 0 & 1 \\ 1 & 0 & 0 & 0 & 0 & 0 \\ 0 & 1 & 0 & 0 & 0 & 0 \end{bmatrix}; \mathbf{G}^{(2)} = \begin{bmatrix} 0 & 0 & 0 & 0 & 1 & 0 \\ 0 & 0 & 0 & 0 & 0 & 1 \\ 0 & 0 & 1 & 0 & 0 & 0 \\ 0 & 0 & 0 & 1 & 0 & 0 \end{bmatrix}; \mathbf{G}^{(3)} = \begin{bmatrix} 1 & 0 & 0 & 0 & 0 & 0 \\ 0 & 1 & 0 & 0 & 0 & 0 \\ 0 & 0 & 1 & 0 & 0 & 0 \\ 0 & 0 & 0 & 1 & 0 & 0 \end{bmatrix};$$

## STEP III

The stiffness matrix for each member is transformed from local to global reference systems.
    Member 1:

$$\Gamma_{ROT}^{(1)} = \begin{bmatrix} 1 & 0 & 0 & 0 \\ 0 & 0 & 1 & 0 \end{bmatrix}$$

The local member stiffness matrix:

$$\bar{\mathbf{K}}^{(1)} = 10^6 \begin{bmatrix} 1 & -1 \\ -1 & 1 \end{bmatrix}$$

The global member stiffness matrix:

$$\mathbf{K}^{(1)} = \Gamma_{ROT}^{(1)}{}^T \bar{\mathbf{K}}^{(1)} \Gamma_{ROT}^{(1)} = 10^6 \begin{bmatrix} 1 & 0 & -1 & 0 \\ 0 & 0 & 0 & 0 \\ -1 & 0 & 1 & 0 \\ 0 & 0 & 0 & 0 \end{bmatrix}$$

Member 2:

$$\Gamma_{ROT}^{(2)} = \begin{bmatrix} 1/\sqrt{2} & -1/\sqrt{2} & 0 & 0 \\ 0 & 0 & 1/\sqrt{2} & -1/\sqrt{2} \end{bmatrix}$$

The local member stiffness matrix:

$$\overline{\mathbf{K}}^{(2)} = 10^6 \begin{bmatrix} 1 & -1 \\ -1 & 1 \end{bmatrix}$$

The global member stiffness matrix:

$$\mathbf{K}^{(2)} = \Gamma_{ROT}^{(2)}{}^T \overline{\mathbf{K}}^{(2)} \Gamma_{ROT}^{(2)} = 0.5 \times 10^6 \begin{bmatrix} 1 & -1 & -1 & 1 \\ -1 & 1 & 1 & -1 \\ -1 & 1 & 1 & -1 \\ 1 & -1 & -1 & 1 \end{bmatrix}$$

Member 3:

$$\Gamma_{ROT}^{(3)} = \begin{bmatrix} 0 & 1 & 0 & 0 \\ 0 & 0 & 0 & 1 \end{bmatrix}$$

The local member stiffness matrix:

$$\overline{\mathbf{K}}^{(3)} = 10^6 \begin{bmatrix} 1 & -1 \\ -1 & 1 \end{bmatrix}$$

The global member stiffness matrix:

$$\mathbf{K}^{(3)} = \Gamma_{ROT}^{(3)}{}^T \overline{\mathbf{K}}^{(3)} \Gamma_{ROT}^{(3)} = 10^6 \begin{bmatrix} 0 & 0 & 0 & 0 \\ 0 & 1 & 0 & -1 \\ 0 & 0 & 0 & 0 \\ 0 & -1 & 0 & 1 \end{bmatrix}$$

## STEP IV

The augmented stiffness matrix for each member is:

Member 1:

$$\widetilde{\overline{\mathbf{K}}}^{(1)} = \mathbf{G}^{(1)}{}^T \mathbf{K}^{(1)} \mathbf{G}^{(1)} = 10^6 \begin{bmatrix} 1 & 0 & 0 & 0 & -1 & 0 \\ 0 & 0 & 0 & 0 & 0 & 0 \\ 0 & 0 & 0 & 0 & 0 & 0 \\ 0 & 0 & 0 & 0 & 0 & 0 \\ -1 & 0 & 0 & 0 & 1 & 0 \\ 0 & 0 & 0 & 0 & 0 & 0 \end{bmatrix}$$

Member 2:

$$\breve{\mathbf{K}}^{(2)} = \mathbf{G}^{(2)^{T}} \mathbf{K}^{(2)} \mathbf{G}^{(2)} = 0.5 \times 10^{6} \begin{bmatrix} 0 & 0 & 0 & 0 & 0 & 0 \\ 0 & 0 & 0 & 0 & 0 & 0 \\ 0 & 0 & 1 & -1 & -1 & 1 \\ 0 & 0 & -1 & 1 & 1 & -1 \\ 0 & 0 & -1 & 1 & 1 & -1 \\ 0 & 0 & 1 & -1 & -1 & 1 \end{bmatrix}$$

Member 3:

$$\breve{\mathbf{K}}^{(3)} = \mathbf{G}^{(3)^{T}} \mathbf{K}^{(3)} \mathbf{G}^{(3)} = 10^{6} \begin{bmatrix} 0 & 0 & 0 & 0 & 0 & 0 \\ 0 & 1 & 0 & -1 & 0 & 0 \\ 0 & 0 & 0 & 0 & 0 & 0 \\ 0 & -1 & 0 & 1 & 0 & 0 \\ 0 & 0 & 0 & 0 & 0 & 0 \\ 0 & 0 & 0 & 0 & 0 & 0 \end{bmatrix}$$

## STEP V

The fixed-end force vector for member 2 is shown in Figure 7.9.

The local fixed-end force vector is:

$$\overline{\mathbf{P}}_{F}^{(2)} = \lfloor 260 \quad -260 \rfloor^{T}$$

The global fixed-end force vector is:

$$\mathbf{P}_{F}^{(2)} = \mathbf{\Gamma}_{ROT}^{(2)}{}^{T} \overline{\mathbf{P}}_{F}^{(2)} = \lfloor 183.85 \quad -183.85 \quad -183.85 \quad 183.85 \rfloor^{T}$$

## STEP VI

The augmented restrained nodal force vector for member 2 is:

$$\breve{\mathbf{R}}^{(2)} = \mathbf{G}^{(2)^{T}} \mathbf{P}_{F}^{(2)} = \lfloor 0 \quad 0 \quad -183.85 \quad 183.85 \quad 183.85 \quad -183.85 \rfloor^{T}$$

Therefore, the complete stiffness matrix of the truss is:

$$\mathbf{K} = \sum_{k=1}^{3} \breve{\mathbf{K}}^{(k)} = \sum_{k=1}^{3} \mathbf{G}^{(k)^{T}} \mathbf{K}^{(k)} \mathbf{G}^{(k)} = 0.5 \times 10^{6} \begin{bmatrix} 2 & 0 & 0 & 0 & -2 & 0 \\ 0 & 2 & 0 & -2 & 0 & 0 \\ 0 & 0 & 1 & -1 & -1 & 1 \\ 0 & -2 & -1 & 3 & 1 & -1 \\ -2 & 0 & -1 & 1 & 3 & -1 \\ 0 & 0 & 1 & -1 & -1 & 1 \end{bmatrix}$$

$$\overline{P}_{F1}^{(2)} = EA\alpha\Delta T = 260 \; kN \qquad\qquad \overline{P}_{F2}^{(2)} = -EA\alpha\Delta T = -260 \; kN$$

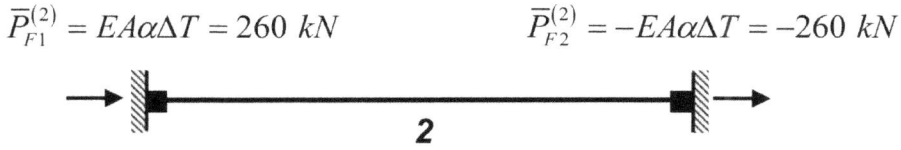

**FIGURE 7.9**   Example 7.2 (Continued).

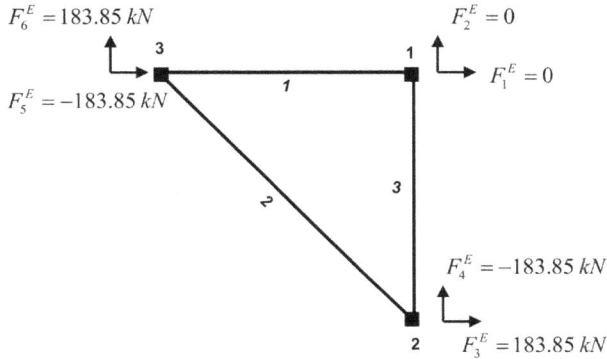

**FIGURE 7.10**   Example 7.2 (Continued).

The restrained nodal force vector for the structure is:

$$\mathbf{R} = \sum_{k=1}^{3} \tilde{\mathbf{R}}^{(k)} = \sum_{k=1}^{3} \mathbf{G}^{(k)^T} \mathbf{P}_F^{(k)} = \begin{bmatrix} 0 & 0 & -183.85 & 183.85 & 183.85 & -183.85 \end{bmatrix}^T$$

The equivalent nodal force vector is shown in Figure 7.10.

$$\mathbf{F}^E = \mathbf{F} - \mathbf{R} = \begin{bmatrix} 0 & 0 & 183.85 & -183.85 & -183.85 & 183.85 \end{bmatrix}^T$$

It is interesting to point out that this set of forces form the self equilibrated system.

## 7.4   DERIVATION OF THE COMPLETE STIFFNESS EQUATIONS OF A FRAME STRUCTURE: MEMBER-CODE ASSEMBLING TECHNIQUE

The assembling algorithm discussed in the previous section is crucial for understanding of the fundamental concepts of the direct stiffness method. Unfortunately, it cannot be implemented efficiently on a computer and consumes large computer storage and computational times for large structural systems. Therefore, it is rarely used in practice.

From Eq. (7.40), it is noticed that each influence coefficient of the structural stiffness matrix can be determined from the global stiffness coefficients of its constituent members, which are placed in the proper slot of the structural stiffness matrix. This way of assembling the structural stiffness matrix as well as the structural nodal force vector can be implemented efficiently on a computer and requires much less computer storage and computational time for large structural systems. This technique is referred to as "the member code technique" and was introduced by S. S. Tezcan in 1963.

Consider the node $k$ of a frame structure shown in Figure 7.11. It is assumed that there are three members ($a$, $b$, and $c$) attached to it. According to nodal degrees of freedom and member connectivity shown in Figure 7.11, mapping vectors for members $a$, $b$, and $c$ are:

$$
LM_a = \begin{Bmatrix} m-3 \\ m-2 \\ m-1 \\ m \\ m+1 \\ m+2 \end{Bmatrix}_a ; \quad
LM_b = \begin{Bmatrix} m+3 \\ m+4 \\ m+5 \\ m \\ m+1 \\ m+2 \end{Bmatrix}_b ; \quad
LM_c = \begin{Bmatrix} m \\ m+1 \\ m+2 \\ m+6 \\ m+7 \\ m+8 \end{Bmatrix}_c ;
\tag{7.43}
$$

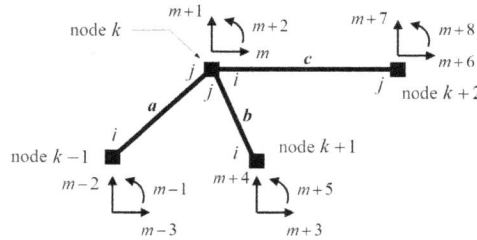

**FIGURE 7.11** Members Connected to Node $k$ of a Frame Structure and Displacement Components of the Connected Nodes.

The global stiffness relations for each member are written in terms of member variables as:

Member $a$:

$$
\begin{Bmatrix} P_1^{(a)} \\ P_2^{(a)} \\ P_3^{(a)} \\ P_4^{(a)} \\ P_5^{(a)} \\ P_6^{(a)} \end{Bmatrix} =
\begin{bmatrix}
K_{11}^{(a)} & K_{12}^{(a)} & K_{13}^{(a)} & K_{14}^{(a)} & K_{15}^{(a)} & K_{16}^{(a)} \\
K_{21}^{(a)} & K_{22}^{(a)} & K_{23}^{(a)} & K_{24}^{(a)} & K_{25}^{(a)} & K_{26}^{(a)} \\
K_{31}^{(a)} & K_{32}^{(a)} & K_{33}^{(a)} & K_{34}^{(a)} & K_{35}^{(a)} & K_{36}^{(a)} \\
K_{41}^{(a)} & K_{42}^{(a)} & K_{43}^{(a)} & K_{44}^{(a)} & K_{45}^{(a)} & K_{46}^{(a)} \\
K_{51}^{(a)} & K_{52}^{(a)} & K_{53}^{(a)} & K_{54}^{(a)} & K_{55}^{(a)} & K_{56}^{(a)} \\
K_{61}^{(a)} & K_{62}^{(a)} & K_{63}^{(a)} & K_{64}^{(a)} & K_{65}^{(a)} & K_{66}^{(a)}
\end{bmatrix}
\begin{Bmatrix} U_1^{(a)} \\ U_2^{(a)} \\ U_3^{(a)} \\ U_4^{(a)} \\ U_5^{(a)} \\ U_6^{(a)} \end{Bmatrix}
\tag{7.44}
$$

Member $b$:

$$
\begin{Bmatrix} P_1^{(b)} \\ P_2^{(b)} \\ P_3^{(b)} \\ P_4^{(b)} \\ P_5^{(b)} \\ P_6^{(b)} \end{Bmatrix} = \begin{bmatrix} K_{11}^{(b)} & K_{12}^{(b)} & K_{13}^{(b)} & K_{14}^{(b)} & K_{15}^{(b)} & K_{16}^{(b)} \\ K_{21}^{(b)} & K_{22}^{(b)} & K_{23}^{(b)} & K_{24}^{(b)} & K_{25}^{(b)} & K_{26}^{(b)} \\ K_{31}^{(b)} & K_{32}^{(b)} & K_{33}^{(b)} & K_{34}^{(b)} & K_{35}^{(b)} & K_{36}^{(b)} \\ K_{41}^{(b)} & K_{42}^{(b)} & K_{43}^{(b)} & K_{44}^{(b)} & K_{45}^{(b)} & K_{46}^{(b)} \\ K_{51}^{(b)} & K_{52}^{(b)} & K_{53}^{(b)} & K_{54}^{(b)} & K_{55}^{(b)} & K_{56}^{(b)} \\ K_{61}^{(b)} & K_{62}^{(b)} & K_{63}^{(b)} & K_{64}^{(b)} & K_{65}^{(b)} & K_{66}^{(b)} \end{bmatrix} \begin{Bmatrix} U_1^{(b)} \\ U_2^{(b)} \\ U_3^{(b)} \\ U_4^{(b)} \\ U_5^{(b)} \\ U_6^{(b)} \end{Bmatrix}
\tag{7.45}
$$

Member $c$:

$$
\begin{Bmatrix} P_1^{(c)} \\ P_2^{(c)} \\ P_3^{(c)} \\ P_4^{(c)} \\ P_5^{(c)} \\ P_6^{(c)} \end{Bmatrix} = \begin{bmatrix} K_{11}^{(c)} & K_{12}^{(c)} & K_{13}^{(c)} & K_{14}^{(c)} & K_{15}^{(c)} & K_{16}^{(c)} \\ K_{21}^{(c)} & K_{22}^{(c)} & K_{23}^{(c)} & K_{24}^{(c)} & K_{25}^{(c)} & K_{26}^{(c)} \\ K_{31}^{(c)} & K_{32}^{(c)} & K_{33}^{(c)} & K_{34}^{(c)} & K_{35}^{(c)} & K_{36}^{(c)} \\ K_{41}^{(c)} & K_{42}^{(c)} & K_{43}^{(c)} & K_{44}^{(c)} & K_{45}^{(c)} & K_{46}^{(c)} \\ K_{51}^{(c)} & K_{52}^{(c)} & K_{53}^{(c)} & K_{54}^{(c)} & K_{55}^{(c)} & K_{56}^{(c)} \\ K_{61}^{(c)} & K_{62}^{(c)} & K_{63}^{(c)} & K_{64}^{(c)} & K_{65}^{(c)} & K_{66}^{(c)} \end{bmatrix} \begin{Bmatrix} U_1^{(c)} \\ U_2^{(c)} \\ U_3^{(c)} \\ U_4^{(c)} \\ U_5^{(c)} \\ U_6^{(c)} \end{Bmatrix}
\tag{7.46}
$$

Based on the mapping vector (7.43) for each member, Eqs. (7.44), (7.45), and (7.46) are expressed in terms of structural variables as:

Member $a$:

$$
LM_a = \begin{Bmatrix} m-3 \\ m-2 \\ m-1 \\ m \\ m+1 \\ m+2 \end{Bmatrix}_a \; ; \quad
\begin{Bmatrix} P_1^{(a)} \\ P_2^{(a)} \\ P_3^{(a)} \\ P_4^{(a)} \\ P_5^{(a)} \\ P_6^{(a)} \end{Bmatrix} = \begin{array}{c} \\ m-3 \\ m-2 \\ m-1 \\ m \\ m+1 \\ m+2 \end{array} \overset{\begin{array}{cccccc} m-3 & m-2 & m-1 & m & m+1 & m+2 \end{array}}{\begin{bmatrix} K_{11}^{(a)} & K_{12}^{(a)} & K_{13}^{(a)} & K_{14}^{(a)} & K_{15}^{(a)} & K_{16}^{(a)} \\ K_{21}^{(a)} & K_{22}^{(a)} & K_{23}^{(a)} & K_{24}^{(a)} & K_{25}^{(a)} & K_{26}^{(a)} \\ K_{31}^{(a)} & K_{32}^{(a)} & K_{33}^{(a)} & K_{34}^{(a)} & K_{35}^{(a)} & K_{36}^{(a)} \\ K_{41}^{(a)} & K_{42}^{(a)} & K_{43}^{(a)} & K_{44}^{(a)} & K_{45}^{(a)} & K_{46}^{(a)} \\ K_{51}^{(a)} & K_{52}^{(a)} & K_{53}^{(a)} & K_{54}^{(a)} & K_{55}^{(a)} & K_{56}^{(a)} \\ K_{61}^{(a)} & K_{62}^{(a)} & K_{63}^{(a)} & K_{64}^{(a)} & K_{65}^{(a)} & K_{66}^{(a)} \end{bmatrix}} \begin{Bmatrix} U_{m-3} \\ U_{m-2} \\ U_{m-1} \\ U_m \\ U_{m+1} \\ U_{m+2} \end{Bmatrix}
\tag{7.47}
$$

Member $b$:

$$
LM_b = \begin{Bmatrix} m+3 \\ m+4 \\ m+5 \\ m \\ m+1 \\ m+2 \end{Bmatrix}_b ; \quad \begin{Bmatrix} P_1^{(b)} \\ P_2^{(b)} \\ P_3^{(b)} \\ P_4^{(b)} \\ P_5^{(b)} \\ P_6^{(b)} \end{Bmatrix} = \begin{matrix} & \overset{m+3}{} & \overset{m+4}{} & \overset{m+5}{} & \overset{m}{} & \overset{m+1}{} & \overset{m+2}{} \\ \begin{matrix} m+3 \\ m+4 \\ m+5 \\ m \\ m+1 \\ m+2 \end{matrix} & \begin{bmatrix} K_{11}^{(b)} & K_{12}^{(b)} & K_{13}^{(b)} & K_{14}^{(b)} & K_{15}^{(b)} & K_{16}^{(b)} \\ K_{21}^{(b)} & K_{22}^{(b)} & K_{23}^{(b)} & K_{24}^{(b)} & K_{25}^{(b)} & K_{26}^{(b)} \\ K_{31}^{(b)} & K_{32}^{(b)} & K_{33}^{(b)} & K_{34}^{(b)} & K_{35}^{(b)} & K_{36}^{(b)} \\ K_{41}^{(b)} & K_{42}^{(b)} & K_{43}^{(b)} & K_{44}^{(b)} & K_{45}^{(b)} & K_{46}^{(b)} \\ K_{51}^{(b)} & K_{52}^{(b)} & K_{53}^{(b)} & K_{54}^{(b)} & K_{55}^{(b)} & K_{56}^{(b)} \\ K_{61}^{(b)} & K_{62}^{(b)} & K_{63}^{(b)} & K_{64}^{(b)} & K_{65}^{(b)} & K_{66}^{(b)} \end{bmatrix} \end{matrix} \begin{Bmatrix} U_{m+3} \\ U_{m+4} \\ U_{m+5} \\ U_m \\ U_{m+1} \\ U_{m+2} \end{Bmatrix} \quad (7.48)
$$

Member $c$:

$$
LM_c = \begin{Bmatrix} m \\ m+1 \\ m+2 \\ m+6 \\ m+7 \\ m+8 \end{Bmatrix}_c ; \quad \begin{Bmatrix} P_1^{(c)} \\ P_2^{(c)} \\ P_3^{(c)} \\ P_4^{(c)} \\ P_5^{(c)} \\ P_6^{(c)} \end{Bmatrix} = \begin{matrix} & \overset{m}{} & \overset{m+1}{} & \overset{m+2}{} & \overset{m+6}{} & \overset{m+7}{} & \overset{m+8}{} \\ \begin{matrix} m \\ m+1 \\ m+2 \\ m+6 \\ m+7 \\ m+8 \end{matrix} & \begin{bmatrix} K_{11}^{(c)} & K_{12}^{(c)} & K_{13}^{(c)} & K_{14}^{(c)} & K_{15}^{(c)} & K_{16}^{(c)} \\ K_{21}^{(c)} & K_{22}^{(c)} & K_{23}^{(c)} & K_{24}^{(c)} & K_{25}^{(c)} & K_{26}^{(c)} \\ K_{31}^{(c)} & K_{32}^{(c)} & K_{33}^{(c)} & K_{34}^{(c)} & K_{35}^{(c)} & K_{36}^{(c)} \\ K_{41}^{(c)} & K_{42}^{(c)} & K_{43}^{(c)} & K_{44}^{(c)} & K_{45}^{(c)} & K_{46}^{(c)} \\ K_{51}^{(c)} & K_{52}^{(c)} & K_{53}^{(c)} & K_{54}^{(c)} & K_{55}^{(c)} & K_{56}^{(c)} \\ K_{61}^{(c)} & K_{62}^{(c)} & K_{63}^{(c)} & K_{64}^{(c)} & K_{65}^{(c)} & K_{66}^{(c)} \end{bmatrix} \end{matrix} \begin{Bmatrix} U_m \\ U_{m+1} \\ U_{m+2} \\ U_{m+6} \\ U_{m+7} \\ U_{m+8} \end{Bmatrix} \quad (7.49)
$$

From Figure 7.12, the equilibrium conditions of node $k$ are written as:

$$
\sum F_X = 0; \quad \Rightarrow \quad F_m^E = P_4^{(a)} + P_4^{(b)} + P_1^{(c)} \quad (7.50)
$$

$$
\sum F_Y = 0; \quad \Rightarrow \quad F_{m+1}^E = P_5^{(a)} + P_5^{(b)} + P_2^{(c)} \quad (7.51)
$$

$$
\sum M_Z = 0; \quad \Rightarrow \quad F_{m+2}^E = P_6^{(a)} + P_6^{(b)} + P_3^{(c)} \quad (7.52)
$$

Equations (7.50), (7.51), and (7.52) are associated with the $mth$, $(m+1)th$, and $(m+2)th$ stiffness equations of the structure, respectively. Using member stiffness relations of Eqs. (7.47), (7.48), and (7.49), the right-hand sides of Eqs. (7.50), (7.51), and (7.52) can be expressed in terms of structural nodal displacements as:

$$
\begin{aligned}
F_m^E = {} & K_{41}^{(a)} U_{m-3} + K_{42}^{(a)} U_{m-2} + K_{43}^{(a)} U_{m-1} + \left( K_{44}^{(a)} + K_{44}^{(b)} + K_{11}^{(c)} \right) U_m + \\
& \left( K_{45}^{(a)} + K_{45}^{(b)} + K_{12}^{(c)} \right) U_{m+1} + \left( K_{46}^{(a)} + K_{46}^{(b)} + K_{13}^{(c)} \right) U_{m+2} + K_{41}^{(b)} U_{m+3} + \\
& K_{42}^{(b)} U_{m+4} + K_{43}^{(b)} U_{m+5} + K_{14}^{(c)} U_{m+6} + K_{15}^{(c)} U_{m+7} + K_{16}^{(c)} U_{m+8}
\end{aligned} \quad (7.53)
$$

$$
\begin{aligned}
F_{m+1}^E = {} & K_{51}^{(a)} U_{m-3} + K_{52}^{(a)} U_{m-2} + K_{53}^{(a)} U_{m-1} + \left( K_{54}^{(a)} + K_{54}^{(b)} + K_{21}^{(c)} \right) U_m + \\
& \left( K_{55}^{(a)} + K_{55}^{(b)} + K_{22}^{(c)} \right) U_{m+1} + \left( K_{56}^{(a)} + K_{56}^{(b)} + K_{23}^{(c)} \right) U_{m+2} + K_{51}^{(b)} U_{m+3} + \\
& K_{52}^{(b)} U_{m+4} + K_{53}^{(b)} U_{m+5} + K_{24}^{(c)} U_{m+6} + K_{25}^{(c)} U_{m+7} + K_{26}^{(c)} U_{m+8}
\end{aligned} \quad (7.54)
$$

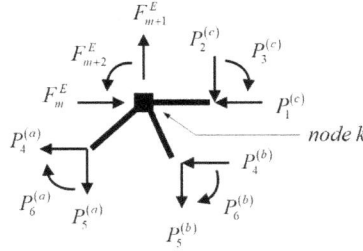

**FIGURE 7.12**   Equilibrium Conditions of Node $k$.

$$F^E_{m+2} = K^{(a)}_{61}U_{m-3} + K^{(a)}_{62}U_{m-2} + K^{(a)}_{63}U_{m-1} + \left(K^{(a)}_{64} + K^{(b)}_{64} + K^{(c)}_{31}\right)U_m \quad +$$

$$\left(K^{(a)}_{65} + K^{(b)}_{65} + K^{(c)}_{32}\right)U_{m+1} + \left(K^{(a)}_{66} + K^{(b)}_{66} + K^{(c)}_{33}\right)U_{m+2} + K^{(b)}_{61}U_{m+3} + \qquad (7.55)$$

$$K^{(b)}_{62}U_{m+4} + K^{(b)}_{63}U_{m+5} + K^{(c)}_{34}U_{m+6} + K^{(c)}_{35}U_{m+7} + K^{(c)}_{36}U_{m+8}$$

From Eq. (7.53), (7.54), and (7.55), it can be concluded that:

| | | |
|---|---|---|
| $K_{(m)(1)} = 0$ | $K_{(m+1)(1)} = 0$ | $K_{(m+2)(1)} = 0$ |
| $K_{(m)(2)} = 0$ | $K_{(m+1)(2)} = 0$ | $K_{(m+2)(2)} = 0$ |
| $\vdots$ | $\vdots$ | $\vdots$ |
| $K_{(m)(m-3)} = K^{(a)}_{41}$ | $K_{(m+1)(m-3)} = K^{(a)}_{51}$ | $K_{(m+2)(m-3)} = K^{(a)}_{61}$ |
| $K_{(m)(m-2)} = K^{(a)}_{42}$ | $K_{(m+1)(m-2)} = K^{(a)}_{52}$ | $K_{(m+2)(m-2)} = K^{(a)}_{62}$ |
| $K_{(m)(m-1)} = K^{(a)}_{43}$ | $K_{(m+1)(m-1)} = K^{(a)}_{53}$ | $K_{(m+2)(m-1)} = K^{(a)}_{63}$ |
| $K_{(m)(m)} = K^{(a)}_{44} + K^{(b)}_{44} + K^{(c)}_{11}$ | $K_{(m+1)(m)} = K^{(a)}_{54} + K^{(b)}_{54} + K^{(c)}_{21}$ | $K_{(m+2)(m)} = K^{(a)}_{64} + K^{(b)}_{64} + K^{(c)}_{31}$ |
| $K_{(m)(m+1)} = K^{(a)}_{45} + K^{(b)}_{45} + K^{(c)}_{12}$ | $K_{(m+1)(m+1)} = K^{(a)}_{55} + K^{(b)}_{55} + K^{(c)}_{22}$ | $K_{(m+2)(m+1)} = K^{(a)}_{65} + K^{(b)}_{65} + K^{(c)}_{32}$ |
| $K_{(m)(m+2)} = K^{(a)}_{46} + K^{(b)}_{46} + K^{(c)}_{13}$ | $K_{(m+1)(m+2)} = K^{(a)}_{56} + K^{(b)}_{56} + K^{(c)}_{23}$ | $K_{(m+2)(m+2)} = K^{(a)}_{66} + K^{(b)}_{66} + K^{(c)}_{33}$ |
| $K_{(m)(m+3)} = K^{(b)}_{41}$ | $K_{(m+1)(m+3)} = K^{(b)}_{51}$ | $K_{(m+2)(m+3)} = K^{(b)}_{61}$ |
| $K_{(m)(m+4)} = K^{(b)}_{42}$ | $K_{(m+1)(m+4)} = K^{(b)}_{52}$ | $K_{(m+2)(m+4)} = K^{(b)}_{62}$ |
| $K_{(m)(m+5)} = K^{(b)}_{43}$ | $K_{(m+1)(m+5)} = K^{(b)}_{53}$ | $K_{(m+2)(m+5)} = K^{(b)}_{63}$ |
| $K_{(m)(m+6)} = K^{(c)}_{14}$ | $K_{(m+1)(m+6)} = K^{(c)}_{24}$ | $K_{(m+2)(m+6)} = K^{(c)}_{34}$ |
| $K_{(m)(m+7)} = K^{(c)}_{15}$ | $K_{(m+1)(m+7)} = K^{(c)}_{25}$ | $K_{(m+2)(m+7)} = K^{(c)}_{35}$ |
| $K_{(m)(m+8)} = K^{(c)}_{16}$ | $K_{(m+1)(m+8)} = K^{(c)}_{26}$ | $K_{(m+2)(m+8)} = K^{(c)}_{36}$ |
| $K_{(m)(n)} = 0 \ \left[n = (m+9), (m+10), \ldots\right]$ | $K_{(m+1)(n)} = 0 \ \left[n = (m+9), (m+10), \ldots\right]$ | $K_{(m+2)(n)} = 0 \ \left[n = (m+9), (m+10), \ldots\right]$ |

Clearly, each stiffness coefficient $K_{ij}$ of the structure is equal to the sum of the stiffness coefficients of its constituent members through the member mapping vectors $LM$. Later, it will be shown that this assembling concept can be implemented efficiently on a computer.

Now, the prototypical frame of Figure 7.1 is revisted. Based on the member-code number technique, the schematic representation of the assembling process of the structural stiffness matrix is shown in Figure 7.13.

Likewise, the schematic representation of the assembling process of the restrained nodal force vector is shown in Figure 7.14.

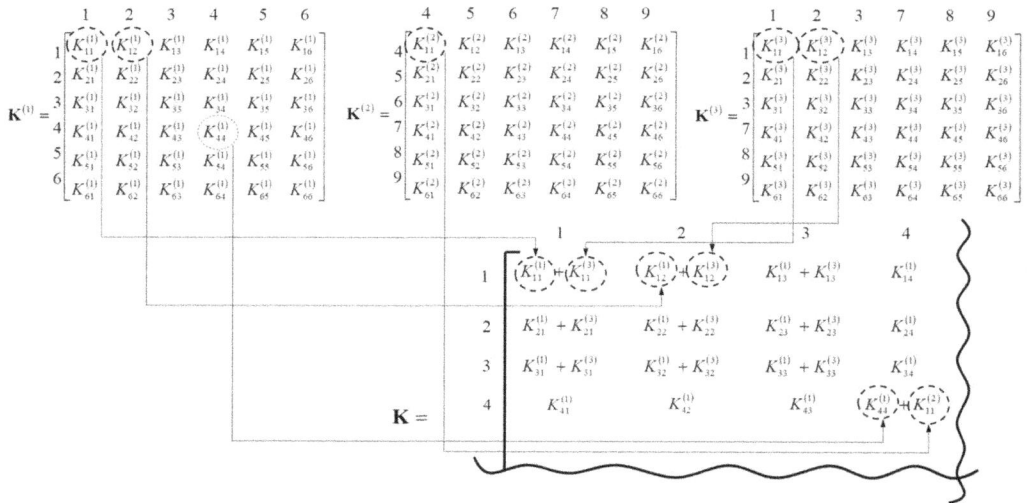

**FIGURE 7.13** Schematic Representation of the Assembling Process of the Structural Stiffness Matrix by Member Code Number.

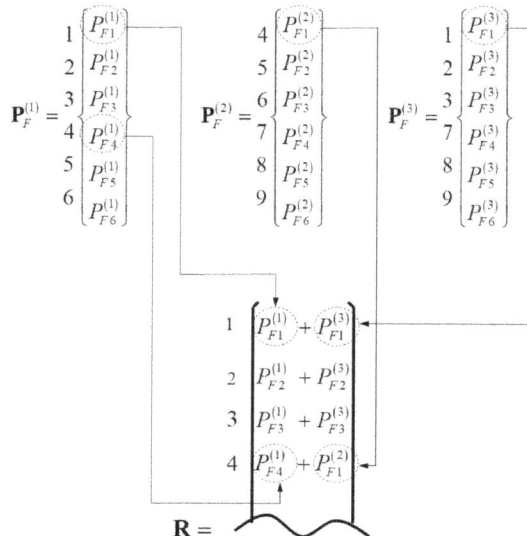

**FIGURE 7.14** Schematic Representation of the Assembling Process of the Restrained Nodal Force Vector by Member Code Number.

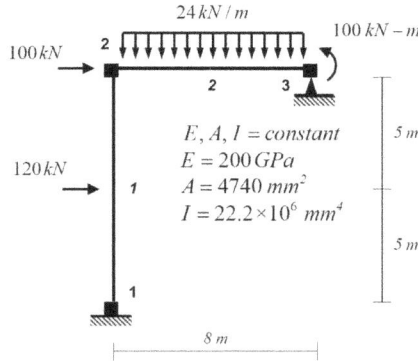

**FIGURE 7.15**  Example 7.3

## EXAMPLE 7.3

For the frame shown in Figure 7.15, set up the structural stiffness matrix and equivalent nodal load vector by the member code technique.

**Solution:** The local reference systems as well as DOF, member, and nodal numbering systems are shown in Figure 7.16.

The global stiffness matrices for each member are:

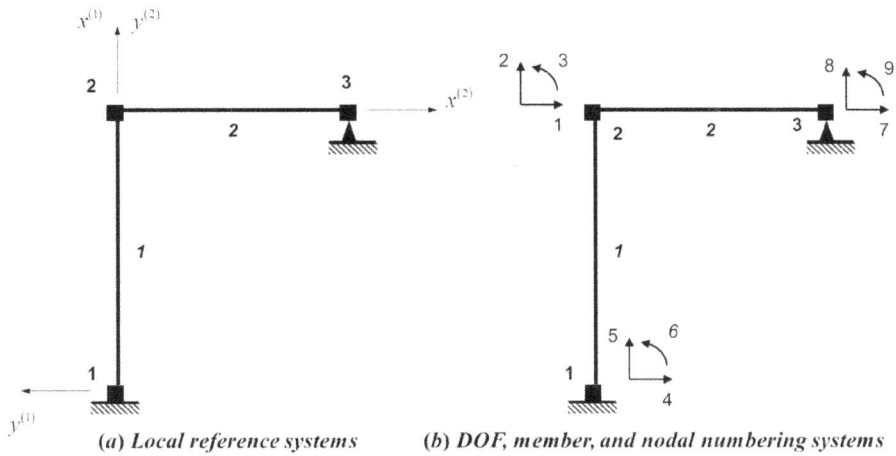

*(a) Local reference systems*       *(b) DOF, member, and nodal numbering systems*

**FIGURE 7.16**  Example 7.3 (Continued).

Member 1:
The global member stiffness matrix is:

$$
\mathbf{K}^{(1)} =
\begin{array}{c}
\begin{array}{cccccc}
\phantom{-}4 & \phantom{-}5 & \phantom{-}6 & \phantom{-}1 & \phantom{-}2 & \phantom{-}3
\end{array} \\
\left[
\begin{array}{cccccc}
53.28 & 0 & -266.4 & -53.28 & 0 & -266.4 \\
0 & 94800 & 0 & 0 & -94800 & 0 \\
-266.4 & 0 & 1776 & 266.4 & 0 & 888 \\
-53.28 & 0 & 266.4 & 53.28 & 0 & 266.4 \\
0 & -94800 & 0 & 0 & 94800 & 0 \\
-266.4 & 0 & 888 & 266.4 & 0 & 1776
\end{array}
\right]
\begin{array}{c}
4 \\ 5 \\ 6 \\ 1 \\ 2 \\ 3
\end{array}
\end{array}
$$

After accounting for the contribution of member 1, the structural stiffness matrix is:

$$
\mathbf{K} =
\begin{array}{c}
\begin{array}{ccccccccc}
1 & 2 & 3 & 4 & 5 & 6 & 7 & 8 & 9
\end{array} \\
\left[
\begin{array}{ccccccccc}
53.28 & 0 & 266.4 & -53.28 & 0 & 266.4 & 0 & 0 & 0 \\
0 & 94800 & 0 & 0 & -94800 & 0 & 0 & 0 & 0 \\
266.4 & 0 & 1776 & -266.4 & 0 & 888 & 0 & 0 & 0 \\
-53.28 & 0 & -266.4 & 53.28 & 0 & -266.4 & 0 & 0 & 0 \\
0 & -94800 & 0 & 0 & 94800 & 0 & 0 & 0 & 0 \\
266.4 & 0 & 888 & -266.4 & 0 & 1776 & 0 & 0 & 0 \\
0 & 0 & 0 & 0 & 0 & 0 & 0 & 0 & 0 \\
0 & 0 & 0 & 0 & 0 & 0 & 0 & 0 & 0 \\
0 & 0 & 0 & 0 & 0 & 0 & 0 & 0 & 0
\end{array}
\right]
\begin{array}{c}
1 \\ 2 \\ 3 \\ 4 \\ 5 \\ 6 \\ 7 \\ 8 \\ 9
\end{array}
\end{array}
$$

Member 2:
The global member stiffness matrix is:

$$
\mathbf{K}^{(2)} =
\begin{array}{c}
\begin{array}{cccccc}
1 & 2 & 3 & 7 & 8 & 9
\end{array} \\
\left[
\begin{array}{cccccc}
118500 & 0 & 0 & -118500 & 0 & 0 \\
0 & 104.06 & 416.25 & 0 & -104.06 & 416.25 \\
0 & 416.25 & 2220 & 266.4 & -416.25 & 1110 \\
-118500 & 0 & 0 & 118500 & 0 & 0 \\
0 & -104.06 & -416.25 & 0 & 104.06 & -416.25 \\
0 & 416.25 & 1110 & 0 & -416.25 & 2220
\end{array}
\right]
\begin{array}{c}
1 \\ 2 \\ 3 \\ 7 \\ 8 \\ 9
\end{array}
\end{array}
$$

After accounting for the contribution of member 2, the structural stiffness matrix is:

$$
\mathbf{K} =
\begin{array}{c}
\begin{array}{ccccccccc}
\;1\; & \;2\; & \;3\; & \;4\; & \;5\; & \;6\; & \;7\; & \;8\; & \;9\;
\end{array} \\
\left[
\begin{array}{ccccccccc}
118500+53.28 & 0 & 266.4 & -53.28 & 0 & 266.4 & -118500 & 0 & 0 \\
0 & 94800+104.06 & 416.25 & 0 & -94800 & 0 & 0 & -104.06 & -416.25 \\
266.4 & 416.25 & 1776+2220 & -266.4 & 0 & 888 & 0 & 416.25 & 1110 \\
-53.28 & 0 & -266.4 & 53.28 & 0 & -266.4 & 0 & 0 & 0 \\
0 & -94800 & 0 & 0 & 94800 & 0 & 0 & 0 & 0 \\
266.4 & 0 & 888 & -266.4 & 0 & 1776 & 0 & 0 & 0 \\
-118500 & 0 & 0 & 0 & 0 & 0 & 118500 & 0 & 0 \\
0 & -104.06 & 416.25 & 0 & 0 & 0 & 0 & 104.06 & -416.25 \\
0 & -416.25 & 1110 & 0 & 0 & 0 & 0 & -416.25 & 2220
\end{array}
\right]
\begin{array}{c}
1 \\ 2 \\ 3 \\ 4 \\ 5 \\ 6 \\ 7 \\ 8 \\ 9
\end{array}
\end{array}
$$

The schematic representation of the member-code assembling technique is shown in Figure 7.17. The fixed end force vector for each member is:

Member 1 (Figure 7.18$a$):

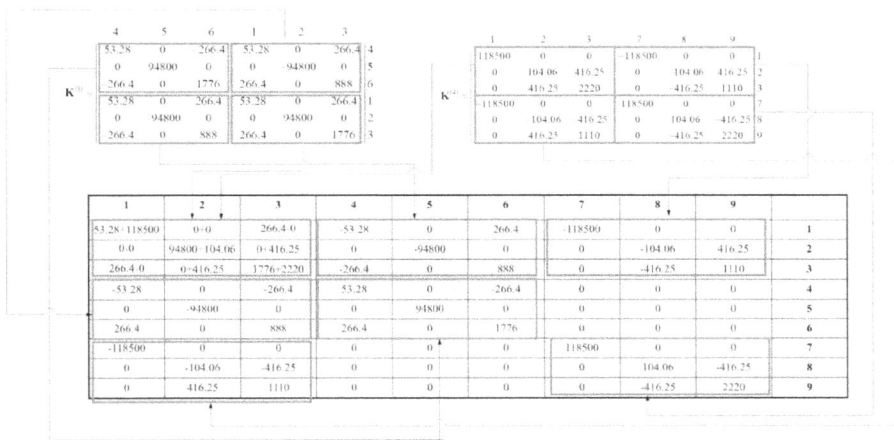

**FIGURE 7.17**  Example 7.3 (Continued).

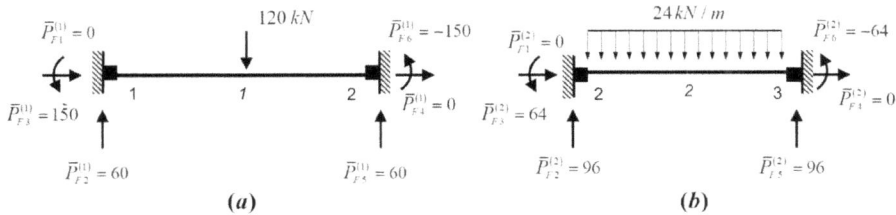

**FIGURE 7.18**  Example 7.3 (Continued).

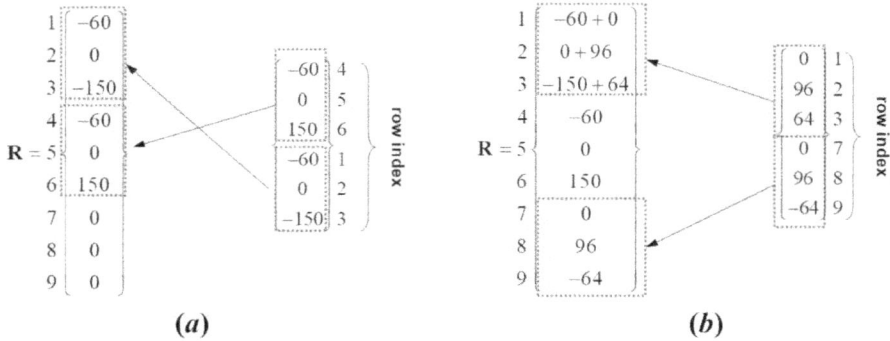

**FIGURE 7.19**  Example 7.3 (Continued).

The local fixed end force vector is:

$$\overline{\mathbf{P}}_F^{(1)} = \lfloor 0 \quad 60 \quad 150 \quad 0 \quad 60 \quad -150 \rfloor^T$$

The global fixed end force vector is:

$$\begin{array}{cccccc} 4 & 5 & 6 & 1 & 2 & 3 \end{array}$$
$$\mathbf{P}_F^{(1)} = {\Gamma_{ROT}^{(1)}}^T \overline{\mathbf{P}}_F^{(1)} = \lfloor -60 \quad 0 \quad 150 \quad -60 \quad 0 \quad -150 \rfloor^T$$

After accounting for the contribution of member 1, the structural fixed-end force vector is shown in Figure 7.19a.

Member 2 (Figure 7.18b):

The local fixed-end force vector is:

$$\overline{\mathbf{P}}_F^{(2)} = \lfloor 0 \quad 96 \quad 64 \quad 0 \quad 96 \quad -64 \rfloor^T$$

The global fixed end force vector is:

$$\begin{array}{cccccc} 1 & 2 & 3 & 7 & 8 & 9 \end{array}$$
$$\mathbf{P}_F^{(2)} = {\Gamma_{ROT}^{(2)}}^T \overline{\mathbf{P}}_F^{(2)} = \lfloor 0 \quad 96 \quad 64 \quad 0 \quad 96 \quad -64 \rfloor^T$$

After accounting for the contribution of member 2, the structural fixed-end force vector is shown in Figure 7.19b.

The nodal load vector is:

$$\mathbf{F} = \lfloor 100 \quad 0 \quad 0 \quad F_4 \quad F_5 \quad F_6 \quad F_7 \quad F_8 \quad 100 \rfloor^T$$

The equivalent nodal load vector is shown in Figure 7.20.

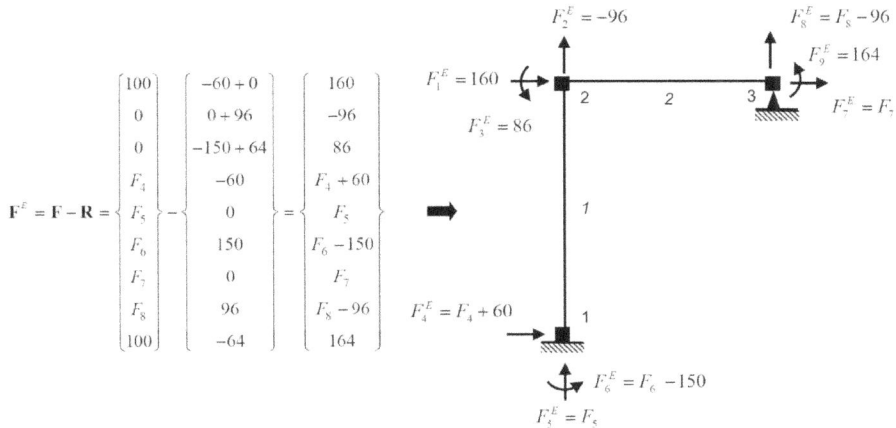

The matrix equation shown:

$$\mathbf{F}^E = \mathbf{F} - \mathbf{R} = \begin{Bmatrix} 100 \\ 0 \\ 0 \\ F_4 \\ F_5 \\ F_6 \\ F_7 \\ F_8 \\ 100 \end{Bmatrix} - \begin{Bmatrix} -60 + 0 \\ 0 + 96 \\ -150 + 64 \\ -60 \\ 0 \\ 150 \\ 0 \\ 96 \\ -64 \end{Bmatrix} = \begin{Bmatrix} 160 \\ -96 \\ 86 \\ F_4 + 60 \\ F_5 \\ F_6 - 150 \\ F_7 \\ F_8 - 96 \\ 164 \end{Bmatrix}$$

**FIGURE 7.20**  Example 7.3 (Continued).

## 7.5  GENERAL PROCEDURE FOR STRUCTURAL ANALYSIS BY THE DIRECT STIFFNESS METHOD

In the last two sections, two assembling schemes for the structural stiffness matrix and (equivalent) nodal force vector were studied. These are the main tasks in the analyzing process. With some additional steps, the whole analyzing process of a frame structure by the direct stiffness method will be completed. The general procedure for structural analysis by the direct stiffness method can be summarized as follows:

1. Prepare a structural model of the actual structure. Usually, this includes:

   - Numbering its degrees of freedom, nodes, and members.
   - Assigning local and global reference systems.
   - Defining boundary conditions.

     Note: this step requires some engineering judgments and experiences.

2. Evaluate the stiffness matrix and fixed-end load vector with respect to both global and local coordinate systems for each member.
3. Assemble member stiffness matrices and member fixed-end load vectors to obtain the structural stiffness matrix and structural fixed-end load vector.
4. Rearrange rows and columns to partition the structural stiffness equations following the free and constrained degrees of freedom.

   Note: this step can be skipped if degrees of freedom are numbered properly.

5. Compute free nodal displacements $\mathbf{U}_{free}$ from Eq. (7.3).
6. Compute member-end displacements, member-end forces, and support reactions. For each member, the following steps are carried out:

I.  Retrieve global member-end displacements $\mathbf{U}^{(k)}$ from the structural nodal displacements $\mathbf{U}$ through the member mapping vector (member code number). This step can be achieved by using the member compatibility mapping relation ($\mathbf{U}^{(k)} = \mathbf{G}^{(k)}\mathbf{U}$).

II. Determine local member-end displacements $\overline{\mathbf{U}}^{(k)}$ through the coordinate transformation relation ($\overline{\mathbf{U}}^{(k)} = \mathbf{\Gamma}_{ROT}^{(k)}\mathbf{U}^{(k)}$).

III. Recover local member-end forces $\overline{\mathbf{P}}^{(k)}$ from the member stiffness equations ($\overline{\mathbf{P}}^{(k)} = \overline{\mathbf{K}}^{(k)}\overline{\mathbf{U}}^{(k)} + \overline{\mathbf{P}}_F^{(k)}$).

IV. Determine global member-end forces $\mathbf{P}^{(k)}$ through the coordinate transformation relation ($\mathbf{P}^{(k)} = \mathbf{\Gamma}_{ROT}^{(k)}{}^T \overline{\mathbf{P}}^{(k)}$).

V.  If the considered member is attached to a support, then use the mapping vector to collect the relevant component of $\mathbf{P}^{(k)}$ in the equivalent nodal load vector $\mathbf{F}^E$.

## EXAMPLE 7.4

Determine the nodal displacements, member-end forces, and member-end displacements of the frame shown in Figure 7.21 due to the combined effect of the loading and a settlement of 0.05 $m$ of the left support.

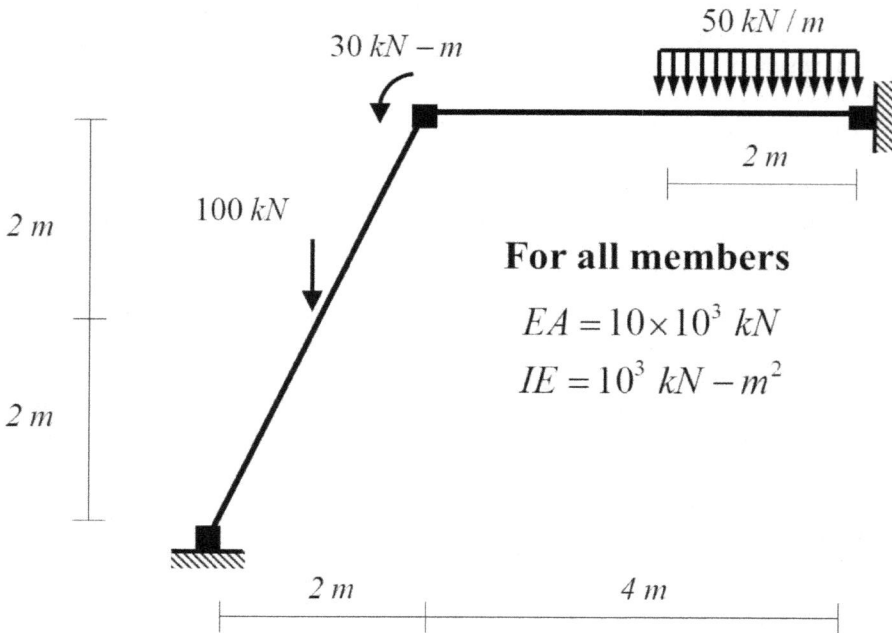

$30 \, kN - m$

$50 \, kN / m$

$2 \, m$

$2 \, m$

$100 \, kN$

**For all members**

$EA = 10 \times 10^3 \, kN$

$IE = 10^3 \, kN - m^2$

$2 \, m$

$2 \, m$

$4 \, m$

**FIGURE 7.21**   Example 7.4.

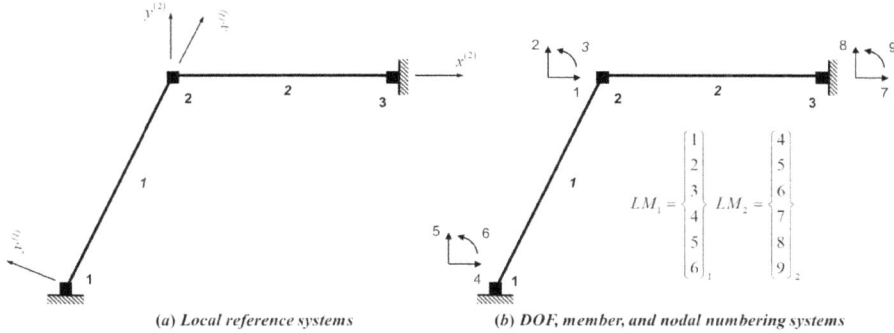

(a) Local reference systems          (b) DOF, member, and nodal numbering systems

**FIGURE 7.22**   Example 7.4 (Continued).

**Solution:** The local reference systems as well as DOF, member, and nodal numbering systems are shown in Figure 7.22.

The local and global stiffness matrices for each member are:

Member 1:

$$\bar{\mathbf{K}}^{(1)} = \begin{bmatrix} 2236.1 & 0 & 0 & -2236.1 & 0 & 0 \\ 0 & 134.2 & 300 & 0 & -134.2 & 300 \\ 0 & 300 & 894.3 & 0 & -300 & 447.2 \\ -2236.1 & 0 & 0 & 2236.1 & 0 & 0 \\ 0 & -134.2 & -300 & 0 & 134.2 & -300 \\ 0 & 300 & 447.2 & 0 & -300 & 894.3 \end{bmatrix}$$

$$\mathbf{K}^{(1)} = {\Gamma_{ROT}^{(1)}}^T \bar{\mathbf{K}}^{(1)} \Gamma_{ROT}^{(1)} = \begin{bmatrix} 554.5 & 840.8 & -268.3 & -554.5 & -840.8 & -268.3 \\ 840.8 & 1815.7 & 134.2 & -840.8 & -1815.7 & 134.2 \\ -268.3 & 134.2 & 894.4 & 268.3 & -134.2 & 447.2 \\ -554.5 & -840.8 & 268.3 & 554.5 & 840.8 & 268.3 \\ -840.8 & -1815.7 & -134.2 & 840.8 & 1815.7 & -134.2 \\ -268.3 & 134.2 & 447.2 & 268.3 & -134.2 & 894.4 \end{bmatrix}$$

Member 2:

$$\bar{\mathbf{K}}^{(2)} = \begin{bmatrix} 2500 & 0 & 0 & -2500 & 0 & 0 \\ 0 & 187.5 & 375 & 0 & -187.5 & 375 \\ 0 & 375 & 1000 & 0 & -375 & 500 \\ -2500 & 0 & 0 & 2500 & 0 & 0 \\ 0 & -187.5 & -375 & 0 & 187.5 & -375 \\ 0 & 375 & 500 & 0 & -375 & 1000 \end{bmatrix}$$

$$\mathbf{K}^{(2)} = {\Gamma_{ROT}^{(2)}}^T \bar{\mathbf{K}}^{(2)} \Gamma_{ROT}^{(2)} = \begin{bmatrix} 2500 & 0 & 0 & -2500 & 0 & 0 \\ 0 & 187.5 & 375 & 0 & -187.5 & 375 \\ 0 & 375 & 1000 & 0 & -375 & 500 \\ -2500 & 0 & 0 & 2500 & 0 & 0 \\ 0 & -187.5 & -375 & 0 & 187.5 & -375 \\ 0 & 375 & 500 & 0 & -375 & 1000 \end{bmatrix}$$

The local and global fixed-end load vectors for each member are:
  Member 1 (Figure 7.23a):
  Local fixed end load vector:

$$\overline{\mathbf{P}}_F^{(1)} = \lfloor 44.72 \quad 22.36 \quad 25 \quad 44.72 \quad 22.36 \quad -25 \rfloor^T$$

Global fixed end load vector:

$$\mathbf{P}_F^{(1)} = \mathbf{\Gamma}_{ROT}^{(1)\,T} \overline{\mathbf{P}}_F^{(1)} = \lfloor 0 \quad 50 \quad 25 \quad 0 \quad 50 \quad -25 \rfloor^T$$

  Member 2 (Figure 7.23b):
  Local fixed-end load vector:

$$\overline{\mathbf{P}}_F^{(2)} = \lfloor 0 \quad 18.75 \quad 20.83 \quad 0 \quad 81.25 \quad -45.83 \rfloor^T$$

Global fixed end load vector:

$$\mathbf{P}_F^{(2)} = \mathbf{\Gamma}_{ROT}^{(2)\,T} \overline{\mathbf{P}}_F^{(2)} = \lfloor 0 \quad 18.75 \quad 20.83 \quad 0 \quad 81.25 \quad -45.83 \rfloor^T$$

Through the member code technique, the structural stiffness matrix is assembled as:

$$\mathbf{K} = \begin{bmatrix} \mathbf{K}_{ff} & \mathbf{K}_{fc} \\ \mathbf{K}_{cf} & \mathbf{K}_{cc} \end{bmatrix}$$

where

$$\mathbf{K}_{ff} = \begin{bmatrix} 3054.5 & 840.8 & 268.3 \\ 840.8 & 2003.2 & 240.8 \\ 268.3 & 240.8 & 1894.4 \end{bmatrix} ;$$

$$\mathbf{K}_{fc} = \mathbf{K}_{cf}^T \begin{bmatrix} -554.5 & -840.8 & 268.3 & -2500 & 0 & 0 \\ -840.8 & -1815.7 & -134.2 & 0 & -187.5 & 375 \\ -268.3 & 134.2 & 447.2 & 0 & -375 & 500 \end{bmatrix} ;$$

$$\mathbf{K}_{cc} = \begin{bmatrix} 554.5 & 840.8 & -268.3 & 0 & 0 & 0 \\ 840.8 & 1815.7 & 134.2 & 0 & 0 & 0 \\ -268.3 & 134.2 & 894.4 & 0 & 0 & 0 \\ 0 & 0 & 0 & 2500 & 0 & 0 \\ 0 & 0 & 0 & 0 & 187.5 & -375 \\ 0 & 0 & 0 & 0 & -375 & 1000 \end{bmatrix} ;$$

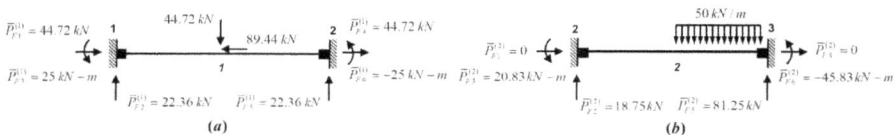

**FIGURE 7.23**  Example 7.4 (Continued).

The structural fixed end load vector is:

$$\mathbf{R} = \lfloor 0 \quad 68.75 \quad -4.17 \quad 0 \quad 50 \quad 25 \quad 0 \quad 81.25 \quad -45.83 \rfloor^T$$

The applied nodal load vector is:

$$\mathbf{F} = \lfloor 0 \quad 0 \quad 30 \quad F_4 \quad F_5 \quad F_6 \quad F_7 \quad F_8 \quad F_9 \rfloor^T$$

The equivalent nodal load vector is:

$$\mathbf{F}^E = \lfloor 0 \quad -68.75 \quad 34.17 \quad F_4 \quad F_5 - 50 \quad F_6 - 25 \quad F_7 \quad F_8 - 81.25 \quad F_9 + 45.83 \rfloor^T$$

In this example, the left support settles by 0.05 $m$. Therefore, constrained displacement degrees of freedom are:

$$\mathbf{U}_{constr} = \lfloor 0 \quad -0.05 \quad 0 \quad 0 \quad 0 \quad 0 \rfloor^T$$

The stiffness equations associated with free degrees of freedom are:

$$\mathbf{F}^E_{free} - \mathbf{K}_{fc}\mathbf{U}_{constr} = \mathbf{K}_{ff}\mathbf{U}_{free}$$

$$\begin{Bmatrix} 0 \\ -68.75 \\ 34.17 \end{Bmatrix} - \begin{Bmatrix} 42.04 \\ 90.79 \\ -6.71 \end{Bmatrix} = \begin{Bmatrix} -42.04 \\ -159.54 \\ 40.88 \end{Bmatrix} = \begin{bmatrix} 3054.5 & 840.8 & 268.3 \\ 840.8 & 2003.2 & 240.8 \\ 268.3 & 240.8 & 1894.4 \end{bmatrix} \begin{Bmatrix} U_1 \\ U_2 \\ U_3 \end{Bmatrix}$$

$$\mathbf{U}_{free} = \lfloor U_1 \quad U_2 \quad U_3 \rfloor^T = \lfloor 0.007272\ m \quad -0.08648\ m \quad 0.03154\ rad \rfloor^T$$

Member end forces and member end displacements for each member are:
   Member 1:
   The global member-end displacements $\mathbf{U}^{(1)}$ are:

$$LM_1 = \begin{Bmatrix} 4 \\ 5 \\ 6 \\ 1 \\ 2 \\ 3 \end{Bmatrix}_1 \Rightarrow \begin{Bmatrix} U_4 \\ U_5 \\ U_6 \\ U_1 \\ U_2 \\ U_3 \end{Bmatrix} \Rightarrow \begin{Bmatrix} U_1^{(1)} \\ U_2^{(1)} \\ U_3^{(1)} \\ U_4^{(1)} \\ U_5^{(1)} \\ U_6^{(1)} \end{Bmatrix} = \begin{Bmatrix} 0 \\ -0.05\ m \\ 0 \\ 0.007272\ m \\ -0.08648\ m \\ 0.03154\ rad \end{Bmatrix}$$

The local member-end displacements $\bar{\mathbf{U}}^{(1)}$ are:

$$\bar{\mathbf{U}}^{(1)} = \mathbf{\Gamma}^{(1)}_{ROT}\mathbf{U}^{(1)} = \lfloor -0.04472\ m \quad -0.02236\ m \quad 0 \quad -0.07409\ m \quad -0.04518\ m \quad 0.03154\ rad \rfloor^T$$

The local member-end forces $\bar{\mathbf{P}}^{(1)}$ are:

$$\bar{\mathbf{P}}^{(1)} = \bar{\mathbf{K}}^{(1)}\bar{\mathbf{U}}^{(1)} + \bar{\mathbf{P}}_F^{(1)} = \lfloor 110.41\,kN \quad 34.88\,kN \quad 45.95\,kN-m \quad -20.97\,kN \quad 9.81\,kN \quad 10.06\,kN-m \rfloor^T$$

The global member-end forces $\mathbf{P}^{(1)}$ are:

$$\mathbf{P}^{(1)} = \Gamma_{ROT}^{(1)}{}^T \bar{\mathbf{P}}^{(1)} = \lfloor 18.18\,kN \quad 114.35\,kN \quad 45.95\,kN-m \quad -18.18\,kN \quad -14.36\,kN \quad 10.06\,kN-m \rfloor^T$$

Member 2:

The global member-end displacements $\mathbf{U}^{(2)}$ are:

$$LM_2 = \begin{Bmatrix} 1 \\ 2 \\ 3 \\ 7 \\ 8 \\ 9 \end{Bmatrix}_2 \Rightarrow \begin{Bmatrix} U_1 \\ U_2 \\ U_3 \\ U_7 \\ U_8 \\ U_9 \end{Bmatrix} \Rightarrow \begin{Bmatrix} U_1^{(2)} \\ U_2^{(2)} \\ U_3^{(2)} \\ U_4^{(2)} \\ U_5^{(2)} \\ U_6^{(2)} \end{Bmatrix} = \begin{Bmatrix} 0.007272\,m \\ -0.08648\,m \\ 0.03154\,rad \\ 0 \\ 0 \\ 0 \end{Bmatrix}$$

The local member-end displacements $\bar{\mathbf{U}}^{(2)}$ are:

$$\bar{\mathbf{U}}^{(2)} = \Gamma_{ROT}^{(2)}\mathbf{U}^{(2)} = \lfloor 0.007272\,m \quad -0.08648\,m \quad 0.03154\,rad \quad 0 \quad 0 \quad 0 \rfloor^T$$

The local member end forces $\bar{\mathbf{P}}^{(2)}$ are:

$$\bar{\mathbf{P}}^{(2)} = \bar{\mathbf{K}}^{(2)}\bar{\mathbf{U}}^{(2)} + \bar{\mathbf{P}}_F^{(2)} = \lfloor 18.18\,kN \quad 14.36\,kN \quad 19.94\,kN-m \quad -18.18\,kN \quad 85.64\,kN \quad -62.49\,kN-m \rfloor^T$$

The global member end forces $\mathbf{P}^{(2)}$ are:

$$\mathbf{P}^{(2)} = \Gamma_{ROT}^{(2)}{}^T \bar{\mathbf{P}}^{(2)} = \lfloor 18.18\,kN \quad 14.36\,kN \quad 19.94\,kN-m \quad -18.18\,kN \quad 85.64\,kN \quad -62.49\,kN-m \rfloor^T$$

In this problem, the effect of the left-support settlement can alternatively be included in the member fixed-end force vector as shown in Figure 7.24.

Local fixed-end load vector due to the left-support settlement:

$$\bar{\mathbf{P}}_F^{(1)} = \lfloor -100 \quad -3 \quad -6.71 \quad 100 \quad 3 \quad -6.71 \rfloor^T$$

Global fixed-end load vector due to the left-support settlement:

$$\mathbf{P}_F^{(1)} = \Gamma_{ROT}^{(1)}{}^T \bar{\mathbf{P}}_F^{(1)} = \lfloor -42.04 \quad -90.78 \quad -6.71 \quad 42.04 \quad 90.78 \quad -6.71 \rfloor^T$$

This global fixed-end force vector due to the left-support settlement is added to that of the one due to the applied point load:

$$\mathbf{P}_F^{(1)} = \lfloor -42.04 \quad -40.78 \quad 18.29 \quad 42.04 \quad 140.78 \quad -31.71 \rfloor^T$$

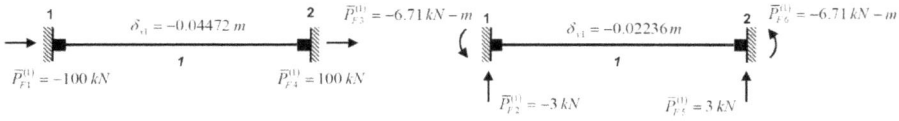

**FIGURE 7.24**  Example 7.4 (Continued).

The structural fixed end load vector is modified accordingly:

$$\mathbf{R} = \lfloor 42.04 \quad 159.54 \quad -10.88 \quad -42.04 \quad -40.78 \quad 18.29 \quad 0 \quad 81.25 \quad -45.83 \rfloor^T$$

The equivalent nodal load vector is:

$$\mathbf{F}^E = \lfloor -42.04 \quad -159.54 \quad 40.88 \quad F_4 + 42.04 \quad F_5 + 40.78 \quad F_6 - 18.29 \quad F_7 \quad F_8 - 81.25 \quad F_9 + 45.83 \rfloor^T$$

Since the effects of the left-support settlement were included in the fixed-end force vector, constrained displacement degrees of freedom have a value of zero ($\mathbf{U}_{constr} = \mathbf{0}$).

The stiffness equations associated with free degrees of freedom are:

$$\mathbf{F}^E_{free} = \mathbf{K}_{ff} \mathbf{U}_{free}$$

$$\begin{Bmatrix} -42.04 \\ -159.54 \\ 40.88 \end{Bmatrix} = \begin{bmatrix} 3054.5 & 840.8 & 268.3 \\ 840.8 & 2003.2 & 240.8 \\ 268.3 & 240.8 & 1894.4 \end{bmatrix} \begin{Bmatrix} U_1 \\ U_2 \\ U_3 \end{Bmatrix}$$

$$\mathbf{U}_{free} = \lfloor U_1 \quad U_2 \quad U_3 \rfloor^T = \lfloor 0.007272 \, m \quad -0.08648 \, m \quad 0.03154 \, rad \rfloor^T$$

Member end forces and member-end displacements for each member are:

Member 1:

The global member-end displacements $\mathbf{U}^{(1)}$ are:

$$LM_1 = \begin{Bmatrix} 4 \\ 5 \\ 6 \\ 1 \\ 2 \\ 3 \end{Bmatrix}_1 \Rightarrow \begin{Bmatrix} U_4 \\ U_5 \\ U_6 \\ U_1 \\ U_2 \\ U_3 \end{Bmatrix} \Rightarrow \begin{Bmatrix} U_1^{(1)} \\ U_2^{(1)} \\ U_3^{(1)} \\ U_4^{(1)} \\ U_5^{(1)} \\ U_6^{(1)} \end{Bmatrix} = \begin{Bmatrix} 0 \\ 0 \\ 0 \\ 0.007272 \, m \\ -0.08648 \, m \\ 0.03154 \, rad \end{Bmatrix}$$

The local member-end displacements $\overline{\mathbf{U}}^{(1)}$ are:

$$\overline{\mathbf{U}}^{(1)} = \mathbf{\Gamma}^{(1)}_{ROT} \mathbf{U}^{(1)} = \lfloor 0 \quad 0 \quad 0 \quad -0.07409 \, m \quad -0.04518 \, m \quad 0.03154 \, rad \rfloor^T$$

The local member-end forces $\overline{\mathbf{P}}^{(1)}$ are:

$$\overline{\mathbf{P}}^{(1)} = \overline{\mathbf{K}}^{(1)}\overline{\mathbf{U}}^{(1)} + \overline{\mathbf{P}}_F^{(1)} = \lfloor 110.41\,kN \quad 34.88\,kN \quad 45.95\,kN - m \quad -20.97\,kN \quad 9.81\,kN \quad 10.06\,kN - m \rfloor^T$$

It is noted that the fixed-end load vector $\overline{\mathbf{P}}_F^{(1)}$ accounts for both mechanical and support settlement actions.

The global member-end forces $\mathbf{P}^{(1)}$ are:

$$\mathbf{P}^{(1)} = \Gamma_{ROT}^{(1)}{}^T \overline{\mathbf{P}}^{(1)} = \lfloor 18.18\,kN \quad 114.35\,kN \quad 45.95\,kN - m \quad -18.18\,kN \quad -14.36\,kN \quad 10.06\,kN - m \rfloor^T$$

**Member 2:**

Since the fixed-end load vector $\overline{\mathbf{P}}_F^{(2)}$ is the same as before, there are no changes in calculations for member-end displacements and member-end forces.

## 7.6   FRAME MEMBERS WITH END FORCE RELEASES

Thus far, two standard plane frame members have been discussed in this textbook: namely the bar and frame members. The bar member is used in the analysis of a truss structure and can resist only axial forces. The frame member is used in the analysis of a frame structure in which all joints are rigid, and both ends of each frame member are rigidly connected to joints. This section considers the case that certain frame members with force releases (displacement discontinuities) exist in frame structures, as shown in Figure 7.25. Common types of force-release devices are symbolized and shown in Figure 7.26. The term "force release" implies that certain force actions cannot be transmitted through these force-release devices. That is, the devices in Figure 7.26a, b, and c cannot transmit thrust, shear, and moment, respectively. These inabilities to transmit force actions result in discontinuities in corresponding displacements (axial displacement, transverse displacement, and rotation).

This section focuses on the development of the stiffness matrix of a frame member with three types of end force releases, namely the thrust release, shear release, and moment release. This can be achieved by modifying the member stiffness matrices developed previously.

### 7.6.1   Bar Member with End Thrust Release

The bar member with a left-end thrust release shown in Figure 7.27a is first considered. Without end thrust releases, the member stiffness equations of a bar are explicitly expressed as:

$$\overline{P}_1 = \frac{EA}{L}\overline{U}_1 - \frac{EA}{L}\overline{U}_4 + \overline{P}_{F1} \tag{7.56}$$

$$\overline{P}_4 = -\frac{EA}{L}\overline{U}_1 + \frac{EA}{L}\overline{U}_4 + \overline{P}_{F4} \tag{7.57}$$

When end $i$ of a bar member is connected to the left-end joint by a thrust-release device, it is obvious that its end force $\overline{P}_1$ has to be zero. By inserting the condition of zero left-end force ($\overline{P}_1 = 0$) into Eq. (7.56), one has:

$$\overline{U}_1 = \overline{U}_4 - \frac{\overline{P}_{F1}L}{EA} \tag{7.58}$$

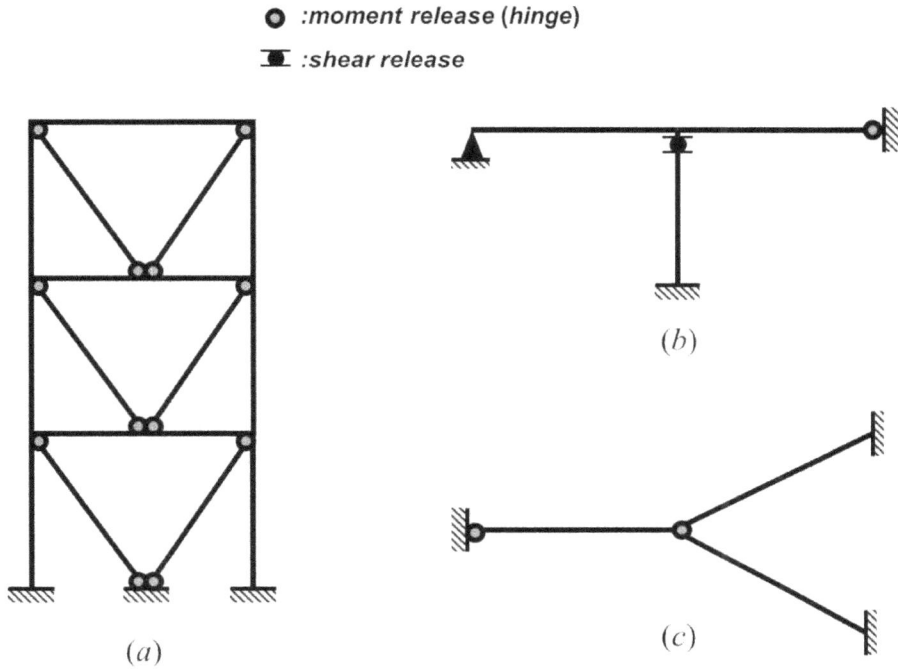

○ :*moment release (hinge)*

■ :*shear release*

(a)

(b)

(c)

**FIGURE 7.25**   Frame Members with End Force Releases.

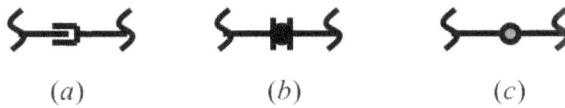

(a)   (b)   (c)

**FIGURE 7.26**   Force Releases: (a) Thrust Release; (b) Shear Release; (c) Moment Release.

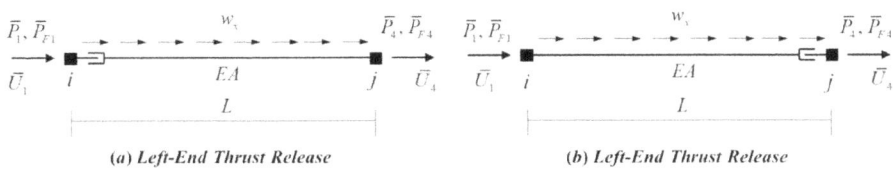

(a) Left-End Thrust Release    (b) Left-End Thrust Release

**FIGURE 7.27**   Bar Member with End Thrust Release.

Substituting Eqs. (7.58) into (7.57), one obtains:

$$\bar{P}_4 = \bar{P}_{F1} + \bar{P}_{F4} \tag{7.59}$$

Eq. (7.59) implies that the member stiffness matrix of a bar with left-end thrust release is null. A similar conclusion can be drawn for a bar with right-end thrust release of Figure 7.27 b.

### 7.6.2  FRAME MEMBER WITH END SHEAR RELEASE

The frame member with left-end shear release shown in Figure 7.28a is first considered. Without end force releases, the member stiffness equations of a frame are explicitly expressed as:

$$\bar{P}_1 = \frac{EA}{L}\bar{U}_1 - \frac{EA}{L}\bar{U}_4 + \bar{P}_{F1} \tag{7.56}$$

$$\bar{P}_2 = 12\frac{IE}{L^3}\bar{U}_2 + 6\frac{IE}{L^2}\bar{U}_3 - 12\frac{IE}{L^3}\bar{U}_5 + 6\frac{IE}{L^2}\bar{U}_6 + \bar{P}_{F2} \tag{7.60}$$

$$\bar{P}_3 = 6\frac{IE}{L^2}\bar{U}_2 + 4\frac{IE}{L}\bar{U}_3 - 6\frac{IE}{L^2}\bar{U}_5 + 2\frac{IE}{L}\bar{U}_6 + \bar{P}_{F3} \tag{7.61}$$

$$\bar{P}_4 = -\frac{EA}{L}\bar{U}_1 + \frac{EA}{L}\bar{U}_4 + \bar{P}_{F4} \tag{7.57}$$

$$\bar{P}_5 = -12\frac{IE}{L^3}\bar{U}_2 - 6\frac{IE}{L^2}\bar{U}_3 + 12\frac{IE}{L^3}\bar{U}_5 - 6\frac{IE}{L^2}\bar{U}_6 + \bar{P}_{F5} \tag{7.62}$$

$$\bar{P}_6 = 6\frac{IE}{L^2}\bar{U}_2 + 2\frac{IE}{L}\bar{U}_3 - 6\frac{IE}{L^2}\bar{U}_5 + 4\frac{IE}{L}\bar{U}_6 + \bar{P}_{F6} \tag{7.63}$$

When end $i$ of a frame member is connected to the left-end joint by a shear-release device, it is obvious that its end force $\bar{P}_2$ has to be zero. By inserting the condition of zero left-end force ($\bar{P}_2 = 0$) into Eq. (7.60), one has:

$$\bar{U}_2 = -\frac{L^3\bar{P}_{F2} - 12IE\bar{U}_5 + 6IEL\left(\bar{U}_3 + \bar{U}_6\right)}{12IE} \tag{7.64}$$

This kinematical relation implies that the left-end transverse displacement $\bar{U}_2$ is no longer an independent degree of freedom but now depends on other displacement degrees of freedom ($\bar{U}_3$, $\bar{U}_5$, and $\bar{U}_6$). With the kinematical relation of Eq. (7.64), Eqs. (7.61), (7.62), and (7.63) are rewritten as:

$$\bar{P}_3 = \frac{IE}{L}\bar{U}_3 - \frac{IE}{L}\bar{U}_6 + \bar{P}_{F3} - \frac{\bar{P}_{F2}L}{2} \tag{7.65}$$

$$\bar{P}_5 = \bar{P}_{F2} + \bar{P}_{F5} \tag{7.66}$$

$$\bar{P}_6 = -\frac{IE}{L}\bar{U}_3 + \frac{IE}{L}\bar{U}_6 + \bar{P}_{F6} - \frac{\bar{P}_{F2}L}{2} \tag{7.67}$$

Therefore, the stiffness equations of a frame with left-end shear release are written in the matrix form as:

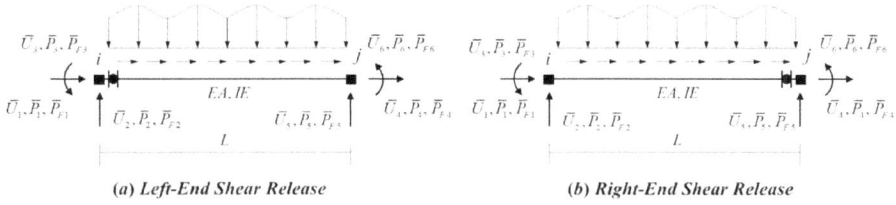

(a) Left-End Shear Release                    (b) Right-End Shear Release

**FIGURE 7.28**   Frame Member with End Shear Release.

$$
\left\{
\begin{array}{c}
\bar{P}_1 \\
\bar{P}_2 \\
\bar{P}_3 \\
\bar{P}_4 \\
\bar{P}_5 \\
\bar{P}_6
\end{array}
\right\}
=
\begin{bmatrix}
\dfrac{EA}{L} & 0 & 0 & -\dfrac{EA}{L} & 0 & 0 \\
0 & 0 & 0 & 0 & 0 & 0 \\
0 & 0 & \dfrac{IE}{L} & 0 & 0 & -\dfrac{IE}{L} \\
-\dfrac{EA}{L} & 0 & 0 & \dfrac{EA}{L} & 0 & 0 \\
0 & 0 & 0 & 0 & 0 & 0 \\
0 & 0 & -\dfrac{IE}{L} & 0 & 0 & \dfrac{IE}{L}
\end{bmatrix}
\left\{
\begin{array}{c}
\bar{U}_1 \\
\bar{U}_2 \\
\bar{U}_3 \\
\bar{U}_4 \\
\bar{U}_5 \\
\bar{U}_6
\end{array}
\right\}
+
\left\{
\begin{array}{c}
\bar{P}_{F1} \\
0 \\
\bar{P}_{F3} - \dfrac{\bar{P}_{F2}L}{2} \\
\bar{P}_{F4} \\
\bar{P}_{F2} + \bar{P}_{F5} \\
\bar{P}_{F6} - \dfrac{\bar{P}_{F2}L}{2}
\end{array}
\right\}
\text{ or } \bar{\mathbf{P}} = \bar{\mathbf{K}}\bar{\mathbf{U}} + \bar{\mathbf{P}}^F
\tag{7.68}
$$

The frame member with right-end shear release shown in Figure 7.28 $b$ is then considered. When end $j$ of a frame member is connected to the right-end joint by a shear-release device, it is obvious that its end force $\bar{P}_5$ has to be zero. By inserting the condition of zero right-end force ($\bar{P}_5 = 0$) into Eq. (7.62), one has:

$$
\bar{U}_5 = -\frac{L^3 \bar{P}_{F5} - 12IE\bar{U}_2 + 6IEL\left(\bar{U}_3 + \bar{U}_6\right)}{12IE}
\tag{7.69}
$$

Similarly, this kinematical relation implies that the right-end transverse displacement $\bar{U}_5$ is no longer an independent degree of freedom but now depends on other displacement degrees of freedom ($\bar{U}_2$, $\bar{U}_5$, and $\bar{U}_6$). With the kinematical relation of Eq. (7.69), Eqs. (7.61), (7.62), and (7.63) are rewritten as:

$$
\bar{P}_2 = \bar{P}_{F2} + \bar{P}_{F5}
\tag{7.70}
$$

$$
\bar{P}_3 = \frac{IE}{L}\bar{U}_3 - \frac{IE}{L}\bar{U}_6 + \bar{P}_{F3} + \frac{\bar{P}_{F5}L}{2}
\tag{7.71}
$$

$$
\bar{P}_6 = -\frac{IE}{L}\bar{U}_3 + \frac{IE}{L}\bar{U}_6 + \bar{P}_{F6} + \frac{\bar{P}_{F5}L}{2}
\tag{7.72}
$$

Therefore, the stiffness equations of a frame with right-end shear release are written in the matrix form as:

$$
\begin{Bmatrix} \bar{P}_1 \\ \bar{P}_2 \\ \bar{P}_3 \\ \bar{P}_4 \\ \bar{P}_5 \\ \bar{P}_6 \end{Bmatrix} = \begin{bmatrix} \dfrac{EA}{L} & 0 & 0 & -\dfrac{EA}{L} & 0 & 0 \\ 0 & 0 & 0 & 0 & 0 & 0 \\ 0 & 0 & \dfrac{IE}{L} & 0 & 0 & -\dfrac{IE}{L} \\ -\dfrac{EA}{L} & 0 & 0 & \dfrac{EA}{L} & 0 & 0 \\ 0 & 0 & 0 & 0 & 0 & 0 \\ 0 & 0 & -\dfrac{IE}{L} & 0 & 0 & \dfrac{IE}{L} \end{bmatrix} \begin{Bmatrix} \bar{U}_1 \\ \bar{U}_2 \\ \bar{U}_3 \\ \bar{U}_4 \\ \bar{U}_5 \\ \bar{U}_6 \end{Bmatrix} + \begin{Bmatrix} \bar{P}_{F1} \\ \bar{P}_{F2} + \bar{P}_{F5} \\ \bar{P}_{F3} + \dfrac{\bar{P}_{F5}L}{2} \\ \bar{P}_{F4} \\ 0 \\ \bar{P}_{F3} + \dfrac{\bar{P}_{F5}L}{2} \end{Bmatrix} \text{ or } \bar{\mathbf{P}} = \bar{\mathbf{K}}\bar{\mathbf{U}} + \bar{\mathbf{P}}^F
$$

$$(7.73)$$

### 7.6.3 FRAME MEMBER WITH END MOMENT RELEASE

The frame member with a left-end moment release (hinge) shown in Figure 7.29a is first considered. When end $i$ of a frame member is connected to the left-end joint by a moment-release device, it is obvious that its end moment $\bar{P}_3$ has to be zero. By inserting the condition of zero left-end moment ($\bar{P}_3 = 0$) into Eq. (7.61), one has:

$$
\bar{U}_3 = -\frac{L^2 \bar{P}_{F3} + 6IE\bar{U}_2 - 6IE\bar{U}_5 + 2IEL\bar{U}_6}{4IEL} \tag{7.74}
$$

This kinematical relation implies that the left-end rotation $\bar{U}_3$ is no longer an independent degree of freedom but now depends on other displacement degrees of freedom ($\bar{U}_2$, $\bar{U}_5$, and $\bar{U}_6$). With the kinematical relation of Eq. (7.74), Eqs. (7.60), (7.62), and (7.63) are rewritten as:

$$
\bar{P}_2 = 3\frac{IE}{L^3}\bar{U}_2 - 3\frac{IE}{L^3}\bar{U}_5 + 3\frac{IE}{L^2}\bar{U}_6 + \bar{P}_{F2} - \frac{3\bar{P}_{F3}}{2L} \tag{7.75}
$$

$$
\bar{P}_5 = -3\frac{IE}{L^3}\bar{U}_2 + 3\frac{IE}{L^3}\bar{U}_5 - 3\frac{IE}{L^2}\bar{U}_6 + \bar{P}_{F5} + \frac{3\bar{P}_{F3}}{2L} \tag{7.76}
$$

$$
\bar{P}_6 = 3\frac{IE}{L^2}\bar{U}_2 - 3\frac{IE}{L^2}\bar{U}_5 + 3\frac{IE}{L}\bar{U}_6 + \bar{P}_{F6} - \frac{\bar{P}_{F3}}{2} \tag{7.77}
$$

Therefore, the stiffness equations of a frame with a left end moment release are written in the matrix form as:

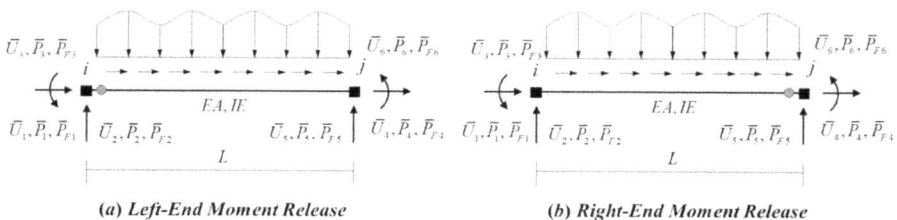

(a) Left-End Moment Release          (b) Right-End Moment Release

**FIGURE 7.29** Frame Member with End Moment Release.

$$\begin{Bmatrix} \bar{P}_1 \\ \bar{P}_2 \\ \bar{P}_3 \\ \bar{P}_4 \\ \bar{P}_5 \\ \bar{P}_6 \end{Bmatrix} = \frac{E}{L} \begin{bmatrix} A & 0 & 0 & -A & 0 & 0 \\ 0 & \dfrac{3I}{L^2} & 0 & 0 & -\dfrac{3I}{L^2} & \dfrac{3I}{L} \\ 0 & 0 & 0 & 0 & 0 & 0 \\ -A & 0 & 0 & A & 0 & 0 \\ 0 & -\dfrac{3I}{L^2} & 0 & 0 & \dfrac{3I}{L^2} & -\dfrac{3I}{L} \\ 0 & \dfrac{3I}{L} & 0 & 0 & -\dfrac{3I}{L} & 3I \end{bmatrix} \begin{Bmatrix} \bar{U}_1 \\ \bar{U}_2 \\ \bar{U}_3 \\ \bar{U}_4 \\ \bar{U}_5 \\ \bar{U}_6 \end{Bmatrix} + \begin{Bmatrix} \bar{P}_{F1} \\ \bar{P}_{F2} - \dfrac{3\bar{P}_{F3}}{2L} \\ 0 \\ \bar{P}_{F4} \\ \bar{P}_{F5} + \dfrac{3\bar{P}_{F3}}{2L} \\ \bar{P}_{F6} - \dfrac{\bar{P}_{F3}}{2} \end{Bmatrix} \quad \text{or } \bar{\mathbf{P}} = \bar{\mathbf{K}}\bar{\mathbf{U}} + \bar{\mathbf{P}}^F \quad (7.78)$$

The frame member with a right-end moment release shown in Figure 7.29$b$ is then considered. When end $j$ of a frame member is connected to the right-end joint by a moment-release device, it is obvious that its end moment $\bar{P}_6$ has to be zero. By inserting the condition of a zero right-end moment ($\bar{P}_6 = 0$) into Eq. (7.63), one has:

$$\bar{U}_6 = -\frac{L^2\bar{P}_{F6} + 6IE\bar{U}_2 - 6IE\bar{U}_5 + 2IEL\bar{U}_3}{4IEL} \tag{7.79}$$

This kinematical relation implies that the right-end rotation $\bar{U}_6$ is no longer an independent degree of freedom but now depends on other displacement degrees of freedom ($\bar{U}_2, \bar{U}_3$, and $\bar{U}_5$). With the kinematical relation of Eq. (7.79), Eqs. (7.60), (7.61), and (7.62) are rewritten as:

$$\bar{P}_2 = 3\frac{IE}{L^3}\bar{U}_2 + 3\frac{IE}{L^2}\bar{U}_3 - 3\frac{IE}{L^3}\bar{U}_5 + \bar{P}_{F2} - \frac{3\bar{P}_{F6}}{2L} \tag{7.80}$$

$$\bar{P}_3 = 3\frac{IE}{L^2}\bar{U}_2 + 3\frac{IE}{L}\bar{U}_3 - 3\frac{IE}{L^2}\bar{U}_5 + \bar{P}_{F3} - \frac{\bar{P}_{F6}}{2} \tag{7.81}$$

$$\bar{P}_5 = -3\frac{IE}{L^3}\bar{U}_2 - 3\frac{IE}{L^2}\bar{U}_3 + 3\frac{IE}{L^3}\bar{U}_5 + \bar{P}_{F5} + \frac{3\bar{P}_{F6}}{2L} \tag{7.82}$$

Therefore, the stiffness equations of a frame with right end moment release are written in the matrix form as:

$$\begin{Bmatrix} \bar{P}_1 \\ \bar{P}_2 \\ \bar{P}_3 \\ \bar{P}_4 \\ \bar{P}_5 \\ \bar{P}_6 \end{Bmatrix} = \frac{E}{L} \begin{bmatrix} A & 0 & 0 & -A & 0 & 0 \\ 0 & \dfrac{3I}{L^2} & \dfrac{3I}{L} & 0 & -\dfrac{3I}{L^2} & 0 \\ 0 & \dfrac{3I}{L} & 3I & 0 & -\dfrac{3I}{L} & 0 \\ -A & 0 & 0 & A & 0 & 0 \\ 0 & -\dfrac{3I}{L^2} & -\dfrac{3I}{L} & 0 & \dfrac{3I}{L^2} & 0 \\ 0 & 0 & 0 & 0 & 0 & 0 \end{bmatrix} \begin{Bmatrix} \bar{U}_1 \\ \bar{U}_2 \\ \bar{U}_3 \\ \bar{U}_4 \\ \bar{U}_5 \\ \bar{U}_6 \end{Bmatrix} + \begin{Bmatrix} \bar{P}_{F1} \\ \bar{P}_{F2} - \dfrac{3\bar{P}_{F6}}{2L} \\ \bar{P}_{F3} - \dfrac{\bar{P}_{F6}}{2} \\ \bar{P}_{F4} \\ \bar{P}_{F5} + \dfrac{3\bar{P}_{F6}}{2L} \\ 0 \end{Bmatrix} \quad \text{or } \bar{\mathbf{P}} = \bar{\mathbf{K}}\bar{\mathbf{U}} + \bar{\mathbf{P}}^F \quad (7.83)$$

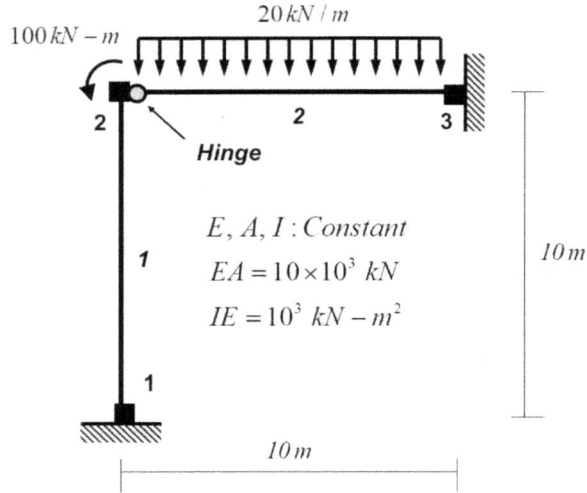

**FIGURE 7.30**   Example 7.5.

## EXAMPLE 7.5

Determine the nodal displacements, member-end forces, and member-end displacements for the frame shown in Figure 7.30.

**Solution:** The local reference systems as well as DOF, member, and nodal numbering systems are shown in Figure 7.31.

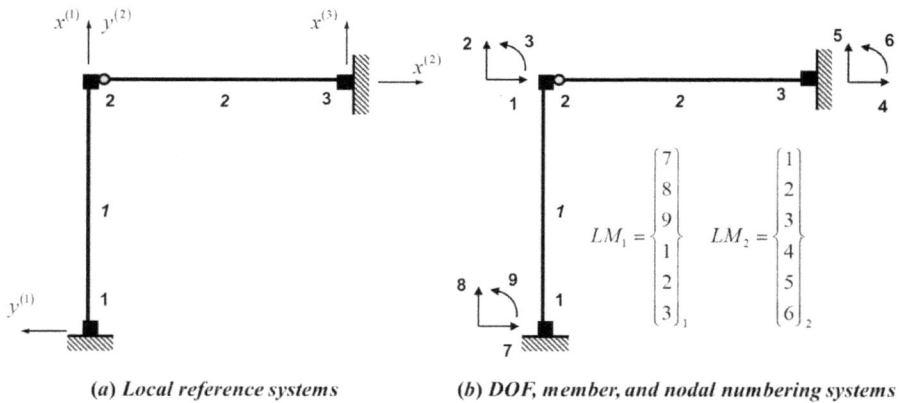

(a) *Local reference systems*          (b) *DOF, member, and nodal numbering systems*

**FIGURE 7.31**   Example 7.5 (Continued).

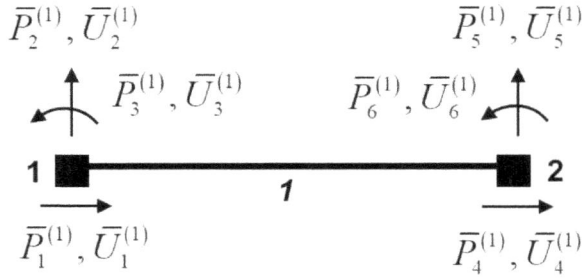

**FIGURE 7.32** Example 7.5 (Continued).

Member stiffness matrices and fixed-end load vectors for each member are:
Member 1 (Figure 7.32):
The local stiffness matrix:

$$\bar{\mathbf{K}}^{(1)} = \begin{bmatrix} 1000 & 0 & 0 & -1000 & 0 & 0 \\ 0 & 12 & 60 & 0 & -12 & 60 \\ 0 & 60 & 400 & 0 & -60 & 200 \\ -1000 & 0 & 0 & 1000 & 0 & 0 \\ 0 & -12 & -60 & 0 & 12 & -60 \\ 0 & 60 & 200 & 0 & -60 & 400 \end{bmatrix}$$

The global stiffness matrix:

$$\mathbf{K}^{(1)} = {\mathbf{\Gamma}_{ROT}^{(1)}}^{T} \bar{\mathbf{K}}^{(1)} \mathbf{\Gamma}_{ROT}^{(1)} = \begin{bmatrix} 12 & 0 & -60 & -12 & 0 & -60 \\ 0 & 1000 & 0 & 0 & -1000 & 0 \\ -60 & 0 & 400 & 60 & 0 & 200 \\ -12 & 0 & 60 & 12 & 0 & 60 \\ 0 & -1000 & 0 & 0 & 1000 & 0 \\ -60 & 0 & 200 & 60 & 0 & 400 \end{bmatrix}$$

Member 2 (Figure 7.33): left-end hinge

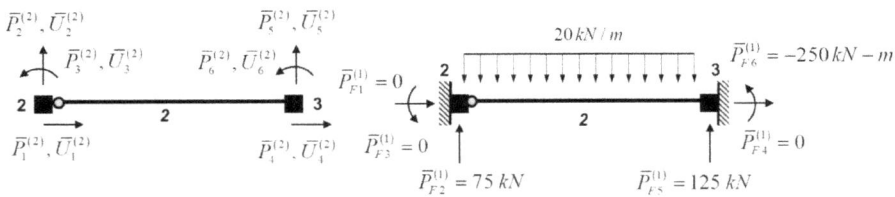

**FIGURE 7.33** Example 7.5 (Continued)

The local and global stiffness matrices:

$$\mathbf{K}^{(2)} = \Gamma_{ROT}^{(2)}{}^{T} \overline{\mathbf{K}}^{(2)} \Gamma_{ROT}^{(2)} = \overline{\mathbf{K}}^{(2)} = \begin{bmatrix} 1000 & 0 & 0 & -1000 & 0 & 0 \\ 0 & 3 & 0 & 0 & -3 & 30 \\ 0 & 0 & 0 & 0 & 0 & 0 \\ -1000 & 0 & 0 & 1000 & 0 & 0 \\ 0 & -3 & 0 & 0 & 3 & -30 \\ 0 & 30 & 0 & 0 & -30 & 300 \end{bmatrix}$$

The global fixed-end force vector:

$$\mathbf{P}_{F}^{(2)} = \Gamma_{ROT}^{(2)}{}^{T} \overline{\mathbf{P}}_{F}^{(2)} = \lfloor 0 \quad 75 \quad 0 \quad 0 \quad 125 \quad -250 \rfloor^{T}$$

Through the member-code technique, the structural stiffness matrix and fixed-end load vector associated with free degrees of freedom are:

$$\mathbf{K}_{ff} = \begin{bmatrix} 1012 & 0 & 60 \\ 0 & 1003 & 0 \\ 60 & 0 & 400 \end{bmatrix} \qquad \text{and} \qquad \mathbf{R}_{free} = \lfloor 0 \quad 75 \quad 0 \rfloor^{T}$$

The applied nodal load vector associated with free degrees of freedom is:

$$\mathbf{F}_{free} = \lfloor 0 \quad 0 \quad 100 \rfloor^{T}$$

The equivalent nodal load vector associated with free degrees of freedom is:

$$\mathbf{F}_{free}^{E} = \mathbf{F}_{free} - \mathbf{R}_{free} = \lfloor 0 \quad -75 \quad 100 \rfloor^{T}$$

The structural stiffness equations corresponding to free degrees of freedom are:

$$\begin{Bmatrix} 0 \\ -75 \\ 100 \end{Bmatrix} = \begin{bmatrix} 1012 & 0 & 60 \\ 0 & 1003 & 0 \\ 60 & 0 & 400 \end{bmatrix} \begin{Bmatrix} U_1 \\ U_2 \\ U_3 \end{Bmatrix} \qquad \text{or} \qquad \mathbf{F}_{free}^{E} = \mathbf{K}_{ff} \mathbf{U}_{free}$$

The free nodal displacements $\mathbf{U}_{free}$ are determined as:

$$\mathbf{U}_{free} = \left[ \mathbf{K}_{ff} \right]^{-1} \mathbf{F}_{free}^{E} = \lfloor -0.01496\,m \quad -0.07478\,m \quad 0.2522\,rad \rfloor^{T}$$

The recovery process of end displacements and end forces will be shown for member 2.

From the local mapping vector $LM_2$ (Figure 7.31$b$), the global member-end displacement vector is:

$$\mathbf{U}^{(2)} = \lfloor -0.01496\,m \quad -0.07478\,m \quad 0 \quad 0 \quad 0 \quad 0 \rfloor^{T}$$

The local member end displacement vector is:

$$\overline{\mathbf{U}}^{(2)} = \Gamma_{ROT}^{(2)} \mathbf{U}^{(2)} = \lfloor -0.01496\,m \quad -0.07478\,m \quad 0 \quad 0 \quad 0 \quad 0 \rfloor^{T}$$

The local member end force vector is:

$$\overline{\mathbf{P}}^{(2)} = \overline{\mathbf{K}}^{(2)}\overline{\mathbf{U}}^{(2)} + \overline{\mathbf{P}}_F^{(2)} = \lfloor -14.95\ kN \quad 74.78\ kN \quad 0 \quad 14.95\ kN \quad 125.22\ kN \quad -252.24\ kN\text{-}m \rfloor^T$$

The global member end force vector is:

$$\mathbf{P}^{(2)} = \mathbf{\Gamma}_{ROT}^{(2)}{}^T \overline{\mathbf{P}}^{(2)} = \lfloor -14.95\ kN \quad 74.78\ kN \quad 0 \quad 14.95\ kN \quad 125.22\ kN \quad -252.24\ kN\text{-}m \rfloor^T$$

End displacements and end forces of member 1 can be recovered in a similar fashion.

## 7.7 EXERCISES

**Problem 7.1:** For the complex truss structure shown in Figure 7.34, use the member-wise assembling technique to construct the structural stiffness matrix and determine all nodal displacements as well as all member forces.

**Problem 7.2:** For the frame structure shown in Figure 7.35, use the member-code assembling technique to construct the structural stiffness matrix as well as equivalent nodal forces and determine all nodal displacements as well as all member end forces.

**Problem 7.3:** Compute the nodal displacements, member-end forces, and support forces for the frame shown in Figure 7.36.

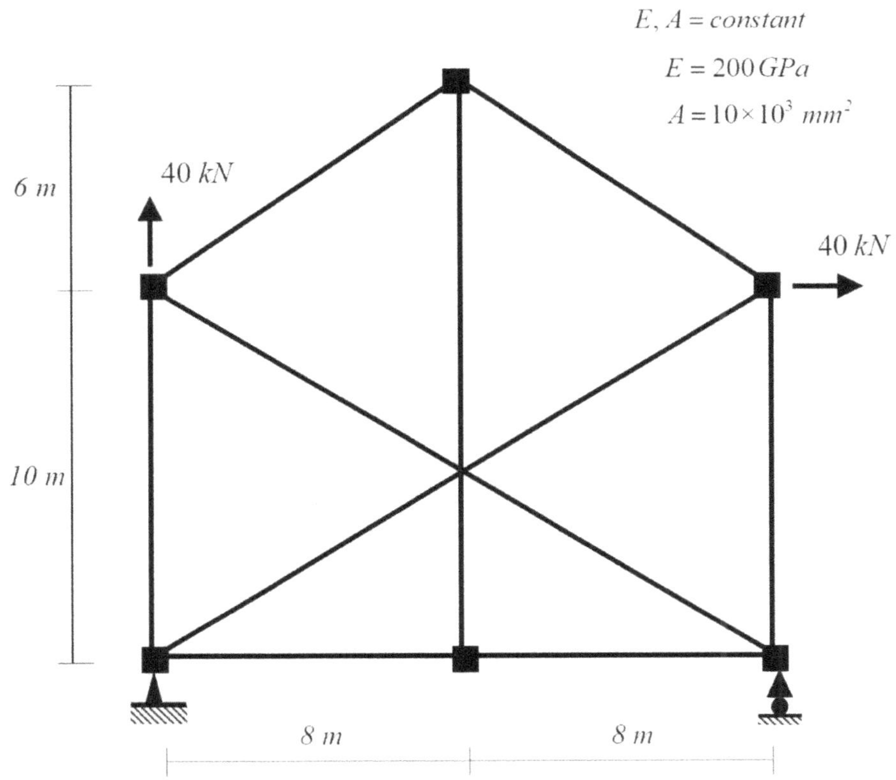

$E, A = constant$

$E = 200\ GPa$

$A = 10 \times 10^3\ mm^2$

$6\ m$

$40\ kN$

$40\ kN$

$10\ m$

$8\ m$

$8\ m$

**FIGURE 7.34** Problem 7.1.

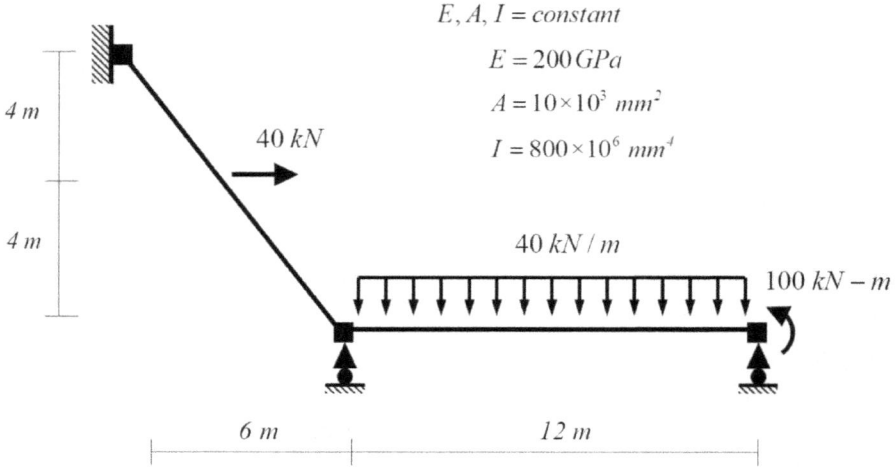

**FIGURE 7.35**   Problem 7.2.

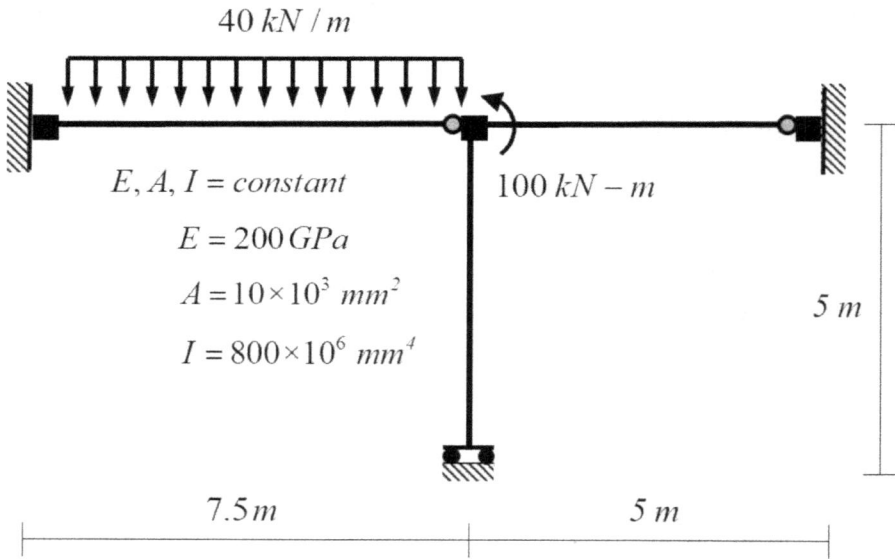

**FIGURE 7.36**   Problem 7.3.

## REFERENCES

Armenakas, A.E. 1991. *Modern structural analysis: The matrix method approach.* McGraw-Hill Inc.

Dawe, D.J. 1984. *Matrix and finite element displacement analysis of structures.* Clarendon Press.

Elias, Z.M. 1986. *Theory and methods of structural analysis.* John Wiley and Son Inc.

Felton, L.P. and R.B. Nelson. 1997. *Matrix structural analysis.* John Wiley & Sons Inc.

Ghali, A. and A.M. Neville. 2017. *Structural analysis: A unified classical and matrix approach* (7th Edition). CRC Press.

Kanchi, M.B. 1993. *Matrix methods of structural analysis* (2nd Edition). Wiley Eastern Limited.

Martin, H.C. 1966. *Introduction to matrix methods of structural analysis.* McGraw-Hill Inc.

McGuire, W., R.H. Gallagher, and R.D. Ziemian. 2000. *Matrix structural analysis* (2nd Edition). John Wiley & Sons Inc.

Pandit, G.S. and S.P. Gupta. 2008. *Structural analysis: A matrix approach* (2nd Edition). Tata McGraw-Hill Publishing Company Limited.

Saouma, V.E. 1997. *CVEN 5525 course notes: Matrix structural analysis.* Department of Civil, Environmental, and Architectural Engineering, University of Colorado.

Spacone, E. 2003. *CVEN 5525 course notes: Matrix structural analysis.* Department of Civil, Environmental, and Architectural Engineering, University of Colorado.

Tezcan, S.S. 1963. Discussion of simplified formulation of stiffness matrices by P.M. Wright. *ASCE Journal of the Structural Division* 89(6): 445–449.

Weaver, W. and J.M. Gere. 1990. *Matrix analysis of framed structures.* Chapman & Hall, Boca Raton.

# 8 The Stiffness Method III
## *Computer-oriented Algorithm*

## 8.1 INTRODUCTION

In this chapter, the analysis procedure of frame structures within the framework of a frame analysis program (e.g. GRASP) is presented in a computerized manner. There are several common aspects between the procedure presented herein and the one employed in a general-purpose finite element program (e.g. ABAQUS) or, more generally, of a computational platform (e.g. FEAP). A detailed discussion of the general layout of such computational frameworks is out of scope and can be found in more advanced classes in the sequence of structural engineering courses as discussed in Chapter 1.

Generally, there are four steps forming the core of any structural computational procedure, and they must be embedded into any structural (finite element) analysis program (Spacone 2003):

(a) Collect input data (e.g. nodal coordinates)
(b) Assemble database containing constant information (e.g. member stiffness matrices)
(c) Perform analysis (linear or nonlinear / statics or dynamics)
(d) Elaborate output data (e.g. graphical representations)

## 8.2 GENERAL ALGORITHM

The general algorithm for implementing the direct stiffness method can be summarized as follows:

### 8.2.1 PRELIMINARY PART

1. Classify the member type used to analyze the actual structure (beam, truss, frame, or grid) and specify
   (a) Number of spatial coordinates (1D, 2D, or 3D).
   (b) Number of degrees of freedom per node (local and global).
   (c) Number of sectional and material properties.
2. Specify the free and restrained degree-of-freedom numbering system.

### 8.2.2 ANALYSIS PART

1. For each member, determine the following quantities:
   (a) Mapping vector $LM_i$ relating member to structural degrees of freedom and vice versa.
   (b) The local member stiffness matrix (complete stiffness matrix $\bar{\mathbf{K}}^{(i)}$ or basic stiffness matrix $\mathbf{k}^{(i)}$).

DOI: 10.1201/9781003595458-8

(c)  Angles between the local and global reference systems.

(d)  Transformation matrices (e.g. $\mathbf{\Gamma}_{ROT}^{(i)}$ , $\mathbf{\Gamma}_{RBM}^{(i)}$ , and $\mathbf{\Gamma}_{REZ}^{(i)}$ ).

(e)  The member stiffness matrix in the global reference system (e.g. $\mathbf{K}^{(i)} = \mathbf{\Gamma}^{(i)^T} \bar{\mathbf{K}}^{(i)} \mathbf{\Gamma}^{(i)}$ ).

The member fixed-end load vector $\mathbf{P}_F^{(i)}$ .

2.  Assemble the complete structural stiffness matrix $\mathbf{K}$.

3.  Extract $\mathbf{K}_{ff}$, $\mathbf{K}_{fc}$, $\mathbf{K}_{cf}$, and $\mathbf{K}_{cc}$ from $\mathbf{K}$.

4.  Invert $\mathbf{K}_{ff}$ (or decompose $\mathbf{K}_{ff}$ as $\mathbf{LL}^T$).

5.  Assemble the equivalent nodal load vector $\mathbf{F}^E$.

6.  Establish the constrained nodal displacement vector $\mathbf{U}_{constr}$.

7.  Determine the free nodal displacements $\mathbf{U}_{free}$ from $\mathbf{U}_{free} = \left[ \mathbf{K}_{ff} \right]^{-1} \left( \mathbf{F}_{free}^E - \mathbf{K}_{fc} \mathbf{U}_{constr} \right)$.

8.  Determine the reactive forces from $\mathbf{F}_{constr} = \mathbf{K}_{cf} \mathbf{U}_{free} + \mathbf{K}_{cc} \mathbf{U}_{constr} + \mathbf{R}_{constr}$. Alternatively, the reactive forces can be determined through the stiffness equations of members attached to supports.

9.  For each member, transform its end displacements from global to local coordinates $\bar{\mathbf{U}}^{(i)} = \mathbf{\Gamma}_{ROT}^{(i)} \mathbf{U}^{(i)}$ and determine its end forces from the member stiffness equation $\bar{\mathbf{P}}^{(i)} = \bar{\mathbf{K}}^{(i)} \bar{\mathbf{U}}^{(i)} + \bar{\mathbf{P}}_F^{(i)}$.

## 8.3  COMPUTER PROGRAM FLOWCHARTS

Referring to the program flowchart in Figure 8.1, the main routine should perform the following tasks:

1.  Read (input data):
    (a)  The job title, analyzer's name, etc.
    (b)  The control parameters.
        1.  Number of nodes.
        2.  Number of members.
        3.  Types of structures.
        4.  Number of different member properties.
        5.  Number of loading cases.
2.  Specify the structural parameters
    (a)  Number of spatial coordinates for the structure.
    (b)  Number of degrees of freedom per node (local and global).
3.  Loop over all members and compute local member stiffness matrices as well as the transformation matrices.
4.  Initialize the global stiffness matrix to zero.
5.  Loop over all members and for each member
    (a) Retrieve the local member stiffness matrix and transform it from local to global systems.
    (b) Establish the local mapping vector.
    (c) Assemble the structural stiffness matrix.
6.  Decompose the global stiffness matrix (e.g. Cholesky's decomposition).
7.  For each loading case
    (a) Loop over associated members and construct member fixed-end load vectors.
    (b) Assemble the load vector.
    (c) Determine the nodal displacements.

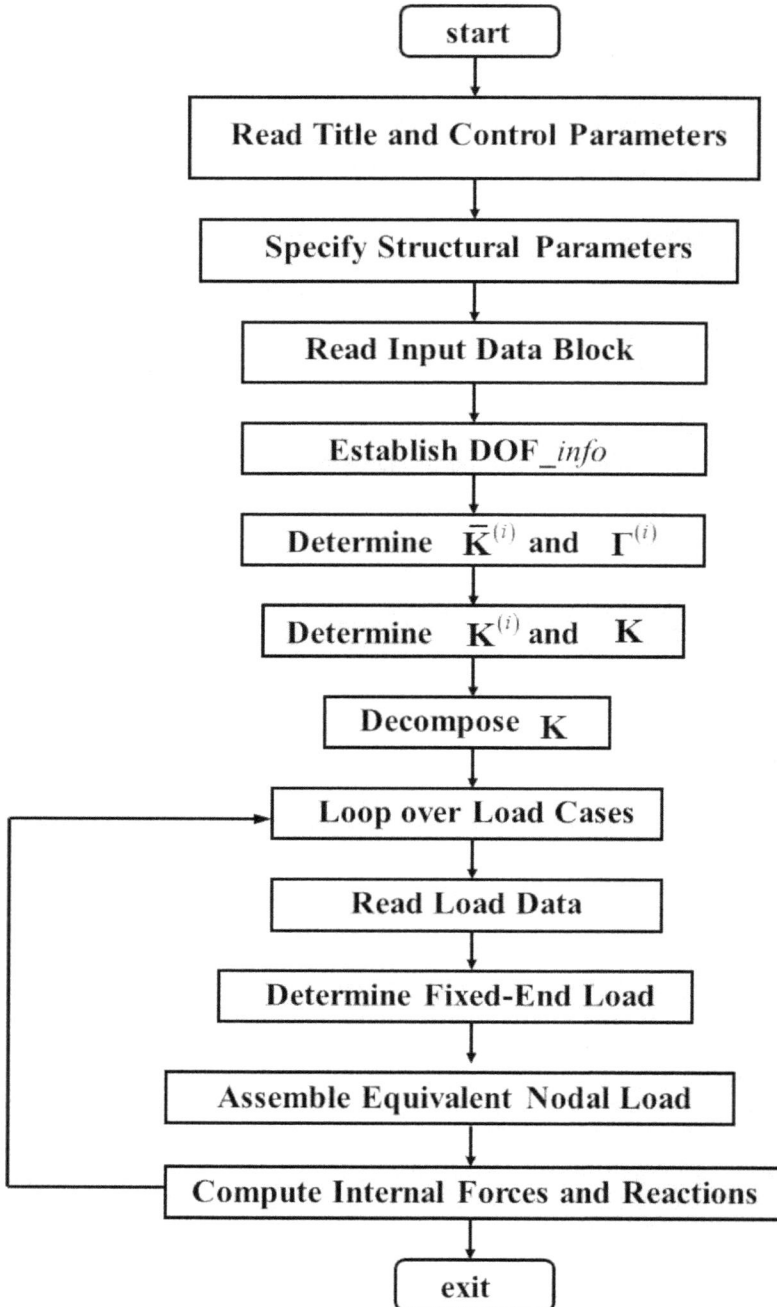

**FIGURE 8.1**   Program Flowchart.

    (d)  Loop over all members and for each member.
8. Determine the end displacements (local and global).
9. Determine the end forces (local and global).

## 8.4   INPUT DATA

Generally, the basic input data for a frame structure must include the following data blocks as suggested by Spacone (2003):

- *GEOMETRY*
- *MEMBERS*
- *MATERIALS*
- *BOUNDARY CONDITIONS*
- *APPLIED NODAL LOADS*
- *APPLIED MEMBER LOADS*
- *APPLIED DISPLACEMENTS*
- *LEVEL OF ANALYSIS*

**GEOMETRY:** this input data block normally collects node numbers and node coordinates in the global reference system.

**MEMBERS:** for a plane (2D) frame, this input data block usually collects member connectivity (starting from node $i$ and ending at node $j$ for each member), member type (bar, beam, etc.), and material sets.

**MATERIALS:** for each material set of a linear elastic 2D frame, this input data block usually collects elastic modulus $E$, shear modulus $G$, section area $A$, and moment of inertia $I$.

**BOUNDARY CONDITIONS:** an appropriate set of boundary conditions must be provided to prevent the structure from forming a mechanism. In general, a structure will possess many more unconstrained degrees of freedom than constrained degrees of freedom. Thus, this input data block usually contains only information on the constrained degrees of freedom (boundary conditions).

**APPLIED NODAL LOADS:** this input data block usually contains information on applied nodal loads.

**APPLIED MEMBER LOADS:** this input data block usually collects information on applied member loads acting on members between nodes.

**APPLIED DISPLACEMENTS:** when supports are subjected to prescribed displacements (e.g. settlements), this input data block contains information regarding such known displacements. Prescribed displacements are usually specified at constrained degrees of freedom only.

**LEVEL OF ANALYSIS:** this input data block contains information on how the loads and/or displacements are imposed and how analysis is performed. In this textbook, this input data block may not be necessary since only a linear elastic analysis is considered.

### 8.4.1   CONSTANT DATABASE

The following arrays are employed to collect the information created from the input data. Usually, this information does not change during analysis and needs to be constructed only once before the analysis process.

**NODE information:** the array **NODE**_*info* contains the coordinates (X, Y) for each node.

$$\mathbf{NODE}\_info \ = \ \big[\,Array\,with\,NODE\,data\,\big]_{\text{num\_nodes} \times \text{num\_coor}}$$

$$\textbf{NODE}\_info = \begin{bmatrix} X_1 & Y_1 \\ \vdots & \vdots \\ X_{num\_nodes} & Y_{num\_nodes} \end{bmatrix} \Rightarrow 2D$$

$$\textbf{NODE}\_info = \begin{bmatrix} X_1 & Y_1 & Z_1 \\ \vdots & \vdots & \vdots \\ X_{num\_nodes} & Y_{num\_nodes} & Z_{num\_nodes} \end{bmatrix} \Rightarrow 3D$$

where num_nodes represents the number of nodes in a structure and num_coor represents the number of spatial coordinates.

Figure 8.2 represents the flowchart for reading and storing geometric data array **NODE**_info.

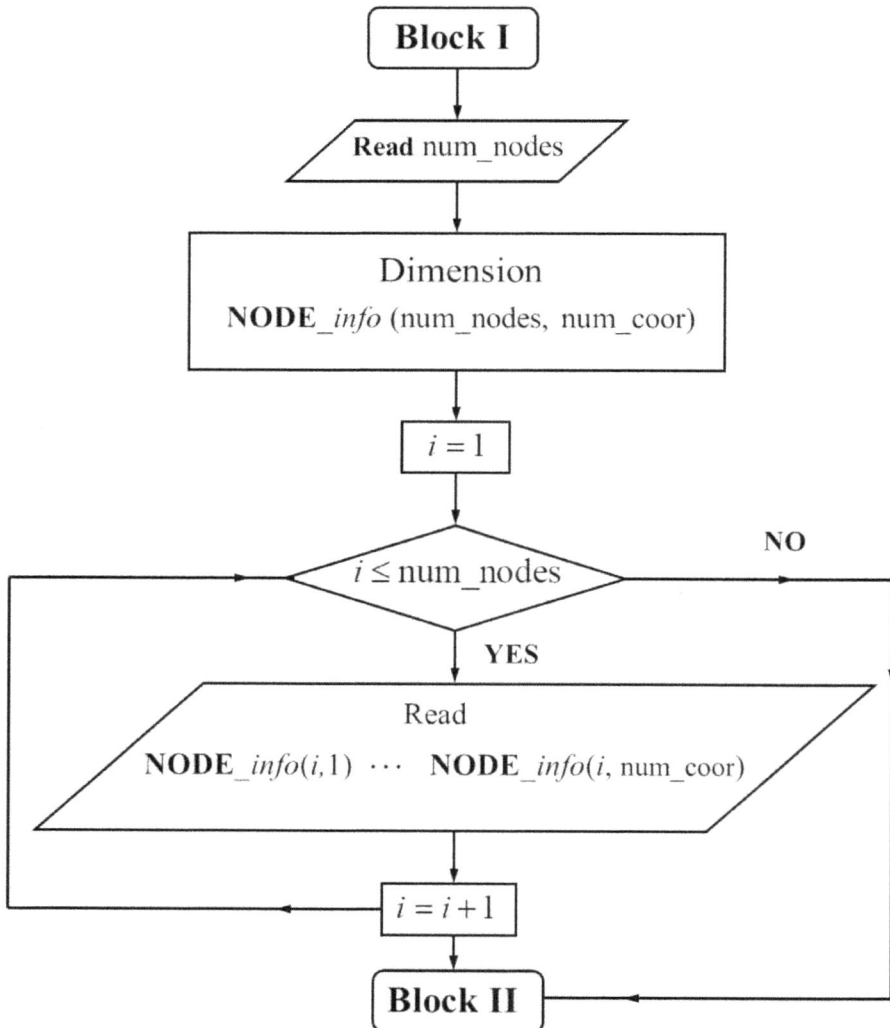

**FIGURE 8.2**   Flowchart for Reading and Storing Geometric Data Block **NODE**_info.

**BOUNDARY information:** due to the support conditions, the total number of free degrees of freedom will be less than the total number of degrees of freedom (num_nodes × Node_DOF). All information of boundary conditions is stored in the array **BOUN**_*info*.

$$\mathbf{BOUN}\_info = \big[ Array\, with\, Boundary\, Conditions \big]_{\text{num\_nodes} \times \text{Node\_DOF}}$$

where Node_DOF denotes the maximum degree-of-freedom number for each node.

**BOUN**_*info* is initialized to zero at the beginning of the input stage. Typically, the restrained code is assigned as:

$$\mathbf{BOUN}\_info\,(i\_DOF, i\_NODE) = \begin{cases} 0 & if \quad free\ DOF \\ 1 & if \quad locked\ DOF \end{cases}$$

After all nodal boundary conditions have been read, equation numbers are assigned incrementally, first to all free degrees of freedom with a positive sign and then to all constrained degrees of freedom with a negative sign.

The maximum positive degree-of-freedom number indicates the size of the square matrix ($\mathbf{K}_{ff}$) needed to be decomposed (e.g. Cholesky's Decomposition).

Consider the frame shown in Figure 8.3. The corresponding array **BOUN**_*info* is

$$\mathrm{DOF} \rightarrow \quad 1 \quad 2 \quad 3$$
$$\mathbf{BOUN}\_info = \begin{bmatrix} 1 & 1 & 0 \\ 0 & 0 & 0 \\ 0 & 1 & 0 \end{bmatrix} \begin{matrix} 1 \\ 2 \\ 3 \end{matrix} \leftarrow \mathrm{Node}$$

The **DOF**_*info* associated with **BOUN**_*info* is

$$\mathbf{DOF}\_info = \begin{bmatrix} -7 & -8 & 1 \\ 2 & 3 & 4 \\ 5 & -9 & 6 \end{bmatrix} \begin{matrix} 1 \\ 2 \\ 3 \end{matrix} \leftarrow \mathrm{Node}$$

The resulting degree-of-freedom numbering system is shown in Figure 8.4.

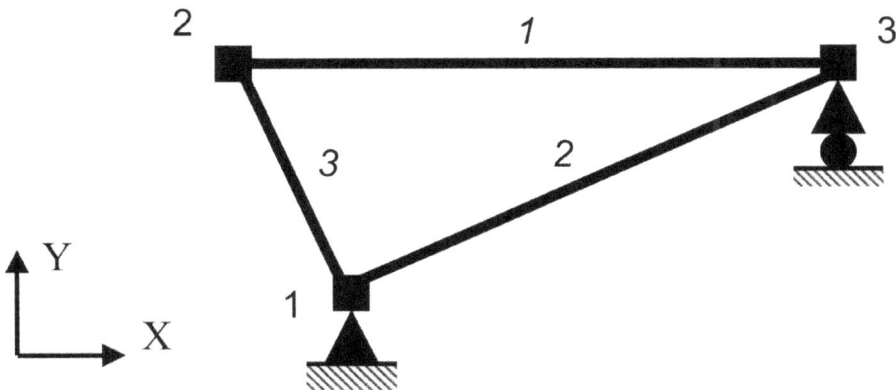

**FIGURE 8.3** Example for Constructing Array **BOUN**_*info*.

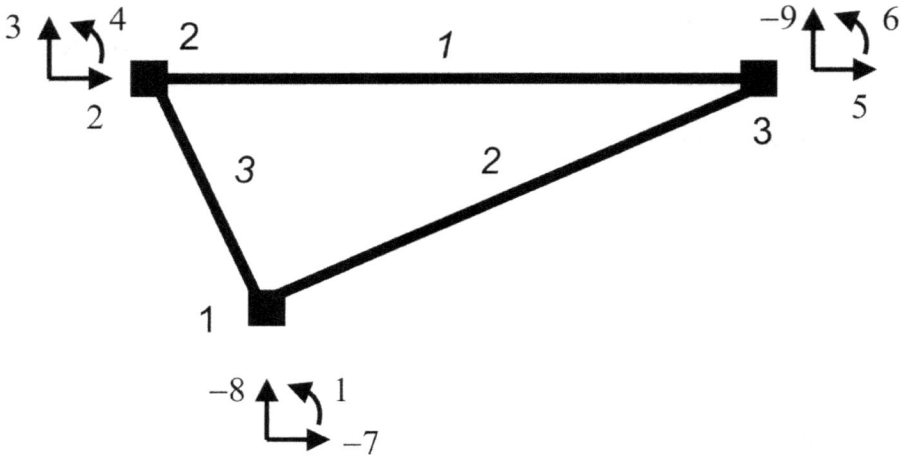

**FIGURE 8.4**  Resulting Degree-of-Freedom Numbering System.

The support codes for some typical supports for plane frames are given in Figure 8.5.

Figure 8.6 represents the flowchart for reading and storing support data array **BOUN**_*info*.

**MATERIAL information:** the array **MATE**_*info* collects given $E$ (or $G$), $A$, and $I$ for each material type.

$$\textbf{MATE}\_info = \begin{bmatrix} Array\, with\, MATERIAL\, data \end{bmatrix}_{\text{num\_mates} \times \text{mate\_para}}$$

$$\textbf{MATE}\_info = \begin{bmatrix} E_1 & A_1 \\ \vdots & \vdots \\ E_{\text{num\_mates}} & A_{\text{num\_mates}} \end{bmatrix} \Rightarrow \text{2D \& 3D Truss}$$

$$\textbf{MATE}\_info = \begin{bmatrix} E_1 & A_1 & I_1 \\ \vdots & \vdots & \vdots \\ E_{\text{num\_mates}} & A_{\text{num\_mates}} & I_{\text{num\_mates}} \end{bmatrix} \Rightarrow \text{2D Euler - Bernoulli Frame}$$

$$\textbf{MATE}\_info = \begin{bmatrix} E_1 & A_1 & I_1 & G_1 \\ \vdots & \vdots & \vdots & \vdots \\ E_{\text{num\_mates}} & A_{\text{num\_mates}} & I_{\text{num\_mates}} & G_{\text{num\_mates}} \end{bmatrix} \Rightarrow \text{2D Timoshenko Frame}$$

where num_mates is denoted for the number of material sets and mate_para for the number of material-set parameters.

Figure 8.7 represents the flowchart for reading and storing material-set data array **MATE**_*info*.

**MEMBER information:** the array **MEM**_*info* collects the starting node $i$, the end node $j$, the member type (e.g. Euler-Bernoulli member, Timoshenko member, bar member, etc.), and material properties for each member.

$$\textbf{MEM}\_info = \begin{bmatrix} Array\, with\, MEMBER\, data \end{bmatrix}_{\text{num\_members} \times \text{member\_para}}$$

| Support Type | | Support Code |
|---|---|---|
| *Free* | 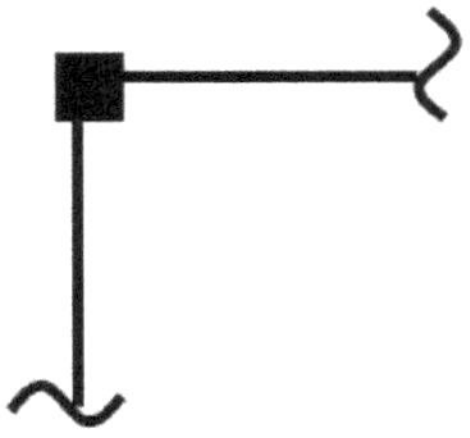 | **0, 0, 0** |
| *Roller* | 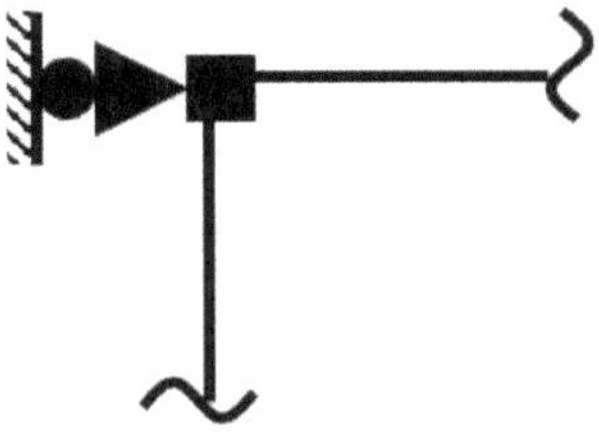 | **1, 0, 0** |
| *Hinge* | 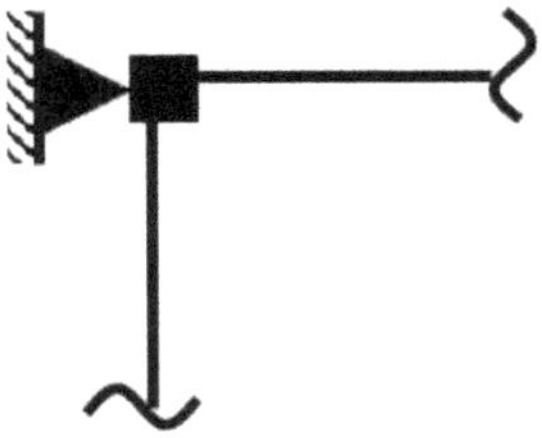 | **1, 1, 0** |
| *Shear Release (Collar)* | 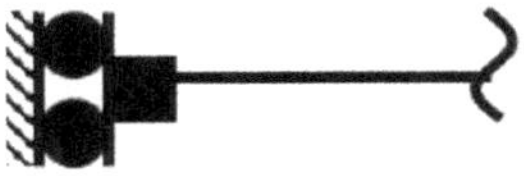 | **1, 0, 1** |
| *Fixed* |  | **1, 1, 1** |

**FIGURE 8.5**    Support Codes for Plane Frames.

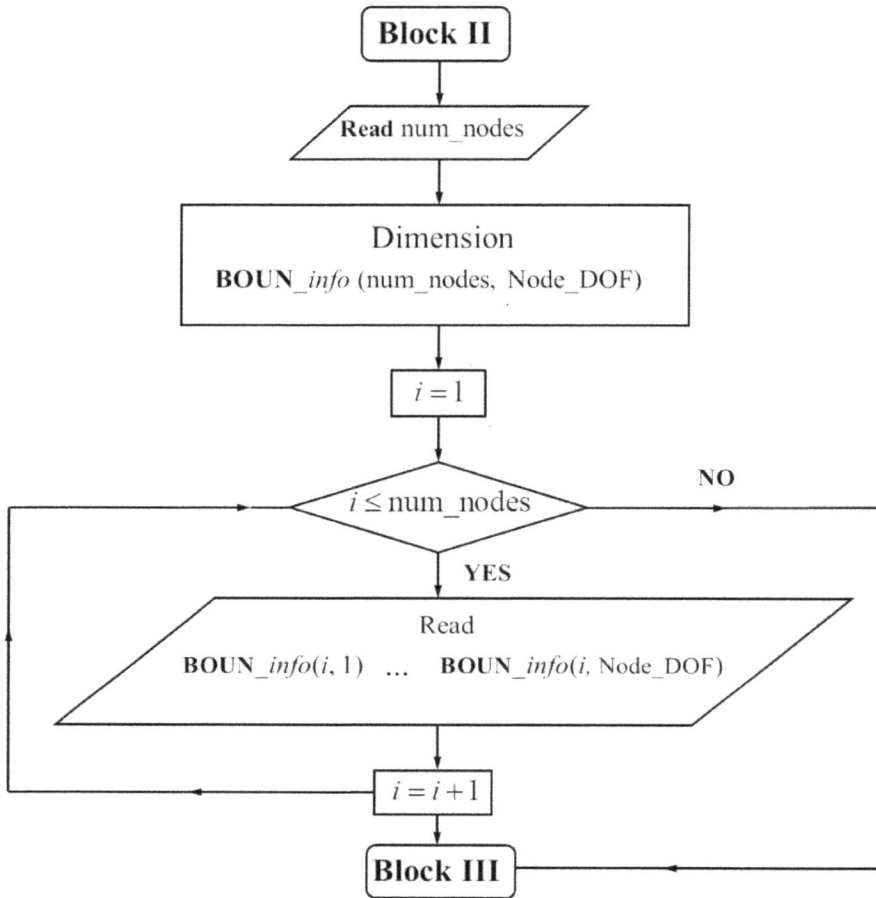

**FIGURE 8.6**    Flowchart for Reading and Storing Support Data Block **BOUN**_*info*.

$$\mathbf{MEM\_}\mathit{info} = \begin{bmatrix} node\_i_1 & node\_j_1 & mem\_type_1 & mate\_type_1 \\ \vdots & \vdots & \vdots & \vdots \\ node\_i_{num\_members} & node\_j_{num\_members} & mem\_type_{num\_members} & mate\_type_{num\_members} \end{bmatrix}$$

where num_members denotes the number of members in a structure and member_para denotes the number of member parameters.

Figure 8.8 represents the flowchart for reading and storing member data array **MEM**_*info*.

**DOF information:** it is essential to provide the number and information about restrained condition (constrained or unconstrained) for each degree of freedom. In this chapter, it can be achieved by assigning a sign to the degree-of-freedom number. Normally, a positive sign defines an unconstrained degree of freedom, and a negative sign defines a constrained degree of freedom. The degree-of-freedom numbering information is stored in the array **DOF**_*info*.

$$\mathbf{DOF\_}\mathit{info} = \begin{bmatrix} Array\ with\ DOF\ data \end{bmatrix}_{num\_nodes \times Node\_DOF}$$

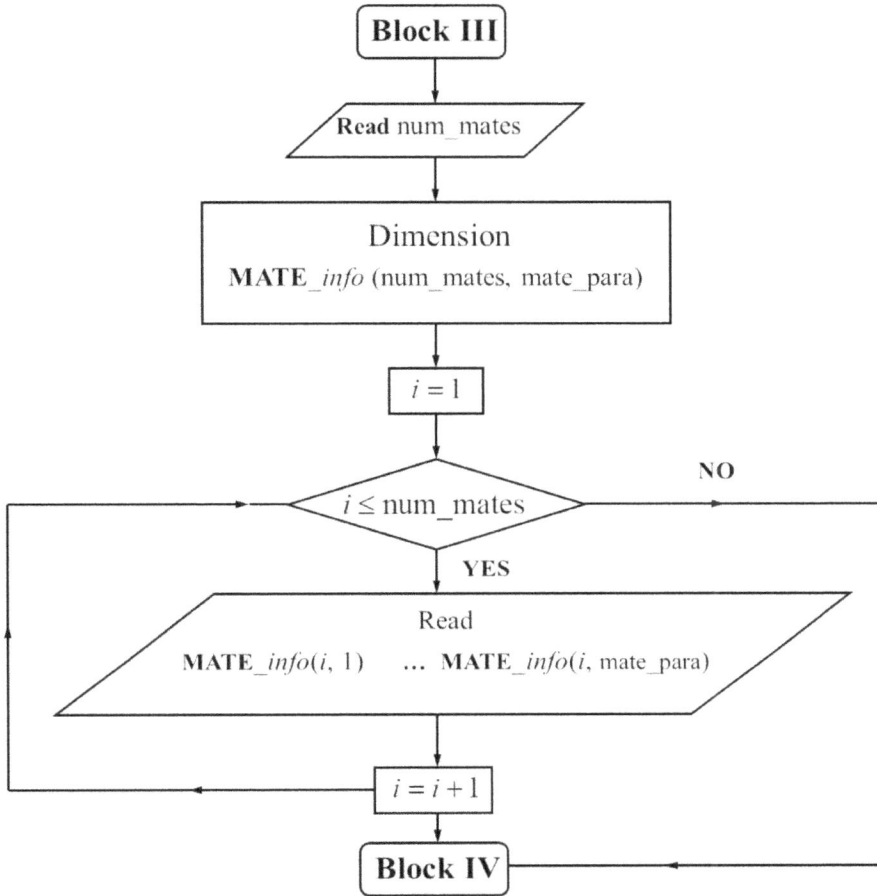

**FIGURE 8.7**    Flowchart for Reading and Storing Material Data Block **MATE**_*info*.

$$\mathbf{DOF\_info} = \begin{bmatrix} DOF\_X_1 & DOF\_Y_1 \\ \vdots & \vdots \\ DOF\_X_{num\_nodes} & DOF\_Y_{num\_nodes} \end{bmatrix} \Rightarrow \text{2D Truss}$$

$$\mathbf{DOF\_info} = \begin{bmatrix} DOF\_X_1 & DOF\_Y_1 & DOF\_\theta_1 \\ \vdots & \vdots & \vdots \\ DOF\_X_{num\_nodes} & DOF\_Y_{num\_nodes} & DOF\_\theta_{num\_nodes} \end{bmatrix} \Rightarrow \text{2D Frame}$$

The array **DOF**_*info* is constructed in conjunction with the array **BOUN**_*info* containing information of boundary conditions.

   **APPLIED NODAL LOAD information:** the nodal loads are stored into a single reference array **NODAL_LOAD**_*info*.

$$\mathbf{NODAL\_LOAD}\_info = \begin{bmatrix} Array\, with\, APPLIED\, NODAL\, LOAD \end{bmatrix}_{num\_nodes \times Node\_DOF}$$

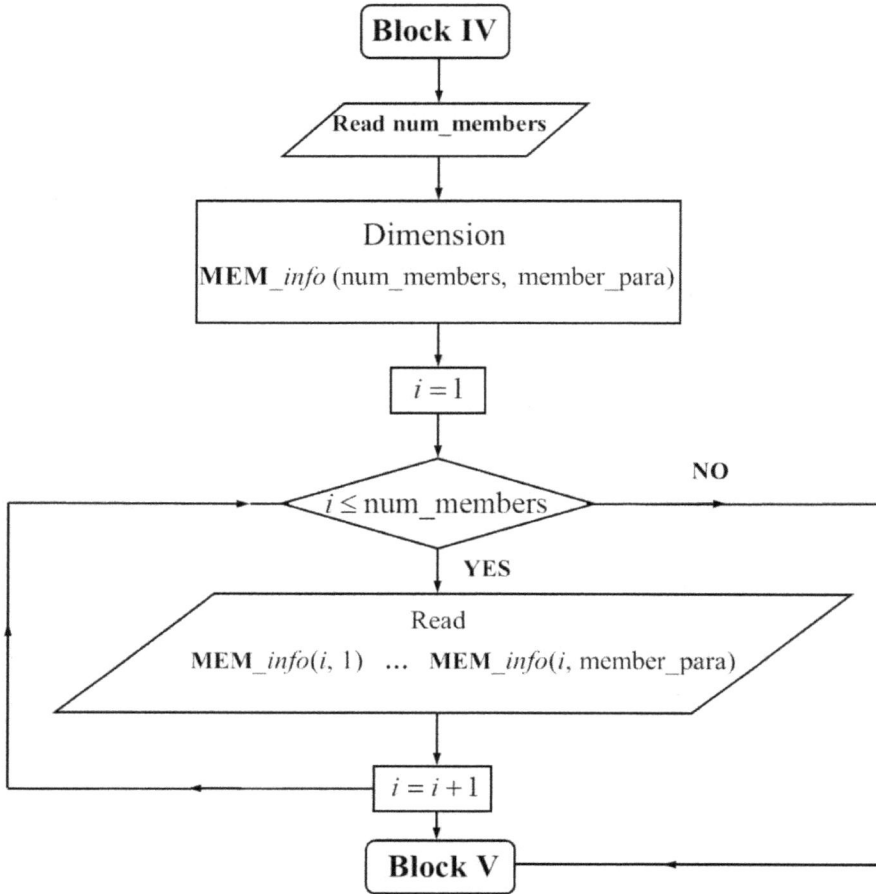

**FIGURE 8.8**    Flowchart for Reading and Storing Member Data Block **MEM_**$info$.

The applied nodal loads associated with free degrees of freedom are extracted from **NODAL _ LOAD _** $info$ and collected in the vector $\mathbf{F}_{free}$ as:

$$\mathbf{F}_{free} = \left[ APPLIED\ NODAL\ LOAD\ AT\ FREE\ DOF \right]_{num\_DOF\_free}$$

where num_DOF_free denotes the number of free degrees of freedom.

Figure 8.9 represents the flowchart for reading and storing the nodal load data array **NODAL _ LOAD _** $info$.

**APPLIED MEMBER LOAD information:** the member loads are stored into a single reference array **MEM _ LOAD _** $info$.

$$\mathbf{MEM \_ LOAD \_} info = \left[ Array\ with\ MEMBER\ LOAD \right]_{num\_members \times ele\_load\_para}$$

where mem_load_para denotes the number of member-load parameters.

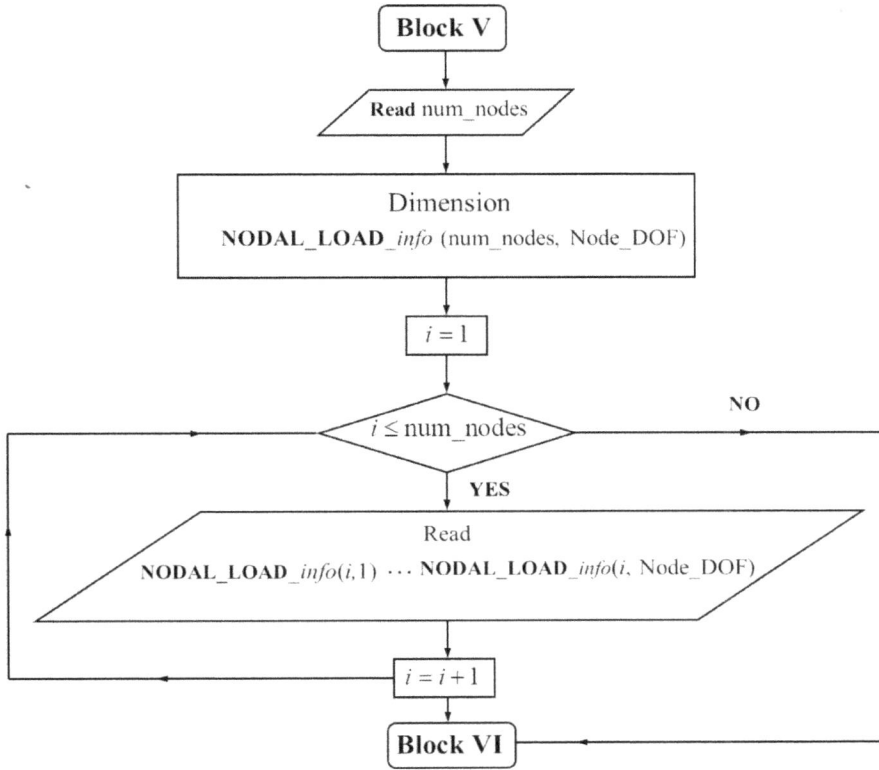

**FIGURE 8.9**   Flowchart for Reading and Storing Nodal Load Data Block **NODAL_LOAD**_info

The fixed-end load vector is constructed in conjunction with member-load data in **MEM _ LOAD _** _info_ and is stored in the vectors $\mathbf{F}^F_{free}$ and $\mathbf{F}^F_{constr}$ as:

$$\mathbf{F}^F_{free} = \big[ FIXED-END\ NODAL\ LOAD\ AT\ FREE\ DOF \big]_{\text{num\_DOF\_free}}$$

$$\mathbf{F}^F_{constr} = \big[ FIXED-END\ NODAL\ LOAD\ AT\ CONSTRAINED\ DOF \big]_{\text{num\_DOF\_constr}}$$

where num_DOF_constr denotes the number of constrained degrees of freedom.

Figure 8.10 represents the flowchart for reading and storing the member load data array **MEM _ LOAD _** _info_.

**APPLIED DISPLACEMENT information:** the applied nodal displacements are stored into a single reference array **NODAL _ DIS _** _info_.

$$\mathbf{NODAL \_ DIS \_} info = \big[ Array\ with\ APPLIED\ NODAL\ DISPLACEMENT \big]_{\text{num\_nodes} \times \text{Node\_DOF}}$$

The applied nodal displacements associated with constrained degrees of freedom are extracted from **NODAL _ DIS _** _info_ and collected in the vector $\mathbf{U}_{constr}$ as:

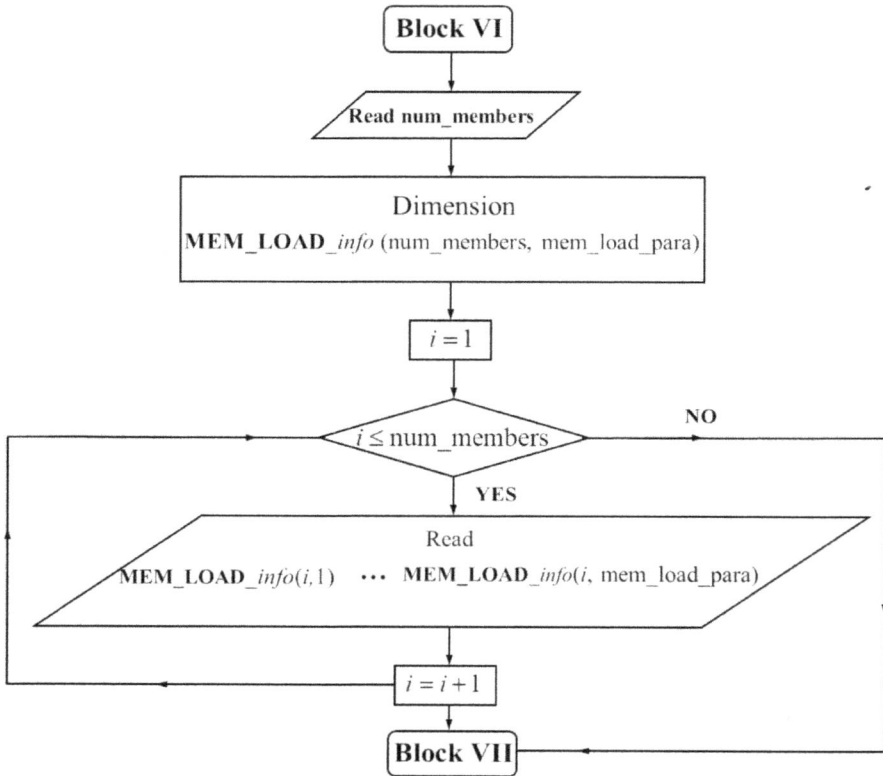

**FIGURE 8.10**     Flowchart for Reading and Storing Member Load Data Block **MEM _ LOAD _** *info.*

$$
\mathbf{U}_{constr} = \left[ \underbrace{Array\,with\,APPLIED\,NODAL\,DISPLACEMENT}_{\text{num\_DOF\_constr}} \right]
$$

It is important to note that the sum of num_DOF_free and num_DOF_constr equals the total degree-of-freedom number (num_DOFs).

Figure 8.11 represents the flowchart for reading and storing the nodal displacement data array **NODAL _ DIS _** *info.*

## 8.4.2  Demonstration: A Small Frame

The above-mentioned input data blocks will be discussed through a very simple frame example. The frame with support conditions, applied nodal and member loads, applied support settlement, and geometric and material properties is shown in Figure 8.12.

The nodes and members, as prepared for the input data, are shown in Figure 8.13.

The procedure for generating the local reference systems from the input data is shown in Figure 8.14. For each member, the local *x*-axis starts from node *i* to node *j*. It is noted that the local *y*-axis is selected in such a way as to form a right reference system with the *z*-axis, which in this example is normal to the sheet plane and directed toward the reader.

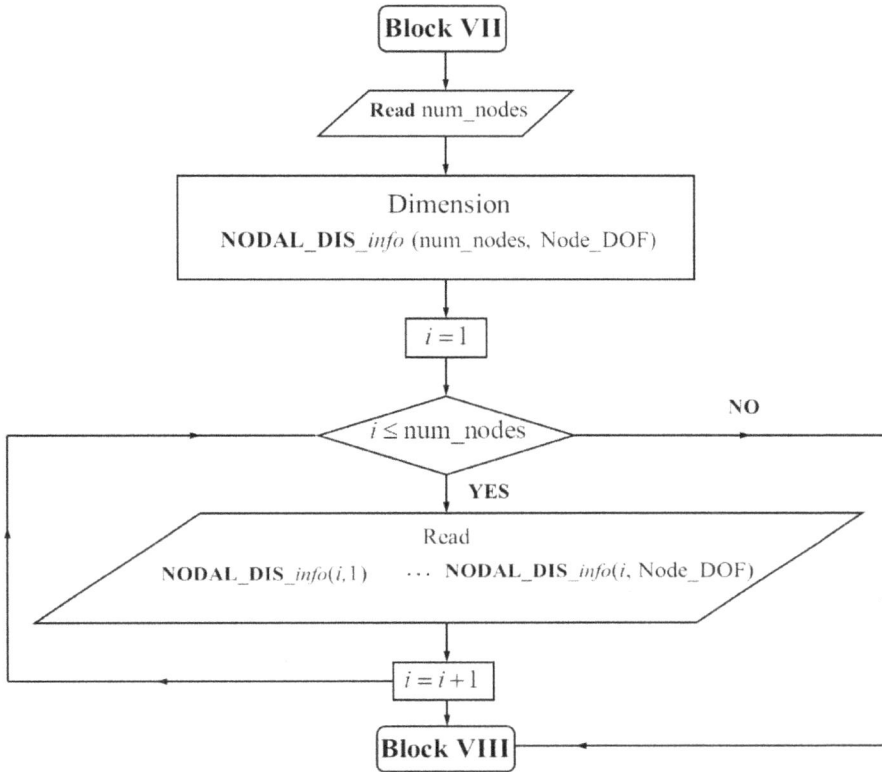

**FIGURE 8.11**   Flowchart for Reading and Storing Nodal Displacement Data Block **NODAL_DIS_*info***

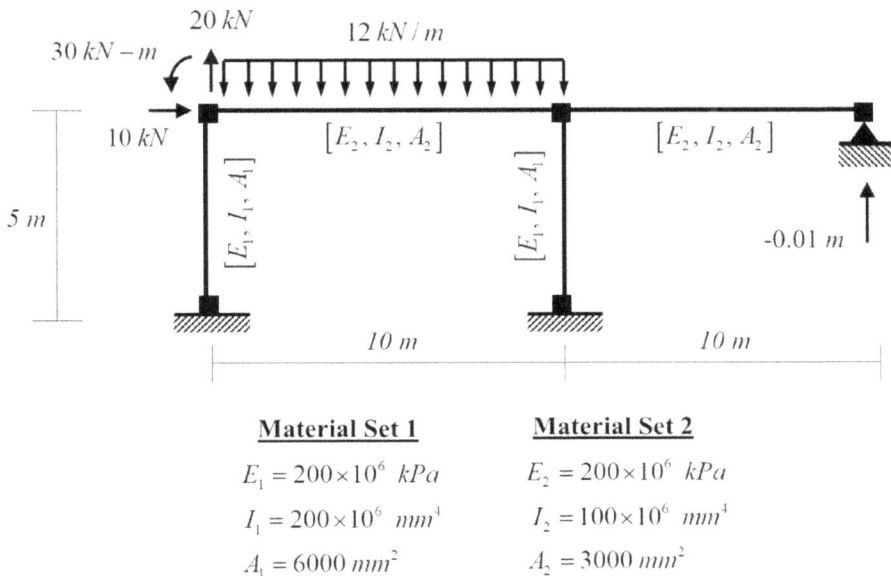

**Material Set 1**

$E_1 = 200 \times 10^6 \ kPa$

$I_1 = 200 \times 10^6 \ mm^4$

$A_1 = 6000 \ mm^2$

**Material Set 2**

$E_2 = 200 \times 10^6 \ kPa$

$I_2 = 100 \times 10^6 \ mm^4$

$A_2 = 3000 \ mm^2$

**FIGURE 8.12**   A Small Frame Example.

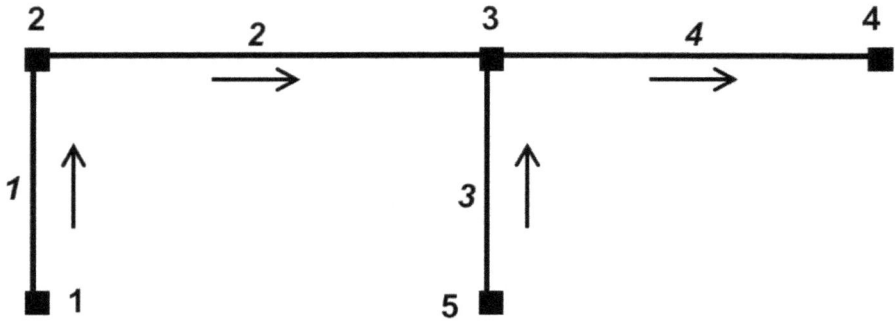

**FIGURE 8.13** A Small Frame: Node and Member Numbering.

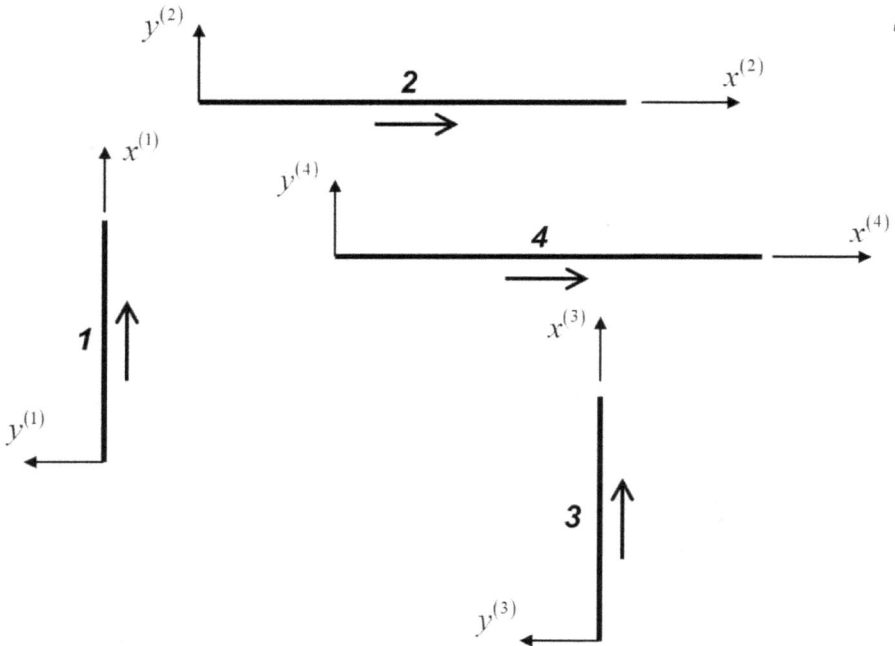

**FIGURE 8.14** A Small Frame: Member in Local Reference System.

The corresponding input data block is shown in Figure 8.15.

The constant database for the small reference frame is comprised of the following arrays:

$$
\textbf{NODE}\_info \;=\; \begin{bmatrix} 0 & 0 \\ 0 & 5 \\ 10 & 5 \\ 20 & 10 \\ 10 & 0 \end{bmatrix}
$$

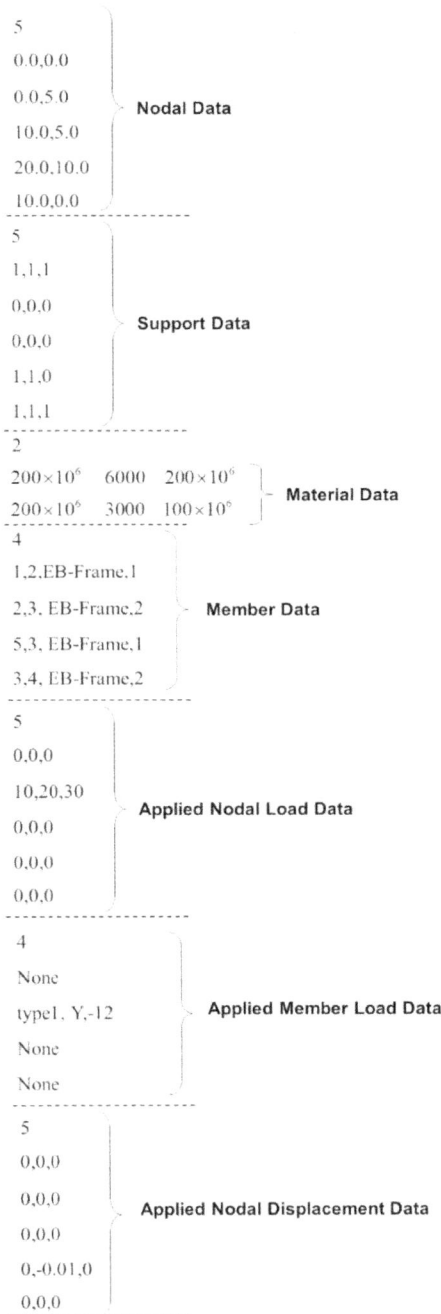

FIGURE 8.15    An Example of Input Data Block.

Each row of **NODE**_*info* contains the coordinates of each node. Node 1 has coordinates {0,0}, Node 2 {0,5}, Node 3 {10,5}, and so on.

The degree-of-freedom numbering system shown in Figure 8.16 is created automatically from the node numbering as prepared for the input data of Figure 8.13. The order of degrees of

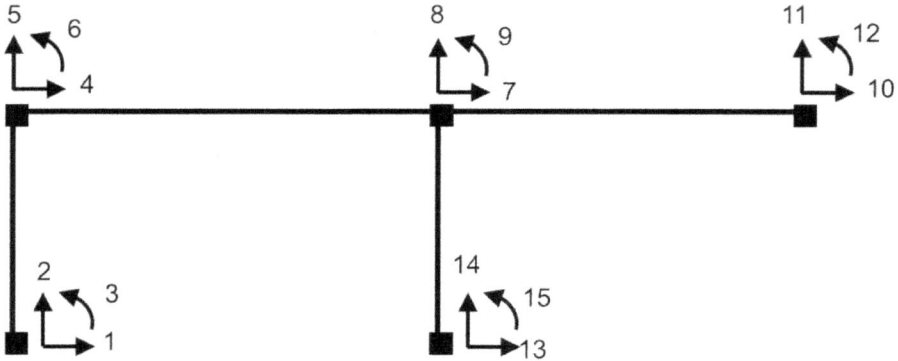

**FIGURE 8.16**  A Small Frame: Initial Degree-of-Freedom Numbering.

freedom is displacement in the global X direction, displacement in the global Y direction, and rotation around the global Z axis.

$$\mathbf{BOUN}\_info = \begin{bmatrix} 1 & 1 & 1 \\ 0 & 0 & 0 \\ 0 & 0 & 0 \\ 1 & 1 & 0 \\ 1 & 1 & 1 \end{bmatrix} \Rightarrow \mathbf{DOF}\_info = \begin{bmatrix} -8 & -9 & -10 \\ 1 & 2 & 3 \\ 4 & 5 & 6 \\ -11 & -12 & 7 \\ -13 & -14 & -15 \end{bmatrix} \Rightarrow Renumbering\ DOF$$

Each row of **BOUN**_*info* contains the support code for each node. Node 1 is attached to a fixed support that prevents it from translating in the X and Y directions as well as rotating about the Z axis. Thus, the support code for node 1 is 1,1,1. Node 2 is free to translate and rotate. Thus, the support code for node 2 is 0,0,0. For nodes 3, 4, and 5, similar rules of support-code numbering are applied. It is left to the reader as his/her term project to improve the storing process of support data. This is due to the fact that in actual structures, most of the structural nodes are free, and only a few are subjected to support restraints. Therefore, it is not efficient to impose support-code numbers on all structural nodes as presented here.

The degree-of-freedom numbering of Figure 8.16 is renumbered in order to distinguish between the unconstrained and constrained degrees of freedom. In this case, the unconstrained degrees of freedom are numbered first. The resulting numbering is shown in Figure 8.17. The negative sign indicates that the degrees of freedom are constrained. In other words, the displacements of these degrees of freedom are prescribed. The total number of degrees of freedom num_DOFs is 15. Of these, eight are constrained (num_DOF_constr = 8) while seven are unconstrained (num_DOF_free = 7).

Each row of **DOF**_*info* contains information for the degrees of freedom of each node. For example, the fourth row **DOF**_*info* indicates that for node 4, the displacement in the global X direction is constrained and corresponds to the global degree of freedom 11, the displacement in the global Y direction is constrained and corresponds to the global degree of freedom 12, and the rotation around the Z axis is unconstrained and corresponds to the global degree of freedom 7.

$$\mathbf{MATE}\_info = \begin{bmatrix} 200 \times 10^6 & 6000 & 200 \times 10^6 \\ 200 \times 10^6 & 3000 & 100 \times 10^6 \end{bmatrix}$$

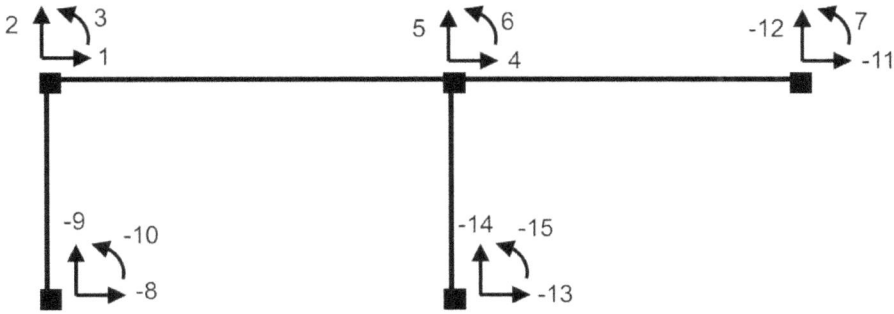

**FIGURE 8.17**   A Small Frame: Degree-of-Freedom Numbering after Separation between Constrained and Unconstrained Degrees of Freedom.

Each row of **MATE**_$info$ contains information for each material set. The first row indicates that for material set 1, the modulus of elasticity $E$ is $200 \times 10^6$ $kPa$; the sectional area $A$ is 6000 $mm^2$; and the moment of inertia $I$ is $200 \times 10^6$ $mm^4$. A similar interpretation is applied to the second row.

$$\textbf{MEM}\_info = \begin{bmatrix} 1 & 2 & \text{EB-Frame} & 1 \\ 2 & 3 & \text{EB-Frame} & 2 \\ 5 & 3 & \text{EB-Frame} & 1 \\ 3 & 4 & \text{EB-Frame} & 2 \end{bmatrix}$$

Each row of **MEM**_$info$ contains information for each member. Member 1 goes from node 1 to node 2, is of EB-Frame type, and has material properties defined as set 1. This last data index indicates that the material properties of member 1 are contained in the first row of **MATE**_$info$. Similar interpretation is applied to other rows.

$$\textbf{NODAL\_LOAD}\_info = \begin{bmatrix} 0 & 0 & 0 \\ 10 & 20 & 30 \\ 0 & 0 & 0 \\ 0 & 0 & 0 \\ 0 & 0 & 0 \end{bmatrix}$$

Each row of **NODAL_LOAD**_$info$ contains applied nodal loads for each node. For example, the second row indicates that the applied nodal loads on node 2 in X, Y, and Z directions are $10\, kN$, $20\, kN$, and $30\, kN - m$, respectively. For other nodes, similar interpretation is applied.

In conjunction with **DOF**_$info$ array, the $\mathbf{F}_{free}$ array containing the applied loads associated with free degrees of freedom is defined as:

$$\mathbf{F}_{free} = \left| \underset{DOF\ 1}{10} \quad \underset{DOF\ 2}{20} \quad \underset{DOF\ 3}{30} \quad \underset{DOF\ 4}{0} \quad \underset{DOF\ 5}{0} \quad \underset{DOF\ 6}{0} \quad \underset{DOF\ 7}{0} \right|^T$$

The length of $\mathbf{F}_{free}$ is 7, that is the number of free degrees of freedom. They correspond to degrees of freedom 1 through 7 in Figure 8.17. It is left to the reader as his/her term project to improve the storing process of applied nodal loads. It would be more efficient to specify nodal loads only for certain nodes instead of specifying nodal loads on all structural nodes as presented here.

$$\mathbf{MEM\_LOAD\_} \textit{info} = \begin{bmatrix} \textit{None} \\ \text{type1} & \text{Y} & -12 \\ \textit{None} \\ \textit{None} \end{bmatrix}$$

Each row of **MEM_LOAD**_*info* contains applied member loads for each member. The first, third, and fourth rows indicate that members 1, 3, and 4 are not subjected to any member load while the second row indicates that member 2 is subjected to the member load of type1 (uniform load) in -Y direction with the intensity of 12 $kN/m$. The way of assigning loading parameters usually is designed by program developers. It is left to the reader as his/her term project to improve the storing process of applied member loads. It would be more efficient to define member loads only for certain members instead of specifying member loads for all members as presented here.

$$\mathbf{NODAL\_DIS\_} \textit{info} = \begin{bmatrix} 0 & 0 & 0 \\ 0 & 0 & 0 \\ 0 & 0 & 0 \\ 0 & -0.01 & 0 \\ 0 & 0 & 0 \end{bmatrix}$$

Each row of **NODAL_DIS**_*info* contains applied nodal displacements for each node. For example, the fourth row indicates that the applied nodal displacement on node 4 in Y direction is $-0.01\ m$. For other nodes, a similar interpretation is applied.

In conjunction with **DOF**_*info* array, the $\mathbf{U}_{constr}$ array containing the applied displacements associated with constrained degrees of freedom is defined as:

$$\mathbf{U}_{constr} = \begin{bmatrix} \underset{DOF\ 8}{0} & \underset{DOF\ 9}{0} & \underset{DOF\ 10}{0} & \underset{DOF\ 11}{0} & \underset{DOF\ 12}{-0.01} & \underset{DOF\ 13}{0} & \underset{DOF\ 14}{0} & \underset{DOF\ 15}{0} \end{bmatrix}^T$$

The length of the array $\mathbf{U}_{constr}$ is 8, which is the number of constrained degrees of freedom. They correspond to degrees of freedom 8 through 15 in Figure 8.17. It is left to the reader as his/her term project to improve the storing process of applied nodal displacements. It would be more efficient to define nodal displacements only for certain nodes instead of specifying nodal displacements for all structural nodes as presented here.

Other database arrays must be computed but not necessarily be stored. This array is essential for assembling the member equations to form the structure equations. This array is the local mapping vector $LM_i$ discussed previously in Chapter 1. It maps the member degrees of freedom in the global reference system to the structure degrees of freedom. $LM_i$ is constructed based on the information already stored in **MEM**_*info* and **DOF**_*info*. For example, referring to the **MEM**_*info* of member 3 (third row) of Figure 8.13, it is observed that the starting and ending nodes of member 3 are 5 and 3, respectively. From **DOF**_*info*, it is clear that 6 member degrees of freedom of member 3 correspond to 6 structure degrees of freedom $\{13\ \ 14\ \ 15\ \ 4\ \ 5\ \ 6\}$.

These numbers map the member degrees of freedom to the structure degrees of freedom. The resulting member mapping array is:

$$LM_3 = \lfloor 13 \quad 14 \quad 15 \quad 4 \quad 5 \quad 6 \rfloor^T$$

Obviously, there is no need to make $LM_i$ a part of the constant database, since it can be constructed from the information contained in **MEM_**$info$ and **DOF_**$info$ as shown above. Similarly, for the other members, the member mapping arrays are:

$$LM_1 = \lfloor 8 \quad 9 \quad 10 \quad 1 \quad 2 \quad 3 \rfloor^T ; \; LM_2 = \lfloor 1 \quad 2 \quad 3 \quad 4 \quad 5 \quad 6 \rfloor^T ; \; LM_4 = \lfloor 4 \quad 5 \quad 6 \quad 11 \quad 12 \quad 7 \rfloor^T$$

## 8.5   ASSEMBLY OF THE STRUCTURAL STIFFNESS MATRIX

### 8.5.1   MEMBER STIFFNESS AND TRANSFORMATION MATRICES

For each member:

1. Retrieve its member properties from the member data block **MEM_**$info$.
2. Determine its length from nodal coordinates and direction cosines.
3. Compute its local stiffness matrix.

The flowchart for determining member stiffness and transformation matrices is shown in Figure 8.18.

### 8.5.2   STRUCTURAL STIFFNESS MATRIX

The assembling steps of the structural stiffness matrix are the following:

1. Initialize the array storing the stiffness matrix to zero.
2. Loop through all members and for each member $k$:
    (a)   Retrieve its local member stiffness matrix $\bar{\mathbf{K}}^{(k)}$ and transformation matrix $\mathbf{\Gamma}^{(k)}$.
    (b)   Compute the global member stiffness matrix from $\mathbf{K}^{(k)} = \mathbf{\Gamma}^{(k)^T} \bar{\mathbf{K}}^{(k)} \mathbf{\Gamma}^{(k)}$.
    (c)   Establish the local mapping vector $LM_k$ fr.om arrays **MEM_**$info$ and **DOF_**$info$.
    (d)   Loop through each row and column of the member stiffness matrix and add the contributions of the member to the structural stiffness matrix.

$$\mathbf{K}\left(LM_k(i), LM_k(j)\right) = \mathbf{K}\left(LM_k(i), LM_k(j)\right) + \mathbf{K}^{(k)}(i,j)$$

    (e)   Extract the submatrices $\mathbf{K}_{ff}$, $\mathbf{K}_{fc}$, $\mathbf{K}_{cf}$, $\mathbf{K}_{cc}$ from $\mathbf{K}$.
3. The flowchart for assembling the member stiffness matrices is shown in Figure 8.19.

## 8.6   ASSEMBLY OF THE EQUIVALENT LOAD VECTOR

The main routine should read and loop through each loading case.

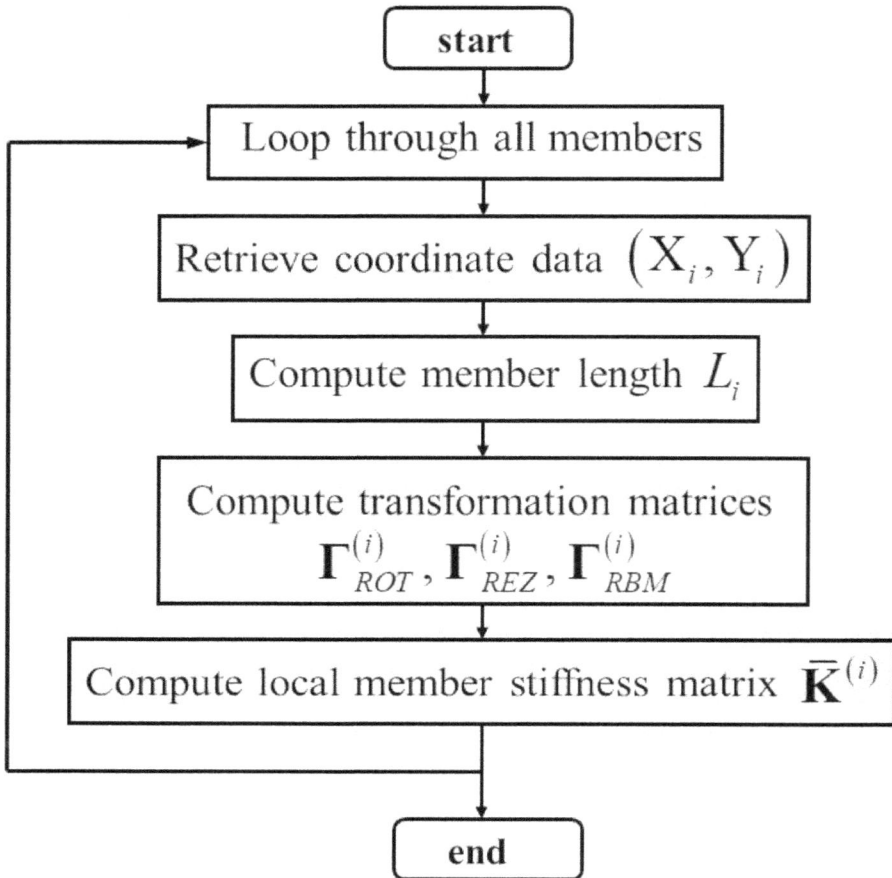

**FIGURE 8.18**    Flowchart for Computing Member Stiffness and Transformation Matrices.

1.  Initialize the array storing the load vector to zero.
2.  Read the number of loaded nodes. For each loaded node, collect the non-zero values in the load vector (using the **DOF**_$info$ to locate the appropriate position).
3.  Loop through all loaded members:
    (a)   Read the member number and load values.
    (b)   Compute the fixed-end loads both in local and global coordinates.
    (c)   Through the local mapping vector $LM_k$, add the fixed-end loads to the applied nodal load vector.
    (d)   Store the fixed-end loads for future use (internal force recovery)
4.  Partition the equivalent nodal load vector $\mathbf{F}^F$ into $\mathbf{F}^F_{free}$ and $\mathbf{F}^F_{constr}$

The flowchart for assembling the load vector is shown in Figure 8.20.

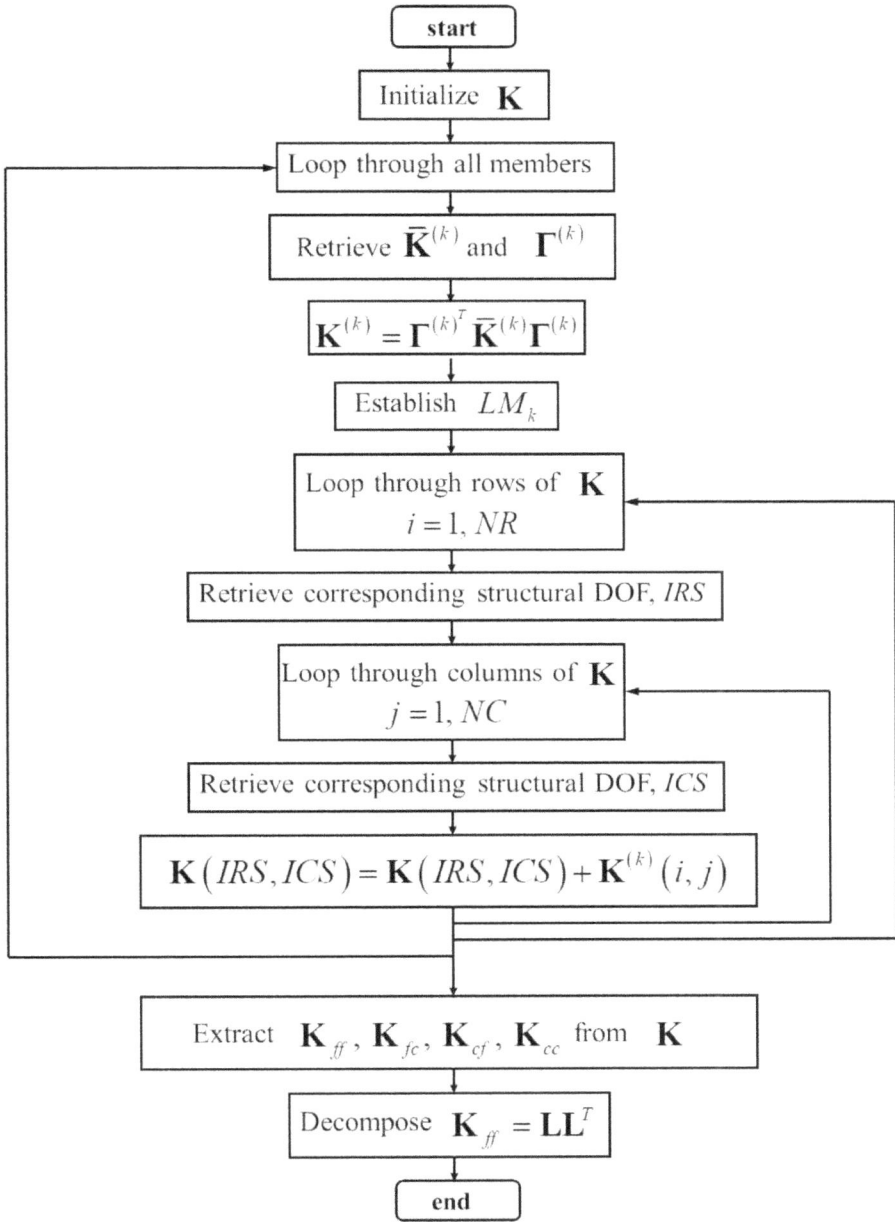

**FIGURE 8.19**   Flowchart for Assembling Member Stiffness Matrices.

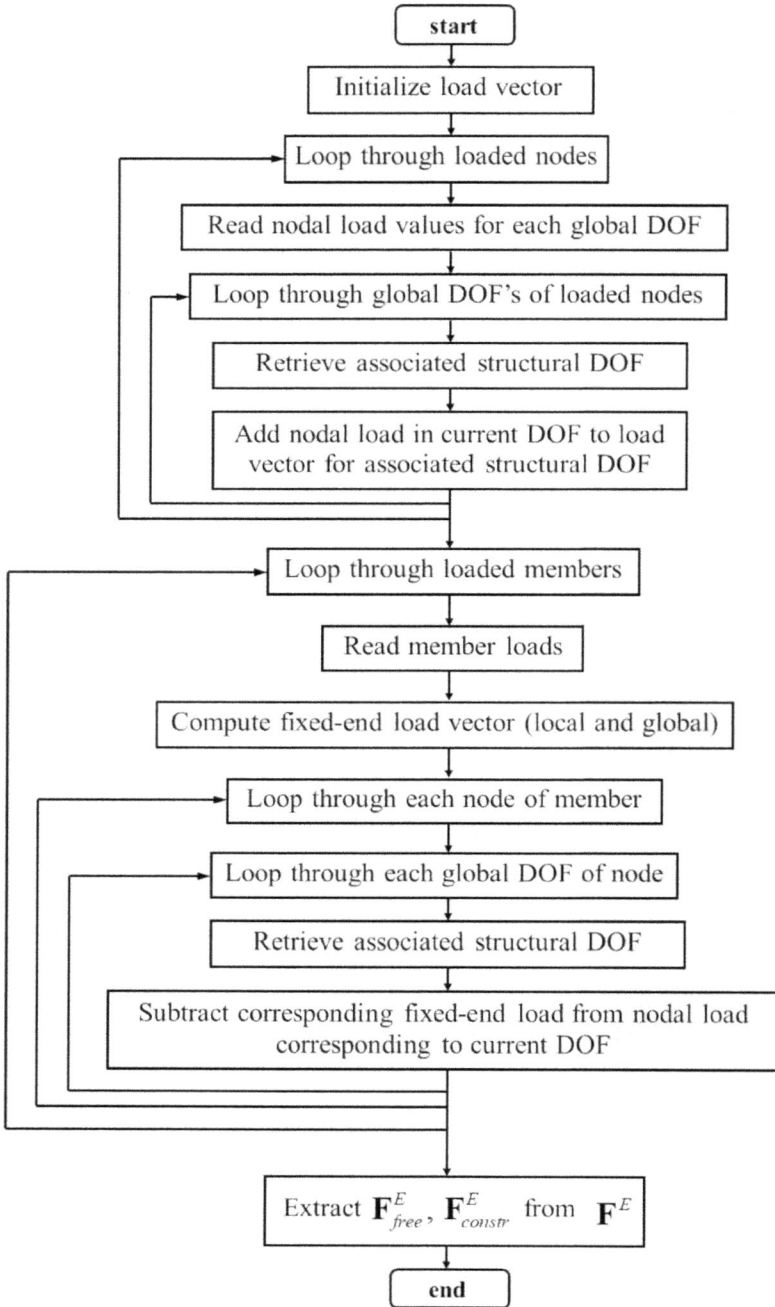

**FIGURE 8.20**   Flowchart for Assembling the Load Vector.

## 8.7 BACK SUBSTITUTION TO DETERMINE NODAL DISPLACEMENTS

Once the stiffness matrix associated with free degrees of freedom has been decomposed and the load vector corresponding to free degrees of freedom has been set up, the nodal displacements are determined through the back substitution process. This is done by multiplying the decomposed stiffness matrix with the load vector.

## 8.8 RECOVERY OF MEMBER-END DISPLACEMENTS, MEMBER-END FORCES, AND SUPPORT REACTIONS

The member-end displacements, member-end forces, and support reactions are recovered through the following procedures:

1. Initialize the reactive force vector to zero.
2. For each member retrieve:
   (a) Transformation matrices.
   (b) The fixed-end load vector.
   (c) The local stiffness matrix.
3. Determine the global member-end displacements $\mathbf{U}^{(k)}$ from the local mapping vector $LM_k$.
4. Compute local member-end displacements $\bar{\mathbf{U}}^{(k)}$ from $\bar{\mathbf{U}}^{(k)} = \mathbf{\Gamma}_{ROT}^{(k)} \mathbf{U}^{(k)}$.
5. Compute local member-end forces $\bar{\mathbf{P}}^{(k)}$ from $\bar{\mathbf{P}}^{(k)} = \bar{\mathbf{K}}^{(k)} \bar{\mathbf{U}}^{(k)} + \bar{\mathbf{P}}_F^{(k)}$.
6. Compute global member-end forces from $\mathbf{P}^{(k)} = \mathbf{\Gamma}_{ROT}^{(k)}{}^T \bar{\mathbf{P}}^{(k)}$ and collect appropriate components for support reactions.

The flowchart for recovering member-end displacements, member-end forces, and support reactions is shown in Figure 8.21.

## 8.9 TERM PROJECT

You are asked to write a computer program that can solve 2D frames and 2D trusses. The program can be written in MATLAB, Mathematica, Fortran90/95, C++, Python, etc.

The program must handle:

(a) *nodal loads*
(b) *nodal displacements (support settlements)*
(c) *constant distributed loads (for 2D frames)*
(d) *temperature effects (extra credits)*
(e) *rigid end zones and rigid floors (extra credits)*
(f) *member-end releases (extra credits)*

***Remark:*** The number of members/degrees of freedom should not be limited.

```
                        ┌─────────────┐
                        │    start    │
                        └─────────────┘
                               │
                 ┌─────────────────────────────┐
                 │ Initialize reactive force vector │
                 └─────────────────────────────┘
                               │
                   ┌───────────────────────┐
                   │ Loop through all members │
                   └───────────────────────┘
                               │
              ┌───────────────────────────────────┐
              │ Loop through each node of member   │
              └───────────────────────────────────┘
                               │
              ┌───────────────────────────────────┐
              │ Loop through global DOF's of node  │
              └───────────────────────────────────┘
                               │
              ┌───────────────────────────────────┐
              │ Retrieve associated structural DOF │
              └───────────────────────────────────┘
                               │
          ┌───────────────────────────────────────┐
          │ Retrieve associated global displacement │
          └───────────────────────────────────────┘
                               │
          ┌───────────────────────────────────────┐
          │ Transform end displacement from global  │
          │      to local reference systems         │
          └───────────────────────────────────────┘
                               │
      ┌─────────────────────────────────────────────┐
      │ Retrieve member stiffness matrices (local)   │
      └─────────────────────────────────────────────┘
                               │
        ┌─────────────────────────────────────────┐
        │ Retrieve member fixed-end load vector    │
        └─────────────────────────────────────────┘
                               │
        ┌─────────────────────────────────────────┐
        │ Retrieve member transformation matrix    │
        └─────────────────────────────────────────┘
                               │
        ┌─────────────────────────────────────────┐
        │ Compute local member-end forces by       │
        └─────────────────────────────────────────┘
                               │
        ┌─────────────────────────────────────────┐
        │ Compute global member-end forces by      │
        └─────────────────────────────────────────┘
                               │
        ┌─────────────────────────────────────────┐
        │ Loop through global DOF's of member      │
        └─────────────────────────────────────────┘
                               │
                          ◇ decision ◇
                   Is global DOF restrained ?
```

Compute local member-end forces by
$$\overline{\mathbf{P}}^{(k)} = \overline{\mathbf{K}}^{(k)}\overline{\mathbf{U}}^{(k)} + \overline{\mathbf{P}}_F^{(k)}$$

Compute global member-end forces by
$$\mathbf{P}^{(k)} = \mathbf{\Gamma}_{ROT}^{(k)}{}^T\,\overline{\mathbf{P}}^{(k)}$$

Is global DOF restrained ?    YES → Add it to reactive force vector

NO

end

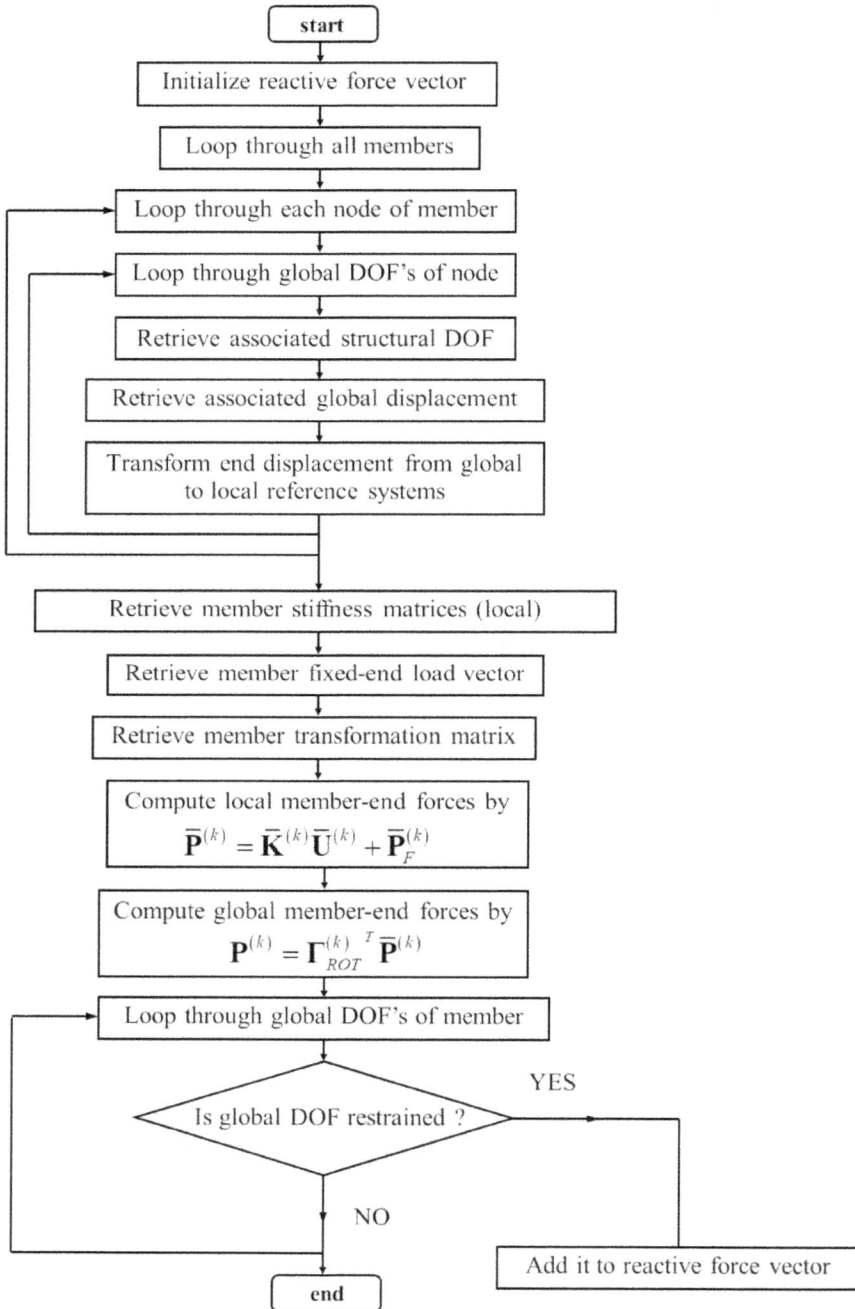

**FIGURE 8.21**    Flowchart for Recovering Member-End Displacements, Member-End Forces, and Support Reactions.

# REFERENCES

Dawe, D.J. 1984. *Matrix and finite element displacement analysis of structures.* Clarendon Press.

Hoit, M. 1995. *Computer-assisted structural analysis and modeling.* Prentice-Hall, Englewood Cliffs.

Kanchi, M.B. 1993. *Matrix methods of structural analysis* (2nd Edition). Wiley Eastern Limited.

Saouma, V.E. 1997. CVEN 5525 course notes: Matrix structural analysis. Department of Civil, Environmental, and Architectural Engineering, University of Colorado.

Spacone, E. 2003. CVEN 5525 course notes: Matrix structural analysis. Department of Civil, Environmental, and Architectural Engineering, University of Colorado.

Weaver, W. and J.M. Gere. 1990. *Matrix analysis of framed structures.* Chapman & Hall, Boca Raton.

# 9 The Flexibility Method

## 9.1 INTRODUCTION

Thus far, the matrix displacement method of structural analysis in which the structural nodal displacements serve as primary unknowns has been discussed. As the equilibrium equations are employed for the determination of unknown nodal displacements, the name "equilibrium method" is also used. Since the stiffness relations are used to express the nodal forces in terms of the nodal displacements, it is also known as the "stiffness method". This chapter aims at presenting the matrix force method of structural analysis. In essence, the force and displacement methods are dual methods of structural analysis which differ only in incorporating the statical and kinematical requirements. In the force method, a set of reactive and/or internal forces known as *redundant forces* are treated as basic unknowns. The number of selected redundant forces is equal to the degree of statical indeterminacy (*SI*) of the structure. Thus, the size of the problem depends on the degree of statical indeterminacy. For each redundant force, there is a relevant condition of compatibility. Consequently, there are as many compatibility equations as the unknown redundant forces. Therefore, the numbers of compatibility equations are sufficient to determine the unknown redundant forces. As the compatibility equations are employed for the determination of the unknown redundant forces, the name "compatibility method" is also used. Since the flexibility relations are used to express the nodal deformations in terms of nodal forces, it is also known as the "flexibility method". As a result, the terms "the force method", "the compatibility method", and "the flexibility method" can be used interchangeably.

Nowadays, the matrix force method is superseded by the matrix displacement method due to its lack of generality and difficulties in establishing the redundant force systems required at the beginning of the analyzing process. Furthermore, the matrix displacement approach possesses several advantages from practical viewpoints of computer implementation as discussed in Chapters 7 and 8. However, from the viewpoints of basic principles and completeness, it is worth pursuing the matrix force method in here.

Before wrapping up this section, it is worth mentioning a brief history of the development of the force method. The first consistent form of the force method for analyzing statically indeterminate structures was proposed by James Clerk Maxwell in 1864. His approach was based on a consideration of deflections, but the presentation was not so illuminating and attracted only little attention (Timoshenko 1953). Ten years later, this analyzing method was refined by Otto Mohr. Thus, the analysis approach of statically indeterminate structures using the deflection calculations is often referred to as the "Maxwell–Mohr" method. The first matrix form of the force method was probably proposed by Samuel Levy in 1947 to analyze the swept wing of aircraft (Martin and Carey 1973). Argyris and Kelsey (1960) subsequently gave a very general formulation of the matrix force method of structural analysis based on energy concepts. For the last 20 years, the matrix force method has reincarnated and its superiority has been discovered in the field of nonlinear frame analysis (Limkatanyu and Spacone 2006; Limkatanyu et al. 2014; Sae-Long et al. 2021).

DOI: 10.1201/9781003595458-9

## 9.2 THE STRUCTURAL FLEXIBILITY EQUATIONS: STATICALLY DETERMINATE SYSTEMS

This section focuses on establishing the structural flexibility equations of a statically determinate frame. A frame structure shown in Figure 9.1 is used as a vehicle for the description. It is worth mentioning that a statically determinate system plays a crucial role as a primary structure in the process of analyzing a statically indeterminate structure by the force method as pursued in undergraduate structural analysis courses.

The equilibrium conditions are first recalled. From Chapter 3, the matrix equilibrium equation of this frame can be expressed as:

$$\begin{Bmatrix} \mathbf{F}_1 \\ \mathbf{F}_2 \\ \mathbf{F}_3 \end{Bmatrix} = \begin{bmatrix} \mathbf{b}_{2B}^{(1)} & \mathbf{b}_{1B}^{(2)} & \mathbf{0} \\ \mathbf{0} & \mathbf{b}_{2B}^{(2)} & \mathbf{b}_{1B}^{(3)} \\ \mathbf{0} & \mathbf{0} & \mathbf{b}_{2B}^{(3)} \end{bmatrix} \begin{Bmatrix} \mathbf{Q}_1 \\ \mathbf{Q}_2 \\ \mathbf{Q}_3 \end{Bmatrix} + \begin{Bmatrix} \mathbf{F}_1^0 \\ \mathbf{F}_2^0 \\ \mathbf{F}_3^0 \end{Bmatrix} \Rightarrow \begin{Bmatrix} \mathbf{F}_1^* \\ \mathbf{F}_2^* \\ \mathbf{F}_3^* \end{Bmatrix} = \begin{Bmatrix} \mathbf{F}_1 \\ \mathbf{F}_2 \\ \mathbf{F}_3 \end{Bmatrix} - \begin{Bmatrix} \mathbf{F}_1^0 \\ \mathbf{F}_2^0 \\ \mathbf{F}_3^0 \end{Bmatrix} = \begin{bmatrix} \mathbf{b}_{2B}^{(1)} & \mathbf{b}_{1B}^{(2)} & \mathbf{0} \\ \mathbf{0} & \mathbf{b}_{2B}^{(2)} & \mathbf{b}_{1B}^{(3)} \\ \mathbf{0} & \mathbf{0} & \mathbf{b}_{2B}^{(3)} \end{bmatrix} \begin{Bmatrix} \mathbf{Q}_1 \\ \mathbf{Q}_2 \\ \mathbf{Q}_3 \end{Bmatrix} \qquad (9.1)$$

or in the concise format as:

$$\mathbf{F}_{free} = \mathbf{B}_{free}\mathbf{Q} + \mathbf{F}_{free}^0 \Rightarrow \mathbf{F}_{free}^* = \mathbf{F}_{free} - \mathbf{F}_{free}^0 = \mathbf{B}_{free}\mathbf{Q} \qquad (9.2)$$

Due to the statical determinacy of the system, the basic member forces $\mathbf{Q}$ (shown in Figure 9.2) can uniquely be expressed in terms of the equivalent nodal applied loads $\mathbf{F}_{free}^*$ associated with free degrees of freedom as:

$$\mathbf{Q} = \begin{bmatrix} \mathbf{B}_{free} \end{bmatrix}^{-1} \mathbf{F}_{free}^* = \mathbf{B}_0 \mathbf{F}_{free}^* \qquad (9.3)$$

It is worth mentioning that the matrix $\mathbf{B}_0$ in Eq. (9.3) can be viewed as a discrete form of force interpolation functions between the force vectors $\mathbf{Q}$ and $\mathbf{F}_{free}^*$. In the flexibility-based frame

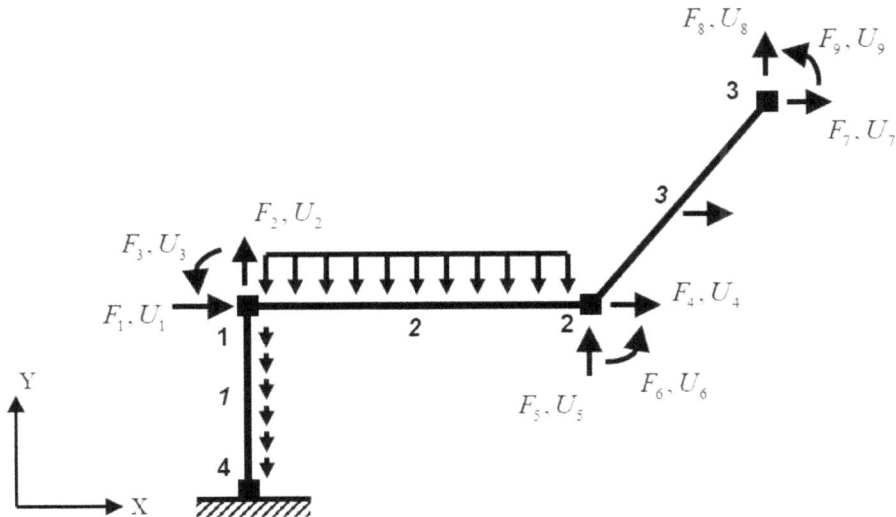

**FIGURE 9.1** A Statically Determinate Frame.

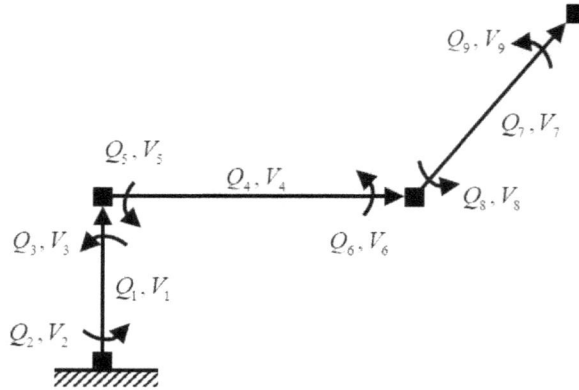

**FIGURE 9.2**  Basic Member Forces and Deformations of the Frame in Figure 9.1.

element formulation, the force interpolations are employed to express internal forces in terms of external forces. More details on the flexibility-based frame element formulation can be found in several papers published by the author as listed in the reference (e.g. Limkatanyu and Spacone 2002, 2006; Limkatanyu et al. 2012).

From Chapter 4, the matrix compatibility equation of this frame can be expressed as:

$$
\begin{Bmatrix} \mathbf{V}_1 \\ \mathbf{V}_2 \\ \mathbf{V}_3 \end{Bmatrix} = \begin{bmatrix} \mathbf{a}_{2B}^{(1)} & \mathbf{0} & \mathbf{0} \\ \mathbf{a}_{1B}^{(2)} & \mathbf{a}_{2B}^{(2)} & \mathbf{0} \\ \mathbf{0} & \mathbf{a}_{1B}^{(3)} & \mathbf{a}_{2B}^{(3)} \end{bmatrix} \begin{Bmatrix} \mathbf{U}_1 \\ \mathbf{U}_2 \\ \mathbf{U}_3 \end{Bmatrix}
\tag{9.4}
$$

or in the concise format as:

$$
\mathbf{V} = \mathbf{V}_{free} = \mathbf{A}_{free}\mathbf{U}_{free}
\tag{9.5}
$$

From Eqs. (9.2) and (9.5), one can observe that

$$
\mathbf{A}_{free} = \mathbf{B}_{free}^{T}
\tag{9.6}
$$

This type of relationship is due to the contragradient nature between the statical and kinematical relations as already proved in Chapter 5 through the virtual work principles.

Due to the statical determinacy of the system, the nodal displacements $\mathbf{U}_{free}$ associated with free degrees of freedom can uniquely be expressed in terms of the basic member deformations $\mathbf{V}$ (shown in Figure 9.2) as:

$$
\mathbf{U}_{free} = \left[\mathbf{A}_{free}\right]^{-1}\mathbf{V} = \mathbf{B}_0^{T}\mathbf{V}
\tag{9.7}
$$

The member flexibility relations are now employed to express the basic member deformations in terms of the basic member forces.

For an individual member, they are written as:

$$\mathbf{v}^{(i)} = \mathbf{f}^{(i)}\mathbf{q}^{(i)} + \mathbf{v}_0^{(i)} + \mathbf{v}_W^{(i)} + \mathbf{v}_s^{(i)} \tag{9.8}$$

where the vector $\mathbf{v}_0^{(i)}$ contains the basic member deformations due to initial deformations; the vector $\mathbf{v}_W^{(i)}$ contains the basic member deformations due to member loads; and $\mathbf{v}_s^{(i)}$ contains the basic member deformations due to support movements. These values of basic member deformations are given in Appendix II.

For all members, they are written together as:

$$\begin{Bmatrix} \mathbf{v}_1 \\ \mathbf{v}_2 \\ \mathbf{v}_3 \end{Bmatrix} = \begin{bmatrix} \mathbf{f}^{(1)} & \mathbf{0} & \mathbf{0} \\ \mathbf{0} & \mathbf{f}^{(2)} & \mathbf{0} \\ \mathbf{0} & \mathbf{0} & \mathbf{f}^{(3)} \end{bmatrix} \begin{Bmatrix} \mathbf{q}_1 \\ \mathbf{q}_2 \\ \mathbf{q}_3 \end{Bmatrix} + \begin{Bmatrix} \mathbf{v}_0^{(1)} \\ \mathbf{v}_0^{(2)} \\ \mathbf{v}_0^{(3)} \end{Bmatrix} + \begin{Bmatrix} \mathbf{v}_W^{(1)} \\ \mathbf{v}_W^{(2)} \\ \mathbf{v}_W^{(3)} \end{Bmatrix} + \begin{Bmatrix} \mathbf{v}_s^{(1)} \\ \mathbf{v}_s^{(2)} \\ \mathbf{v}_s^{(3)} \end{Bmatrix}$$

$$\Downarrow \tag{9.9}$$

$$\begin{Bmatrix} \mathbf{V}_1 \\ \mathbf{V}_2 \\ \mathbf{V}_3 \end{Bmatrix} = \begin{bmatrix} \mathbf{f}^{(1)} & \mathbf{0} & \mathbf{0} \\ \mathbf{0} & \mathbf{f}^{(2)} & \mathbf{0} \\ \mathbf{0} & \mathbf{0} & \mathbf{f}^{(3)} \end{bmatrix} \begin{Bmatrix} \mathbf{Q}_1 \\ \mathbf{Q}_2 \\ \mathbf{Q}_3 \end{Bmatrix} + \begin{Bmatrix} \mathbf{V}_0^1 \\ \mathbf{V}_0^2 \\ \mathbf{V}_0^3 \end{Bmatrix} + \begin{Bmatrix} \mathbf{V}_W^1 \\ \mathbf{V}_W^2 \\ \mathbf{V}_W^3 \end{Bmatrix} + \begin{Bmatrix} \mathbf{V}_s^1 \\ \mathbf{V}_s^2 \\ \mathbf{V}_s^3 \end{Bmatrix}$$

or in the concise format as:

$$\mathbf{V} = \mathbf{f}\mathbf{Q} + \mathbf{V}_0 + \mathbf{V}_W + \mathbf{V}_s \tag{9.10}$$

All of the fundamental ingredients ((9.3), (9.7), and (9.10)) can be written together through the virtual force principle as follows.

From the invariant property of the work discussed in Chapter 5, it can be concluded that:

$$\underbrace{\delta \mathbf{Q}^T \mathbf{V}}_{\delta W_{int}^*} = \underbrace{\delta \mathbf{F}_{free}^{*T} \mathbf{U}_{free}}_{\delta W_{ext}^*} \tag{9.11}$$

From Eq. (9.3), one has:

$$\delta \mathbf{Q}^T = \delta \mathbf{F}_{free}^{*T} \mathbf{B}_0^T \tag{9.12}$$

It is noted that the equilibrium relations in Eq. (9.12) guarantee the satisfaction of equilibrium requirements of the virtual forces $\delta \mathbf{Q}$ and $\delta \mathbf{F}_{free}^*$. This is an essential aspect of the virtual force system as discussed in Chapter 5.

Substituting Eqs. (9.12) into (9.11), one has:

$$\delta \mathbf{F}_{free}^{*T} \mathbf{B}_0^T \mathbf{V} = \delta \mathbf{F}_{free}^{*T} \mathbf{U}_{free} \quad \Rightarrow \quad \delta \mathbf{F}_{free}^{*T} \left( \mathbf{B}_0^T \mathbf{V} - \mathbf{U}_{free} \right) = 0 \tag{9.13}$$

Due to the arbitrariness of $\delta \mathbf{F}_{free}^*$, the following matrix compatibility relation is obtained:

$$\mathbf{U}_{free} = \mathbf{B}_0^T \mathbf{V} \tag{9.14}$$

Based on the unassembled flexibility equations of Eq. (9.10), one has:

$$\mathbf{U}_{free} = \mathbf{B}_0^{\ T}\mathbf{V} = \mathbf{B}_0^{\ T}\left(\mathbf{fQ} + \mathbf{V}_0 + \mathbf{V}_w + \mathbf{V}_s\right) = \mathbf{B}_0^{\ T}\mathbf{fQ} + \mathbf{B}_0^{\ T}\mathbf{V}_0 + \mathbf{B}_0^{\ T}\mathbf{V}_w + \mathbf{B}_0^{\ T}\mathbf{V}_s \quad (9.15)$$

Substituting Eqs. (9.3) into (9.15), the structural flexibility equation is obtained:

$$\mathbf{U}_{free} = \left(\mathbf{B}_0^{\ T}\mathbf{fB}_0\right)\mathbf{F}_{free}^* + \mathbf{B}_0^{\ T}\mathbf{V}_0 + \mathbf{B}_0^{\ T}\mathbf{V}_w + \mathbf{B}_0^{\ T}\mathbf{V}_s = \bar{\mathbf{F}}\mathbf{F}_{free}^* + \mathbf{U}_{free}^0 + \mathbf{U}_{free}^w + \mathbf{U}_{free}^s \quad (9.16)$$

where the structural flexibility matrix $\bar{\mathbf{F}}$ is defined as:

$$\bar{\mathbf{F}} = \mathbf{B}_0^{\ T}\mathbf{fB}_0 \quad (9.17)$$

The structural nodal displacements $\mathbf{U}_{free}^0$ due to the presence of initial basic deformations are defined as:

$$\mathbf{U}_{free}^0 = \mathbf{B}_0^{\ T}\mathbf{V}_0 \quad (9.18)$$

The structural nodal displacements $\mathbf{U}_{free}^w$ due to the presence of member loads are defined as:

$$\mathbf{U}_{free}^w = \mathbf{B}_0^{\ T}\mathbf{V}_w \quad (9.19)$$

The structural nodal displacements $\mathbf{U}_{free}^s$ due to the support movements are defined as:

$$\mathbf{U}_{free}^s = \mathbf{B}_0^{\ T}\mathbf{V}_s \quad (9.20)$$

The schematic representation of the flexibility method is presented in Figure 9.3.

## EXAMPLE 9.1

For the frame shown in Figure 9.4, use the flexibility method to determine its nodal displacements.

**Solution:** Member, node, DOFs, basic-member force, and basic-member deformation numbering systems are shown in Figure 9.5.

From the statical considerations described in Chapter 3, all nodal equilibrium equations can be written in the matrix form as:

$$\mathbf{F}_{free}^* = \mathbf{B}_{free}\mathbf{Q} \quad \text{or} \quad \begin{Bmatrix} 0 \\ -\dfrac{wL}{2} \\ 0 \\ 0 \\ -\dfrac{wL}{2} \\ 0 \\ wL \\ 0 \\ 0 \end{Bmatrix} = \begin{bmatrix} 0 & \dfrac{1}{L} & \dfrac{1}{L} & -1 & 0 & 0 & 0 & 0 & 0 \\ 1 & 0 & 0 & 0 & \dfrac{1}{L} & \dfrac{1}{L} & 0 & 0 & 0 \\ 0 & 0 & 1 & 0 & 1 & 0 & 0 & 0 & 0 \\ 0 & 0 & 0 & 1 & 0 & 0 & 0 & \dfrac{2}{L} & \dfrac{2}{L} \\ 0 & 0 & 0 & 0 & -\dfrac{1}{L} & -\dfrac{1}{L} & 1 & 0 & 0 \\ 0 & 0 & 0 & 0 & 0 & 1 & 0 & 0 & 1 \\ 0 & 0 & 0 & 0 & 0 & 0 & 0 & -\dfrac{2}{L} & -\dfrac{2}{L} \\ 0 & 0 & 0 & 0 & 0 & 0 & -1 & 0 & 0 \\ 0 & 0 & 0 & 0 & 0 & 0 & 0 & 1 & 0 \end{bmatrix}\begin{Bmatrix} Q_1 \\ Q_2 \\ Q_3 \\ Q_4 \\ Q_5 \\ Q_6 \\ Q_7 \\ Q_8 \\ Q_9 \end{Bmatrix}$$

$$Q = B_0 F_{free}^s$$

Equilibrium

Q: Basic Forces

$$V = fQ + V_0 + V_w + V_s$$ Members

Structure

V: Basic Deformations

Compatibility

$$U_{free} = \bar{F}F_{free}^s + U_{free}^0 + U_{free}^w + U_{free}^s$$

$$U_{free} = B_0^T V$$

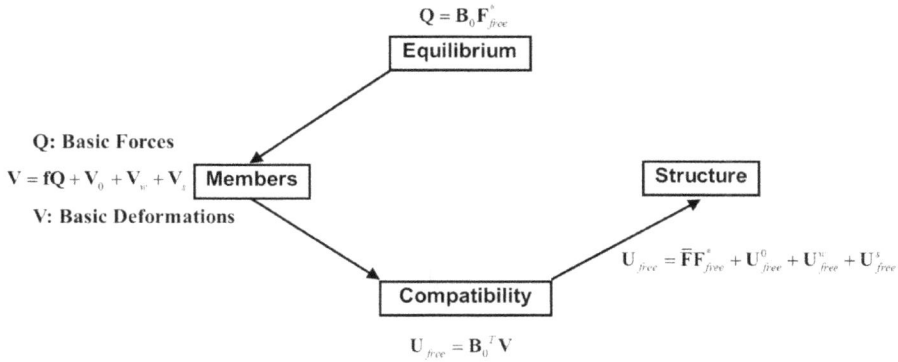

**FIGURE 9.3**  Schematic Representation of the Flexibility Method.

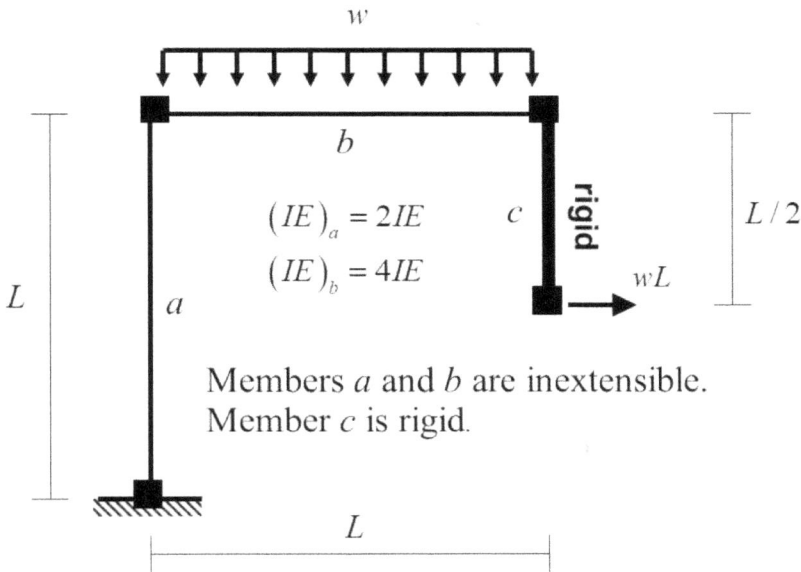

$$(IE)_a = 2IE$$

$$(IE)_b = 4IE$$

**rigid**

$$L/2$$

$$wL$$

Members *a* and *b* are inextensible.
Member *c* is rigid.

**FIGURE 9.4**  Example 9.1.

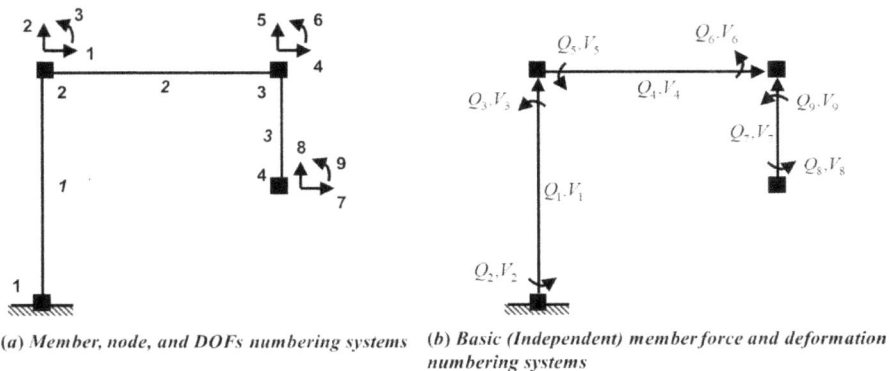

*(a) Member, node, and DOFs numbering systems*      *(b) Basic (Independent) member force and deformation numbering systems*

**FIGURE 9.5**  Example 9.1 (Continued).

The member basic forces $\mathbf{Q}$ can be expressed in terms of the equivalent nodal applied loads $\mathbf{F}^*_{free}$ as:

$$\mathbf{Q} = \mathbf{B}_0\mathbf{F}^*_{free} \quad \text{or} \quad
\begin{Bmatrix} Q_1 \\ Q_2 \\ Q_3 \\ Q_4 \\ Q_5 \\ Q_6 \\ Q_7 \\ Q_8 \\ Q_9 \end{Bmatrix} =
\begin{bmatrix}
0 & 1 & 0 & 0 & 1 & 0 & 0 & 1 & 0 \\
L & 0 & -1 & L & -L & -1 & \dfrac{L}{2} & -L & -1 \\
0 & 0 & 1 & 0 & L & 1 & \dfrac{L}{2} & L & 1 \\
0 & 0 & 0 & 1 & 0 & 0 & 1 & 0 & 0 \\
0 & 0 & 0 & 0 & -L & -1 & -\dfrac{L}{2} & -L & -1 \\
0 & 0 & 0 & 0 & 0 & 1 & \dfrac{L}{2} & 0 & 1 \\
0 & 0 & 0 & 0 & 0 & 0 & 0 & -1 & 0 \\
0 & 0 & 0 & 0 & 0 & 0 & 0 & 0 & 1 \\
0 & 0 & 0 & 0 & 0 & 0 & -\dfrac{L}{2} & 0 & -1
\end{bmatrix}
\begin{Bmatrix} 0 \\ -\dfrac{wL}{2} \\ 0 \\ 0 \\ \dfrac{wL}{2} \\ 0 \\ wL \\ 0 \\ 0 \end{Bmatrix}$$

The unassembled flexibility matrix (Figure 9.6) is:

$$\mathbf{V} = \mathbf{f}\mathbf{Q} + \mathbf{V}_W + \underbrace{\mathbf{V}_0}_{0} + \underbrace{\mathbf{V}_S}_{0} \quad \text{or} \quad
\begin{Bmatrix} \mathbf{V}_1 \\ \mathbf{V}_2 \\ \mathbf{V}_3 \end{Bmatrix} =
\begin{bmatrix} \mathbf{f}_1 & \mathbf{0} & \mathbf{0} \\ \mathbf{0} & \mathbf{f}_2 & \mathbf{0} \\ \mathbf{0} & \mathbf{0} & \mathbf{f}_3 \end{bmatrix}
\begin{Bmatrix} \mathbf{Q}_1 \\ \mathbf{Q}_2 \\ \mathbf{Q}_3 \end{Bmatrix} +
\begin{Bmatrix} \mathbf{V}^1_W \\ \mathbf{V}^2_W \\ \mathbf{V}^3_W \end{Bmatrix}$$

For members 1 and 3, $\mathbf{V}^1_W = \mathbf{V}^3_W = \mathbf{0}$ and for member 2, $\mathbf{V}^2_W = \begin{bmatrix} 0 & -\dfrac{wL^3}{96IE} & \dfrac{wL^3}{96IE} \end{bmatrix}^T$.

$$
\begin{Bmatrix} V_1 \\ V_2 \\ V_3 \\ V_4 \\ V_5 \\ V_6 \\ V_7 \\ V_8 \\ V_9 \end{Bmatrix} =
\begin{bmatrix}
0 & 0 & 0 & 0 & 0 & 0 & 0 & 0 & 0 \\
0 & \dfrac{L}{6IE} & -\dfrac{L}{12IE} & 0 & 0 & 0 & 0 & 0 & 0 \\
0 & -\dfrac{L}{12IE} & \dfrac{L}{6IE} & 0 & 0 & 0 & 0 & 0 & 0 \\
0 & 0 & 0 & 0 & 0 & 0 & 0 & 0 & 0 \\
0 & 0 & 0 & 0 & \dfrac{L}{12IE} & -\dfrac{L}{24IE} & 0 & 0 & 0 \\
0 & 0 & 0 & 0 & -\dfrac{L}{24IE} & \dfrac{L}{12IE} & 0 & 0 & 0 \\
0 & 0 & 0 & 0 & 0 & 0 & 0 & 0 & 0 \\
0 & 0 & 0 & 0 & 0 & 0 & 0 & 0 & 0 \\
0 & 0 & 0 & 0 & 0 & 0 & 0 & 0 & 0
\end{bmatrix}
\begin{Bmatrix} Q_1 \\ Q_2 \\ Q_3 \\ Q_4 \\ Q_5 \\ Q_6 \\ Q_7 \\ Q_8 \\ Q_9 \end{Bmatrix} +
\begin{Bmatrix} 0 \\ 0 \\ 0 \\ 0 \\ -\dfrac{wL^3}{96IE} \\ \dfrac{wL^3}{96IE} \\ 0 \\ 0 \\ 0 \end{Bmatrix}
$$

The structural flexibility equation is:

$$\mathbf{U}_{free} = \overline{\mathbf{F}}\mathbf{F}^*_{free} + \mathbf{U}^W_{free}$$

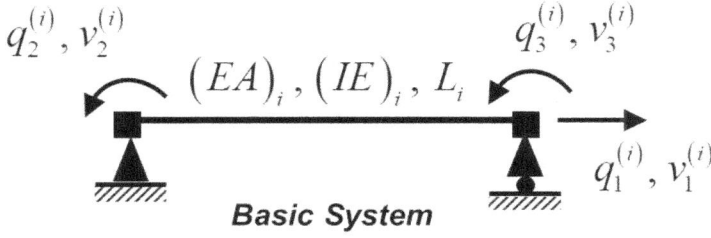

**Basic System**

**FIGURE 9.6**   Example 9.1 (Continued).

where the structural flexibility matrix $\bar{\mathbf{F}}$ is:

$$\bar{\mathbf{F}} = \mathbf{B}_0{}^T \mathbf{f} \mathbf{B}_0 = \frac{L}{IE}
\begin{bmatrix}
\dfrac{L^2}{6} & 0 & -\dfrac{L}{4} & \dfrac{L^2}{6} & -\dfrac{L^2}{4} & -\dfrac{L}{4} & \dfrac{L^2}{24} & -\dfrac{L^2}{4} & -\dfrac{L}{4} \\[2mm]
0 & 0 & 0 & 0 & 0 & 0 & 0 & 0 & 0 \\[2mm]
-\dfrac{L}{4} & 0 & \dfrac{1}{2} & -\dfrac{L}{4} & \dfrac{L}{2} & \dfrac{1}{2} & 0 & \dfrac{L}{2} & \dfrac{1}{2} \\[2mm]
\dfrac{L^2}{6} & 0 & -\dfrac{L}{4} & \dfrac{L^2}{6} & -\dfrac{L^2}{4} & -\dfrac{L}{4} & \dfrac{L^2}{24} & -\dfrac{L^2}{4} & -\dfrac{L}{4} \\[2mm]
-\dfrac{L^2}{4} & 0 & \dfrac{L}{2} & -\dfrac{L^2}{4} & \dfrac{7L^2}{12} & \dfrac{5L}{8} & \dfrac{L^2}{16} & \dfrac{7L^2}{12} & \dfrac{5L}{8} \\[2mm]
-\dfrac{L}{4} & 0 & \dfrac{1}{2} & -\dfrac{L}{4} & \dfrac{5L}{8} & \dfrac{3}{4} & \dfrac{L}{8} & \dfrac{5L}{8} & \dfrac{3}{4} \\[2mm]
\dfrac{L^2}{24} & 0 & 0 & \dfrac{L^2}{24} & \dfrac{L^2}{16} & \dfrac{L}{8} & \dfrac{5L^2}{48} & \dfrac{L^2}{16} & \dfrac{L}{8} \\[2mm]
-\dfrac{L^2}{4} & 0 & \dfrac{L}{2} & -\dfrac{L^2}{4} & \dfrac{7L^2}{12} & \dfrac{5L}{8} & \dfrac{L^2}{16} & \dfrac{7L^2}{12} & \dfrac{5L}{8} \\[2mm]
-\dfrac{L}{4} & 0 & \dfrac{1}{2} & -\dfrac{L}{4} & \dfrac{5L}{8} & \dfrac{3}{4} & \dfrac{L}{8} & \dfrac{5L}{8} & \dfrac{3}{4}
\end{bmatrix}$$

The structural nodal displacements $\mathbf{U}^W_{free}$ due to the presence of member loads are:

$$\mathbf{U}^W_{free} = \mathbf{B}_0{}^T \mathbf{V}_w = \left\lfloor 0 \;\; 0 \;\; 0 \;\; 0 \;\; \frac{wL^4}{96IE} \;\; \frac{wL^3}{48IE} \;\; \frac{wL^4}{96IE} \;\; \frac{wL^4}{96IE} \;\; \frac{wL^3}{48IE} \right\rfloor^T$$

The structural nodal displacements $\mathbf{U}_{free}$ are:

$$\mathbf{U}_{free} = \left\lfloor U_1 \;\; U_2 \;\; U_3 \;\; U_4 \;\; U_5 \;\; U_6 \;\; U_7 \;\; U_8 \;\; U_9 \right\rfloor^T$$

$$= \left\lfloor \frac{wL^4}{6IE} \;\; 0 \;\; -\frac{wL^3}{4IE} \;\; \frac{wL^4}{6IE} \;\; -\frac{7wL^4}{32IE} \;\; -\frac{wL^3}{6IE} \;\; \frac{wL^4}{12IE} \;\; -\frac{7wL^4}{32IE} \;\; -\frac{wL^3}{6IE} \right\rfloor^T$$

It can be observed that both inextensibility and inflexibility constraints can be easily imposed in the flexibility method when compared to the stiffness method.

## EXAMPLE 9.2

Consider the truss structure supported by springs shown in Figure 9.7. Use the flexibility method to determine its nodal displacements. For simplicity, it is assumed that all members have the same $EA$.

**Solution:** Member, node, DOFs, basic-member force, and basic-member deformation numbering systems are shown in Figure 9.8.

From the statical considerations described in Chapter 3, the matrix equilibrium equation is:

$$\mathbf{F} = \mathbf{BQ} \quad \text{or} \quad \begin{Bmatrix} F_1 \\ F_2 \\ F_3 \\ F_4 \\ F_5 \\ F_6 \\ F_7 \\ F_8 \end{Bmatrix} = \begin{bmatrix} 0 & 0 & 0 & -1 & -0.6 & 1 & 0 & 0 \\ -1 & 0 & 0 & 0 & -0.8 & 0 & 1 & 0 \\ 0 & -1 & 0 & 0 & 0 & 0 & 0 & 0 \\ 1 & 0 & 0 & 0 & 0 & 0 & 0 & 0 \\ 0 & 1 & -0.6 & 0 & 0.6 & 0 & 0 & 0 \\ 0 & 0 & 0.8 & 0 & 0.8 & 0 & 0 & 0 \\ 0 & 0 & 0.6 & 1 & 0 & 0 & 0 & 0 \\ 0 & 0 & -0.8 & 0 & 0 & 0 & 0 & 1 \end{bmatrix} \begin{Bmatrix} Q_1 \\ Q_2 \\ Q_3 \\ Q_4 \\ Q_5 \\ Q_6 \\ Q_7 \\ Q_8 \end{Bmatrix}$$

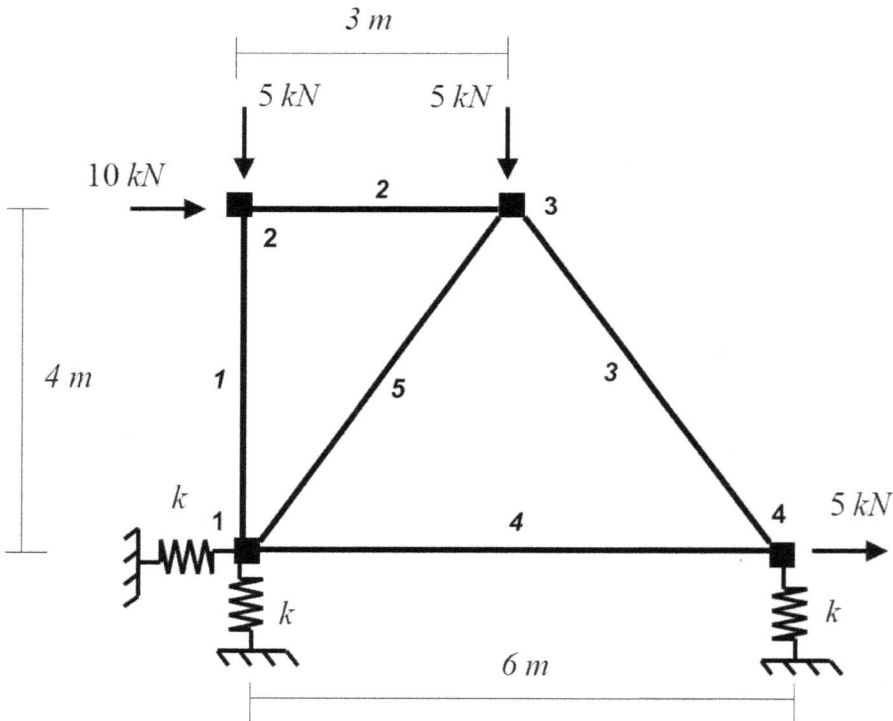

**FIGURE 9.7** Example 9.2.

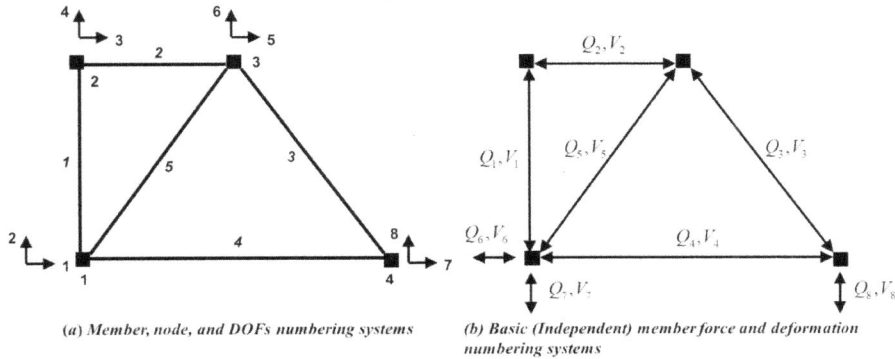

*(a) Member, node, and DOFs numbering systems*

*(b) Basic (Independent) member force and deformation numbering systems*

**FIGURE 9.8**   Example 9.2 (Continued).

The member basic forces $\mathbf{Q}$ can be expressed in terms of the nodal applied loads $\mathbf{F}$ as:

$$\mathbf{Q} = \mathbf{B}_0\mathbf{F} \quad \text{or} \quad \begin{Bmatrix} Q_1 \\ Q_2 \\ Q_3 \\ Q_4 \\ Q_5 \\ Q_6 \\ Q_7 \\ Q_8 \end{Bmatrix} = \begin{bmatrix} 0 & 0 & 0 & 1 & 0 & 0 & 0 & 0 \\ 0 & 0 & -1 & 0 & 0 & 0 & 0 & 0 \\ 0 & 0 & -0.833 & 0 & -0.833 & 0.625 & 0 & 0 \\ 0 & 0 & 0.5 & 0 & 0.5 & -0.375 & 1 & 0 \\ 0 & 0 & 0.833 & 0 & 0.833 & 0.625 & 0 & 0 \\ 1 & 0 & 1 & 0 & 1 & 0 & 1 & 0 \\ 0 & 1 & 0.667 & 1 & 0.667 & 0.5 & 0 & 0 \\ 0 & 0 & -0.667 & 0 & -0.667 & 0.5 & 0 & 1 \end{bmatrix} \begin{Bmatrix} F_1 \\ F_2 \\ F_3 \\ F_4 \\ F_5 \\ F_6 \\ F_7 \\ F_8 \end{Bmatrix}$$

The unassembled flexibility matrix is:

$$\mathbf{V} = \mathbf{f}\mathbf{Q} \quad \text{or} \quad \begin{Bmatrix} V_1 \\ V_2 \\ V_3 \\ V_4 \\ V_5 \\ V_6 \\ V_7 \\ V_8 \end{Bmatrix} = \frac{1}{EA}\begin{bmatrix} 4 & 0 & 0 & 0 & 0 & 0 & 0 & 0 \\ 0 & 3 & 0 & 0 & 0 & 0 & 0 & 0 \\ 0 & 0 & 5 & 0 & 0 & 0 & 0 & 0 \\ 0 & 0 & 0 & 6 & 0 & 0 & 0 & 0 \\ 0 & 0 & 0 & 0 & 5 & 0 & 0 & 0 \\ 0 & 0 & 0 & 0 & 0 & \dfrac{EA}{k} & 0 & 0 \\ 0 & 0 & 0 & 0 & 0 & 0 & \dfrac{EA}{k} & 0 \\ 0 & 0 & 0 & 0 & 0 & 0 & 0 & \dfrac{EA}{k} \end{bmatrix} \begin{Bmatrix} Q_1 \\ Q_2 \\ Q_3 \\ Q_4 \\ Q_5 \\ Q_6 \\ Q_7 \\ Q_8 \end{Bmatrix}$$

The structural flexibility equation is:

$$\mathbf{U} = \overline{\mathbf{F}}\mathbf{F}$$

where the structural flexibility matrix $\bar{\mathbf{F}}$ is:

$$\bar{\mathbf{F}} = \mathbf{B}_0{}^T \mathbf{f} \mathbf{B}_0 = \begin{bmatrix} \dfrac{1}{k} & 0 & \dfrac{1}{k} & 0 & \dfrac{1}{k} & 0 & \dfrac{1}{k} & 0 \\[2mm] 0 & \dfrac{1}{k} & \dfrac{0.667}{k} & \dfrac{1}{k} & \dfrac{0.667}{k} & \dfrac{0.5}{k} & 0 & 0 \\[2mm] \dfrac{1}{k} & \dfrac{0.667}{k} & \dfrac{1.889}{k}+\dfrac{11.444}{EA} & \dfrac{0.667}{k} & \dfrac{1.889}{k}+\dfrac{8.444}{EA} & -\dfrac{1.125}{EA} & \dfrac{1}{k}+\dfrac{3}{EA} & -\dfrac{0.667}{k} \\[2mm] 0 & \dfrac{1}{k} & \dfrac{0.6667}{k} & \dfrac{1}{k}+\dfrac{4}{EA} & \dfrac{0.6667}{k} & \dfrac{0.5}{k} & 0 & 0 \\[2mm] \dfrac{1}{k} & \dfrac{0.667}{k} & \dfrac{1.889}{k}+\dfrac{8.444}{EA} & \dfrac{0.667}{k} & \dfrac{1.889}{k}+\dfrac{8.444}{EA} & -\dfrac{1.125}{EA} & \dfrac{1}{k}+\dfrac{3}{EA} & -\dfrac{0.667}{k} \\[2mm] 0 & \dfrac{0.5}{k} & -\dfrac{1.125}{EA} & \dfrac{0.5}{k} & -\dfrac{1.125}{EA} & \dfrac{0.5}{k}+\dfrac{4.75}{EA} & -\dfrac{2.25}{EA} & \dfrac{0.5}{k} \\[2mm] \dfrac{1}{k} & 0 & \dfrac{1}{k}+\dfrac{3}{EA} & 0 & \dfrac{1}{k}+\dfrac{3}{EA} & -\dfrac{2.25}{EA} & \dfrac{1}{k}+\dfrac{6}{EA} & 0 \\[2mm] 0 & 0 & -\dfrac{0.667}{k} & 0 & -\dfrac{0.667}{k} & \dfrac{0.5}{k} & 0 & \dfrac{1}{k} \end{bmatrix}$$

The nodal displacements are:

$$\begin{Bmatrix} U_1 \\ U_2 \\ U_3 \\ U_4 \\ U_5 \\ U_6 \\ U_7 \\ U_8 \end{Bmatrix} = \begin{bmatrix} \dfrac{1}{k} & 0 & \dfrac{1}{k} & 0 & \dfrac{1}{k} & 0 & \dfrac{1}{k} & 0 \\[2mm] 0 & \dfrac{1}{k} & \dfrac{0.667}{k} & \dfrac{1}{k} & \dfrac{0.667}{k} & \dfrac{0.5}{k} & 0 & 0 \\[2mm] \dfrac{1}{k} & \dfrac{0.667}{k} & \dfrac{1.889}{k}+\dfrac{11.444}{EA} & \dfrac{0.667}{k} & \dfrac{1.889}{k}+\dfrac{8.444}{EA} & -\dfrac{1.125}{EA} & \dfrac{1}{k}+\dfrac{3}{EA} & -\dfrac{0.667}{k} \\[2mm] 0 & \dfrac{1}{k} & \dfrac{0.6667}{k} & \dfrac{1}{k}+\dfrac{4}{EA} & \dfrac{0.6667}{k} & \dfrac{0.5}{k} & 0 & 0 \\[2mm] \dfrac{1}{k} & \dfrac{0.667}{k} & \dfrac{1.889}{k}+\dfrac{8.444}{EA} & \dfrac{0.667}{k} & \dfrac{1.889}{k}+\dfrac{8.444}{EA} & -\dfrac{1.125}{EA} & \dfrac{1}{k}+\dfrac{3}{EA} & -\dfrac{0.667}{k} \\[2mm] 0 & \dfrac{0.5}{k} & -\dfrac{1.125}{EA} & \dfrac{0.5}{k} & -\dfrac{1.125}{EA} & \dfrac{0.5}{k}+\dfrac{4.75}{EA} & -\dfrac{2.25}{EA} & \dfrac{0.5}{k} \\[2mm] \dfrac{1}{k} & 0 & \dfrac{1}{k}+\dfrac{3}{EA} & 0 & \dfrac{1}{k}+\dfrac{3}{EA} & -\dfrac{2.25}{EA} & \dfrac{1}{k}+\dfrac{6}{EA} & 0 \\[2mm] 0 & 0 & -\dfrac{0.667}{k} & 0 & -\dfrac{0.667}{k} & \dfrac{0.5}{k} & 0 & \dfrac{1}{k} \end{bmatrix} \begin{Bmatrix} 0 \\ 0 \\ 10 \\ -5 \\ 0 \\ -5 \\ 5 \\ 0 \end{Bmatrix}$$

$$\mathbf{U} = \begin{bmatrix} U_1 & U_2 & U_3 & U_4 & U_5 & U_6 & U_7 & U_8 \end{bmatrix}^T$$

$$= \begin{bmatrix} \dfrac{15}{k} & \dfrac{-0.83}{k} & \dfrac{20.56}{k}+\dfrac{135.07}{EA} & \dfrac{-0.83}{k}-\dfrac{20}{EA} & \dfrac{20.56}{k}+\dfrac{105.07}{EA} & -\dfrac{5}{k}-\dfrac{46.25}{EA} & \dfrac{15}{k}+\dfrac{71.25}{EA} & -\dfrac{9.17}{k} \end{bmatrix}^T$$

## 9.3 THE FLEXIBILITY APPROACH: STATICALLY INDETERMINATE SYSTEMS

The main objective of this chapter is to present the flexibility approach for analyzing a statically indeterminate system. By introducing a sufficient number of force releases, any given indeterminate structure can be transformed into a determinate one. Generally, the required number of force releases is equal to the degree of statical indeterminacy (*SI*) of the system. The structure becomes unstable when the number of force releases is larger than the degree of statical indeterminacy. Common types of force releases are shown in Figure 9.9. The term "force release" implies that certain force actions cannot be transmitted through these force-release devices. That is, the devices in Figure 9.9*a*, *b*, and *c* cannot transmit thrust, shear, and moment, respectively. Figure 9.9*d* represents the complete release through which no force can be transmitted. These inabilities to transmit force actions result in discontinuities in corresponding displacement components (axial displacement, transverse displacement, and rotation).

In general, the forces associated with the releases are called *redundants*, and they serve as the basic unknowns in the force method. The resulting released structure is statically determinate and is referred to as the primary structure. Due to its statical determinacy, the internal forces can simply be expressed in terms of the given applied loads and the redundants through the equilibrium conditions. As a result, the associated internal deformations can be obtained through the member flexibility equations. Consequently, the associated internal deformations can be written in terms of the given applied loads and the redundants. The virtual force principle can be employed to determine displacements at any desired location, especially the relative displacements associated with the released locations. As mentioned earlier, the force releases usually result in the discontinuities (relative displacements) in their associated displacements (axial displacement, transverse displacement, and rotation). For example, when the shear (Figure 9.9*b*) and moment (Figure 9.9*c*) are released, there are relative displacement and rotation, respectively. However, no such discontinuities exist in the given structure. Thus, the continuity conditions can be imposed at the released locations and form the compatibility equations to solve for the unknown redundants. Once the redundants are obtained, the analysis of the primary system (statically determinate) yields the internal forces, internal deformations, and nodal displacements as a superposition of the contributions from the given loads and the redundants.

With the forgoing description of the flexibility method, the general approach to the flexibility method can be summarized as:

1. Construct the degree-of-freedom, node, and member numbering systems.
2. Convert the statically indeterminate structure to a determinate one (primary structure) by inserting a sufficient number of force releases. The associated released forces serve as basic unknowns of the system. It should be noted that a poor choice of force releases may lead to an ill-conditioned system. More details of this issue can be found in Felton and Nelson (1997).

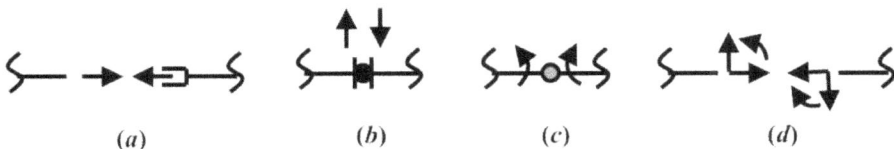

**FIGURE 9.9** Force-Releases: (*a*) Thrust Release; (*b*) Shear Release; (*c*) Moment Release; (*d*) Complete Release.

3.  Express the internal (basic member) forces $\mathbf{Q}$ of the primary structure in terms of the given applied loads $\mathbf{F}^*_{free}$ and the redundants $\mathbf{R}$ through the equilibrium conditions. Generally, these statical relations can be expressed as:

$$\mathbf{Q} = \mathbf{Q}' + \mathbf{Q}'' = \mathbf{B}_0 \mathbf{F}^*_{free} + \mathbf{B}_R \mathbf{R} \tag{9.21}$$

4.  It is worth observing that $\mathbf{Q}'$ represents the internal (basic member) forces due to the given applied loads $\mathbf{F}^*_{free}$ and $\mathbf{Q}''$ represents the internal (basic member) forces due to the redundants $\mathbf{R}$. Furthermore, it should be pointed out that the matrices $\mathbf{B}_0$ and $\mathbf{B}_R$ in Eq. (9.21) can be viewed as discrete forms of force interpolation functions between the internal force vectors $\mathbf{Q}$ and the applied load $\mathbf{F}^*_{free}$ and redundant $\mathbf{R}$ vectors, respectively. The author has used this form of force interpolation functions to formulate the flexibility-based frame model with tangential interfaces (Limkatanyu 2002) and flexibility-based frame model with lateral interfaces (Limkatanyu and Spacone 2006). The reader who has a keen interest in the flexibility-based frame model is urged to explore several papers published by the author as well as doctoral dissertations by students of Prof. F.C. Filippou (University of California, Berkeley, USA) and of Prof. E. Spacone (University "G. D'Annunzio" of Chieti, Pescara, Italy).

5.  Using the member flexibility relations of Eq. (9.10), determine the internal (basic member) deformations $\mathbf{V}$. Essentially, the basic member deformations are expressed in terms of the given applied loads $\mathbf{F}^*_{free}$ and the redundants $\mathbf{R}$.

6.  Use the virtual force principle to determine the relative displacements at the assumed released locations.

7.  Since in the given structure, such relative displacements do not exist. Thus, the compatibility equations associated with the released locations are obtained. The compatibility equations are expressed in terms of the internal (basic member) deformations $\mathbf{V}$ and essentially, in terms of the given applied loads $\mathbf{F}^*_{free}$ and the redundants $\mathbf{R}$. Solve the compatibility equations for the redundants $\mathbf{R}$.

8.  Substituting $\mathbf{R}$ into step (3), the basic member forces $\mathbf{Q}$ are obtained.

9.  Based on Eq. (9.10), the basic member deformations $\mathbf{V}$ are obtained.

10. Use the virtual force principle to obtain the nodal displacements $\mathbf{U}$.

A statically indeterminate frame shown in Figure 9.10 is used as a vehicle for the description of the aforementioned steps. From the statical consideration, it is noticed that the degree of statical indeterminacy ($SI$) is equal to $9-7 = 2$. Therefore, two force releases are required in order to render this frame statically determinate. Several choices of primary structures are shown in Figure 9.11. In Figure 9.11$b$–$g$, some acceptable choices of primary structures are shown. The primary structure shown in Figure 9.11$h$ is unacceptable since it cannot resist applied horizontal loads.

The released structure of Figure is selected as the primary structure. Referring to Figure 9.2, the basic member forces and deformations of this frame are written in the matrix form as:

$$\mathbf{Q} = \lfloor \mathbf{Q}_1 \quad \mathbf{Q}_2 \quad \mathbf{Q}_3 \rfloor^T \text{ and } \mathbf{V} = \lfloor \mathbf{V}_1 \quad \mathbf{V}_2 \quad \mathbf{V}_3 \rfloor^T$$

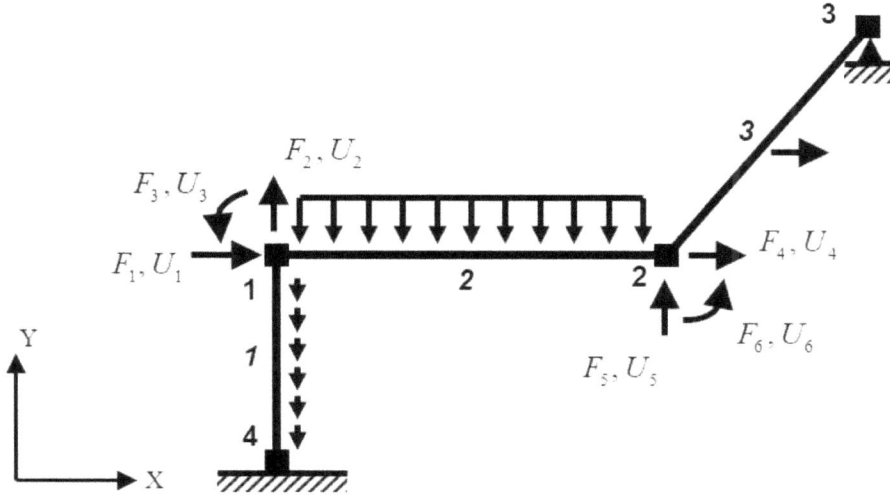

**FIGURE 9.10** A Statically Indeterminate Frame ($SI = 2$).

In this case, the basic member forces $Q_2$ and $Q_5$ are redundants. Thus, the redundant force vector **R** is:

$$\mathbf{R} = \lfloor Q_5 \quad Q_2 \rfloor^T$$

The statical relation of Eq. (9.21) is recalled:

$$\mathbf{Q} = \mathbf{Q}' + \mathbf{Q}'' = \mathbf{B}_0 \mathbf{F}_{free}^* + \mathbf{B}_R \mathbf{R} \tag{9.22}$$

The unassembled flexibility equations are:

$$\begin{Bmatrix} \mathbf{V}_1 \\ \mathbf{V}_2 \\ \mathbf{V}_3 \end{Bmatrix} = \begin{bmatrix} \mathbf{f}^{(1)} & \mathbf{0} & \mathbf{0} \\ \mathbf{0} & \mathbf{f}^{(2)} & \mathbf{0} \\ \mathbf{0} & \mathbf{0} & \mathbf{f}^{(3)} \end{bmatrix} \begin{Bmatrix} \mathbf{Q}_1 \\ \mathbf{Q}_2 \\ \mathbf{Q}_3 \end{Bmatrix} + \begin{Bmatrix} \mathbf{V}_0^1 \\ \mathbf{V}_0^2 \\ \mathbf{V}_0^3 \end{Bmatrix} + \begin{Bmatrix} \mathbf{V}_w^1 \\ \mathbf{V}_w^2 \\ \mathbf{V}_w^3 \end{Bmatrix} + \begin{Bmatrix} \mathbf{V}_s^1 \\ \mathbf{V}_s^2 \\ \mathbf{V}_s^3 \end{Bmatrix} \text{ or } \mathbf{V} = \mathbf{f}\mathbf{Q} + \mathbf{V}_0 + \mathbf{V}_w + \mathbf{V}_s \tag{9.23}$$

Substituting Eqs. (9.21) into (9.23), one has:

$$\mathbf{V} = \mathbf{f}\left(\mathbf{B}_0 \mathbf{F}_{free}^* + \mathbf{B}_R \mathbf{R}\right) + \mathbf{V}_0 + \mathbf{V}_w + \mathbf{V}_s \tag{9.24}$$

So far, the equilibrium conditions and flexibility properties have been considered. Next, the compatibility conditions are derived based on the virtual force principle.

$$\delta W_{ext}^* = \delta W_{int}^* \tag{9.25}$$

Internal complementary virtual work is:

$$\delta W_{int}^* = \delta \mathbf{Q}^T \mathbf{V} \tag{9.26}$$

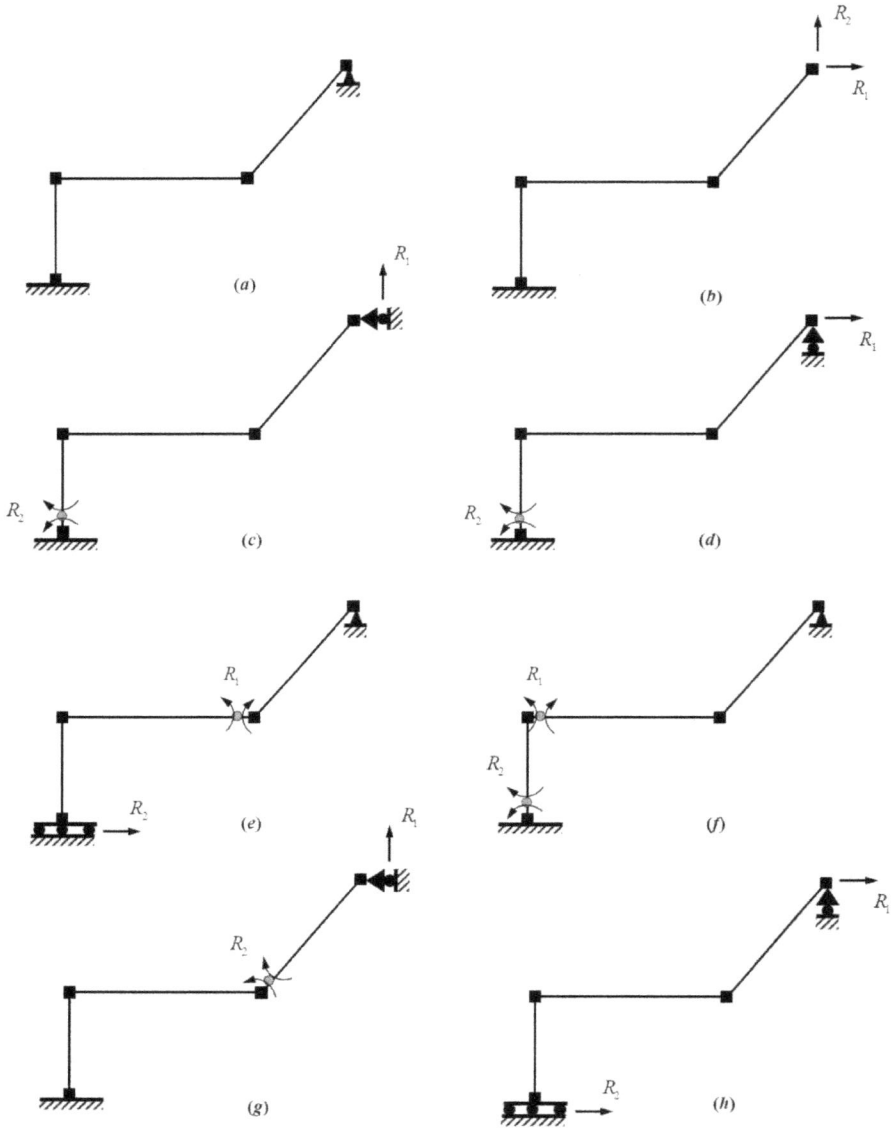

**FIGURE 9.11** (*a*) Two-Degree Indeterminate Frame; (*b*)–(*g*) Acceptable Choices of Primary Structures; (*h*) Improper Choice of Primary Structures.

It is noted that:

$$\delta \mathbf{Q}^T = \delta \mathbf{F}_{free}^{*}{}^T \mathbf{B}_0^T + \delta \mathbf{R}^T \mathbf{B}_R^T \tag{9.27}$$

Substituting Eqs. (9.27) into (9.26), one has:

$$\delta W_{int}^{*} = \delta \mathbf{Q}^T \mathbf{V} = \left( \delta \mathbf{F}_{free}^{*}{}^T \mathbf{B}_0^T + \delta \mathbf{R}^T \mathbf{B}_R^T \right) \mathbf{V} = \delta \mathbf{F}_{free}^{*}{}^T \mathbf{B}_0^T \mathbf{V} + \delta \mathbf{R}^T \mathbf{B}_R^T \mathbf{V} \tag{9.28}$$

External complementary virtual work is:

$$\delta W_{ext}^* = \delta \mathbf{F}_{free}^{*T} \mathbf{U}_{free} \tag{9.29}$$

Imposing the virtual force principle, one has:

$$\delta W_{int}^* = \delta W_{ext}^* \rightarrow \delta \mathbf{F}_{free}^{*T} \left( \mathbf{B}_0^T \mathbf{V} - \mathbf{U}_{free} \right) + \delta \mathbf{R}^T \mathbf{B}_R^T \mathbf{V} = 0 \tag{9.30}$$

From arbitrariness of $\delta \mathbf{F}_{free}^*$ and $\delta \mathbf{R}$, it can be concluded that:

$$\mathbf{B}_0^T \mathbf{V} - \mathbf{U}_{free} = \mathbf{0} \tag{9.31}$$

and

$$\mathbf{B}_R^T \mathbf{V} = \mathbf{0} \tag{9.32}$$

Eq. (9.31) provides the nodal displacements for a given set of member deformations $\mathbf{V}$ and Eq. (9.32) provides the compatibility conditions. Substituting Eqs. (9.24) into (9.32), one has:

$$\mathbf{B}_R^T \mathbf{f} \mathbf{B}_0 \mathbf{F}_{free}^* + \mathbf{B}_R^T \mathbf{f} \mathbf{B}_R \mathbf{R} + \mathbf{B}_R^T \mathbf{V}_0 + \mathbf{B}_R^T \mathbf{V}_W + \mathbf{B}_R^T \mathbf{V}_S = \mathbf{0} \tag{9.33}$$

or

$$\overline{\mathbf{F}}_{R0} \mathbf{F}_{free}^* + \overline{\mathbf{F}}_{RR} \mathbf{R} + \mathbf{U}_{R0} + \mathbf{U}_{RW} + \mathbf{U}_{RS} = \mathbf{0} \tag{9.34}$$

where

$$\overline{\mathbf{F}}_{R0} = \mathbf{B}_R^T \mathbf{f} \mathbf{B}_0; \quad \overline{\mathbf{F}}_{RR} = \mathbf{B}_R^T \mathbf{f} \mathbf{B}_R; \quad \mathbf{U}_{R0} = \mathbf{B}_R^T \mathbf{V}_0; \quad \mathbf{U}_{RW} = \mathbf{B}_R^T \mathbf{V}_W; \text{and } \mathbf{U}_{RS} = \mathbf{B}_R^T \mathbf{V}_S \tag{9.35}$$

The compatibility conditions of Eq. (9.34) can be solved for the redundants $\mathbf{R}$. The physical interpretation of Eq. (9.34) is that the summation of relative displacements at released locations induced by the given applied loads $\mathbf{F}_{free}^*$, the redundants $\mathbf{R}$, the initial deformations, the member loads, and the support movements is equal to zero.

## EXAMPLE 9.3

Use the flexibility method to analyze the inextensible frame shown in Figure 9.12 and determine all nodal displacements.

   **Solution:** Member, node, DOFs, basic-member force, and basic-member deformation numbering systems are shown in Figure 9.13.

   From the statical considerations described in Chapter 3, the matrix equilibrium equation is:

$$\mathbf{F}_{free} = \mathbf{B}_{free} \mathbf{Q} + \mathbf{F}_{free}^0 \quad \text{or} \quad \begin{Bmatrix} F_1 \\ F_2 \\ F_3 \\ F_4 \end{Bmatrix} = \begin{bmatrix} \dfrac{1}{\sqrt{2}} & \dfrac{1}{2L} & \dfrac{1}{2L} & -1 & 0 & 0 \\ \dfrac{1}{\sqrt{2}} & -\dfrac{1}{2L} & -\dfrac{1}{2L} & 0 & \dfrac{1}{L} & \dfrac{1}{L} \\ 0 & 0 & 1 & 0 & 1 & 0 \\ 0 & 0 & 0 & 1 & 0 & 0 \end{bmatrix} \begin{Bmatrix} Q_1 \\ Q_2 \\ Q_3 \\ Q_4 \\ Q_5 \\ Q_6 \end{Bmatrix} + \begin{Bmatrix} 0 \\ \dfrac{wL}{2} \\ 0 \\ 0 \end{Bmatrix}$$

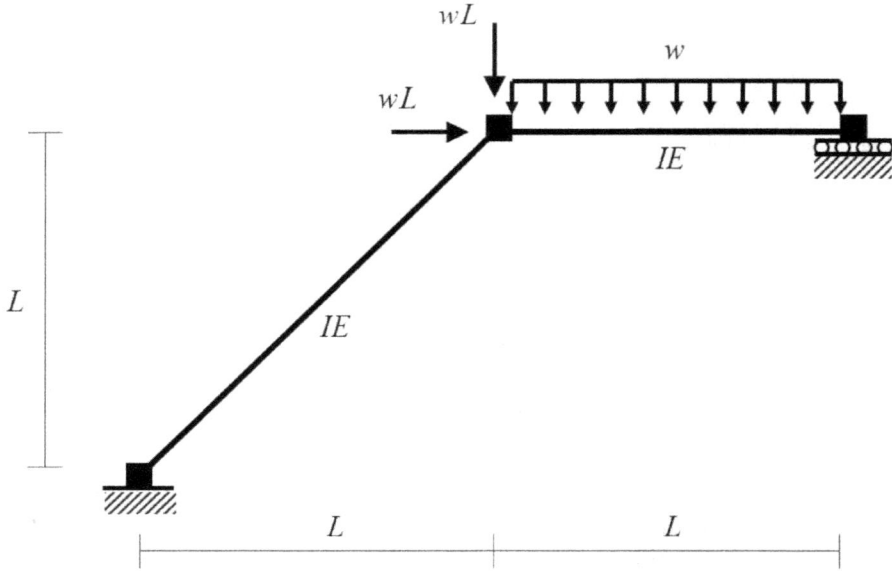

**FIGURE 9.12** Example 9.3.

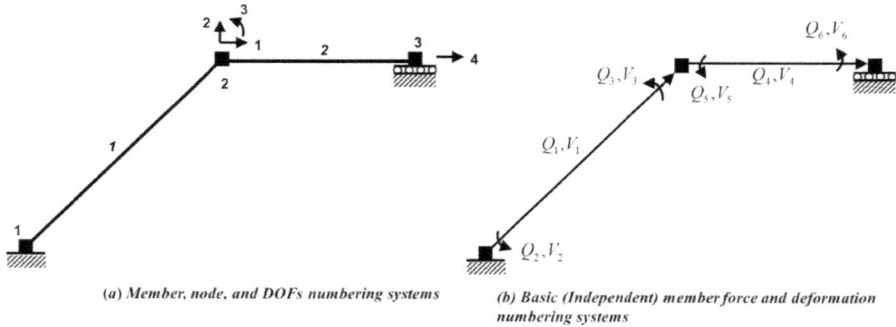

*(a) Member, node, and DOFs numbering systems*   *(b) Basic (Independent) member force and deformation numbering systems*

**FIGURE 9.13** Example 9.3 (Continued).

From the statical consideration, it is noticed that the degree of statical indeterminacy (*SI*) of this system is equal to 6–4 = 2. Thus, two force releases are needed in order to render this frame statically determinate. Basic member forces $Q_5$ and $Q_6$ are selected as redundants.

$$\mathbf{R} = \lfloor R_1 \quad R_2 \rfloor^T = \lfloor Q_5 \quad Q_6 \rfloor^T$$

The primary structure is shown in Figure 9.14.

Thus, the basic member forces $\mathbf{Q}$ can be expressed in terms of the given applied loads $\mathbf{F}^0_{free}$ and redundants $\mathbf{R}$ as:

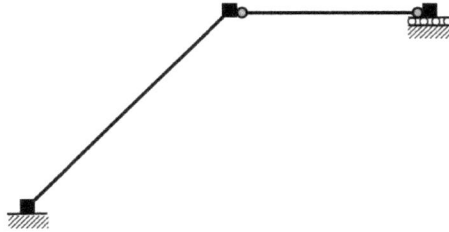

**FIGURE 9.14** Example 9.3 (Continued).

$$\mathbf{Q} = \mathbf{B}_0 \mathbf{F}_{free} + \mathbf{B}_R \mathbf{R} \rightarrow \begin{Bmatrix} Q_1 \\ Q_2 \\ Q_3 \\ Q_4 \\ Q_5 \\ Q_6 \end{Bmatrix} = \begin{bmatrix} \dfrac{1}{\sqrt{2}} & \dfrac{1}{\sqrt{2}} & 0 & \dfrac{1}{\sqrt{2}} \\ L & -L & -1 & L \\ 0 & 0 & 1 & 0 \\ 0 & 0 & 0 & 1 \\ 0 & 0 & 0 & 0 \\ 0 & 0 & 0 & 0 \end{bmatrix} \begin{Bmatrix} F^*_{free1} \\ F^*_{free2} \\ F^*_{free3} \\ F^*_{free4} \end{Bmatrix} + \begin{bmatrix} -\dfrac{1}{\sqrt{2}L} & -\dfrac{1}{\sqrt{2}L} \\ 2 & 1 \\ -1 & 0 \\ 0 & 0 \\ 1 & 0 \\ 0 & 1 \end{bmatrix} \begin{Bmatrix} R_1 \\ R_2 \end{Bmatrix}$$

The physical interpretation of matrices $\mathbf{B}_0$ and $\mathbf{B}_R$ is shown in Figures 9.15 to 9.17. The unassembled flexibility matrix is:

$$\mathbf{V} = \mathbf{f}\mathbf{Q} + \mathbf{V}_w \qquad \text{or} \qquad \begin{Bmatrix} \mathbf{V}_1 \\ \mathbf{V}_2 \end{Bmatrix} = \begin{bmatrix} \mathbf{f}_1 & \mathbf{0} \\ \mathbf{0} & \mathbf{f}_2 \end{bmatrix} \begin{Bmatrix} \mathbf{Q}_1 \\ \mathbf{Q}_2 \end{Bmatrix} + \begin{Bmatrix} \mathbf{V}_w^1 \\ \mathbf{V}_w^2 \end{Bmatrix}$$

For member 2, one has:

$$\mathbf{V}_w^2 = \begin{bmatrix} 0 & -\dfrac{wL^3}{24IE} & \dfrac{wL^3}{24IE} \end{bmatrix}^T$$

$$\begin{Bmatrix} V_1 \\ V_2 \\ V_3 \\ V_4 \\ V_5 \\ V_6 \end{Bmatrix} = \begin{bmatrix} 0 & 0 & 0 & 0 & 0 & 0 \\ 0 & \dfrac{\sqrt{2}L}{3IE} & -\dfrac{L}{3\sqrt{2}IE} & 0 & 0 & 0 \\ 0 & -\dfrac{L}{3\sqrt{2}IE} & \dfrac{\sqrt{2}L}{3IE} & 0 & 0 & 0 \\ 0 & 0 & 0 & 0 & 0 & 0 \\ 0 & 0 & 0 & 0 & \dfrac{L}{3IE} & -\dfrac{L}{6IE} \\ 0 & 0 & 0 & 0 & -\dfrac{L}{6IE} & \dfrac{L}{3IE} \end{bmatrix} \begin{Bmatrix} Q_1 \\ Q_2 \\ Q_3 \\ Q_4 \\ Q_5 \\ Q_6 \end{Bmatrix} + \begin{Bmatrix} 0 \\ 0 \\ 0 \\ 0 \\ -\dfrac{wL^3}{24IE} \\ \dfrac{wL^3}{24IE} \end{Bmatrix}$$

The matrix compatibility equation is:

$$\underbrace{\overline{\mathbf{F}}_{R0}\mathbf{F}^*_{free}}_{\mathbf{U}_{RP}} + \underbrace{\overline{\mathbf{F}}_{RR}\mathbf{R}}_{\mathbf{U}_{RR}} + \mathbf{U}_{RW} = \mathbf{0}$$

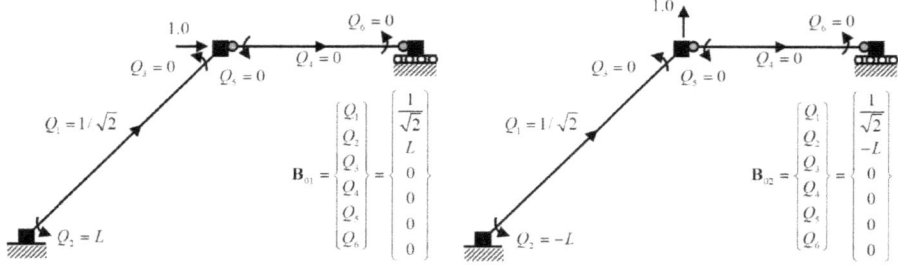

**FIGURE 9.15**    Example 9.3 (Continued).

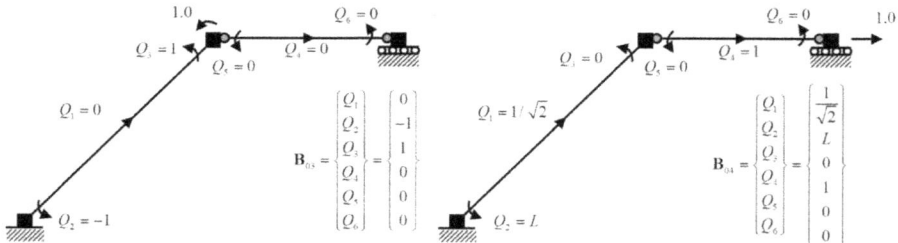

**FIGURE 9.16**    Example 9.3 (Continued).

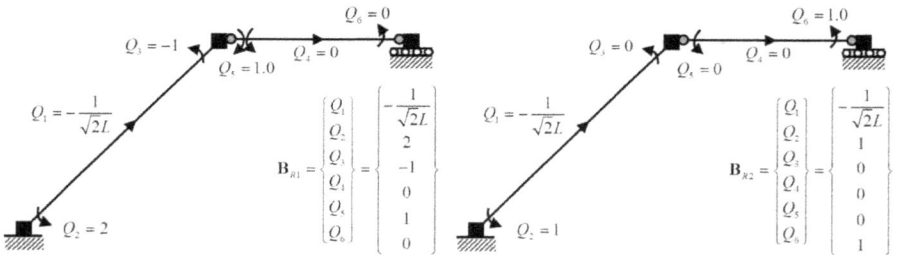

**FIGURE 9.17**    Example 9.3 (Continued).

where

$$\overline{\mathbf{F}}_{RR} = \mathbf{B}_R^T \mathbf{f} \mathbf{B}_R = \frac{L}{IE}\begin{bmatrix} 3.633 & 1.012 \\ 1.012 & 0.805 \end{bmatrix};$$

$$\overline{\mathbf{F}}_{R0} = \mathbf{B}_R^{\ T} \mathbf{f} \mathbf{B}_0 = \frac{L}{IE}\begin{bmatrix} 1.179L & -1.179L & -2.12132 & 1.179L \\ 0.471L & -0.471L & -0.707 & -0.471L \end{bmatrix};$$

$$\mathbf{F}_{free}^* = \begin{bmatrix} wL & -\dfrac{3}{2}wL & 0 & 0 \end{bmatrix}^T; \ \mathbf{U}_{RP} = \overline{\mathbf{F}}_{R0}\mathbf{F}_{free}^* = \begin{bmatrix} \dfrac{2.946\,wL^3}{IE} & \dfrac{1.179\,wL^3}{IE} \end{bmatrix}^T;$$

$$\mathbf{U}_{RW} = \mathbf{B}_R{}^T\mathbf{V}_w = \left| -\frac{wL^3}{24IE} \quad \frac{wL^3}{24IE} \right|^T$$

The physical interpretation of the matrix compatibility equation is that the summation of relative rotations at released locations induced by the given applied loads $\mathbf{F}_{free}^*$, the redundants $\mathbf{R}$, and the member load is equal to zero.

$$\mathbf{R} = -\left[\overline{\mathbf{F}}_{RR}\right]^{-1}\left(\mathbf{U}_{RP} + \mathbf{U}_{RW}\right) \quad \Rightarrow \quad \mathbf{R} = \begin{Bmatrix} R_1 \\ R_2 \end{Bmatrix} = \begin{Bmatrix} Q_5 \\ Q_6 \end{Bmatrix} = \begin{Bmatrix} -0.580\,wL^2 \\ -0.786\,wL^2 \end{Bmatrix}$$

The basic member forces $\mathbf{Q}$ are:

$$\begin{Bmatrix} Q_1 \\ Q_2 \\ Q_3 \\ Q_4 \\ Q_5 \\ Q_6 \end{Bmatrix} = \begin{bmatrix} \dfrac{1}{\sqrt{2}} & \dfrac{1}{\sqrt{2}} & 0 & \dfrac{1}{\sqrt{2}} \\ L & -L & -1 & L \\ 0 & 0 & 1 & 0 \\ 0 & 0 & 0 & 1 \\ 0 & 0 & 0 & 0 \\ 0 & 0 & 0 & 0 \end{bmatrix} \begin{Bmatrix} wL \\ \dfrac{3wL}{2} \\ 0 \\ 0 \end{Bmatrix} + \begin{bmatrix} -\dfrac{1}{\sqrt{2}L} & -\dfrac{1}{\sqrt{2}L} \\ 2 & 1 \\ -1 & 0 \\ 0 & 0 \\ 1 & 0 \\ 0 & 1 \end{bmatrix} \begin{Bmatrix} -0.580wL^2 \\ -0.786wL^2 \end{Bmatrix} = \begin{Bmatrix} 0.613wL \\ 0.553wL^2 \\ 0.580wL^2 \\ 0 \\ -0.580wL^2 \\ -0.786wL^2 \end{Bmatrix}$$

The basic member deformations $\mathbf{V}$ are:

$$\begin{Bmatrix} V_1 \\ V_2 \\ V_3 \\ V_4 \\ V_5 \\ V_6 \end{Bmatrix} = \begin{bmatrix} 0 & 0 & 0 & 0 & 0 & 0 \\ 0 & \dfrac{\sqrt{2}L}{3IE} & -\dfrac{L}{3\sqrt{2}IE} & 0 & 0 & 0 \\ 0 & -\dfrac{L}{3\sqrt{2}IE} & \dfrac{\sqrt{2}L}{3IE} & 0 & 0 & 0 \\ 0 & 0 & 0 & 0 & 0 & 0 \\ 0 & 0 & 0 & 0 & \dfrac{L}{3IE} & -\dfrac{L}{6IE} \\ 0 & 0 & 0 & 0 & -\dfrac{L}{6IE} & \dfrac{L}{3IE} \end{bmatrix} \begin{Bmatrix} 0.613wL \\ 0.553wL^2 \\ 0.580wL^2 \\ 0 \\ -0.580wL^2 \\ -0.786wL^2 \end{Bmatrix} + \begin{Bmatrix} 0 \\ 0 \\ 0 \\ 0 \\ -\dfrac{wL^3}{24IE} \\ \dfrac{wL^3}{24IE} \end{Bmatrix} = \begin{Bmatrix} 0 \\ \dfrac{0.124wL^3}{IE} \\ \dfrac{0.143wL^3}{IE} \\ 0 \\ -\dfrac{0.104L^3}{IE} \\ -\dfrac{0.124wL^3}{IE} \end{Bmatrix}$$

The nodal displacements $\mathbf{U}_{free}$ are:

$$\mathbf{U}_{free} = \mathbf{B}_0{}^T\mathbf{V} = \left| \frac{0.124wL^4}{IE} \quad -\frac{0.124wL^4}{IE} \quad \frac{0.0196wL^3}{IE} \quad \frac{0.124wL^4}{IE} \right|^T$$

The complete member end forces can be determined from the matrix equilibrium equation of each member.

For example, the complete end forces of member 1 are:

$$
\mathbf{P}^{(1)} = \mathbf{b}_B^{(1)}\mathbf{Q}_1 \quad \text{or} \quad
\begin{Bmatrix} P_1^{(1)} \\ P_2^{(1)} \\ P_3^{(1)} \\ P_4^{(1)} \\ P_5^{(1)} \\ P_6^{(1)} \end{Bmatrix}
=
\begin{bmatrix}
-\dfrac{1}{\sqrt{2}} & -\dfrac{1}{2L} & -\dfrac{1}{2L} \\[2mm]
-\dfrac{1}{\sqrt{2}} & \dfrac{1}{2L} & \dfrac{1}{2L} \\[2mm]
0 & 1 & 0 \\[2mm]
\dfrac{1}{\sqrt{2}} & \dfrac{1}{2L} & \dfrac{1}{2L} \\[2mm]
\dfrac{1}{\sqrt{2}} & -\dfrac{1}{2L} & -\dfrac{1}{2L} \\[2mm]
0 & 0 & 1
\end{bmatrix}
\begin{Bmatrix} 0.613wL \\ 0.553wL^2 \\ 0.580wL^2 \end{Bmatrix}
=
\begin{Bmatrix} -wL \\ 0.133wL \\ 0.553wL^2 \\ wL \\ -0.133wL \\ 0.580wL^2 \end{Bmatrix}
$$

## EXAMPLE 9.4

Use the flexibility method to determine the forces in all members of the truss shown due to lack of fits as given in Figure 9.18. For all members, $E = 210$ GPa.

| Member | $L(\mathbf{m})$ | Area $(\mathbf{m^2})$ | Lack of Fit $(\mathbf{m})$ |
|--------|------|--------|-------------|
| 1 | 2 | 0.0016 | 0.01 |
| 2 | 1.5 | 0.0016 | −0.004 |
| 3 | 2 | 0.0016 | 0 |
| 4 | 1.5 | 0.0016 | 0.002 |
| 5 | 2.5 | 0.0008 | −0.002 |
| 6 | 2.5 | 0.0008 | 0 |

**Solution:** The basic force and deformation numbering systems are shown in Figure 9.19a.

From the equilibrium consideration, it is noticed that the degree of statical indeterminacy is equal to $6-5 = 1$. Thus, one release is needed to render this truss statically determinate. The basic member force $Q_6$ is selected as the redundant. Therefore, the primary structure is shown in Figure 9.19b. The matrices $\mathbf{B}_0$ and $\mathbf{B}_R$ are shown in Figure 9.20.

The statical relations between $\mathbf{Q}$ and $\mathbf{R}$ can be written as:

$$
\mathbf{Q} = \mathbf{B}_R \mathbf{R} \quad \Rightarrow \quad
\begin{Bmatrix} Q_1 \\ Q_2 \\ Q_3 \\ Q_4 \\ Q_5 \\ Q_6 \end{Bmatrix}
=
\begin{bmatrix} -0.8 \\ -0.6 \\ -0.8 \\ -0.6 \\ 1 \\ 1 \end{bmatrix}
\{R_1\}
$$

The initial basic member deformations $\mathbf{V}_0$ due to lack of fit are:

$$
\mathbf{V}_0 = \begin{bmatrix} V_0^1 & V_0^2 & V_0^3 & V_0^4 & V_0^5 & V_0^6 \end{bmatrix}^T = \begin{bmatrix} 0.01 & -0.004 & 0 & 0.002 & -0.002 & 0 \end{bmatrix}^T
$$

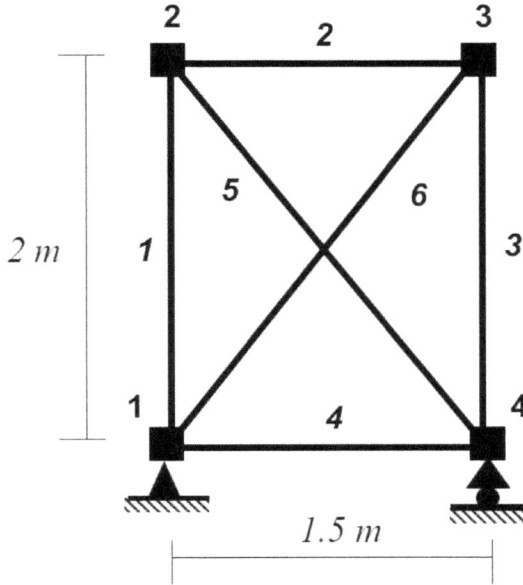

**FIGURE 9.18** Example 9.4.

The unassembled flexibility equations are:

$$\mathbf{V} = \mathbf{f}\mathbf{Q} + \mathbf{V}_0 \qquad \text{or} \qquad \begin{Bmatrix} V_1 \\ V_2 \\ V_3 \\ V_4 \\ V_5 \\ V_6 \end{Bmatrix} = \begin{bmatrix} f_1 & 0 & 0 & 0 & 0 & 0 \\ 0 & f_2 & 0 & 0 & 0 & 0 \\ 0 & 0 & f_3 & 0 & 0 & 0 \\ 0 & 0 & 0 & f_4 & 0 & 0 \\ 0 & 0 & 0 & 0 & f_5 & 0 \\ 0 & 0 & 0 & 0 & 0 & f_6 \end{bmatrix} \begin{Bmatrix} Q_1 \\ Q_2 \\ Q_3 \\ Q_4 \\ Q_5 \\ Q_6 \end{Bmatrix} + \begin{Bmatrix} V_0^1 \\ V_0^2 \\ V_0^3 \\ V_0^4 \\ V_0^5 \\ V_0^6 \end{Bmatrix}$$

where

$$f_1 = \frac{L_1}{EA_1} = \frac{1.25}{210000}; \quad f_2 = \frac{L_2}{EA_2} = \frac{0.9375}{210000}; \quad f_3 = \frac{L_3}{EA_3} = \frac{1.25}{210000};$$

$$f_4 = \frac{L_4}{EA_4} = \frac{0.9375}{210000}; \quad f_5 = \frac{L_5}{EA_5} = \frac{3.75}{210000}; \quad f_6 = \frac{L_6}{EA_6} = \frac{3.75}{210000};$$

The matrix compatibility equation is:

$$\bar{\mathbf{F}}_{RR}\mathbf{R} + \mathbf{U}_{R0} = \mathbf{0}$$

where

$$\bar{\mathbf{F}}_{RR} = \mathbf{B}_R^T \mathbf{f} \mathbf{B}_R = \begin{bmatrix} 0.04655 \times 10^{-3} \end{bmatrix} \quad \text{and} \quad \mathbf{U}_{R0} = \mathbf{B}_R^T \mathbf{V}_0 = -8.8 \times 10^{-3} \ m$$

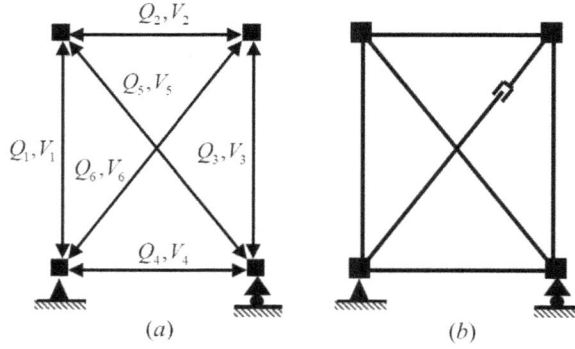

**FIGURE 9.19**  Example 9.4 (Continued).

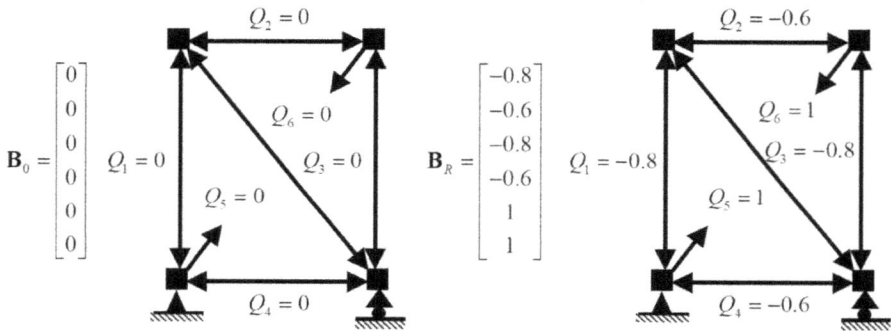

**FIGURE 9.20**  Example 9.4 (Continued)

The redundant force $R_1$ is:

$$\mathbf{R} = -\left[\overline{\mathbf{F}}_{RR}\right]^{-1}\mathbf{U}_{R0} \quad \Rightarrow \quad Q_6 = R_1 = -\frac{-8.8\times10^{-3}}{0.04655\times10^{-3}} = 189.04 \; kN$$

The axial forces in all members are:

$$\mathbf{Q} = \mathbf{B}_R\mathbf{R} \quad \Rightarrow \quad \begin{Bmatrix} Q_1 \\ Q_2 \\ Q_3 \\ Q_4 \\ Q_5 \\ Q_6 \end{Bmatrix} = \begin{bmatrix} -0.8 \\ -0.6 \\ -0.8 \\ -0.6 \\ 1 \\ 1 \end{bmatrix}\{189.04\} = \begin{Bmatrix} -151.23 \\ -113.42 \\ -151.23 \\ -113.42 \\ 189.04 \\ 189.04 \end{Bmatrix} kN$$

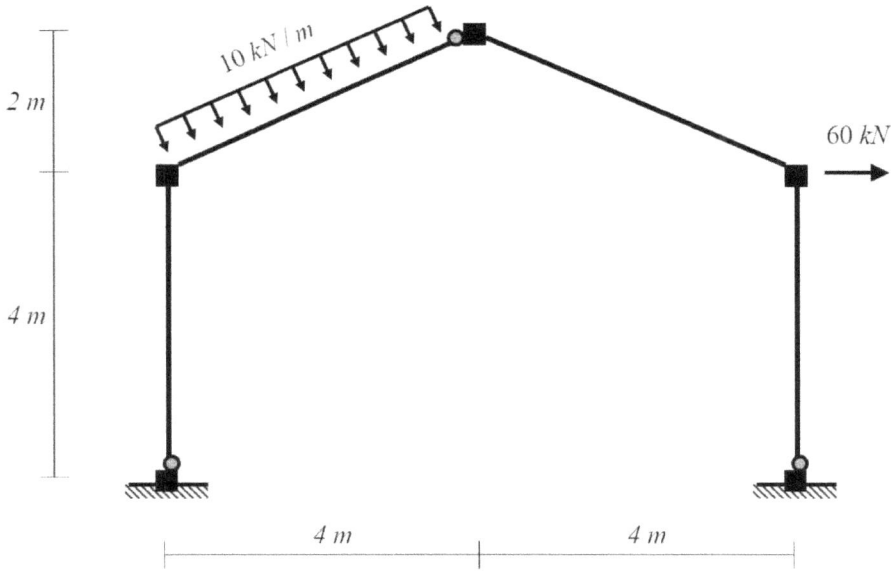

**FIGURE 9.21**    Problem 9.1.

## 9.4   EXERCISES

**Problem 9.1:** Use the flexibility method to compute all nodal displacements of the frame structure shown in Figure 9.21. All members of the frame are made of the same material ( $E = 210$ GPa ) and have the same constant cross section ( $I = 60 \times 10^6$ mm$^4$ and $A = 20 \times 10^3$ mm$^2$).

**Problem 9.2:** Use the flexibility method to analyze the truss shown in Figure 9.22. The support at node 1 settles 25 mm. Assume that the sectional area $A$ is 2000 mm$^2$ for all members and the elastic modulus $E$ is 200 GPa.

**Problem 9.3:** Use the flexibility method to analyze the coupling beam system shown in Figure 9.23 and determine all nodal displacements.

**Problem 9.4:** Use the flexibility method to analyze the beam shown in Figure 9.24 and determine all nodal displacements.

**Problem 9.5:** The vertical deflection at $b$ in the king post bridge shown in Figure 9.25 is too large. Therefore, it is decided to shorten bars $ad$ and $dc$ by equal amounts until there is no vertical deflection at $b$.

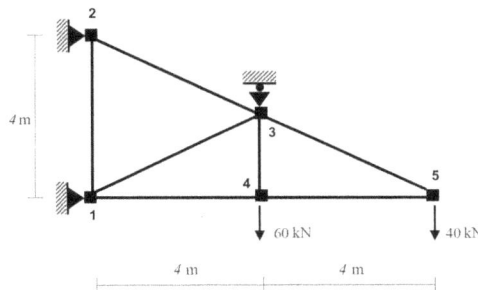

**FIGURE 9.22**    Problem 9.2.

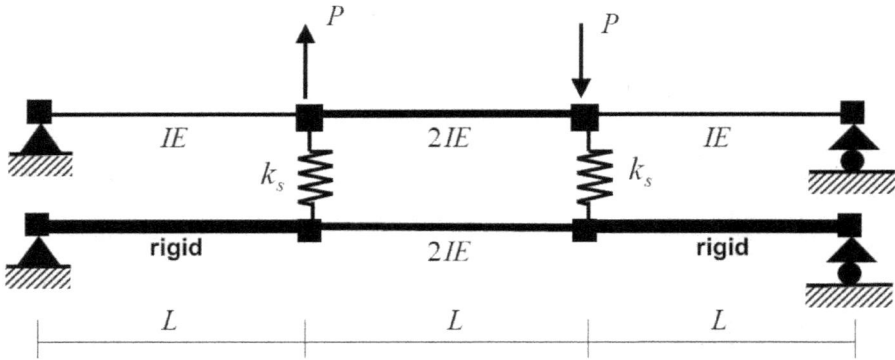

**FIGURE 9.23**   Problem 9.3.

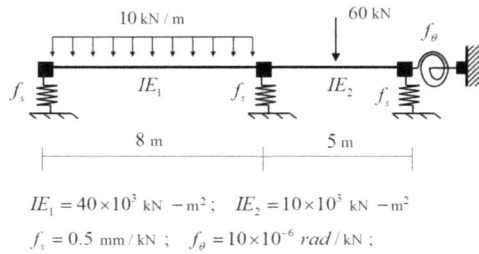

$IE_1 = 40 \times 10^3$ kN $-$ m$^2$; $IE_2 = 10 \times 10^3$ kN $-$ m$^2$

$f_s = 0.5$ mm / kN ; $f_\theta = 10 \times 10^{-6}$ rad / kN ;

**FIGURE 9.24**   Problem 9.4.

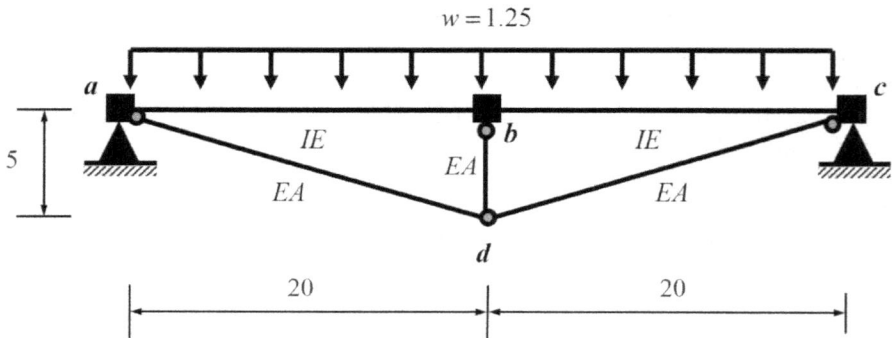

**FIGURE 9.25**   Problem 9.5.

(a) What are the forces in *ad*, *dc*, and *bd* when the vertical deflection at *b* is zero?

(b) Draw the bending moment diagram.

(c) How much shortening is required in *ad* and *dc* to make the vertical deflection at *b* zero.

Assume that $IE = 50 \times 10^3$ and $EA = 5000$.

# REFERENCES

Argyris, J.H. and S. Kelsey. 1960. *Energy theorems and atructural analysis.* Butterworths & Co Ltd.

Felton, L.P. and R.B. Nelson. 1997. *Matrix structural analysis.* John Wiley & Sons Inc.

Kanchi, M.B. 1993. *Matrix methods of structural analysis* (2nd Edition). Wiley Eastern Limited.

Limkatanyu, S., K. Kuntiyawichai, E. Spacone, and M. Kwon. 2012. Natural stiffness matrix for beams on Winkler foundation: Exact force-based derivation. *Structural Engineering and Mechanics* 42(1): 39–-53.

Limkatanyu, S., W. Prachasaree, G. Kaewkulchai, and E. Spacone. 2014. Unification of mixed Euler-Bernoulli-von Karman planar frame model and corotational approach. *Mechanics Based Design of Structures and Machines: An International Journal* 42(4): 419–441.

Limkatanyu, S. and E. Spacone. 2002. R/C frame element with bond interfaces: Parts I & II. *ASCE Journal of Structural Engineering* 128(3): 346–364.

Limkatanyu, S. and E. Spacone. 2006. Frame element with lateral deformable supports: Formulation and numerical validation. *Computers and Structures* 84(13–14): 942–954.

Martin, H.C. 1966. *Introduction to matrix methods of structural analysis.* McGraw-Hill Inc.

Martin, H.C. and G.F. Carey. 1973. *Introduction to finite element analysis: Theory and application.* McGraw-Hill Inc.

Meek, J.L. 1971. *Matrix structural analysis.* McGraw-Hill Inc.

Pandit, G.S. and S.P. Gupta. 2008. *Structural analysis: A matrix approach* (2nd Edition). Tata McGraw-Hill Publishing Company Limited.

Przemieniecki, J.S. 1985. *Theory of matrix structural analysis.* Dover Publications Inc.

Sae-Long, W., S. Limkatanyu, C. Hansapinyo, W. Prachasaree, J. Rungamornrat, and M. Kwon. 2021. Nonlinear flexibility-based beam element on Winkler-Pasternak foundation. *Geomechanics and Engineering* 24(4): 371–388.

Timoshenko, S.P. 1953. *History of strength of materials.* McGraw-Hill Inc.

# Appendix I
## *Fixed-end Forces and Moments*

1. Concentrated transverse force $P_y$

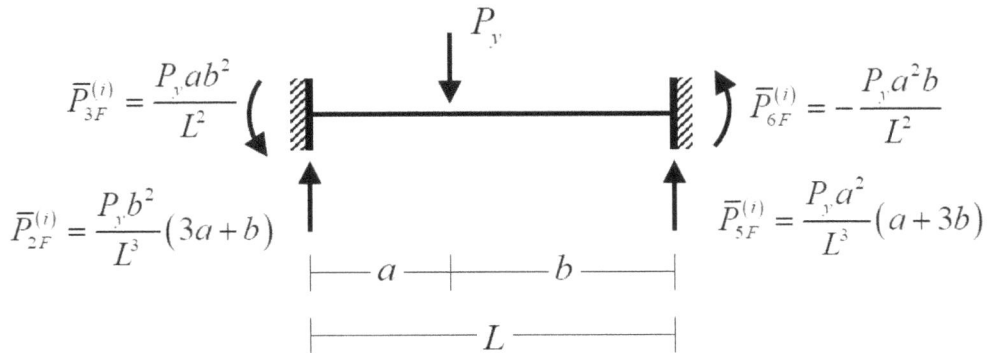

$$\overline{P}_{3F}^{(i)} = \frac{P_y ab^2}{L^2}$$

$$\overline{P}_{6F}^{(i)} = -\frac{P_y a^2 b}{L^2}$$

$$\overline{P}_{2F}^{(i)} = \frac{P_y b^2}{L^3}(3a+b)$$

$$\overline{P}_{5F}^{(i)} = \frac{P_y a^2}{L^3}(a+3b)$$

**FIGURE A1.1**

2. Concentrated transverse moment $M_z$

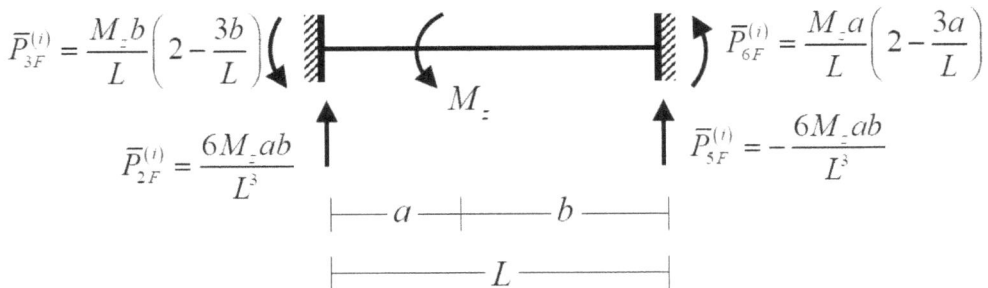

$$\overline{P}_{3F}^{(i)} = \frac{M_z b}{L}\left(2 - \frac{3b}{L}\right)$$

$$\overline{P}_{6F}^{(i)} = \frac{M_z a}{L}\left(2 - \frac{3a}{L}\right)$$

$$\overline{P}_{2F}^{(i)} = \frac{6M_z ab}{L^3}$$

$$\overline{P}_{5F}^{(i)} = -\frac{6M_z ab}{L^3}$$

**FIGURE A1.2**

3. Uniform transverse force/length $w_{y0}$ over the entire length

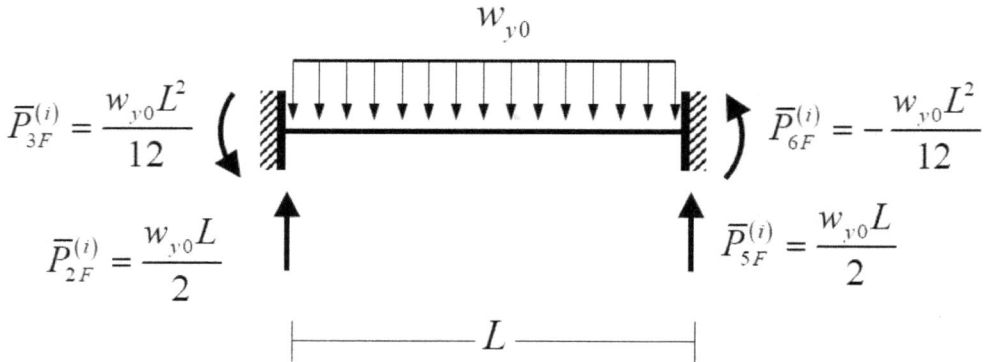

$$\overline{P}_{3F}^{(i)} = \frac{w_{y0}L^2}{12}$$

$$\overline{P}_{6F}^{(i)} = -\frac{w_{y0}L^2}{12}$$

$$\overline{P}_{2F}^{(i)} = \frac{w_{y0}L}{2}$$

$$\overline{P}_{5F}^{(i)} = \frac{w_{y0}L}{2}$$

**FIGURE A1.3**

4. Uniform transverse force/length $w_{y0}$ over portion of the length

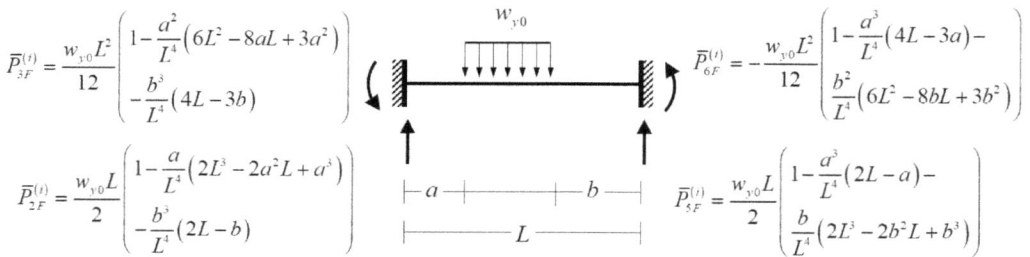

$$\overline{P}_{3F}^{(i)} = \frac{w_{y0}L^2}{12}\left(\begin{array}{c} 1 - \dfrac{a^2}{L^4}(6L^2 - 8aL + 3a^2) \\ -\dfrac{b^3}{L^4}(4L - 3b) \end{array}\right)$$

$$\overline{P}_{6F}^{(i)} = -\frac{w_{y0}L^2}{12}\left(\begin{array}{c} 1 - \dfrac{a^3}{L^4}(4L - 3a) - \\ \dfrac{b^2}{L^4}(6L^2 - 8bL + 3b^2) \end{array}\right)$$

$$\overline{P}_{2F}^{(i)} = \frac{w_{y0}L}{2}\left(\begin{array}{c} 1 - \dfrac{a}{L^4}(2L^3 - 2a^2L + a^3) \\ -\dfrac{b^3}{L^4}(2L - b) \end{array}\right)$$

$$\overline{P}_{5F}^{(i)} = \frac{w_{y0}L}{2}\left(\begin{array}{c} 1 - \dfrac{a^3}{L^4}(2L - a) - \\ \dfrac{b}{L^4}(2L^3 - 2b^2L + b^3) \end{array}\right)$$

**FIGURE A1.4**

5. Concentrated axial force at $x = a$

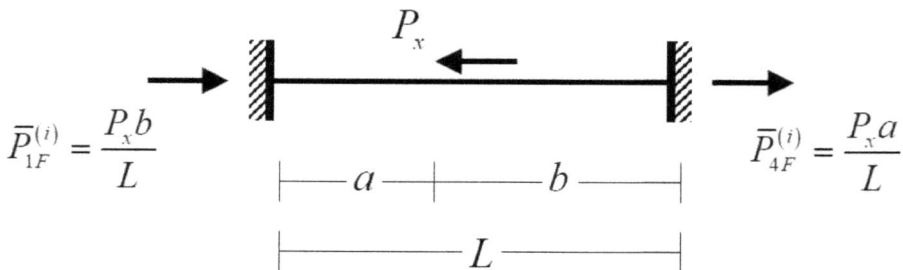

$$\overline{P}_{1F}^{(i)} = \frac{P_x b}{L}$$

$$\overline{P}_{4F}^{(i)} = \frac{P_x a}{L}$$

**FIGURE A1.5**

6. Uniform axial force/length along a portion of the length

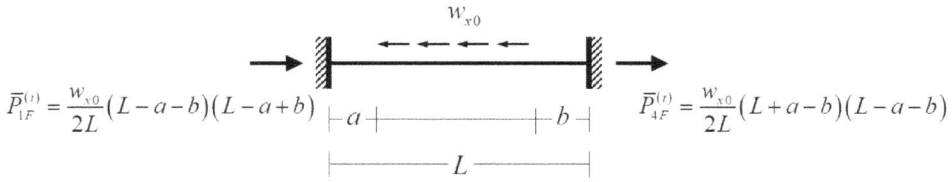

$$\bar{P}_{1F}^{(i)} = \frac{w_{x0}}{2L}\left(L-a-b\right)\left(L-a+b\right)$$

$$\bar{P}_{4F}^{(i)} = \frac{w_{x0}}{2L}\left(L+a-b\right)\left(L-a-b\right)$$

**FIGURE A1.6**

7. Linear varying temperature change through depth $h$ over the entire length

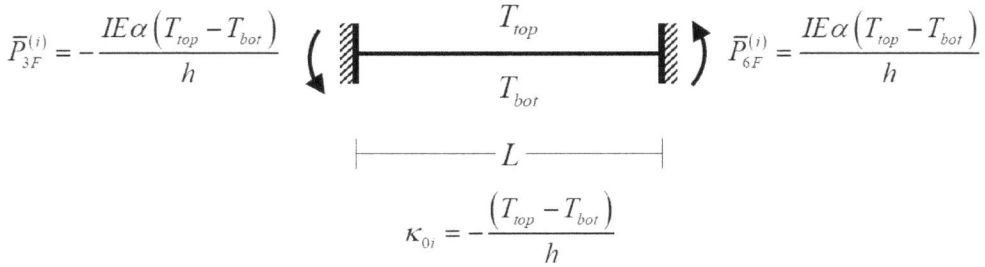

$$\bar{P}_{3F}^{(i)} = -\frac{IE\alpha\left(T_{top}-T_{bot}\right)}{h}$$

$$\bar{P}_{6F}^{(i)} = \frac{IE\alpha\left(T_{top}-T_{bot}\right)}{h}$$

$$\kappa_{0i} = -\frac{\left(T_{top}-T_{bot}\right)}{h}$$

**FIGURE A1.7**

8. Initial axial displacement $u_{x0}^{(i)}$

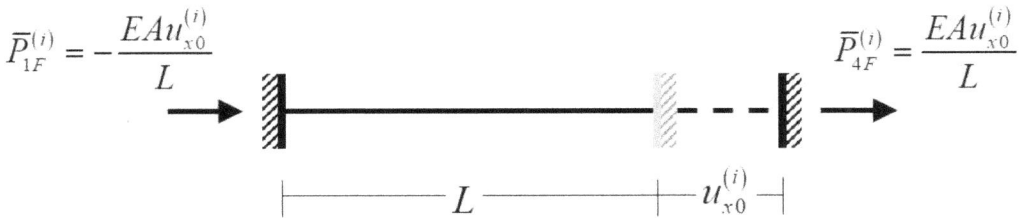

$$\bar{P}_{1F}^{(i)} = -\frac{EAu_{x0}^{(i)}}{L}$$

$$\bar{P}_{4F}^{(i)} = \frac{EAu_{x0}^{(i)}}{L}$$

**FIGURE A1.8**

9. Initial transverse displacement $u_{y0}^{(i)}$

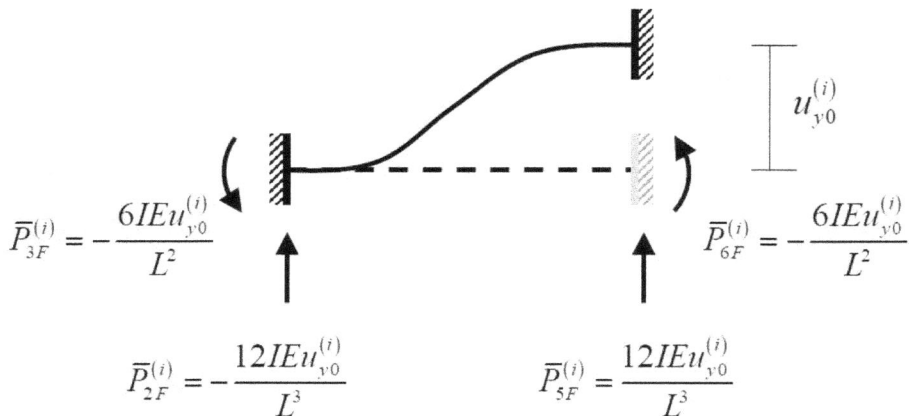

$$\overline{P}_{3F}^{(i)} = -\frac{6IEu_{y0}^{(i)}}{L^2}$$

$$\overline{P}_{6F}^{(i)} = -\frac{6IEu_{y0}^{(i)}}{L^2}$$

$$\overline{P}_{2F}^{(i)} = -\frac{12IEu_{y0}^{(i)}}{L^3}$$

$$\overline{P}_{5F}^{(i)} = \frac{12IEu_{y0}^{(i)}}{L^3}$$

**FIGURE A1.9**

10. Initial transverse displacement $\theta_{z0}^{(i)}$

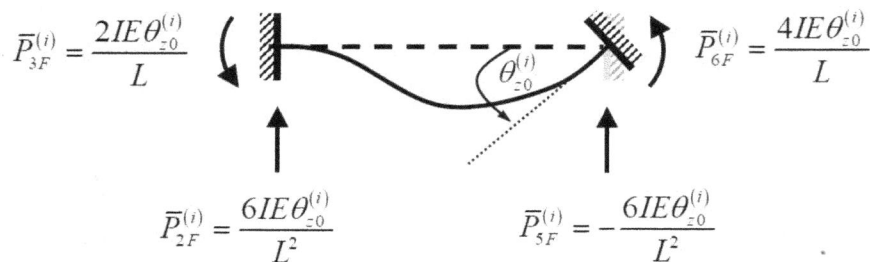

$$\overline{P}_{3F}^{(i)} = \frac{2IE\theta_{z0}^{(i)}}{L}$$

$$\overline{P}_{6F}^{(i)} = \frac{4IE\theta_{z0}^{(i)}}{L}$$

$$\overline{P}_{2F}^{(i)} = \frac{6IE\theta_{z0}^{(i)}}{L^2}$$

$$\overline{P}_{5F}^{(i)} = -\frac{6IE\theta_{z0}^{(i)}}{L^2}$$

**FIGURE A1.10**

# Appendix II
## Basic Member Deformations due to Member Loads and Initial Strains

1. Uniform axial force/length $w_{x0}$ and uniform transverse force/length $w_{y0}$ over the entire length: Simply supported system

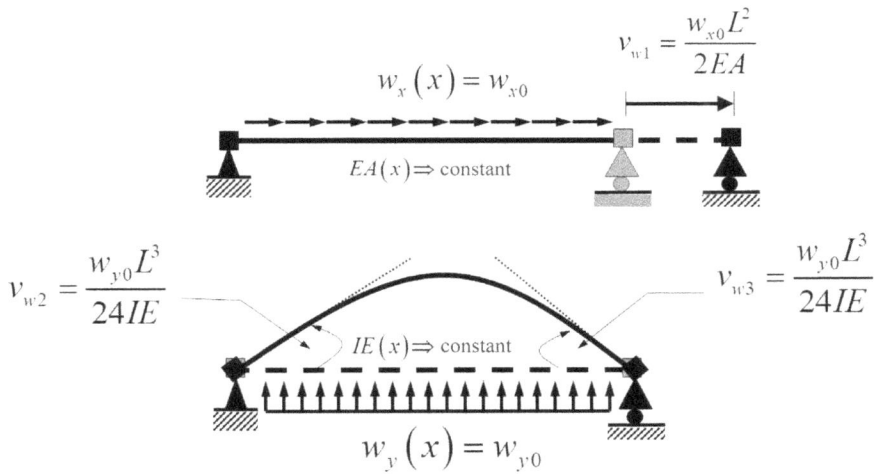

$$v_{w1} = \frac{w_{x0}L^2}{2EA}$$

$$w_x(x) = w_{x0}$$

$$EA(x) \Rightarrow \text{constant}$$

$$v_{w2} = \frac{w_{y0}L^3}{24IE}$$

$$IE(x) \Rightarrow \text{constant}$$

$$v_{w3} = \frac{w_{y0}L^3}{24IE}$$

$$w_y(x) = w_{y0}$$

**FIGURE A2.1**

2. Uniform axial strain $\varepsilon_0 i$ and uniform bending curvature $\kappa_{0i}$: Simply supported system

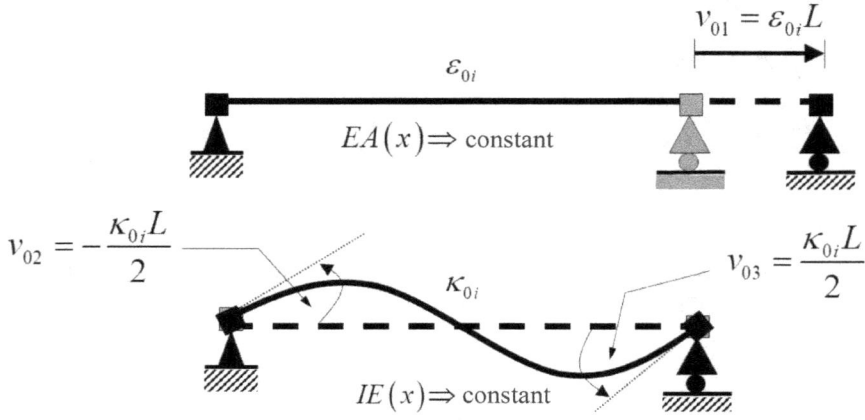

$$v_{01} = \varepsilon_{0i} L$$

$$\varepsilon_{0i}$$

$$EA(x) \Rightarrow \text{constant}$$

$$v_{02} = -\frac{\kappa_{0i} L}{2}$$

$$\kappa_{0i}$$

$$v_{03} = \frac{\kappa_{0i} L}{2}$$

$$IE(x) \Rightarrow \text{constant}$$

**FIGURE A2.2**

# Index

For Product Safety Concerns and Information please contact our EU
representative  GPSR@taylorandfrancis.com
Taylor & Francis Verlag GmbH, Kaufingerstraße 24, 80331 München, Germany

www.ingramcontent.com/pod-product-compliance
Lightning Source LLC
Chambersburg PA
CBHW080709220326
41598CB00033B/5355

* 9 7 8 1 0 3 2 9 7 6 6 6 2 *